AF566948

CLASSICAL FEEDBACK CONTROL

CONTROL ENGINEERING

A Series of Reference Books and Textbooks

Editor

NEIL MUNRO, PH.D., D.SC.

Professor
Applied Control Engineering
University of Manchester Institute of Science and Technology
Manchester, United Kingdom

1. Nonlinear Control of Electric Machinery, *Darren M. Dawson, Jun Hu, and Timothy C. Burg*
2. Computational Intelligence in Control Engineering, *Robert E. King*
3. Quantitative Feedback Theory: Fundamentals and Applications, *Constantine H. Houpis and Steven J. Rasmussen*
4. Self-Learning Control of Finite Markov Chains, *A. S. Poznyak, K. Najim, and E. Gómez-Ramírez*
5. Robust Control and Filtering for Time-Delay Systems, *Magdi S. Mahmoud*
6. Classical Feedback Control: With MATLAB, *Boris J. Lurie and Paul J. Enright*

Additional Volumes in Preparation

CLASSICAL FEEDBACK CONTROL
With MATLAB

Boris J. Lurie
Paul J. Enright
Jet Propulsion Laboratory
California Institute of Technology

MARCEL DEKKER, INC. NEW YORK • BASEL

Library of Congress Cataloging-in-Publication Data

Lurie, B. J.
Classical feedback control with MATLAB / Boris J. Lurie, Paul J. Enright.
p. cm. — (Control engineering ; 6)
ISBN 0-8247-0370-7
1. Feedback control systems. I. Enright, Paul J. II. Title. III. Control engineering (Marcel Dekker) ; 6.

TJ216 .L865 2000
629.8'3—dc21 99-087832

This book is printed on acid-free paper.

Headquarters
Marcel Dekker, Inc.
270 Madison Avenue, New York, NY 10016
tel: 212-696-9000; fax: 212-685-4540

Eastern Hemisphere Distribution
Marcel Dekker AG
Hutgasse 4, Postfach 812, CH-4001 Basel, Switzerland
tel: 41-61-261-8482; fax: 41-61-261-8896

World Wide Web
http://www.dekker.com

The publisher offers discounts on this book when ordered in bulk quantities. For more information, write to Special Sales/Professional Marketing at the headquarters address above.

Current printing (last digit)
10 9 8 7 6 5 4 3

PRINTED IN THE UNITED STATES OF AMERICA

PREFACE

Classical Feedback Control describes design and implementation of high-performance feedback controllers for engineering systems. The book emphasizes the frequency-domain approach which is widely used in practical engineering. It presents design methods for linear and nonlinear high-order controllers for single-input, single-output and multi-input, multi-output, analog and digital control systems.

Modern technology allows implementation of high-performance controllers at a very low cost. Conversely, several analysis tools which were previously considered an inherent part of control system courses limit the design to low-order (and therefore low-performance) compensators. Among these are the root-locus method, the detection of right-sided polynomial roots using the Routh-Hurwitz criterion, and manual calculations using the Laplace and Fourier transforms. These methods have been rendered obsolete by computers and are granted only a brief treatment in the book, making room for loop shaping, Bode integrals, structural simulation of complex systems, multiloop systems, and nonlinear controllers, all of which are essential for good design practice.

In the design philosophy adopted by *Classical Feedback Control*, Bode integral relations play a key role. The integrals are employed to estimate the available system performance and to determine the frequency responses that maximize the disturbance rejection and the feedback bandwidth. This ability to quickly estimate the attainable performance is critical for system-level trades in the design of complex engineering systems, of which the controller is one of many subsystems. Only at the final design stage and only for the finally selected option of the system configuration do the compensators need to be designed in detail, by approximation of the already found optimal frequency responses.

Nonlinear dynamic compensation is employed to provide global and process stability, and to improve transient responses. The nearly-optimal high-order compensators are then economically implemented using analog and digital technology.

The first six chapters support a one-semester course in linear control. The rest of the book considers the issues of complex system simulation, robustness, global stability, and nonlinear control. Throughout the book, MATLAB ® and Simulink ® from Mathworks Inc.* and SPICE are used for simulation and design. Student versions of the software are sufficient for studying control with this book, and even for some professional work; no preliminary experience with this software is required. The student should have some knowledge of the Laplace transform and frequency responses; the required theory is reviewed in Appendix 2.

Appendix 1 is an elementary treatment of feedback control, which can be used as an introduction to the course.

It is the authors' intention to make *Classical Feedback Control* not only a textbook but also a reference for students as they become engineers, enabling them to design high-performance controllers and easing the transition from school to the competitive industrial environment. The methods described in this book were used by the authors and their colleagues as the major design tools for feedback loops of aerospace and telecommunication systems.

We would be grateful for any comments, corrections, and criticism our readers may take the trouble to communicate to us, via E-mail b.j.lurie@jpl.nasa.gov or luriecontrol.com, or addressed to B. J. Lurie, 198/326, JPL, 4800 Oak Grove Drive, Pasadena CA 91109.

---

* The MathWorks, Inc., 3 Apple Hill Drive, Natick, MA, 01760-2098 USA
Tel: 508-647-7000 Fax: 508-647-7101 E-mail: info@mathworks.com
Web: www.mathworks.com

Acknowledgment. We thank Alla Lurie for technical editing and acknowledge the generous help of Asif Ahmed. We greatly appreciate previous discussions on many control issues with Professor Isaac Horowitz, and collaboration, comments and advice of our colleagues at the Jet Propulsion Laboratory, and especially of Edward Kopf (who told the authors about the jump-resonance in the attitude control loop of the Mariner 10 spacecraft) and Drs. Alexander Abramovich, Thomas Bak, David Bayard (who helped edit the chapter on adaptive systems), Dimitrius Boussalis, Daniel Chang, Gun-Shing Chen (who contributed Appendix 7), Ali Ghavimi (who contributed to Appendix A13.14), Fred Hadaegh (who co-authored several papers on which Chapter 13 is based), John Hench (who contributed the digital signal profiling function in Section 5.11), Kenneth Lau, Wei Min Liu, Mehran Mesbahi, Gregory Neat, John O'Brien, Samuel Sirlin, John Spanos, and Michael Zak. Suggestions and corrections made by Professors Randolph Beard, Osita Nwokah, and Roy Smith allowed us to improve the manuscript. Dr. Jason Modisette read the manuscript and suggested many changes and corrections. Alan Schier contributed the example of a mechanical snake control in Appendix A13.15. To all of them we extend our sincere gratitude.

Boris J. Lurie
Paul J. Enright

CONTENTS

Chapter 7

Chapter 8

Chapter 9

Chapter 10

Chapter 11

TO INSTRUCTORS

The book presents the design techniques which the authors found most useful in designing control systems for industry, telecommunication, and space programs. It also prepares the reader for research in the area of high-performance nonlinear controllers.

Plant and compensator. In classical control, the object of control (plant) is characterized either by its measured frequency response over the range of effective feedback or by a rather simple input/output mathematical model.

Classical design does not utilize the plant's internal variables and/or their estimates for compensation, unlike the full-state feedback approach. The appropriate loop responses are achieved by a stand-alone high-order compensator. These are the reasons this book starts with feedback, disturbance rejection, loop shaping, and compensator design, and not with extensive plant modeling.

Book architecture. The material contained in this book is organized as a sequence of, roughly speaking, four design level layers. Each layer considers linear and nonlinear systems, feedback, modeling and simulation. The layers are the following:

1. Control system *analysis*: elementary linear feedback theory, a short description of the effects of nonlinearities, and elementary simulation methods (Chapters 1–2).
2. Control system *design*: feedback theory and design of linear single-loop systems developed in depth (Chapters 3–4) followed by implementation methods (Chapter 5–6).
   This completes the first one-semester course in control.
3. Integration of linear and nonlinear subsystem models into the system model, utilization of the effects of feedback on impedances, various simulation methods (Chapter 7) followed by a brief survey of alternative controller design methods and of adaptive systems (Chapters 8 and 9).
4. Nonlinear systems study with practical design methods (Chapters 10 and 11), methods of elimination or reduction of process instability (Chapter 12), and composite nonlinear controllers (Chapter 13).

Each consecutive layer is based on the preceding layers. For example, introduction of absolute stability and Nyquist stable systems in the second layer is preceded by a primitive treatment of saturation effects in the first layer; global stability and absolute stability are then treated more precisely in the fourth layer. Treatment of the effects of links' input and output impedances on the plant uncertainty in the third layer is based on the elementary feedback theory of the first layer and the effects of plant tolerances on the available feedback developed in the second layer.

This architecture reflects the multifaceted character of real life design, and allows illustration of the theory by real system examples without excessive idealization.

Design examples. The following are examples of high-performance controllers. These controllers are chosen among those designed by the authors of the book at the Jet Propulsion Laboratory for various space robotic missions.

- A prototype of the controller for a retroreflector carriage and several other controllers of the Chemistry Spacecraft is described in Section 4.2.3. All these controllers are high-order and nonlinear, control plants with structural modes, and include a high-order linear part with a Bode step.

- A nonlinear digital controller for the Mars Pathfinder high-gain antenna pointing to Earth, Section 5.10.5.
- Switched-capacitor controller for the STRV spacecraft cryogenic cooler vibration rejection, Section 6.4.2.
- Vibration damping in the model of a space stellar interferometer, Example 2 in Section 7.1.1 and Example 2 in Section 7.7.
- Mars Global Surveyor attitude control, briefly described in Example 1, Section 13.6.
- Microgravity accelerometer analog feedback loop, described in detail in Example 2, Section 11.9.
- Cassini Narrow View Camera thermal control, described in detail in Section 7.1.2 and in Example 3, Section 13.6.
- More design examples are described in Appendix 13; in particular, Cassini mapping spectrometer controller, Section A13.2; high-voltage power source for Deep Space Network klystron radar, Section 13.6; Cassini computer clock PLL, Section A13.10; a feedback-feedforward digital attitude controller for DS-1 spacecraft solar panels, Section A13.11; Microwave Limb Sounder antenna attitude control of the Chemistry spacecraft, Section A13.12; and Space Interferometer optical delay line control, Section A13.13.

(The book's cover depicts Chemistry, Cassini, and DS-1 spacecraft.)

Software. Most simulation and design examples in the book use MATLAB . Some examples can use Simulink, but the introductory course can be given using MATLAB alone. No preliminary knowledge of MATLAB is assumed, and only a small subset of MATLAB commands is used, listed below in the order of their introduction: `logspace`, `bode`, `conv`, `tf2zp`, `zp2tf`, `step`, `gtext`, `title`, `set`, `grid`, `hold on`, `hold off`, `rlocus`, `plot(x,y)`, `inv`, `linspace`, `lp2lp`, `lp2bp`, `format`, `roots`, `poly`, `inv`, `bilinear`, `residue`, `ezplot`, `linmod`, `laplace`, `invlaplace`, `impulse`. Additional MATLAB functions described in Appendix 14 may be downloaded from <luriecontrol.com> and complement the design methods described in the book. If Simulink is used as a major design tool, students should be taught how to plot Bode diagrams from the block diagrams (see pp. 228, 250).

EE majors should also be shown how to make simulations in SPICE (SPICE examples given in this book are listed in the Index). These examples can be bypassed in teaching other specialties. Some simple C code examples are given in Chapter 5.

Frequency responses. The design methods taught in this book are based on frequency responses. The *s*-plane is used only for proving several theorems essential for frequency response methods. No previous experience with practical applications of the Laplace transform is required, and in the feedback system design as described in this book, the transforms are performed numerically with MATLAB commands `step` and `impulse`.

EE students know the frequency responses from the *Signals and Systems* course, the prerequisite to the control course. Mechanical, chemical, and aerospace engineering majors know frequency responses from the courses on dynamic responses. If needed, frequency responses for these specialties can be taught using Appendix 2, either before or in parallel with Chapter 3. Appendix 2 contains a number of problems on the Laplace transform and frequency responses.

Undergraduate course. The first 6 chapters, which constitute the first course in control, also include some material better suited for a graduate course. This material should be omitted from a one-semester course, especially when the course is taught to mechanical/aerospace/chemical engineering majors when extra time is needed to teach frequency responses. The sections to be bypassed are listed in the abstracts at the beginning of each chapter.

The material from Appendix 1 can be used for an introductory lecture.

Digital controllers. MATLAB, Simulink, and C make digital controller design simple. The best way of designing a good discrete controller is to design a high-order continuous controller, break it properly into several links, and then convert each link to a digital form; thus, two small tables of formulas or one MATLAB command is all that is needed. The *links* of digital controllers *must* be low-order, so there is no need for conversion formulas for high-order functions. The accuracy of the Tustin transform is adequate, so there is no need for pre-warping.

The digital control design can serve as a prerequisite for the following special courses on DSP, estimation, and adaptive digital control.

Analog controller implementation. The input of a modern mechanical actuator is most frequently electrical, the output signal of a sensor is typically electrical, and the compensators are therefore almost always electrical. Very often, the signals are analog, which allows implementation of the summer and compensator as analog electrical circuits in electrical and as well mechanical/hydraulic/thermal control systems.

Analog controllers are easier and cheaper to design, implement, troubleshoot, and manufacture than digital in many applications. Therefore, Chapter 6 is important for the engineers of all specialties. Nonetheless, the chapter need not be fully covered during a one-semester course, and can be used for self-study or as a reference later, when the need for practical design arises.

Second one-semester course. Chapters 7–13 can be used for the second one-semester course (it can be a graduate course), *and*, for self-study and as a reference for engineers who took only the first course.

Chapter 7 describes structural design and simulation systems with drivers, motors, and sensors. In particular, it shows that tailoring the output impedance of the actuator is important to reduce the plant tolerances and to increase the feedback in the outer loop.

Chapter 8 gives a short introduction to quantitative feedback theory and H_∞ control and the time-domain control based on state variables.

Undergraduate students need not be taught adaptive systems design since practical control systems rarely need to be designed as adaptive. But the need for some adaptive systems does exist. Therefore, the engineer should be aware of the major concepts, advantages and limitations of adaptive control. He must be able to recognize the need for such control, and, at the same time, not to waste time on trying to achieve the impossible. The material in Chapter 9 will enable him either to figure out how to design an adaptive system himself, or to understand the language of the specialized literature.

The design of high-order nonlinear controllers is covered in Chapters 10–13. These design methods have been proven very effective in practice, but are far from being finalized. Further research needs to be done to advance these methods.

Problems. Design problems with mechanical plants are suitable for both ME and EE majors. Additional problems for EE majors can be found in [9]. Some problems

EE majors. Additional problems for EE majors can be found in [9]. Some problems convey insights not presented in the text explicitly. A booklet with solutions to the problems is available for instructors from the publisher.

CLASSICAL FEEDBACK CONTROL

Chapter 1

FEEDBACK AND SENSITIVITY

Chapter 1 introduces the basics of feedback control. The purpose of feedback is to make the output insensitive to plant parameter variations and disturbances. Negative, positive, and large feedback are defined and discussed along with sensitivity and disturbance rejection. The notion of frequency response, the Nyquist diagram, and the Nichols chart are introduced. (The Nyquist stability criterion is presented in Chapter 3.)

Feedback control and block diagram algebra are explained at an elementary level in Appendix 1, which can be used as an introduction to this chapter. Laplace transfer functions are described in Appendix 2.

1.1 Feedback control system

It is best to begin with an example. Fig. 1.1(a) depicts a ***servomechanism*** regulating the elevation of an antenna. Fig. 1.1(b) shows a block diagram for this control system made of cascaded elements, i.e. ***links***. The capital letters stand for the signals' Laplace transforms and also for the transfer functions of the linear links. This block diagram shows a ***single-input single-output*** (***SISO***) system. There is one input command, U_1, which is the commanded elevation angle, and just one output, U_2, which is the actual elevation of the antenna. Evidently there is one feedback loop, and so the system is also referred to as ***single-loop***.

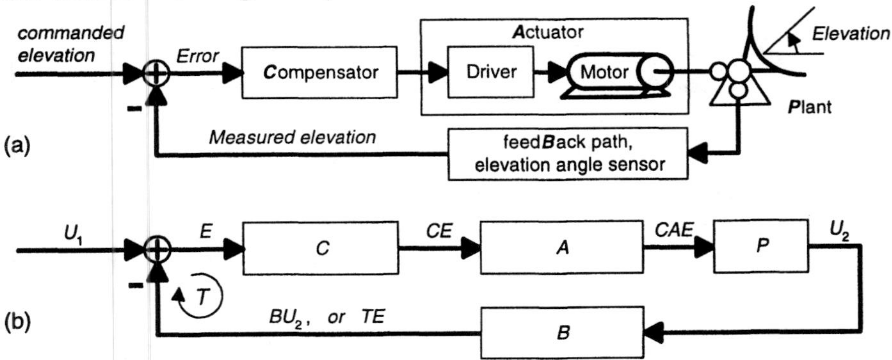

Fig. 1.1 Single-loop feedback system

The ***feedback path*** contains some sort of sensor for the output variable and has the transfer function B. Ideally, the measured output value BU_2 equals the commanded value U_1, and the ***error*** $E = U_1 - BU_2$ at the output of the summer is zero. In practice, most of the time the error is nonzero but small.

The error amplified by the ***compensator*** C is applied to the ***actuator*** A, in this case a motor regulator (driver) and a motor. The motor rotates the ***plant*** P, the antenna itself, which is the object of the control. The compensator, actuator and plant make up the ***forward path*** with the transfer function CAP. If the feedback path were not present, the system would be referred to as ***open-loop***, and the output U_2 would simply equal the product $CAPU_1$.

The ***return signal*** which goes into the summer from the feedback path is TE, where the product $T = CAPB$ is called the ***loop transfer function***, or the ***return ratio***.

The output of the summer is

$$E = U_1 - ET \tag{1.1}$$

so that the error

$$E = \frac{U_1}{T+1} = \frac{U_1}{F}, \tag{1.2}$$

where $F = T + 1$ is the ***return difference*** and its magnitude $|F|$ is the ***feedback***. It is seen that *when the feedback is large, the error is small*.

Example 1. A servomechanism for steering a toy car (using wires) is shown in Fig. 1.2. The command voltage U_1 is regulated by a joystick potentiometer. Another identical potentiometer (angle sensor) placed on the shaft of the motor produces voltage U_{angle} proportional to the shaft rotation angle. The feedback makes the error small so that the sensor voltage approximates the input voltage, and therefore the motor shaft angle tracks the joystick commanded angle.

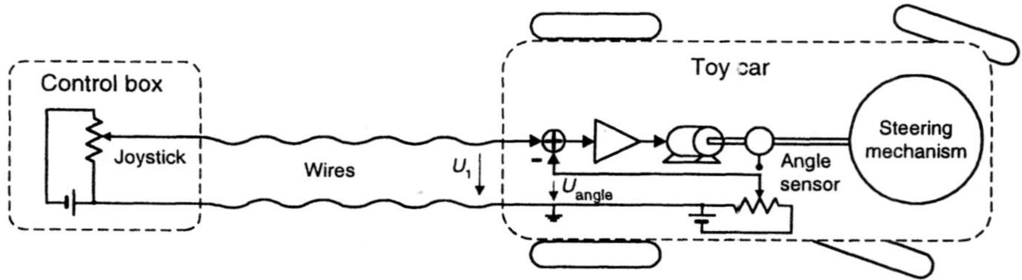

Fig. 1.2 Joystick control of a steering mechanism

The arrangement of a motor with an angle sensor is often called servomotor, or simply ***servo***. Similar servos are used for animation purposes in movie production.

The system of regulating aircraft control surfaces using joysticks and servos was termed "fly by wire" when it was first introduced to replace bulky mechanical gears and cables. The required high reliability was achieved by using four independent parallel analog electrical circuits.

The telecommunication link between the control box and the servo can certainly also be wireless.

Example 2. A ***phase-locked loop*** (***PLL***) is shown in Fig. 1.3. The plant here is a ***voltage-controlled oscillator*** (***VCO***). The VCO is an ac generator whose frequency is proportional to the voltage applied to its input. The ***phase detector*** combines the functions of phase sensors and input summer: its output is proportional to the phase difference between the input signal and the output of the VCO. Large feedback makes the phase difference (phase error) small, so that the output signal has only a small phase difference compared with the input signal and, therefore, the same frequency. In other words, the PLL synchronizes the VCO with the input periodic signal.

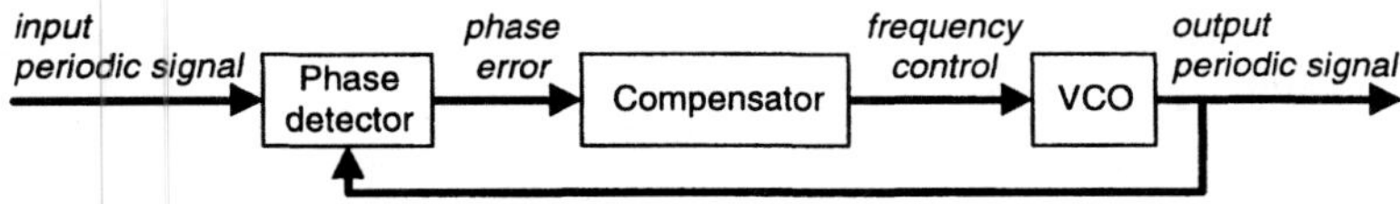

Fig. 1.3 Phase-locked loop

Analog and digital PLLs are widely used in telecommunications (for tuning receivers and for recovering the computer clock from a string of digital data), for synchronizing several motors' angular positions and velocities, and for many other purposes.

1.2 Feedback: positive and negative

The output signal in Fig. 1.1(b) is $U_2 = ECAP$, and from (1.2) $U_1 = EF$, so the input-to-output transfer function of the system with the feedback loop closed, commonly referred to as the ***closed-loop transfer function***, is:

$$\frac{U_2}{U_1} = \frac{ECAP}{EF} = \frac{CAP}{F} \ . \tag{1.3}$$

It is clear that <u>*the feedback reduces the input-output signal transmission by the factor*</u> $|F|$.

The system is said to have "***negative***" feedback when $|F| > 1$ (although the expression $|F|$ is certainly positive). This definition was developed in the 1920s and has to do with the fact that "negative" feedback *reduces* the error $|E|$ and the output $|U_2|$, i.e., produces a *negative increment* in the output *level* when the level is expressed in logarithmic values (in dB, for example) preferred by engineers.

The feedback is said to be "***positive***" if $|F| < 1$, which makes $|E| > |U_1|$. "Positive" feedback *increases* the error and the level of the output.

We will adhere to these definitions of "negative" and "positive" feedback (and use these terms without quotation marks) since very important theoretical developments, to be studied in Chapters 3 and 4, are based on these definitions.

Whether the feedback is positive or negative depends on the amplitude and phase of the return ratio (and *not* only on the sign at the feedback summer as is stated sometimes in elementary treatments of feedback).

Let's consider several numerical examples.

Example 1. The forward path gain coefficient CAP is 100 and the feedback path coefficient B is -0.003. The return ratio T is -0.3. Hence, the return difference F is 0.7, the feedback is positive, and the closed-loop gain coefficient 100/0.7 = 143 is greater than the open-loop gain coefficient.

Example 2. The forward path gain coefficient is 100 and the feedback path coefficient is 0.003. The return ratio T is 0.3. Hence, the return difference F is 1.3, the feedback is negative, and the closed-loop gain coefficient 100/1.3 = 77 is less than the open-loop gain coefficient.

It is seen that <u>*when T is small, whether the feedback is positive or negative depends on the sign of the transfer function about the loop*</u>.

When $|T| > 2$, then $|T + 1| > 1$ and the feedback is negative. I.e. *when $|T|$ is large, the feedback is always negative*.

Example 3. The forward path gain coefficient is 1000 and $B = 0.1$. The return ratio is therefore 100. The return difference is 101, the feedback is negative, and the closed-loop gain coefficient is 9.9.

Example 4. In the previous example, the forward path transfer function is changed to –1000, and the return ratio becomes –100. The return difference is –99, the feedback is still negative, and the closed-loop gain coefficient is 10.1.

1.3 Large feedback

Multiplying the numerator and denominator of (1.3) by B yields another meaningful formula:

$$\frac{U_2}{U_1} = \frac{1}{B}\frac{T}{F} = \frac{1}{B}M, \tag{1.4}$$

where

$$M = \frac{T}{F} = \frac{T}{T+1}. \tag{1.5}$$

Equation (1.4) indicates that *the closed-loop transfer function is the inverse of the feedback path transfer function multiplied by the coefficient M*. When the feedback is ***large***, i.e. when $|T| >> 1$, the return difference $F \approx T$, the coefficient $M \approx 1$, and the output becomes

$$U_2 \approx \frac{1}{B}U_1. \tag{1.6}$$

One result of large feedback is that *the closed-loop transfer function depends nearly exclusively on the feedback path which can usually be constructed of precise components*. This feature is of fundamental importance since the parameters of the actuator and the plant in the forward path typically have large uncertainties. In a system with large feedback, the affect of these uncertainties on the closed-loop characteristics is small. The larger the feedback, the smaller the error expressed by (1.2).

Manufacturing an actuator that is sufficiently powerful and precise to handle the plant without feedback can be prohibitively expensive or impossible. An imprecise actuator may be much cheaper, and a precise sensor may also be relatively inexpensive. Using feedback the cheaper actuator and the sensor can be combined to form a powerful, precise, and reasonably inexpensive system.

According to (1.6), the antenna elevation angle in Fig. 1.1 equals the command divided by B. If the elevation angle is required to be q, then the command should be Bq.

If $B = 1$, as shown in Fig. 1.4(a), then the closed-loop transfer function is just M and $U_2 \approx U_1$, i.e. the output U_2 follows (tracks) the commanded input U_1. Such ***tracking systems*** are widely used. Examples are: a telescope tracking a star or a planet, an antenna on the roof of a vehicle tracking the position of a knob rotated by the operator inside the vehicle, and a cutting tool following a probe on a model to be copied.

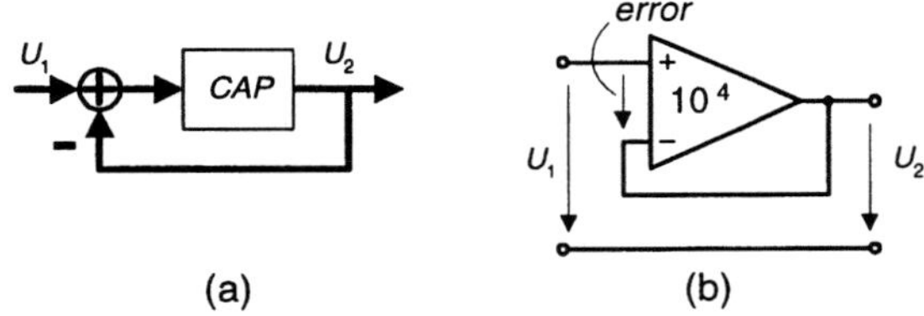

Fig. 1.4 (a) Tracking system, (b) voltage follower

Example 1. Fig. 1.4(b) shows an amplifier with unity feedback. The error voltage is the difference between the input and output voltages. If the amplifier gain coefficient is 10^4, the error voltage constitutes only 10^{-4} of the output voltage. Since the output voltage nearly equals the input voltage, this arrangement is commonly called a ***voltage follower***.

Example 2. Suppose that $T = 100$, so that $M = T/(T + 1) = 0.9901$. If P were to deviate from its nominal value by +10%, then T would become 110. This would make $M = 0.991$, an increase of 0.1%, which is reflected in the output signal. Without the feedback, the variation of the output signal would be 10%. Therefore, introduction of negative feedback in this case reduces the output signal variations 100 times. Introducing positive feedback would do just the opposite – it would increase the variations in the closed-loop input-output transfer function.

Example 3. Consider the voltage regulator shown in Fig. 1.5(a) with its block diagram shown in Fig. 1.5(b). Here, the differential amplifier with transimpedance (ratio of output current I to input voltage E) 10 A/V and high input and output impedances plays the dual role of compensator and actuator. The power supply voltage is VCC. The plant is the load resistor R_L. The potentiometer with the voltage division ratio B constitutes the feedback path.

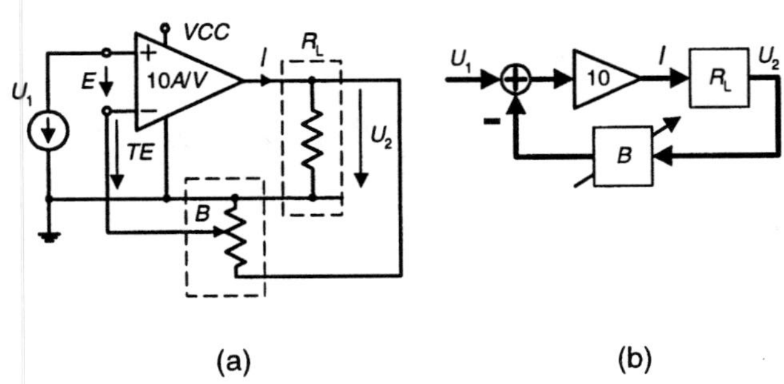

Fig. 1.5 Voltage regulator: (a) schematic diagram, (b) block diagram

The amplifier input voltage is the error $E = U_1 - TE$, and the return ratio is $T = 10BR_L$. Assume that the load resistor is 1 kΩ and the potentiometer is set to $B = 0.5$. Consequently, the return ratio is $T = 5000$.

The command is the 5 V input voltage (when the command is constant, as in this case, it is commonly called a ***reference***, and the control system is called a ***regulator***). Hence, the output voltage according to (1.4) is $10 \times 5000/5001 = 9.998$ V. The VCC should be higher than this value, 12 V to 30 V would be appropriate.

When the load resistance is reduced by 10%, without the feedback the output voltage will be 10% less. With the feedback, T decreases by 10% and the output voltage is $10\times 4500/4501 \approx 9.99778$ i.e. only 0.002% less. The feedback reduces the output voltage variations 10%/0.002% = 5000 times.

This example also illustrates another feature of feedback. Insensitivity of the output voltage to the loading indicates that the regulator output resistance is very low. The feedback dramatically alters the output impedance from very high to very low. (The same is true for the follower shown in Fig. 1.4(b).) The effects of feedback on impedance will be studied in detail in Chapter 7.

1.4 Loop gain and phase frequency responses

1.4.1 Gain and phase responses

The sum (or the difference) of sinusoidal signals of the same frequency is a sinusoidal signal with the same frequency. The summation is simplified when the sinusoidal signals are represented by phasors, vectors on a complex plane. The modulus of a vector equals the signal amplitude and the phase of the vector equals the phase shift of the signal.

Signal $u_1 = |U_1|\sin(\omega t + \varphi_1)$ is represented by the vector $U_1 = |U_1|\angle\varphi_1$, i.e. by the complex number $U_1 = |U_1|\cos\varphi_1 + j|U_1|\sin\varphi_1$.

Signal $u_2 = |U_2|\sin(\omega t + \varphi_2)$ is represented by the vector $U_2 = |U_2|\angle\varphi_2$, i.e. by the complex number $U_2 = |U_2|\cos\varphi_2 + j|U_2|\sin\varphi_2$.

The sum of these two signals is

$$u = |U|\sin(\omega t + \varphi) = (|U_1|\cos\varphi_1 + |U_2|\cos\varphi_2)\sin\omega t + (|U_1|\sin\varphi_1 + |U_2|\sin\varphi_2)\cos\omega t\,,$$

i.e. $\text{Re}U = \text{Re}U_1 + \text{Re}U_2$ and $\text{Im}U = \text{Im}U_1 + \text{Im}U_2$, i.e. $U = U_1 + U_2$. Thus, the vector for the sum of the signals equals the sum of the vectors for the signals.

Example 1. Signal $u_1 = 4\sin(\omega t + \pi/6)$ is represented by the vector $4\angle\pi/6$, or $3.464 + j2$. Signal $u_2 = 6\sin(\omega t + \pi/4)$ is represented by the vector $6\angle\pi/4$, or $4.243 + j4.243$. The sum of these two signals $u_1 + u_2$ is represented by the vector (complex number) $7.707 + j6.243 = 9.92\angle 0.681$ rad or $9.92\angle 39°$.

Example 2. Fig. 1.6 shows four possible phasor diagrams $U_1 = E + TE$ of the signals at the feedback summer at some frequency. In cases (a) and (b), the presence of fed back signal TE makes $|E| > |U_1|$; therefore, $|F| < 1$ and the feedback at this frequency is positive. In cases (c) and (d), $|E| < |U_1|$, and the feedback at this frequency is negative.

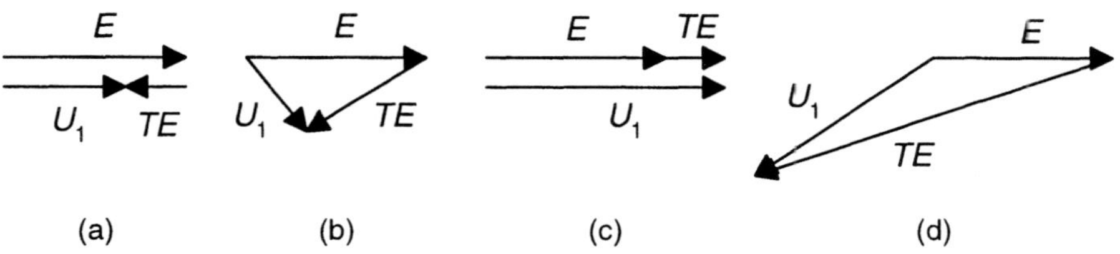

Fig. 1.6 Examples of phasor diagrams for (a,b) positive and (c,d) negative feedback

Replacing the Laplace variable s by $j\omega$ in a Laplace transfer function, where ω is understood to be the frequency in rad/sec, results in the frequency-dependent transfer function. The transfer function is the ratio of the signals at the link's output and input, and generally depends on the signal frequency. Therefore, the return ratio and the return difference are functions of frequency. To accurately track rapidly varying commands, i.e., commands with substantial high-frequency content, one has to make the feedback large over a sufficiently wide frequency range.

Plots of the gain and phase of a transfer function vs. frequency are referred to generically as the ***frequency response***.

The ***loop frequency response*** is defined by the complex function $T(j\omega)$. The magnitude $|T(j\omega)|$ expressed in decibels (dB) is referred to as the ***loop gain***. The angle of $T(j\omega)$ is the ***loop phase shift*** which is usually expressed in degrees. Plots of the gain (and sometimes the phase) of the loop transfer function with logarithmic frequency scale are often called ***Bode diagrams***, in honor of Hendrik W. Bode, who, although he did not invent the diagrams, did develop an improved methodology of using them for feedback system design, to be explained in detail in Chapters 3 and 4. The plots can be drawn using angular frequency ω or $f = \omega/(2\pi)$ in Hz.

Example 3. Let

$$T(s) = \frac{num}{den} = \frac{5000}{s(s+5)(s+50)} = \frac{5000}{s^3 + 55s^2 + 250s + 0}$$

where *num*, *den* are the numerator and denominator polynomials.

The frequency responses for the loop gain in dB, $20\log|T(j\omega)|$, and the loop phase shift in degrees, $(180/\pi)\arg T(j\omega)$, can be plotted over the 0.1 to 100 rad/sec range with the software package MATLAB ® from Mathworks Inc. with the following script:

```
w = logspace(-1, 2);
% log scale of angular
% frequency w
num = 5000;
den = [1 55 250 0];
bode(num, den, w)
```

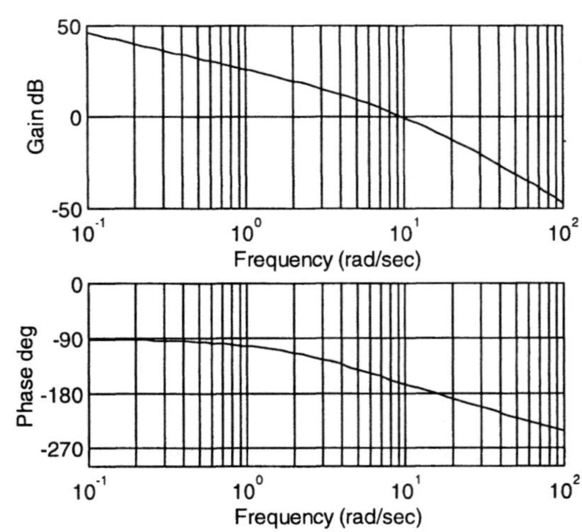

Fig. 1.7 MATLAB plots of gain and phase loop responses

The plots are shown in Fig. 1.7. The loop gain rapidly decreases with frequency, and the slope of the gain response gets even steeper at higher frequencies; this is typical for practical control systems.

The loop gain is 0 dB at 9 rad/sec, i.e., at $9/(2\pi) \approx 1.4$ Hz. The phase shift gradually changes from -90° toward -270°, i.e., the phase lag increases from 90° to 270°.

MATLAB function `conv` can be used to multiply the polynomials s, $(s + 5)$, and

$(s+50)$ in the denominator:

```
a = [1 0]; b = [1 5]; c = [1 50];
ab = conv(a,b); den = conv(ab,c)
```

More information about the MATLAB functions used above can be obtained by typing `help bode`, `help logspace`, and `help conv` in the MATLAB working window, and from the MATLAB manual. Conversions from one to another form of a rational function can be also done using MATLAB functions `tf2zp` (transfer function to zero-pole form) and `zp2tf` (zero-pole form to transfer function).

Example 4. The return ratio from Example 1, explicitly expressed as a function of $j\omega$, is

$$T(j\omega) = \frac{5000}{j\omega(j\omega+5)(j\omega+50)}.$$

At frequency $\omega = 2$, $T \approx 10\angle -110^\circ$ and $F \approx 9\angle -115^\circ$, so the feedback is negative. $|T|$ reduces with frequency. At frequency $\omega = 9$, $T \approx 1\angle -160^\circ$ and $F \approx 0.2\angle -70^\circ$, so the feedback is positive.

The Bode plot of $|T|$ for a typical tracking system is shown in Fig. 1.8 by the solid line. The loop gain decreases with increasing frequency. The diagram crosses the 0 dB line at the ***crossover frequency*** f_b where, by definition, $|T(f_b)| = 1$.

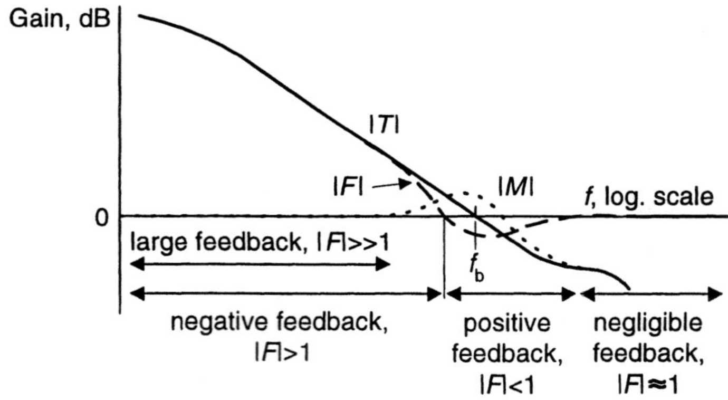

Fig. 1.8 Typical frequency responses for *T*, *F*, and *M*

The frequency response of the feedback $|F|$ is shown by the dashed line. It can be seen that *the feedback is negative (i.e., 20 log$|F| > 0$) up to a certain frequency, becomes positive in the neighborhood of f_b, and then becomes negligible at higher frequencies where $F \to 1$ and 20 log$|F| \to 0$.*

The input-output closed-loop system response is shown by the dotted line. The gain is 0 dB (i.e., the gain coefficient is 1) over the entire bandwidth of large feedback. The hump near the crossover frequency is a result of the positive feedback. This hump, as will be demonstrated in Chapter 3, results in an oscillatory closed-loop transient response, and should therefore be bounded.

In general, *feedback improves the tracking system's accuracy for commands whose dominant Fourier components belong to the area of negative feedback, but degrades the system's accuracy for commands whose frequency content is in the area of positive feedback*.

1.4.2 Nyquist diagram

To visualize the transition from negative to positive feedback, it is helpful to look at the plot of T on the T-plane as the frequency varies from 0 to ∞. This plot is referred to as the ***Nyquist diagram*** and is shown in Fig. 1.9. Either Cartesian (ReT, ImT) or polar coordinates ($|T|$ and arg T) can be used.

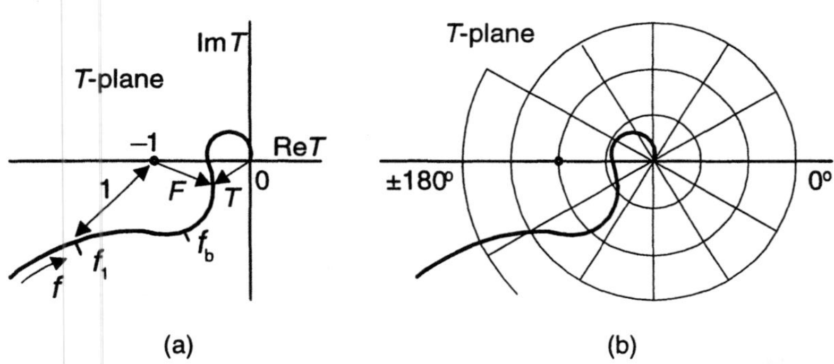

Fig. 1.9 Nyquist diagram with (a) Cartesian and (b) polar coordinates; the feedback is negative at frequencies up to f_1

The Nyquist diagram is a major tool in feedback system design and will be discussed in detail in Chapter 3. Here, we use the diagram only to show the locations of the frequency bands of negative and positive feedback in typical control systems.

At each frequency, the distance to the diagram from the origin is $|T|$, and the distance from the −1 point is $|F|$. It can be seen that *$|F|$ becomes less than 1 at higher frequencies, which means the feedback is positive there*.

The Nyquist diagram should not pass excessively close to the ***critical point*** *−1* or else the closed-loop gain at this frequency will be unacceptably large.

In practice, Nyquist diagrams are commonly plotted on the logarithmic ***L-plane*** with rectangular coordinate axes for the phase and the gain of T, as shown in Fig. 1.10(a). Notice that the critical point −1 of the T-plane maps to point (−180°, 0 dB) of the L-plane. The Nyquist diagram should avoid this point by a certain margin.

Example 1. Fig. 1.10(b) shows the L-plane Nyquist diagram for $T = (20s + 10)/(s^4 + 10s^3 + 20s^2 + s)$ charted with MATLAB script

```
num = [20 10]; den = [1 10 20 1 0];
[mag, phase] = bode(num, den);
plot(phase, 20*log10(mag), 'r', -180, 0, 'wo')
title('L-plane Nyquist diagram')
set(gca,'XTick',[-270 -240 -210 -180 -150 -120 -90])
grid
```

It is recommended for the reader to run this program for modified transfer functions and to observe the effects of the polynomial coefficient variations on the shape of the Nyquist diagram.

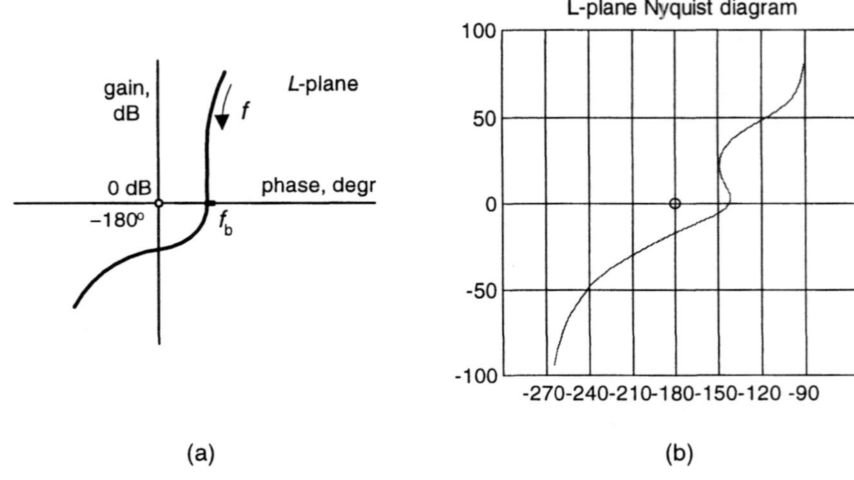

Fig. 1.10 Nyquist diagrams on the *L*-plane,
(a) typical for a well-designed system and (b) MATLAB generated for Example 1

1.4.3 Nichols chart

The ***Nichols chart*** is an *L*-plane template for the mapping from T to M, according to (1.5), and is shown in Fig. 1.11. When the Nyquist diagram for T is drawn on this template, the curves indicate the tracking system gain $20 \log |M|$.

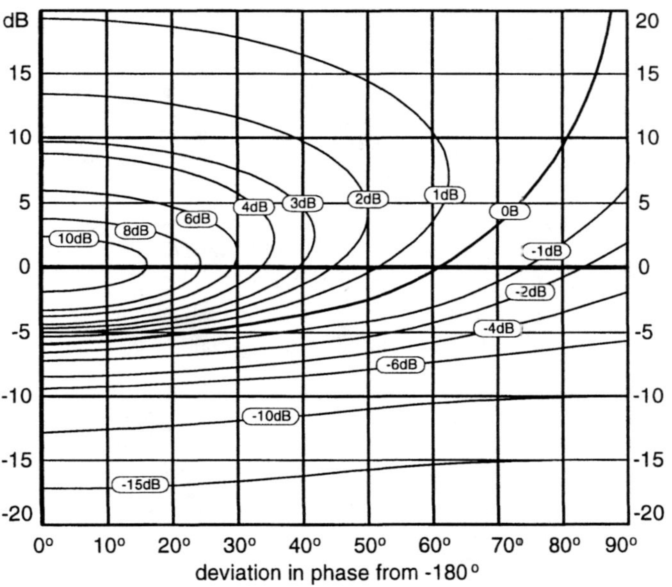

Fig. 1.11 Nichols chart

It is seen that the closer the Nyquist diagram approaches the critical point (−180°, 0 dB), the larger |*M*| is, and therefore, the higher is the peak of the closed-loop frequency response in Fig. 1.8. The limiting case has |*M*| approaching infinity, indicating that the system goes unstable. Typically, |*M*| is allowed to increase not more than two times, i.e., not to exceed 6 dB.

Therefore, *the Nyquist diagram should not penetrate into the area bounded by the line marked "6 dB"*.

Consider several examples which make use of the Nichols chart.

Example 1. The loop gain is 15 dB, the loop phase shift is −150°. From the Nichols chart, the closed-loop gain is −1.4 dB. The feedback is 15 − 1.4 = 13.6 dB. The feedback is negative.

Example 2. The loop gain is 1 dB, the loop phase shift is −150°. From the Nichols chart, the closed-loop gain is 6 dB. The feedback is 1 − 6 = −5 dB. The feedback is positive.

Example 3. The loop gain is −10 dB, the loop phase shift is −170°. From the Nichols chart, the closed-loop gain is −7 dB. The feedback is −10 − (−7) = −3 dB. The feedback is positive.

1.5 Disturbance rejection

Disturbances are signals which enter the feedback system at the input or output of the plant or actuator, as shown in Fig. 1.12, and cause undesirable signal at the system output. In the antenna pointing control system, disturbances might be due to wind, gravity, temperature changes, and imperfections in the motor, the gearing, and the driver. The disturbances can be characterized either by their time history, or, in frequency domain, by the Fourier transform of this time history which gives the disturbance ***spectral density***.

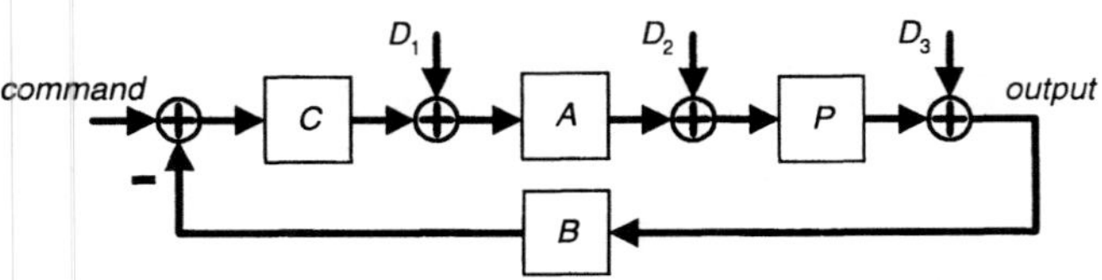

Fig. 1.12 Disturbance sources in a feedback system

The frequency response of the effect of a disturbance at the system's output can be calculated in the same way that the output frequency response to a command is calculated: it is the open-loop effect (D_1AP, for example) divided by the return difference F.

In Fig. 1.12, three disturbance sources are shown. Since in linear systems, the combined effect at the output of several different input signals is the sum of the effects of each separate signal, the disturbances produce the output effect

$$\frac{D_1 AP}{F} + \frac{D_2 P}{F} + \frac{D_3}{F} = \frac{D_1 AP + D_2 P + D_3}{F}.$$

The effects of the disturbances on the output are reduced when the feedback is negative and increased when the feedback is positive. Disturbance rejection is the major purpose for using negative feedback in most control systems.

There exist systems where there is no command at all, and the disturbance rejection is the only purpose of introducing feedback. Such systems are called ***homing systems***. A typical example is a homing missile which is designed to follow the target. No explicit command is given to the missile. Rather, the missile receives only an error signal which is the deviation from the target. The feedback causes the vehicle's aerodynamic surfaces to reduce the error. This error can be considered as a disturbance, and large feedback reduces the error effectively. Another popular type of a system without an explicit command is the active suspension, which uses motors or solenoids to attenuate the vibration propagating from the base to the payload.

Example 1. The feedback in a temperature control loop of a chamber is 100. Without feedback, when the temperature outside the chamber changes, the temperature within the chamber changes by 6°. With feedback, the temperature within the chamber changes by only 0.06°.

Example 2. Gusty winds disturb the orientation of a radio telescope. The winds contain various frequency components, some slowly varying in time and others rapidly oscillating. The feedback in the antenna attitude control loop is 200 at very low frequencies, but drops with frequency (since motors cannot move the huge antenna rapidly), and at 0.1 Hz the feedback is only 5. The disturbance components are attenuated by the feedback accordingly, 200 times for the effect of steady wind, and 5 times for the 0.1 Hz gust components. Detailed calculations for a similar example will be given in the next section.

To further reduce the higher frequency disturbances, an additional feedback loop might be introduced that will adjust the position not of the entire antenna dish but of some smaller mirror in the optical path from the antenna dish to the receiver front-end (or from the power amplifier of the transmitter).

1.6 Example of system analysis

We proceed now with the analysis of the simplified antenna elevation control system shown in Fig. 1.1(b). Assume that the elevation angle sensor function is 1 V/rad, the feedback path coefficient is $B = 1$, the actuator transfer function (the ratio of the output torque to the input voltage) is $A = 5000/(s + 10)$ Nm/V, and the antenna is a rigid body with the moment of inertia $J = 5000\,\text{kgm}^2$. The plant's input variable is torque, the output variable is the elevation angle, i.e., the plant is a double integrator with gain coefficient $1/J$. Since the Laplace transform of an integrator is $1/s$, the transfer function of the plant is

$$P(s) = 1/(Js^2)\,.$$

As shown in Fig. 1.13, the torque τ applied to the antenna is the sum of the torque produced by the actuator and the disturbance wind torque τ_W.

It is known that for large antennas, the wind torque spectral density is approximately proportional to

$$\frac{1}{(s+0.1)(s+2)}\ .$$

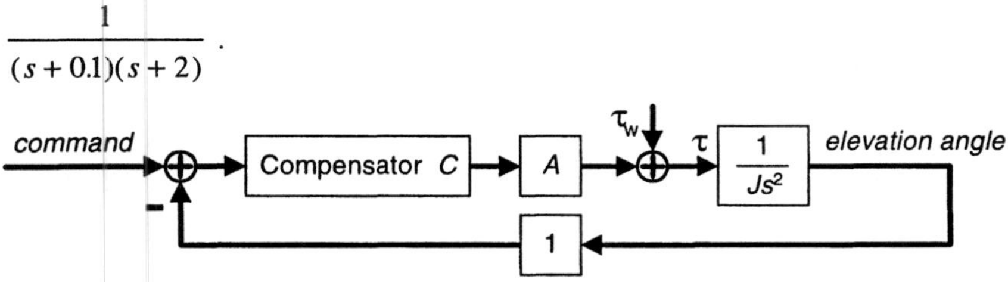

Fig. 1.13 Elevation control system block diagram

The spectral density of the disturbance in the antenna elevation angle is therefore proportional to

$$\frac{1}{(s+0.1)(s+2)}\frac{1}{s^2}=\frac{1}{s^4+2.1s^3+0.2s^2}\ . \tag{1.7}$$

The spectral density plot of the disturbance in the elevation angle is shown in Fig. 1.14, normalized to 100 dB at ω = 0.01. The spectral density is larger at lower frequencies. Large feedback must be introduced at these frequencies to reject the disturbance.

The frequency response of (1.7) can be plotted using MATLAB with:

```
w = logspace(-2,1);
% frequency range
% 0.01 to 10 rad/sec
num = 1;
den = [1 2.1 0.2 0 0];
bode(num, den, w)
```

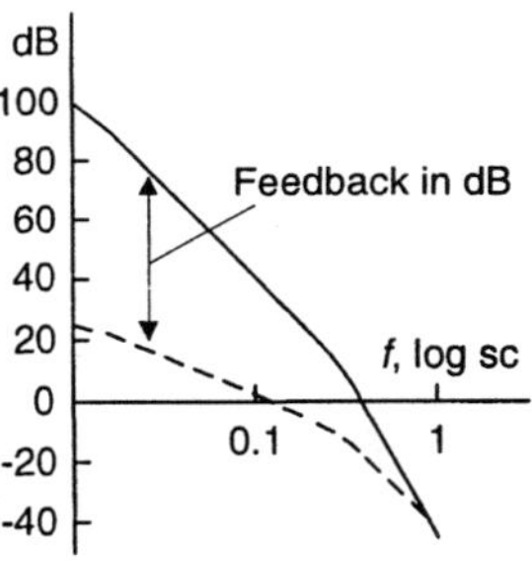

Fig. 1.14 Spectral density of the elevation angle disturbance, in relative units: before the feedback was introduced, solid line; with the feedback, dashed line

The compensator transfer function is

$$C(s)=\frac{50(s+0.05)(s+0.5)}{s(s+5)}$$

(this simple compensator makes the system work reasonably well, although not optimally). The loop transfer function is

$$T(s)=CAP=\frac{50(s+0.05)(s+0.5)}{s(s+5)}\times\frac{5000}{s+10}\times\frac{1}{5000s^2},$$

i.e.

$$T(s)=\frac{num}{den}=\frac{50(s+0.05)(s+0.5)}{s^3(s+5)(s+10)}=\frac{50s^2+27.5s+0.25}{s^5+15s^4+50s^3}\ .$$

The return difference is

$$F(s) = T(s) + 1 = (num + den)/den\ . \tag{1.8}$$

The closed-loop transfer function $M(s) = T/F = num/(num + den)$.

The plots of the gain and phase for the loop transfer function $T(j\omega)$, for $F(j\omega)$, and for $M(j\omega)$ can be made in MATLAB with:

```
w = logspace(-1,1);              % freq range 0.1 to 10 rad/sec
den = [1 15 50 0 0 0];
num = [0 0 0 50 27.5 0.25];      % equal length of the vectors
g = num + den;                   % makes the addition allowable
bode(num, den, w)                % for T
hold on
bode(g, den, w)                  % for F
bode(num, g, w)                  % for M
hold off
```

The plots are shown in Fig. 1.15(a). The labels are placed with mouse and cursor, one at a time, using MATLAB command `gtext('label')`. The feedback is large at low frequencies and is negative up 0.8 rad/sec.

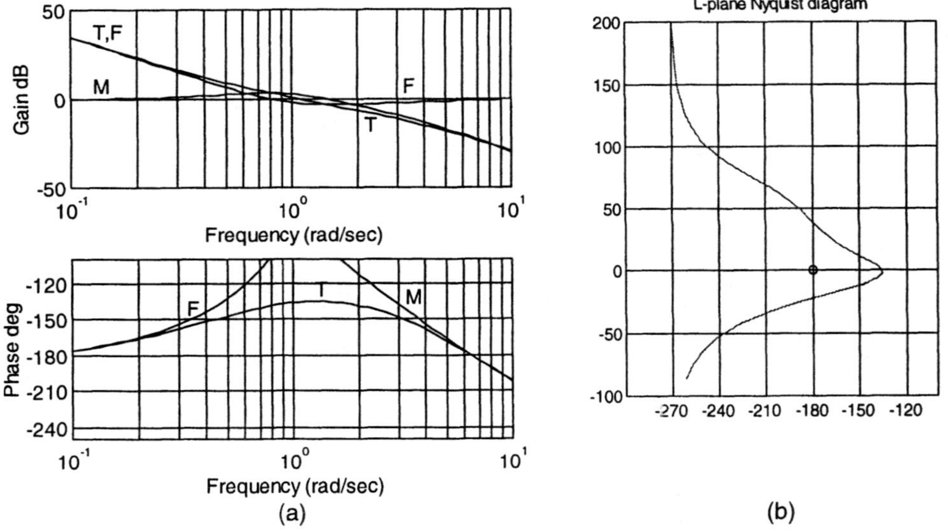

Fig. 1.15 (a) Loop frequency response for the elevation control system: at lower frequencies, T and F overlap; at higher frequencies, T and M overlap. (b) L-plane Nyquist diagram

The closed-loop gain response $20 \log |M|$ is nearly flat up to 1.4 rad/sec, i.e., up to 0.2 Hz. The gain is peaking at 0.8 rad/sec. The hump on the gain response does not exceed 6 dB, which satisfies the design rule mentioned in Section 1.4.3. More precise design methods will be studied in the following chapters.

The plot for disturbances in the system with feedback, the dashed line in Fig. 1.14, is obtained by subtracting the feedback response (in dB) from the disturbance spectral density response, or directly by dividing (1.7) by (1.8). The disturbances are greatly reduced by the feedback.

The mean square of the output error is proportional to the integral of the squared spectral density with linear scales of the axes. The plots required to calculate the reduction in the mean square error can be generated with MATLAB and the areas under the responses found graphically, or the mean square error can be directly calculated using MATLAB functions.

The L-plane Nyquist diagram is shown in Fig. 15(b). The diagram avoids the critical point by significant margins: by 20 dB from below, by 40 dB from above, and by 42 degrees from the right.

Example 1. If the compensator gain coefficient in the system with the Nyquist diagram shown in Fig. 1.15(a) is increased 5 times, i.e. by 14 dB, the Nyquist diagram shifts up by 14 dB and the margin from below decreases from 20 dB to 6 dB.

If the loop gain is increased by (approximately) 20 dB, the Nyquist diagram shifts up by 20 dB, the return ratio becomes 1 at a certain frequency, and the closed-loop gain at this frequency therefore becomes infinite. As we already mentioned in Section 1.4.3, this is a condition for the system to become unstable. Similarly, if the loop gain is reduced by 40 dB, the Nyquist diagram shifts down, at some frequency the return ratio becomes 1, and the system becomes unstable. Using the Nyquist diagram for stability analysis will be discussed in detail in Chapter 3.

The transient response of the closed-loop system to the 1 radian ***step command*** ("increase instantly the elevation angle by 1 radian") is found with:

```
num = [0 0 0 50 27.5 0.25];
den = [1 15 50 0 0 0];
g = num + den; step(num, g)
grid
```

The step-response is shown in Fig. 1.16, nominal case. The output doesn't rise instantly by 1 radian as would be ideal: it rises by 1 radian in less than 2 seconds but then overshoots by 30%, then slightly undershoots, and settles to 1 radian with reasonable accuracy in about 10 seconds. This shape of the closed-loop transient response is typical and, commonly, acceptable.

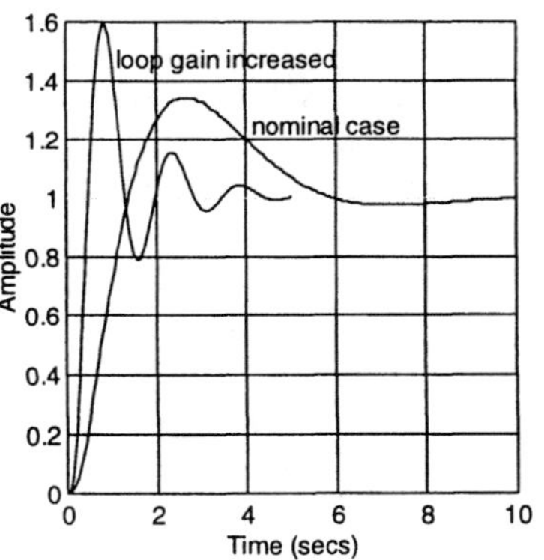

Fig. 1.16 Output response to 1 radian step command

Example 2. The effect of the Nyquist diagram passing closer to the critical point can be seen on the closed-loop transient response. The response generated with `den = 0.2*den` is shown Fig. 1.16, curve "loop gain increased". The response is faster, but it overshoots much more and is quite oscillatory.

1.7 Effect of feedback on the actuator nonlinearity

Actuators can be relatively expensive, bulky, heavy, and power-hungry. Economy requires that the actuator be as small as possible. However, when using small actuators, we have to accept the fact that these actuators will not be able to reproduce signals of relatively large amplitudes without distortion. The controllers must be designed such that this effect will not cause a catastrophic failure (and such a danger is quite real). We will consider these issues in Chapter 3 and later, in Chapters 9–13.

Because the output power of any actuator is limited, the input-to-output relationship invariably saturates when the input amplitude is large enough. ***Saturation*** limits the amplitude of the output signal. The ideal (***hard***) saturation characteristic is shown in Fig. 1.17(a) by the dashed line. In many actuators, the saturation is ***soft*** as shown by the solid curve.

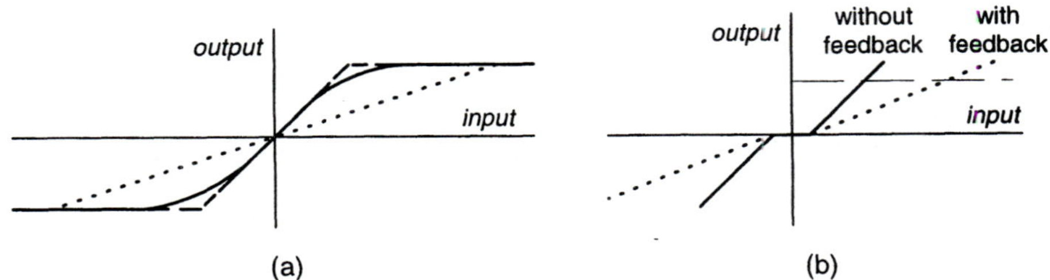

Fig. 1.17 Input-output characteristic of the actuator with (a) soft saturation and (b) dead-zone

When the amplitude of the input increases, the ratio of the output to the input decreases, or, loosely speaking, the gain coefficient of the saturation link decreases. (This gain reduction will be studied in Chapter 11.)

Large feedback about the actuator changes the shape of the input-output characteristic. If the input signal level is such that the slope (differential gain) of the saturation curve is not yet too flattened out, then the differential feedback may remain large and the closed-loop differential transfer function can be quite close to B^{-1}. The closed-loop amplitude characteristic shown by the dotted line is therefore a segment of a nearly straight line. (The slope of the line is shallow since the feedback reduces the input-output differential gain.) Therefore, *in a system with soft saturation, the input-output curve appears as hard saturation when the feedback is large.*

The ***dead-zone*** characteristic is shown in Fig. 1.17(b). Large feedback reduces the differential input-output gain coefficient and therefore makes the input-output characteristic shallower as shown by the dotted curve. Therefore, at any specified amplitude of the output signal, and particularly for the output amplitude marked by the thin dashed line, the input signal amplitude with the feedback is larger. Relative to this input amplitude, the part of the input signal that causes no response in the output due to the dead zone, decreases.

In other words, *the feedback reduces the relative width of the dead zone*. This feature allows the achievement of high resolution and linearity in control systems which use actuators and drivers with rather large dead zones (such actuators and drivers may be less expensive or consume less power from the power supply line, like push-pull class B amplifiers or hydraulic spool valve amplifiers, briefly described in Section 7.1.3, or the motors with mechanical gears).

Next, consider the output signal distortions caused by a small deviation of the actuator from linearity. In response to a sinusoidal input with frequency f, the output of the nonlinear forward path consists of a fundamental component with amplitude U_2 and additional Fourier components called ***nonlinear products***. The ratio of the amplitude of a nonlinear product to the amplitude of the fundamental is the ***nonlinear product coefficient***.

Consider one of these products having frequency nf and amplitude $U_{2(n)}$. If the forward path is approximately linear, the nonlinear product can be viewed as a disturbance source added to the output as shown in Fig. 1.18.

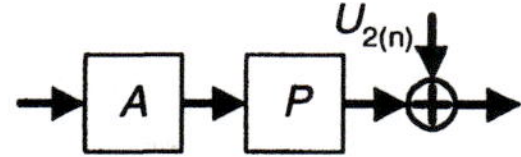

Fig. 1.18 Equivalent representation of nonlinear distortions in the actuator

Now, compare two cases: first, the system without feedback, and second, the system with feedback and with the input signal increased so that the output signal amplitude U_2 is preserved. In the second case, the disturbance, i.e., the amplitude of the nonlinear product at the system's output, is reduced by the value of the feedback. Therefore, *the nonlinear product coefficient for a closed-loop system is reduced by the value of feedback at the frequency of the product.*

Example 1. Without feedback, the nonlinear distortion coefficient in an audio amplifier is 5%. When feedback of 100 is introduced over the entire frequency band of interest, the coefficient becomes 0.05%.

Example 2. An amplifier is used to amplify signals of several TV channels. The feedback in the amplifier is very large at lower frequencies but drops to 5 at 200 MHz. A third harmonic of the 67 MHz signal produces an undesirable nonlinear product with frequency 200 MHz that falls within the band of a higher frequency channel. The feedback reduces the amplitude of this product 5 times.

Low nonlinear distortion in the telecommunication feedback amplifiers invented by Harold Black at the Bell Laboratories in the 1920s, allowed development of long-haul multi-channel telecommunication systems. The works of Black and the Bell Laboratories' scientists Harry Nyquist and Hendrik Bode established the basis for the frequency domain design of feedback systems and feedback maximization. This enabled an increase in the number of telephone channels over expensive telecommunication cables. Later, these methods were applied to feedback control systems to maximize the accuracy and speed of operation.

1.8 Sensitivity

Sensitivity functions are generally used to quantify the undesirable effects of some parameters' deviations from normal of a transfer function. Sensitivity is not a transfer function, although numerically can happen to be equal to one. For control engineers, of particular interest is the sensitivity of the closed-loop transfer function to the plant parameter variations.

An infinitesimal relative change in the plant transfer function dP/P causes an infinitesimal relative change $d(U_2/U_1)/(U_2/U_1)$ in the closed-loop system transfer function U_2/U_1. The feedback system **sensitivity** is defined as the ratio of these changes:

$$S = \frac{d(U_2/U_1)/(U_2/U_1)}{dP/P} = \frac{d[(\log(U_2/U_1)]}{d(\log P)}. \tag{1.9}$$

The smaller the sensitivity, the better. From (1.3),

$$\frac{U_2}{U_1} = \frac{CAP}{CAPB+1};$$

then

$$S = \frac{P}{U_2 / U_1} \frac{d(U_2 / U_1)}{dP} = \frac{P(CAPB+1)}{CAP} \frac{[CA(CAPB+1) - C^2 A^2 PB]}{(CAPB+1)^2} = \frac{1}{(CAPB+1)}.$$

Therefore, as was established by Harold Black,

$$S = \frac{1}{F}. \tag{1.10}$$

The sensitivity is small when the feedback is large, and *the feedback reduces small variations in the output variables, expressed either in percents or in logarithmic values, $|F|$ times*.

Example 1. When the feedback is 10 and the plant magnitude $|P|$ changes by 10%, the closed-loop transfer function changes by only 1%.

Example 2. When the feedback is 10 and the plant magnitude $|P|$ changes by 3 dB, the closed-loop gain changes by 0.3 dB.

When the plant transfer function depends on some parameter q (temperature, pressure, power supply voltage, etc.), this dependence can be characterized by the sensitivity $S_{Pq} = (dP/P)/(dq/q)$. The chain rule can then be used to find the sensitivity of a closed-loop transfer function to q, i.e., the product of the sensitivities: $S \times S_{Pq}$.

Example 3. The feedback is 10, the plant's sensitivity to one of its elements is 0.5. When the value of this element changes by 20%, the closed-loop transfer function changes by 1%.

Since the actuator and compensator transfer functions enter the equation (1.3) in the same way as P, similar formulas describe the effects of variations in the actuator and compensator transfer functions. Since the sensitivity of a large-feedback closed-loop response to the feedback path is nearly 1, *lesser accuracy of the compensator and actuator implementation is acceptable than the accuracy of the feedback path.*

Example 4. The feedback is 100, the plant gain uncertainty is 3 dB. When the actuator and compensator implementation accuracy (together) is 0.2 dB, the total uncertainty of the forward path gain is 3.2 dB. Due to the feedback, the closed-loop uncertainty will be 0.032 dB.

If the feedback path uncertainty is, say, 0.01 dB, the total closed-loop response uncertainty is 0.042 dB.

1.9 Effect of finite plant parameter variations

Plant parameters often vary widely. For example, payloads in a temperature-controlled furnace can be quite different. The mass of a rocket changes while the propellant is being used up. The mass of a cart whose position or velocity must be controlled depends on what is placed in the cart. The moment of inertia of an antenna about one axis might depend on the angle of rotation about another axis.

Sensitivity analysis is a convenient tool for calculation of closed-loop response error and provides sufficient accuracy when plant parameter variations are small. When the plant P deviates from the nominal plant P_o by ΔP which is more than 50% of P_o, for better accuracy the ***Horowitz sensitivity*** (after Isaac M. Horowitz) can be used:

$$S_H = \frac{[\Delta(U_2/U_1)]/(U_2/U_1)_o}{[\Delta P]/P_o} \tag{1.11}$$

This is the ratio of finite relative changes, where the plant perturbed value $P = P_o + \Delta P$. With the input U_1 kept the same for both cases and U_{2o} the nominal output, the perturbed output $U_2 = U_{2o} + \Delta U_2$. It can be shown that the Horowitz sensitivity is the inverse of the feedback for the perturbed plant (see Problem 8):

$$S_H = \frac{1}{CA(P_o + \Delta P)B + 1} = \frac{1}{F} \ . \tag{1.12}$$

Again it is seen that *large negative feedback renders the closed-loop transfer function insensitive to plant parameter variations*.

Example 1. If the nominal value of the plant is $P_o = 100$, the perturbed plant is $P = 200$, and the feedback is $F = 50$ (with this P), then the relative output variation is found to be

$$\frac{\Delta U_2}{U_{2o}} = \frac{\Delta P}{P_o} F^{-1} = \frac{100}{100}\frac{1}{50} = 0.02 \,. \tag{1.13}$$

1.10 Automatic signal level control

The block diagram of automatic volume control in an AM receiver is shown in Fig. 1.19. The goal for the control system is to maintain the carrier level constant at the AM detector, in spite of variations of the signal strength at the antenna.

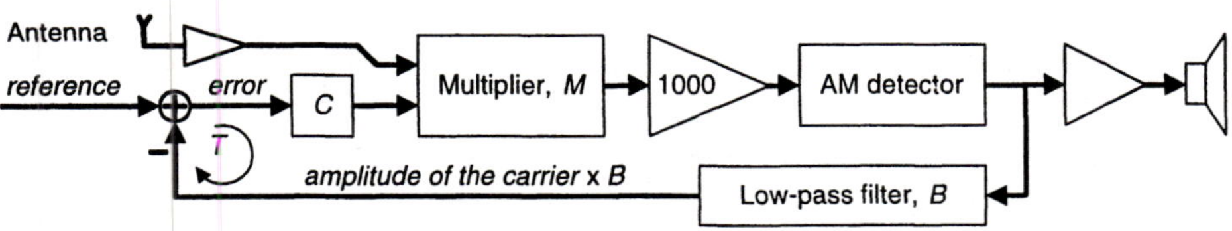

Fig. 1.19 Automatic carrier level control

The antenna signal is amplified and applied to the multiplier M. The signal from the multiplier is further amplified 1000 times and applied to the AM detector. The signal that appears at the output of the detector is the sum of the audio component and the very-low-frequency component that is proportional to the carrier. The low-pass filter B with corner frequency 0.5 Hz removes the audio signal, and the amplitude of the carrier multiplied by B is then compared with the reference voltage.

The error signal processed by the compensator is applied to the second input of the multiplier, changing its transfer function for the RF signal. When the feedback is large, the error gets small and the output carrier level approximates *reference* /B.

Example 1. If, for example, *reference* = 0.5 V, $B = 0.5$, $C = 100$, and the RF carrier amplitude at the modulator input is 0.001V, the return ratio $T = 50$. The carrier amplitude at the detector equals to *reference* × $T/[(T + 1)B]$ = 0.98 V which is very close to the desired carrier amplitude *reference*/B = 1 V.

Since RF signals can vary over a very large range, the gain in the feedback loop can change accordingly. The feedback must be large even when the RF signal is the smallest acceptable (when the signal level is only slightly larger than the level of the always present noise), and, on the other hand, the system must perform well when the loop gain increases by an additional 60 to 80 dB due to the increase of the signal at the antenna.

1.11 Lead and *PID* compensators

Compensator design will be discussed in Chapters 3–6. However, to make this chapter a short self-contained course on servo design, we present below simple design rules for two most frequently used types of the compensators. We assume ω_b has been already chosen (it must be, typically, at least 10 times lower than the frequency of the plant structural mode, and the sensor imperfection and noise at ω_b must not be excessive).

A ***lead compensator*** $C(s) = k(s + a)/(s + b)$ is often used when the plant transfer function is close to a double integrator. The zero $a \approx 0.3\omega_b$, and the pole $b \approx 3\omega_b$. The coefficient k is adjusted for the loop gain to be 0 dB at ω_b, $k \approx b/[\omega_b \times Plant(\omega_b)]$. Lead compensators are used in the Problems 42–44.

A ***PID compensator*** transfer function $C(s) = P + I/s + Dqs/(s + q)$ is tuned by adjusting three real coefficients: P (proportional), I (integral), and D (derivative). The coefficient (not to confuse with the plant) $P \approx 1/Plant(\omega_b)$. The coefficient $I \leq 0.2P\omega_b$. (If $I = 0$, the compensator becomes a lead.) The coefficient D is 0 for a single-integrator plant and, approximately, $0.3P/\omega_b$ for a double-integrator plant. The pole $q \approx 4\omega_b$.

The compensator parameters are fine-tuned either experimentally or using a mathematical plant model and plotting computer simulated open-loop and closed-loop frequency responses and the closed-loop step-response. The loop phase shift at ω_b [i.e., arg $T(\omega_b)$] must be kept between $-120°$ and $-150°$ degrees (as shown in Fig. 1.15). The compensator hardware and software implementation is described in Chapters 5 and 6.

1.12 Conclusion and a look ahead

The material presented in this chapter enables some analysis of single-loop linear control systems. The reader can probably even design some control systems and also demonstrate that feedback vastly improves the system performance. The examples in Appendix 13 should start to become interesting and comprehensible. Still, in order to design systems with performance close to the best possible, the following topics need to be mastered:

- addition of prefilters and feedforward paths to improve the closed-loop responses;
- analysis and design of multiloop control systems – since, for example, in a real antenna attitude control system there also exists a feedback loop stabilizing the motor rate which is "nested" within the main loop;
- design of feedback systems with maximum available accuracy – since the feedback is limited, as we will learn, by some fundamental laws;
- implementation of controllers with analog and digital technology;
- building mathematical models of plants and control systems to evaluate system performance;
- analysis of the effects of the link's nonlinearities, and design of nonlinear controllers. (The controller for the system having the Bode and Nyquist diagrams

when the actuator becomes overloaded.)

1.13 Problems

1 Using proper names for the actuators, plants, and sensors, draw block diagrams for feedback systems controlling (a) temperature and (b) pressure in a chamber, (c) angular velocity of a rotating machine element, (d) luminescence of an illuminated surface, (e) frequency of an oscillator, (f) pitch, yaw and roll of an airplane.

2 When a control system is being designed, conversions from numbers to dB and back need be performed fast. To do this, the following table must be memorized:

decibel	0	1	3	6	10	20	$20n$
number	1	1.12	1.4	2	3.16	10	10^n

Then, say, what is 30 dB? 30 dB = 10 dB + 20 dB, i.e. 31.6 ≈ 30 times.
Or, what is 54 dB? 54 dB = 60 dB – 6 dB, i.e. 1000/2 = 500 times.

Convert to dB: (a) 100; (b) 200; (c) 3000; (d) 5, (e) 8.
Convert to numbers: (f) 110 dB; (g) 63 dB; (h) 12 dB; (i) 18 dB; (j) 30 dB.

3 T is equal to (a) 0.01; (b) –0.01; (c) 0.1; (d) –0.1; (e) 2.72; (f) –0.9; (g) 10; (h) –10; (i) $1.5\angle 150^\circ$.
Calculate F and M, and conclude whether the feedback is positive or negative, large or negligible.

4 For $T = 99$ and B equal to (a) 0.01; (b) 0.1; (c) 1; (d) 0.05; (e) 2.72; (f) 3, calculate the closed-loop transfer functions, and find the error E and the command U_1 to make the output $U_2 = 10$.

5 The open-loop gain coefficient is 3000, the closed-loop gain coefficient is (a) 100, (b) 200, (c) 3000, (d) 5, (e) 2.72.
What are the feedback, return difference, and return ratio? Is this the case of positive or negative feedback?

6 The open-loop gain coefficient is 5000, the closed-loop gain coefficient is (a) 100, (b) 200, (c) 3000, (d) 5, (e) 2.72.
Is the feedback large? How much will the closed-loop gain coefficient change when, because of changes in the plant, the open-loop gain coefficient becomes 6000?

7 In an antenna elevation control system, the feedback is large, the antenna moment of inertia is 430 kgm^2, the three-phase 10 kW motor with gear ratio 120:1 is painted green, and the angle sensor gain coefficient $B = 0.1$ V/degree. What must be the command for the elevation angle to be
(a) 30.5°; (b) 15.5°; (c) 3.5°; (d) 1.5°; (e) 2.72°; (f) 30 mrad?

8 Use the MATLAB commands `conv` and/or `zp2tf` to calculate the coefficients of the numerator and the denominator polynomials for the functions having the following coefficient k, zeros, and poles:
(a) $k = 10$; zeros: –1, –3, –8; poles: –4, –35, –100, –200;
(b) $k = 20$; zeros: –3, –3, –9; poles: –4, –65, –100, –400;
(c) $k = 13$; zeros: –1, –5, –8; poles: –6, –35, –100, –600;
(d) $k = 25$; zeros: –3, –7, –9; poles: –5, –65, –300, –400;
(e) $k = 2.72$; zeros: –1, –1, –2.72; poles: –1, –10, –100, –100;

(f) $k = 20$; zeros: –3, –4, –12; poles: –4, –165, –150, –500;
(g) $k = 1300$; zeros: –1, –50, –80; poles: –6, –35, –300, –500;
(h) $k = 150$; zeros: –3, –70, –90; poles: –5, –150, –300, –400.

9 Use the MATLAB command `root` and/or `tf2zp` to calculate the poles and zeros of the function
(a) $(20s^2 + 30s + 40)/(2s^4 + 10s^3 + 100s^2 + 900)$;
(b) $(s^2 + 3s + 4)/(s^4 + 2s^3 + 20s^2 + 300)$;
(c) $(10s^2 + 10s + 40)/(2s^4 + 20s^3 + 100s^2 + 2000)$;
(d) $(s^2 + 20s + 200)/(s^4 + 5s^3 + 50s^2 + 300)$;
(e) $(2.72s^2 + 27.2s + 200)/(s^4 + 2.72s^3 + 50s^2 + 272)$;
(f) $(s^2 + 10s + 8)/(s^4 + 2s^3 + 12s^2 + 150)$.

10 Use the MATLAB command `bode` to plot the frequency response for the first-order, second-order, and third-order functions
(a) $10/(s + 10)$;
(b) $100/(s + 10)^2$;
(c) $1000/(s + 10)^3$.
Describe the correlation between the slope of the gain-frequency response and the phase shift.
Plot the step time-response. Describe the correlation between the slope of the gain response at higher frequencies and the curvature of the time-response at small times.

11 Use MATLAB to plot the frequency response and step time-response for the first- and second-order functions
(a) $10/(s + 10)$;
(b) $100/(s^2 + 4s + 100)$;
(c) $100/(s^2 + 2s + 100)$;
(d) $100/(s^2 + s + 100)$.
Describe the correlation between the step time-responses and the shapes of the frequency responses.
Find the denominator roots. Describe the correlation between the denominator polynomial roots, the shapes of the frequency responses, and the step-responses.

12 Use MATLAB to convert the function to a ratio of polynomials and plot the frequency response for the function
(a) $50(s + 3)(s + 12)/[(s + 30)(s + 55)(s + 100)(s + 1000)]$;
(b) $60(s + 3)(s + 16)/[(s + 33)(s + 75)(s + 200)(s + 2000)]$;
(c) $10(s + 2)(s + 22)/[(s + 40)(s + 65)(s + 150)]$;
(d) $-20(s + 2)(s + 26)/[(s + 43)(s + 85)(s + 250)(s + 2500)]$;
(e) $2.72(s + 7)(s + 20)/[(s + 10)(s + 100)(s + 1000)]$;
(f) $-25(s + 2)(s + 44)/[(s + 55)(s + 66)(s + 77)(s + 8800)]$.

13 For Example 2 in Section 1.3.1, find $F(s)$ and $M(s)$, and plot their frequency responses using MATLAB.

14 When $|T|$ monotonically decreases in relation to frequency (as in most feedback control systems), what happens to F and M?

15 With the return ratio equal to the function in Problem 12, and $B = 1$, plot the closed-loop frequency responses for (a), (b), (c), (d), (e), (f).

16 With the return ratio equal to the function in Problem 12, and $B = 4$, plot the closed-

16 With the return ratio equal to the function in Problem 12, and $B = 4$, plot the closed-loop frequency responses for (a), (b), (c), (d), (e), (f).

17 Considering functions in Problem 12 as the return ratio, and $B = 10$, plot the closed-loop frequency responses and the return ratio responses for (a), (b), (c), (d), (e), (f).

18 Plot Nyquist diagrams on the L-plane for the functions
(a) $(20s^2 + 30s + 40)/(2s^4 + s^3 + s^2 + 3)$;
(b) $(s^2 + 30s + 4)/(s^4 + 2s^3 + 2s^2 + 3)$;
(c) $(10s^2 + 10s + 40)/(2s^4 + 2s^3 + s^2 + 3)$;
(d) $(s^2 + 20s + 5)/(s^4 + 5s^3 + s^2 + 3)$.

19 What feedback needs to be introduced in an amplifier with harmonic coefficient 5%, for the resulting harmonic coefficient to be 0.02%?

20 Find the third harmonic coefficient in a feedback amplifier with an open-loop harmonic coefficient of 5%, if the feedback at the frequency of the fundamental is (a) 100; (b) 200; (c) 300; (d) 400; (e) 272, and the return ratio is inversely proportional to the frequency.

21 Before introduction of feedback, the actuator dead zone was (a) 5 N; (b) 200 mrad; what is the dead zone after the feedback of 50 was introduced?

22 Before introduction of feedback, the maximum actuator output signal was 100 m/sec. What is this parameter after the feedback of 30 was introduced?

23 The open-loop gain is (a) 80 dB; (b) 100 dB; (c) 120 dB. The closed-loop gain is 20 dB. Because of plant parameter variations, the open-loop gain reduced by 1 dB. What is the change in the closed-loop gain?

24 The feedback is 80 dB. Plant gain is uncertain within ±1.5 dB. What is the uncertainty in the closed-loop gain?

25 In an amplifier, differential gain variations constitute 0.1 dB. What will these variations be when 40 dB of feedback is introduced?

26 The open-loop gain coefficient is (a) 1000; (b) 2000; (c) 50,000. The closed-loop gain coefficient is 20. Because of plant parameter variations, the open-loop gain is reduced by 5%. What is the change in the closed-loop gain?

27 The feedback is (a) 200; (b) 100; (c) 1; (d) 0.5. Plant gain is uncertain within ±15%. What is the uncertainty in the closed-loop gain coefficient?

28 The feedback is (a) 10; (b) 100; (c)1; (d) 0.5. The plant's sensitivity to the temperature is 0.1 dB/degree. When the temperature changes by −6 degrees, by how much will the output change?

29 Tubes were expensive in the early days of tube radio receivers, and positive (regenerative) feedback was used to increase the amplifier gain. The positive feedback also improved the selectivity of the regenerative receiver (but narrowed the bandwidth of the received signal).

Illustrate the above using the following example. The forward path consists of a resonance contour tuned at the signal frequency 100 kHz and an amplifier. The forward path transfer function is $10000/(s^2 + 12.5s + 628^2)$, where the angular velocity is expressed in krad/sec. The feedback path coefficient B $= -0.0077$.

Plot the gain response of the receiver without and with regenerative feedback. By how much does the feedback increase the receiver gain? What is the sensitivity at the frequency of the resonance? How do small deviations in the amplifier's gain affect the output signal? What needs to be done in the feedback path to keep the closed-loop gain constant?

What, in your opinion, will happen when the amplifier gain coefficient increases by 2% but the feedback path is not adjusted? By 2.5%?

30 In the previous example, introduce negative feedback (degenerative feedback) with $B = 0.001$ and then $B = 0.01$. Explain the effect of the feedback on the gain and the selectivity.

31 Prove that Horowitz sensitivity equals $1/F$.

32 The nominal value of the plant is 30, the perturbed value is 50. The perturbed value of the feedback is 20. With the nominal plant, the value of the output would be 10. Using Horowitz sensitivity, find the change in the value of the output when the plant changes from nominal to perturbed.

33 The perturbed value of the plant is 30, the nominal value is 50. The nominal value of the feedback is 20. With the perturbed plant, the value of the output is 10. Using Horowitz sensitivity, find the change in the value of the output when the plant changes from nominal to perturbed.

34 Derive the expression for the Bode sensitivity for the following links: *P*, *A*, *C*. Give numerical examples. What is the required implementation accuracy for these links?

For example, the plant sensitivity (1.10) can be derived by using (1.3) while keeping U_1 constant:

$$S = \frac{dU_2/U_2}{dP/P} = \frac{PdU_2}{U_2 dP} = \frac{P}{\dfrac{CAPBU_1}{1+CAPB}} \frac{d}{dP} \frac{U_1 CAPB}{1+CAPB} = \frac{1}{F}.$$

Therefore, the plant implementation accuracy can be $|F|$ times worse than the required accuracy for the closed-loop system transfer function.

35 Derive an expression for the Bode sensitivity of the output to variations in the transfer function of the feedback path link *B*. Give numerical examples, including cases where the feedback is large (negative), positive, and negligible.

36 From the frequency hodograph of *T* plotted on the Nichols chart in Fig. 1.20, find the closed-loop gain frequency response of a tracking system (assuming $B = 1$) and plot this frequency response with logarithmic frequency scale.

37 Prove that if *T* is plotted upside-down on the Nichols chart, the curvilinear coordinates give $20 \log|F|$.

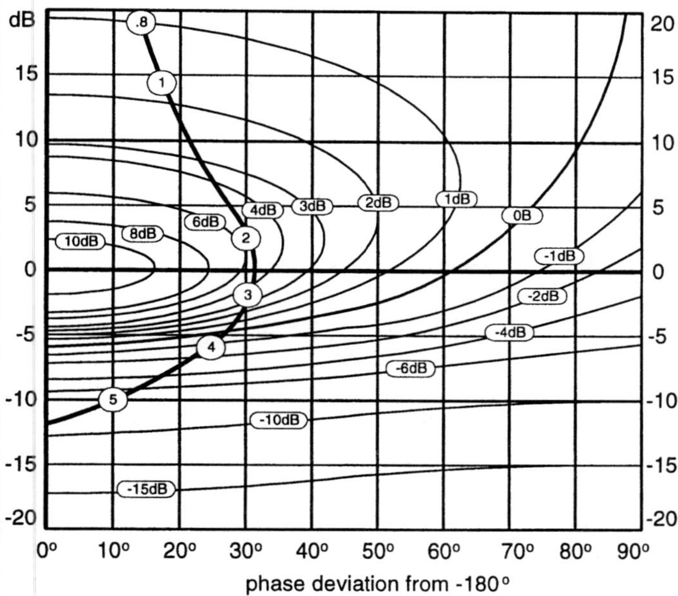

Fig. 1.20 Locus of T on the Nichols chart

38 A periodic disturbance with frequency 0.05 Hz and amplitude 2 mrad in pointing of a spacecraft is caused by magnetometer boom oscillation. Find the value of feedback in the pointing control loop at this frequency required for reducing the disturbance to less than (a) 0.1 mrad; (b) 34 μrad; (c) 12 nrad; (d) 0.2 mrad; (e) 2.72 arcsec.

39 The disturbance spectral density is described by the function (a) $k/[(s+1)s]$; (b) $k/[(s+2)s]$; (c) $k/[(s+5)s]$; (d) $k/[(s+10)s]$; (e) $k/[(s+2.72)s]$.

The return ratio is $1000(s+20)/[(s+1)s]$. Plot the relative disturbance spectral density without and with the feedback.

40 For the voltage regulator depicted in Fig. 1.4, derive the dependence of the feedback and the output voltage on the load resistor R_L. By comparison with the known formula for the voltage at a source terminal: $V = \text{emf} \times R_L/(R_L + R_S)$, find the equivalent output impedance R_s of the voltage regulator.

41 Consider a current regulator with the schematic diagram shown in Fig. 1.21. Here, B is the current sensing resistor, i.e. the system has current feedback at the output. Therefore, the output *current* in the load R_L is stabilized by the feedback, and the regulator output represents a current source. Hence, the output impedance of the regulator must be high.

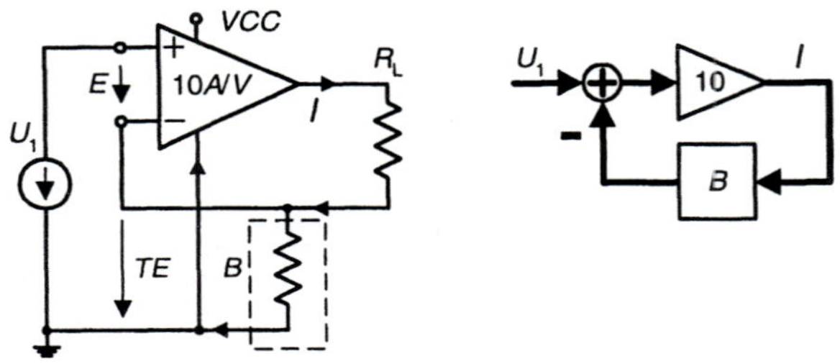

Fig. 1.21 Current regulator

The reference voltage is 0.5 V. The output impedance of the amplifier is much larger than $R_L + B$. Set B for the output current to be (a) 0.1 A; (b) 0.25 A; (c) 0.5 A; (d) 0.4 A; (e) 2.72 A; (f) 0.66 A;
In each case find the return ratio and the output impedance of the regulator.

42 A temperature control loop is shown in Fig. 1.22. The dimensionality of the signals is shown at the block joints. The heater is a voltage-controlled power source with the gain coefficient 2 kW/V. The transfer function (in °C/kW) of the loaded furnace was measured and the experimental response was approximated by transfer function $P(s) = 500/[(s + 0.1)(s + 25)]$. (At dc, when $s = 0$, $P(0) = 200\,^{\circ}$C/kW.) The compensator transfer function $C = 8(s + 1.6)/(s + 0.2)$. The thermometer transfer function is 0.01 V/°C. The command is 4 V.

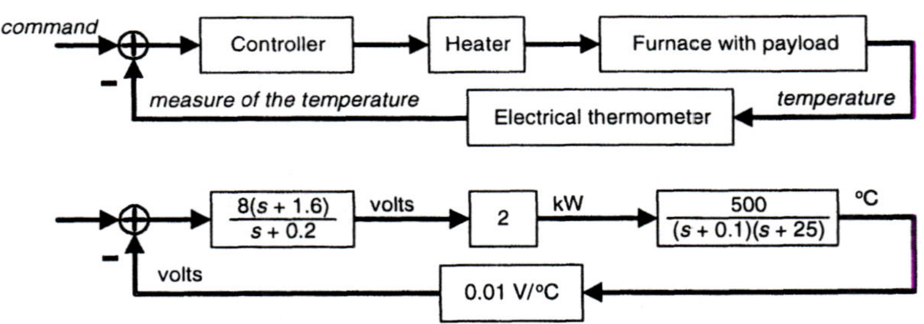

Fig. 1.22 Temperature regulator

(a) Find the return ratio and the feedback at dc (in a stationary regime).

Find the loop transfer function and input-output transfer function. Plot Bode diagrams for the loop transfer function and for the input-output transfer function. Plot the L-plane Nyquist diagram. Check whether the diagram enters the area surrounded by the "6 dB" line on the Nichols diagram.

Plot the output time-response for the 0.01 V step command.

Plot the output time-response to the step disturbance −100 W applied to the input of the furnace that represents some undesired cooling effects.

(b) Create a SIMULINK model. Include in the heater model a saturation with threshold 2 kW. Plot output responses to command steps of different amplitudes. Explain the results.

43 A scanning mirror with an angular velocity drive is shown in Fig. 1.23.

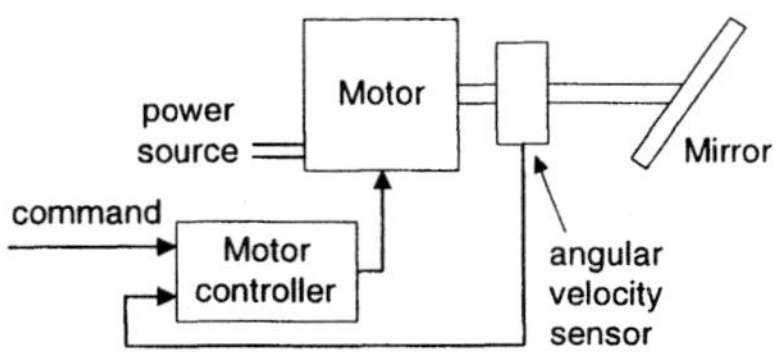

Fig. 1.23 Mirror drive

The mirror angular velocity control loop is shown in Fig. 1.24. The controller transfer function is $C(s) = 10(s + 30)/[(s + 300)s]$. The motor (with driver) is a voltage-controlled velocity source with transfer function 300 (rad/sec)/V (the actuator might have an internal angular velocity control loop). The mirror angular velocity differs from the angular velocity of the motor because of flexibility of the motor shaft and the mirror inertia. The plant transfer function (the ratio of the mirror angular velocity to the motor angular velocity) is $P(s) = 640000/(s^2 + 160s + 640000)$. At dc, the plant transfer function is 1. The gain coefficient of the angular velocity sensor is 0.01 V/(rad/sec).

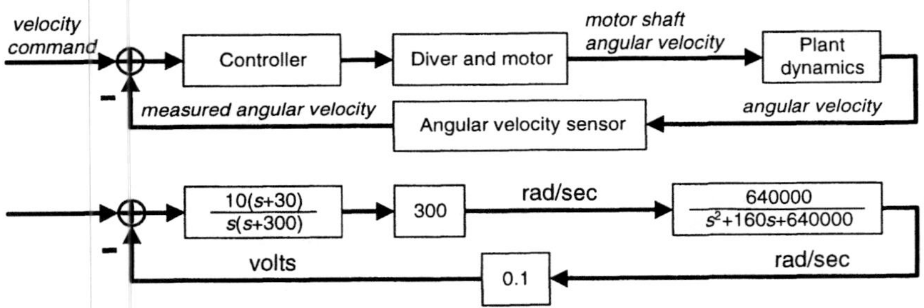

Fig. 1.24 Angular velocity control

(a) Find the loop transfer function and input-output transfer function. Plot Bode diagrams for the loop transfer function and for the input-output transfer function. Plot an *L*-plane Nyquist diagram. Determine whether the diagram enters the area surrounded by the "6 dB" line on the Nichols diagram.

Plot the output time-response for the 1 V step command.

Plot the output time-response to step disturbance −0.1 rad/sec applied to the input of the plant that represents the inaccuracy of the dc permanent magnet motor caused by switching between the stator windings.

(b) Create a SIMULINK model. Include a saturation with threshold 60 W in the driver. Plot output responses to command steps of different amplitude. Explain the results.

44 A pulley with a torque drive is shown in Fig. 1.25. The goal is to maintain the prescribed profile of the torque of the pulley so that the force in the cable lifting a load will be as desired. The motor torque control loop is shown in Fig. 1.26. The plant dynamics, i.e., the transfer function from the torque to the angle, is out of the feedback loop since the torque sensor is measuring the pulley torque directly at the motor.

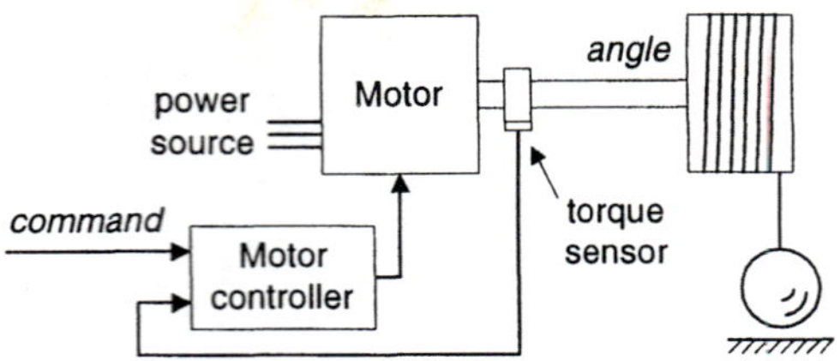

Fig. 1.25 Torque regulation in a pulley

The torque sensor gain coefficient is 0.1 V/Nm. The controller transfer function is $C(s) = 0.8(s + 15)/(s + 100)$. The actuator (a motor with an appropriate driver) is a voltage-controlled torque source with transfer function 30 Nm/V. The disturbance torque due to the motor imperfections is 0.1 Nm.

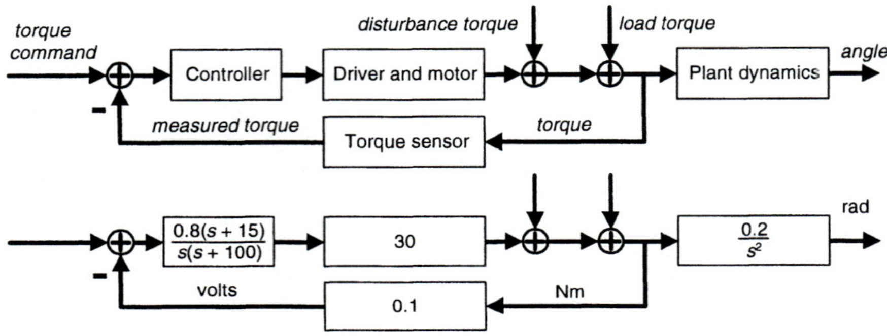

Fig. 1.26 Torque control

(a) Find the loop transfer function and the input-output transfer function. Plot Bode diagrams for the loop transfer function and for the input-output transfer function. Plot an L-plane Nyquist diagram. Check whether the diagram enters the area surrounded by the "6 dB" line on the Nichols diagram.

Plot the torque time-response for the 0.01 V step command.

Plot the torque time-response to the disturbance torque as a step function.

(b) Assuming that the pulley radius is 0.1 m and the load mass is 2 kg, i.e., the plant moment of inertia is 0.02 kg × m^2, i.e., the plant is a double integrator and its transfer function is $50/s^2$, plot the position time history in response to a step torque command (assume that at zero time the load is on the ground).

(c) Create a SIMULINK model. Include in the driver model a saturation with threshold 2 kW. Plot output responses to command steps of different amplitude. Explain the results.

(d) With SIMULINK plot position time history in response to the following command: 1 V for the duration of 2 seconds, 0 V for the duration of the next 2 seconds, and −1 V for the duration of the next 2 seconds.

(e) Make a schematic drawing of torque control in a drilling rig.

(f) Make a schematic drawing of torque control in a lathe, for keeping constant the force applied to the cutting tool (there could be several possible kinematic schemes).

45 An x-positioner is shown in Fig. 1.27. A ball screw converts the motor rotational motion to translational motion of a table along the x-axis, with gear ratio 0.5 mm/rad.

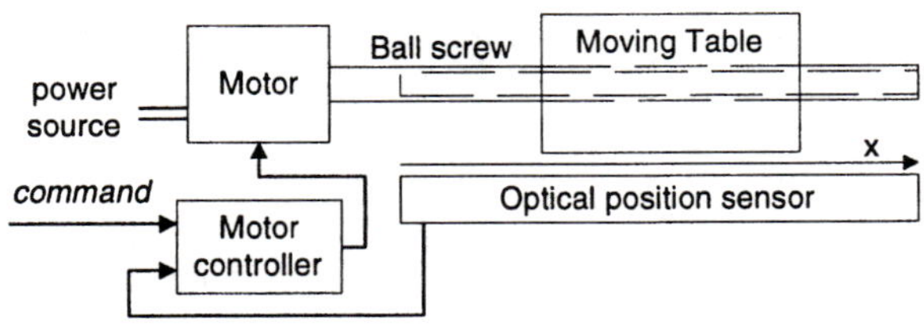

Fig. 1.27 A positioner with an optical position sensor

The positioner control loop is shown in Fig. 1.28. The controller transfer function $C(s) = 0.03(s + 20)/(s + 70)$. The actuator (the driver with the motor) is a voltage controlled angular velocity source with transfer function 500 (rad/sec)/V. The plant transfer function is the ratio of the position to the velocity, i.e., $(1/s)0.5$. The gain coefficient of the optical position sensor is 0.2 V/mm. The disturbance in the angle represents the ball screw imperfections.

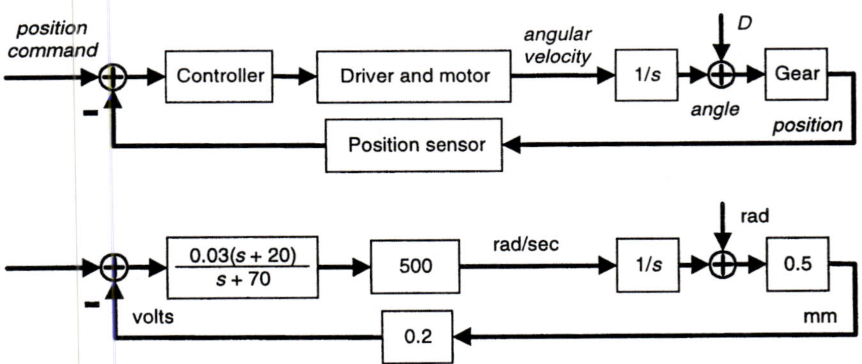

Fig. 1.28 Position control

(a) Find the loop transfer function and input-output transfer function. Plot Bode diagrams for the loop transfer function and for the input-output transfer function. Plot an L-plane Nyquist diagram. Check whether the diagram enters the area surrounded by the "6 dB" line on the Nichols diagram.

Plot the output time-response for the 0.01 V step command.

Plot the output time-response response to step disturbance – 0.1 mm.

If the gear has a dead zone of 0.02 mm, what is the resulting dead zone of the entire closed-loop system?

(b) Create a SIMULINK model. Include a saturation with threshold 3000 rpm in the driver and motor model. Plot output responses to command steps of different amplitude. Explain the results.

46 The signal at an antenna of an AM receiver varies 100 times, from station to station and due to changing reception conditions. The linearized (differential) loop gain of automatic level control at the largest signal level is 1000. What is the range of the output signal carrier variations?

47 The dependence of M on T and many other formulas of this chapter are bilinear functions. It is known that a bilinear function maps a circle (or a straight line) from the complex plane of the variable onto a circle (or a straight line) in the complex plane of the function. Why then are the coordinate curves in the Nichols chart not circles?

48 What is the reason for using voltage followers?

49 A common electric heating pad has thermal control. What would happen if it is left under the blankets? What would happen if there were no thermal control?

Answers to selected problems

1 (a) Heater, furnace with payload, thermometer.

3 (b) $F = 0.99$, $M = -0.01$, the feedback is positive but negligible (i.e. $|F| \approx 1$).

21 (a) 0.1 N.

31 Given the values P and P_o for the nominal and the perturbed plant transfer functions, from (1.9):

$$S_H = \frac{P_o}{P - P_o} \frac{\left[\dfrac{CAPB}{1+CAPB} - \dfrac{CAP_oB}{1+CAP_oB}\right]}{\dfrac{CAP_oB}{1+CAP_oB}} = \frac{1}{1+CAPB} = \frac{1}{F}.$$

Notice the peculiarity of this formula: the changes in the left side (in S_H) are relative to the *nominal* plant value, while the feedback in the right side is calculated for the *perturbed* plant value. The opposite is also true (since the nominal and perturbed (current) values can be swapped).

48 Due to the feedback, the voltage follower has high input and low output impedances, and it relieves the preceding signal source from its voltage being reduced when the load is connected.

49 The temperature will be kept safely low by reducing the consumed power. Without the thermal control, the consumed power will be nearly constant, and the temperature under the blanket may reach unsafe levels.

Chapter 2

FEEDFORWARD, MULTILOOP, AND MIMO SYSTEMS

The command feedforward scheme is equivalent to using a prefilter, or using a feedback path with a specified transfer function. All of these methods enable to obtain the desired input-output closed-loop transfer function with any arbitrary compensator transfer function. Therefore, the compensator transfer function can be chosen as required to maximize the disturbance rejection, i.e., the feedback, while the desired closed-loop transfer function is obtained by using an appropriate feedback path, a command feedforward, or a prefilter.

While the command feedforward scheme and its equivalents do not affect disturbance rejection and plant sensitivity, the error feedforward scheme and Black's feedforward increase disturbance rejection and reduce plant sensitivity.

Multiloop feedback systems are defined, following Bode, as those having a nonlinear (saturation) element in each loop. This definition reflects the importance of taking into account the large uncertainty in the signal transmission of the saturation link caused by the signal amplitude changes.

Major practical types of multiloop systems are studied: local and common loops, nested loops, crossed loops, and the main/vernier loop configuration.

The methods for equivalent transformation of block diagrams are described.

The chapter ends with an introduction of multi-input multi-output (MIMO) systems, coupling, and decoupling matrices in the forward and in the feedback path. Typical MIMO systems are discussed, with examples.

Example 1 from Section 2.1 and Sections 2.3 and 2.4 can be omitted from an introductory control course.

2.1 Command feedforward

The accuracy of the system transfer function can be improved not only by the feedback but also by feeding certain signals forward and combining the signals at the load, or by some combination of feedback and ***feedforward***. There are several feedforward schemes, each having decisive advantages for specific applications; the schemes are briefly described below.

The first type is referred to as ***command feedforwarding*** and is shown in Fig. 2.1. The transfer function of the command feedforward path is

$$(AP_o)^{-1}$$

where P_o is the ***nominal plant transfer function***.

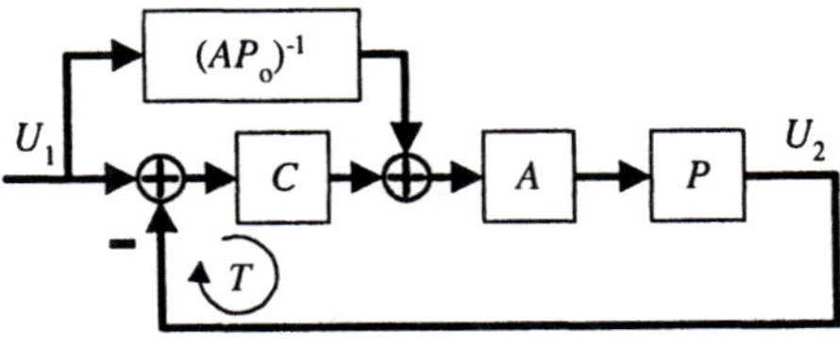

Fig. 2.1 Command feedforwarding

The input-output transfer function of the system with command feedforward is found by summing the transmission functions of parallel paths of the input-to-output signal propagation. In each path, the feedback reduces the transmission F times:

$$\frac{U_2}{U_1} = \left(\frac{1}{AP_o} AP\right)\frac{1}{F} + \frac{T}{F} = \frac{\frac{P}{P_o} + T}{F} = \frac{1+T_o^{-1}}{1+T^{-1}}.$$

This transfer function approaches 1 in two cases: first, when $C \to \infty$ and both T^{-1} and T_o^{-1} vanish, and second, when the plant transfer function does not deviate much from the known nominal plant transfer function P_o, i.e., $P_o \approx P$ and, therefore, $T_o \approx T$. For example, if $C = 0$ (open-loop case), it follows directly from the block diagram (and, certainly, from the formula) that the input-output transfer function is $(P_o/P) \approx 1$.

The input-output transfer function can also be expressed as

$$\frac{T_o(T+1)}{T(T_o+1)} = \frac{M_o}{M}$$

where $M_o = T_o/(T_o + 1)$. If the deviations of the plant P from the nominal value P_o are small, then $M \approx M_o$ and the system transfer function approaches 1, even when the feedback is not large.

The command feedforward can substantially improve the accuracy of the output response to the command, especially over the frequency band where the feedback cannot be made large. There are three limitations however on using the command feedforward:

- The plant must be known pretty well. If the uncertainty in the plant transfer function is large, then P_o cannot be made close to P, and the advantages of the command feedforward decrease drastically.
- The power of the numerator polynomial of the feedforward path transfer function should be smaller than the power of the denominator, for the transfer function to be feasible.

  There is also a limit on the bandwidth of the command feedforward: at higher frequencies $|P|$ typically decreases; however, making $|P_o|$ too small and, therefore, making the feedforward path gain too large would produce an excessively large signal at the input of the actuator, and the actuator would become saturated.
- A plant with substantial pure delay would imply using feedforward with substantial phase advance which is not feasible.

The command feedforward does not change the feedback about the plant, the sensitivity to the plant parameter variations, or the disturbance rejection.

Example 1. Biological, robotic, and many other engineering systems must perform well both in slow and fast modes of operation. In the slow mode, high accuracy must be available, as when surgery is performed. The same actuators (muscles, motors) must also be able to act fast and provide high acceleration for the plant, as when hitting a ball or jumping. The accuracy of the fast components of the motion must be reasonably good but need not be as good as the accuracy of the slow mode action.

High-accuracy slow motion can be achieved with closed-loop control using eyes or other position sensors. The feedback configuration, however, doesn't suit the fast action mode since the speed of the feedback action is bounded by the delays in the feedback loop. To make the actuators move the plant with the maximum speed and acceleration they are capable of while maintaining a reasonable accuracy, the actuators must be commanded directly.

The block diagram of a feedback/feedforward system to answer these conflicting requirements is shown in Fig. 2.2. The ***commander*** transforms the general input command into commands for the individual actuators. The actuators are equipped with wideband (fast) feedback loops using local sensors (***inner loops***) which make them accurate. The plant output variables (average speed, direction of motion, etc.) are controlled by the ***outer feedback loops***.

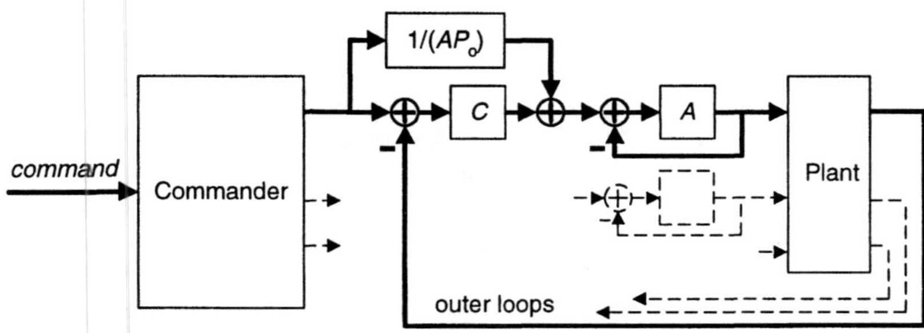

Fig. 2.2 Feedforward system with fast actuator loops and slow outer loops

At low frequencies (i.e., for slow motion), the compensators' gains are high and the system operates closed-loop. At higher frequencies (for fast motion), the compensators' gains roll down and the command feedforward paths become dominant. For each of the controlled output variables, the system input-output transfer function

$$\frac{1+T_o^{-1}}{1+T^{-1}}$$

approximates 1 over the entire frequency range of operation, although much more accurately at lower frequencies where the feedback is large. The transition in accuracy from the slow to the fast mode is gradual, since the gain of the compensator gradually decreases with frequency.

When the motion comprises both slow and fast components, the outer feedback loops control the slow components (as when adjusting the general direction of running) while for the fast components, only the inner loops are closed.

2.2 Prefilter and the feedback path equivalent

The block with transfer function R in the command path preceding the ***command summer*** (same as the feedback summer) in Fig. 2.3 is called the ***prefilter***. This system has return ratio T and closed-loop response $RT/(T+1)$. The system in Fig. 2.4 has return ratio T and closed-loop response $B^{-1}T/(T+1)$. These two systems are equivalent if $B = 1/R$.

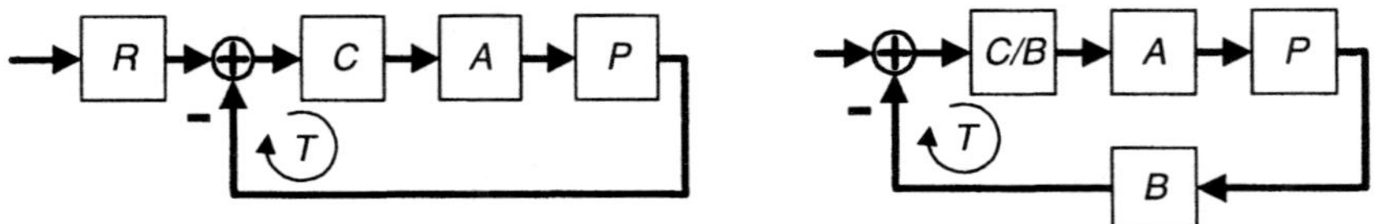

Fig. 2.3 System with prefilter

Fig. 2.4 System with feedback path

With $R = 1/B = 1/M_0$, the systems in Figs. 2.3 and 2.4 have the closed-loop transfer function M/M_0, the same as that for the system in Fig. 2.1. Therefore, all three systems are potentially equivalent. A system having all three links: feedforward path, prefilter, and feedback path, can always be equivalently transformed to any one of the systems in Figs. 2.1, 2.3, 2.4.

The three equivalent methods modify the input-output transfer function and can make it closer to the desired, compared with the system in Fig. 1.1(b), *but do not change the feedback or the sensitivity*.

The system design is performed in two stages: first, the compensator C is defined; then, R, or B, or the feedforward path $(AP_0)^{-1}$.

The compensator transfer function must be chosen so as to maximize the feedback over the bandwidth of interest – as will be shown in Chapters 4 and 5. Once designed, the compensator should not be compromised during the next stage of the design, which is the implementation of a suitable nominal closed-loop response. This goal can be achieved by a proper choice of B, or R, or the feedforward path.

Since the feedback reduces the effects of the compensator parameter variations, compensators need not be precise. The tolerances in the prefilter and the feedback link directly contribute to the output error, so these links do need to be precise. The required accuracy of the feedforward path implementation depends on the accuracy of the knowledge of the plant transfer function, and may be different at different frequencies.

It might seem attractive to integrate the prefilter, the compensator, and the feedback link into a generalized linear subsystem which can be designed using some universal performance index. However, this is not recommended since the sensitivities and the accuracies with which the blocks should be implemented are quite different; the design of these blocks is to a large extent independent; and it is much easier to design these blocks one at a time.

2.3 Error feedforward

Fig. 2.5(a) describes an entirely different scheme known as ***error feedforwarding***.

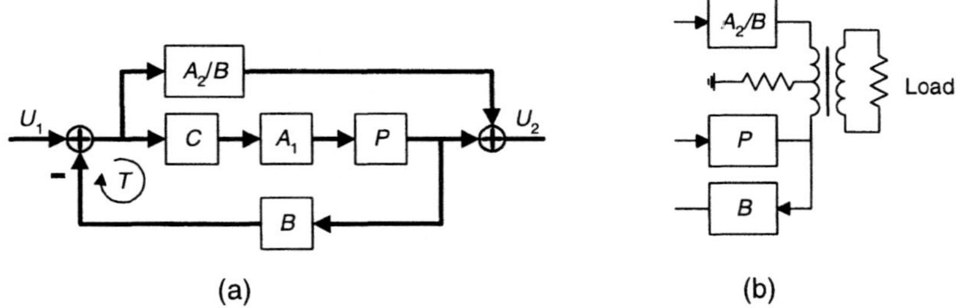

Fig. 2.5 Error feedforward

The input-output transfer function is

$$\frac{U_2}{U_1} = \frac{CA_1 P}{1 + CA_1 PB} + \frac{A_2}{1 + CA_1 PB}, \tag{2.1}$$

and if $A_2 = 1$, then $U_2 = U_1/B$. The sensitivity of the system input-output transfer function to the plant parameter variations (and to variations in C and A_1) can be calculated as $(1 - A_2)/F$. If A_2 is made close to 1, the sensitivity approaches zero.

Practical applications of this method of sensitivity reduction are restricted by the difficulties in the design of the output summer. For an electrical amplifier system with a known load, the output summer can be made using a bridge-type signal combiner as shown in Fig. 2.5(b); the bridge prevents the output signal from the upper path from going into the feedback path. It is more difficult to implement a mechanical system with such properties.

2.4 Black's feedforward method

Finally, consider ***Black's feedforward*** method for sensitivity reduction, which was invented by Harold Black, around the same time he invented the feedback method. Fig. 2.6(a) depicts the method. Note that no feedback appears in this system. The upper signal path is the main one, and the lower one is the error compensation path. The error signal is the difference between the command U_1 and the output of the main path, measured via the B-path. The error is amplified by the error path and added to the system's output, so as to compensate for the initial error in the main path.

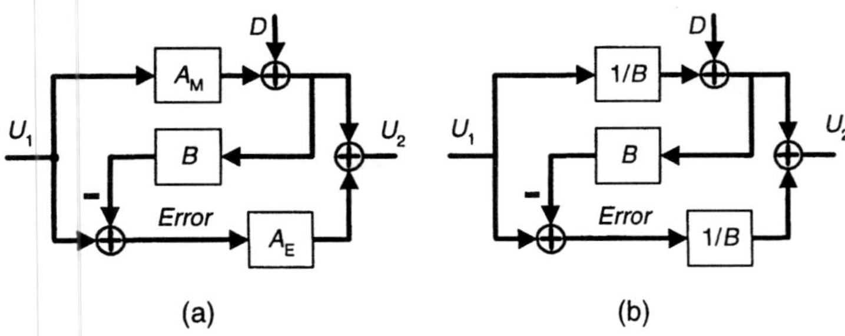

Fig. 2.6 Black's feedforward system: (a) general, (b) ideal case

The input-output transfer function is

$A_M + A_E - A_M A_E B$.

If either A_M or A_E or both equal $1/B$, the input-output transfer function is $1/B$.

The sensitivity of the output to variations in A_M is

$$S = \frac{\dfrac{dU_2}{U_2}}{\dfrac{dA_M}{A_M}} = \frac{1 - A_E B}{1 + \dfrac{A_E}{A_M} - A_E B} \tag{2.2}$$

(derivation of the formula is requested in Problem 11).

If $B \approx 1/A_M$, the error signal is the difference between two nearly equal signals, and the error is small. In this case the actuator A_E can be low-power. Such an actuator can be made very precise. When the gain coefficient of this actuator is $A_E = 1/B$, the sensitivity (2.2) becomes zero. In this case the output effect of the disturbance D is also zero.

In order to make A_M and A_E each equal to $1/B$, and preserve these conditions in spite of variations in these actuators' parameters, both actuators are commonly stabilized by internal feedback or by some adaptive automatic gain adjustment. Also, in the case where both of the transfer functions are $1/B$ as in Fig. 2.6(b), the feedforward scheme provides redundancy: if one of the actuators fails, the remaining one takes the full load and the input-output gain remains unchanged.

In some physical systems, the links A_M, B, and A_E incorporate substantial delays, respectively τ_M, τ_B, and τ_E as shown in Fig. 2.7. These delays do not prevent the use of feedforward if they are properly compensated by insertion of delay link $\tau_M + \tau_B$ in the signal path to the first summer, and delay link $\tau_B + \tau_E$ at the output of the main channel. Then, the phase difference between the signals reaching the summers remains the same, and the only difference in the resulting input-output transfer function of the system is the extra delay $\tau_M + \tau_B + \tau_E$.

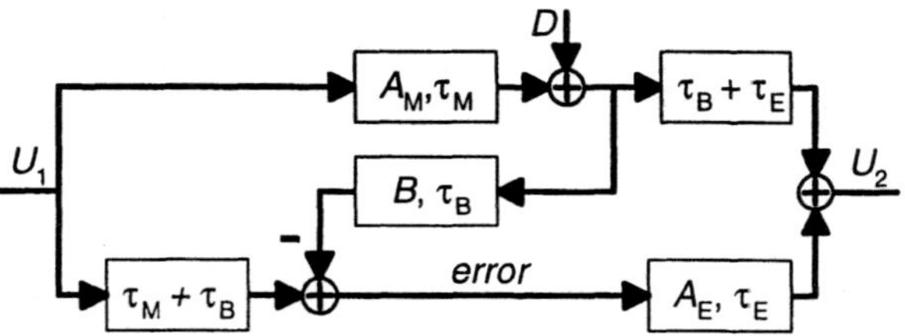

Fig. 2.7 Black's feedforward with delay compensation

This method is often employed in low-distortion amplifiers for telecommunication systems, for signals with frequencies from hundreds of MHz to tens of GHz, but it is not common in control systems. There may be applications to control systems when extreme accuracy is required.

2.5 Multiloop feedback systems

Linear systems can always be transformed to another configuration with a different number of loops, as illustrated in Fig. 2.8.

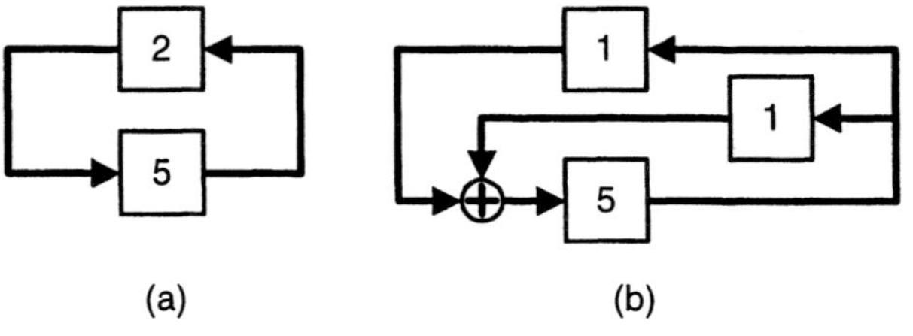

Fig. 2.8 Modifications of a linear system

In accordance with Bode's definition of physical multiloop systems, only the loops which comprise *nonlinear* saturation-type elements are counted. For example, in Fig. 2.9, system (a) is a single-loop system, and system (b) is a three-loop system. This definition is related to the problem of stability analysis of practical systems whose actuators are always nonlinear. Such systems will be studied in Chapters 4–13.

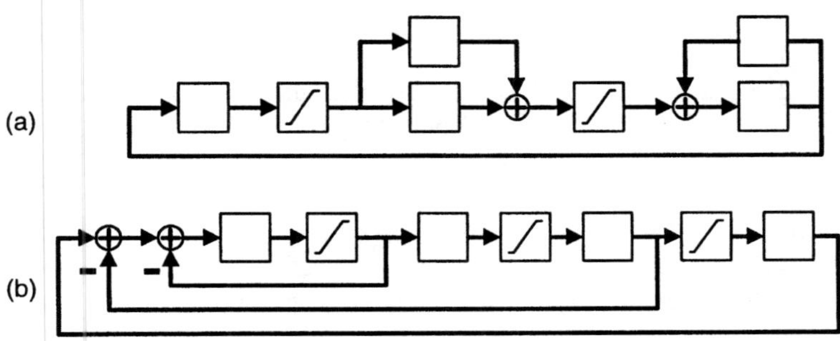

Fig. 2.9 Single-loop (a) and three-loop (b) feedback systems

In this chapter, we will analyze feedback systems in only the linear state of operation, i.e., for small amplitude signals, with saturation links equivalently replaced by unity links.

2.6 Local, common, and nested loops

Fig. 2.1(a) depicts ***local feedback loops***. The sensitivities to each link's parameter variations depend on the feedback in the local loop. The total gain of the chain of these links is reduced by the product of all these feedback values. This gain reduction effect is much larger in this arrangement than in the ***common-loop arrangement***, shown in Fig. 2.10(b).

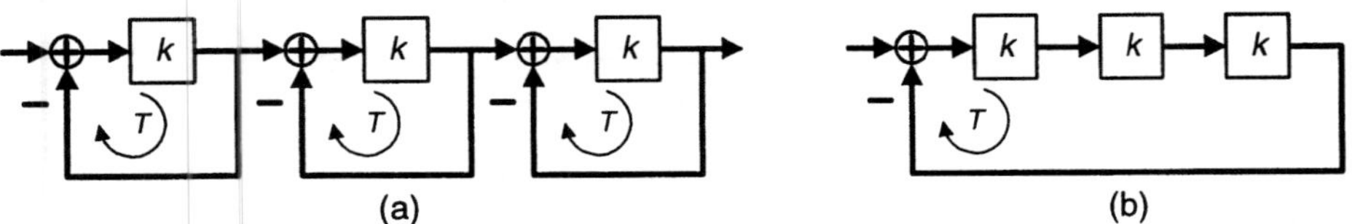

Fig. 2.10 Local loops (a) and the common loop arrangement (b)

If each of the three links with nominal gain coefficient k has the same tolerances, then to provide the same degree of accuracy, it is required to enloop each stage by the same value of feedback. In this case the closed-loop gain $k^3/(T+1)$ of the single-loop system is much higher than the closed-loop gain $k^3/(T+1)^3$ of the system with local loops.

This is why common-loop feedback is preferred in amplification techniques where the resulting gain is important. For control systems, this consideration is not important since the gain can be increased by adding inexpensive gain blocks, but this effect does need to be taken into account when designing analog compensators.

Example 1. Each stage of an amplifier has gain coefficient of 50. The feedback about each stage needs to be at least 10 to make the gain coefficient stable in time. Then, when local feedback is used, the total amplifier gain coefficient will be 125. When the common loop configuration is used, the gain coefficient will be much higher, 12500.

Local loops are often made about links with large parameter variations. These loops can be ***nested*** as shown in Fig. 2.11. Here, the driver amplifier is enlooped by large feedback to make its gain accurate and stable in time, and also to manipulate the output impedance of the driver. The actuator loop makes the actuator-plus-driver subsystem accurate and stable in time. The outer loop improves the accuracy of all links in the forward path, including the plant. The nested loop arrangement is employed for several reasons, to be discussed in detail in Chapter 7, primarily because the feedback bandwidth in the outer loop cannot be made arbitrarily large. The wideband inner loop is about the electrical amplifier (driver), the intermediate bandwidth loop is about the actuator (motor), and the rather narrow bandwidth outer loop, about the plant.

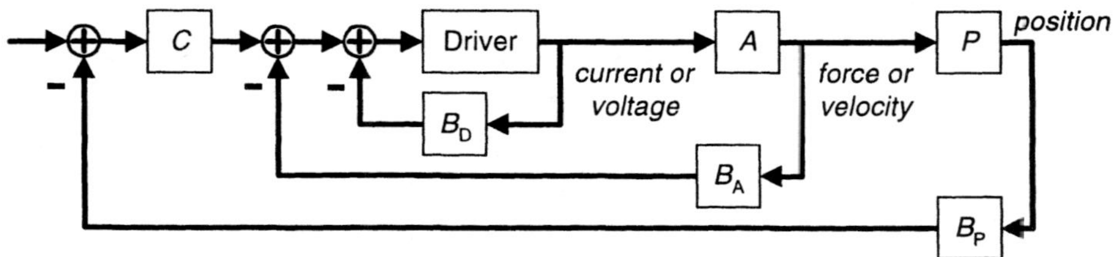

Fig. 2.11 Nested feedback loops

For example, consider the typical case of a driver implemented with an operational amplifier. The op-amps without feedback have very large gain uncertainty and variations due to power supply voltage and temperature changes. An easily implemented large local feedback loop about the driver amplifiers will reduce the tolerances of the forward path to only that of the compensator, plant, and actuator. If op-amps are used in the compensator, they must also have large local feedback.

The variables that are fed back in the inner loops can be different: at the output of the driver, the variable can be the voltage or the current, and at the output of the actuator, the velocity (rate) or the force. The choice of these variables alters the plant transfer function P which is the ratio of the output to the input variables of the plant. For example, in a position control system, force feedback about the actuator makes a rigid body plant a double integrator while the rate feedback makes the plant a single integrator. When the actuator is an electromagnetic motor, rate feedback about the motor is typically accompanied by voltage feedback about the driver. (These issues will be studied in more detail in Chapters 4 – 7.)

2.7 Crossed loops and main/vernier loops

Crossed feedback loops are shown in Fig. 2.12. Such loops are often formed by parasitic coupling. Crossed dc feedback loops are frequently used in bias stabilization circuitry in amplifiers, as in the amplifier illustrated in Fig. 2.12(b).

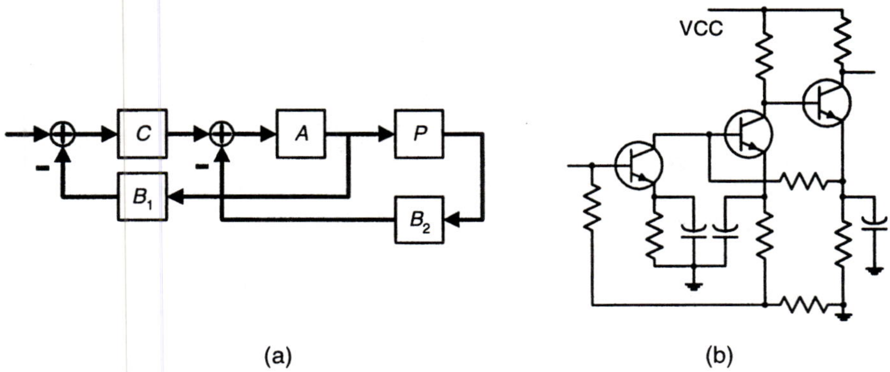

Fig. 2.12 (a) Crossed feedback loops, (b) crossed dc loops in an amplifier

For high control accuracy over a large dynamic range, the actuator must be fast and powerful. If such an actuator is not available, and if large changes in the output variable need not be fast, then an arrangement of two complementary actuators can be employed: the ***main actuator*** and the ***vernier actuator*** which is orders-of-magnitude faster but also orders-of-magnitude less powerful. Fig. 2.13(a) and (b) show two equivalent block diagrams for the ***main/vernier*** loop arrangement.

The main actuator provides most of the action applied to the plant (force, voltage etc.). However, due to its large inertia which is represented in the block diagram by the low-pass link LP, the main actuator cannot render fast signal components. These components, smaller in amplitude but rapidly changing, are provided by the vernier actuator.

From the diagram in Fig. 2.13(a), it is apparent what the actuators are doing, but it is rather difficult to figure out how to design the compensators in the main and vernier channels. For this purpose, the diagram is modified as shown in Fig. 2.13(b).

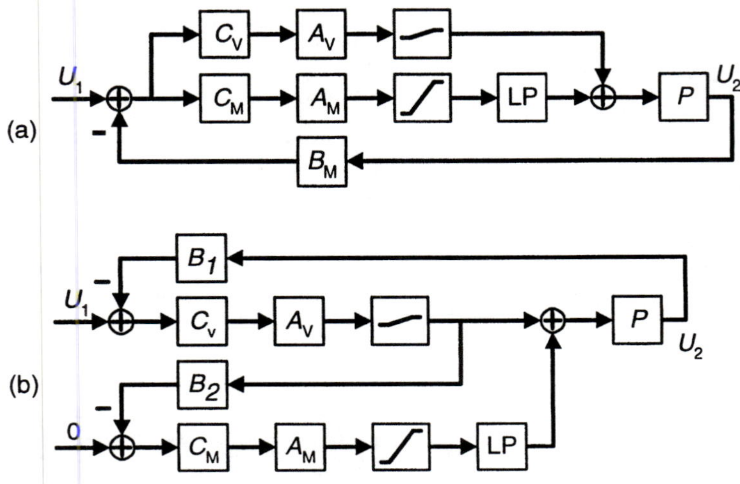

Fig. 2.13 Feedback systems with the main and vernier loops

Now, it is clear how the system operates. The command is given to the vernier loop summer, and the vernier actuator tries to reduce the error rapidly. However, when the error is large, the vernier actuator becomes saturated and cannot compensate high-frequency disturbances in the system. This situation is corrected – the vernier is desaturated – by the main loop. The output signal of the vernier actuator is applied to the feedback summer of the main loop. The command for the main loop is zero since the desired value of the vernier actuator output for slowly varying signal components is zero. (Therefore, there is no physical command summer in the main loop in Fig. 2.13(a); the command summer in the main loop in Fig. 2.13(b) is shown only to simplify the explanation of how the system works.) The slow but powerful main actuator unloads the vernier actuator from slow but large amplitude commands and disturbances. Two examples of such a system are described in Appendix 13.

By the same principle, the system can be extended to a three-loop configuration, etc. Each extra loop provides an economical way to improve the control accuracy by a few orders of magnitude. The feedback bandwidth of each subsequent loop increases. Due to the difference in the loop bandwidth, loop coupling is rather easy to account for during the system stability analysis, both in linear and nonlinear modes of operation.

Example 1. In the orbiting stellar interferometer (a high resolution optical instrument to be placed on an orbit about the earth), the lengths of the optical paths from the two primary mirrors to the summing point must be kept equal to each other. The optical path lengths are measured with laser interferometers, and must be adjusted with nanometer accuracy. For the purpose of this adjustment, in one of the paths a variable delay is introduced by bouncing the light between additional mirrors. The position of one of these mirrors is regulated by three means.

One mirror is mounted on a piezoelectric actuator. The piezoelectric actuator can be controlled with nanometer accuracy, but its maximum displacement (stroke) is only 50 μm. The second mirror is moved by a voice coil. (A voice coil is an electromechanical actuator based on a coil placed in a field of a permanent magnet; voice coils are widely employed in loudspeakers and hard disk drives where they position the reading/writing heads.) The accuracy of the voice coil control loop is lower since its feedback is limited by some mechanical structural resonances, but the actuator maximum stroke is much longer, 1 cm. The voice coil is placed on a cart that can be moved on wheels along a set of rails.

The voice coil is the vernier for the cart, and the piezoelement is the vernier for the voice coil. The voice coil desaturates the piezoactuator, and the cart desaturates the voice coil. The entire control system is able to adjust the optical path length rapidly and very accurately. (The control system is described in detail in Appendix 13.13.)

2.8 Manipulations of block diagrams and calculations of transfer functions

Equivalence block diagram transformations facilitate the conversions of various configurations to standard ones for the purpose of analysis. For example, the diagram in Fig. 2.14(a) can be transformed into diagrams (b) and (c) by changing the node from which the signal is taken while preserving the signal value at the branch output. In this transformation, the forward path transfer function and the feedback loop return ratio are preserved.

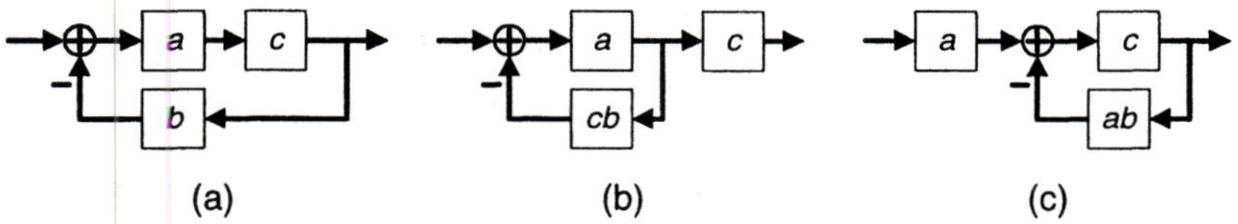

Fig. 2.14 Feedback system equivalent transformations

For the transfer function calculation, the following evident rules apply:

(1) *Transmission along a forward path is reduced by the value of the feedback in the loop that includes links in the path.*
(2) *When there are several parallel forward paths, the total transfer function can be found by superposition of the signals propagating along the paths, i.e., summing the path transfer functions.*

Example 1. The diagram in Fig. 2.16 is obtained from the diagram in Fig. 2.15 by equivalence transformations. The signals taken at different points are multiplied by additional blocks' coefficients so that the signals at the outputs of the branches remain the same.

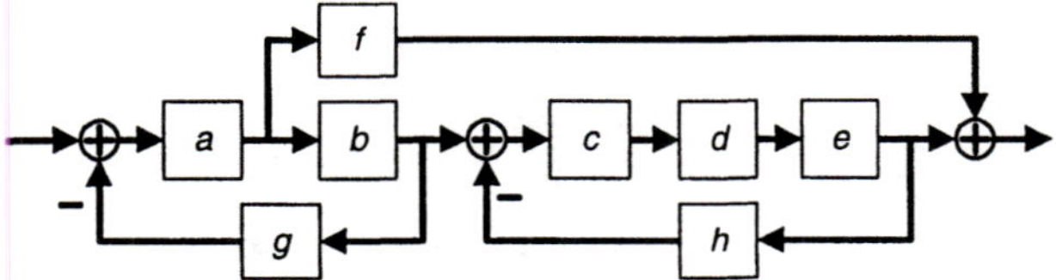

Fig. 2.15 Block diagram of a feedback system

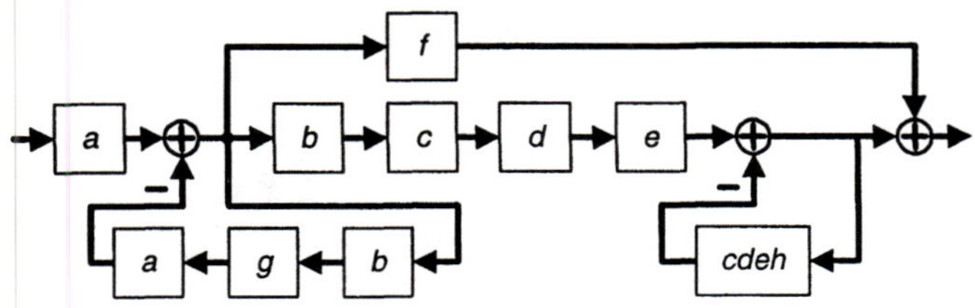

Fig. 2.16 Feedback system with tangent loops

The diagram in Fig. 2.16 has ***tangent loops***, i.e., loops with unity forward paths. According to (1.3), each tangent loop reduces the signal transmission by the value of feedback in the loop.

There are two forward paths and two loops with return rations *bga* and *cdeh*, so that the transfer function is

$$\frac{af}{1+bga}+\frac{abcde}{(1+bga)(1+cdeh)}. \qquad (2.3)$$

Often, instead of the block diagram representations, systems are described by the signal flowchart exemplified in Fig. 2.17.

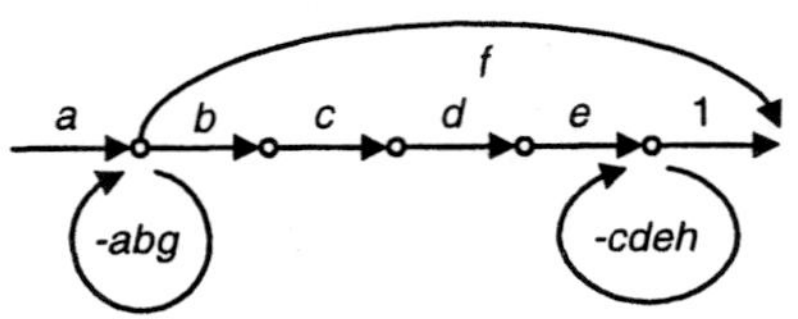

Fig. 2.17 System flowchart representation

Example 2. The gain coefficient of the graph shown in Fig. 2.18 can be calculated as the sum of transmissions along two parallel paths, divided by the feedback in the tangent loop:

$$\frac{(3-2\times 5)10}{1+5\times 6}=-2.26\,.$$

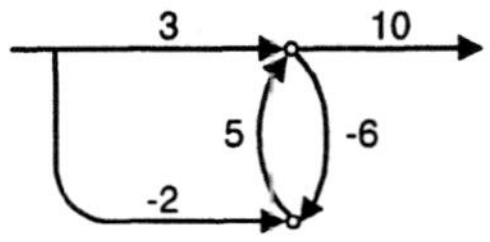

Fig. 2.18 Flowchart

Next, consider nested loops. As shown in Fig. 2.19, by converting all nested loops to the loops between the same nodes, we obtain several parallel loops. The equivalent single loop has the loop transfer function equal to the sum of all the nested loop transfer functions.

Thus, the third rule can be formulated:

(3) *When the loops are nested, the input-output transfer function is the forward path transmission divided by the sum of all loop return ratios and 1*.

The three rules constitute ***Mason's rule***.

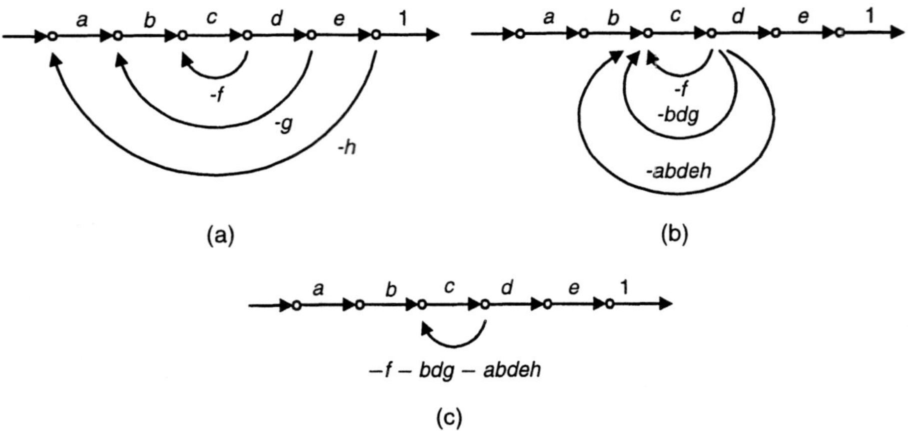

Fig. 2.19 Transformation of (a) nested loops to (b) parallel loops and further to (c) a single equivalent loop

Example 3. The transfer function for the system with nested loops shown in Fig. 2.11 is

$$\frac{CDAP}{DB_D + DAB_A + CDAPB_P + 1} \tag{2.4}$$

where D is the driver transfer function.

With the block diagram manipulations, it is often possible to prove the equivalence of different control schemes. For example, position command in a single-input, single-output system is sometimes split into several paths to form position, velocity, and acceleration command, and these three signals are separately feedforwarded into three different summing points; the sensor output is often passed through a low-pass filter to attenuate the sensor noise (as well as some components of the signal), and then the filtered and the unfiltered sensor signals are fed to different summing points; linear filters are used to estimate the output position, velocity, and acceleration, and these signals are combined linearly to form the signal driving the actuator; some block diagrams include linear time-invariable links named predictors, plant models, and estimators. If the equivalence of these block diagrams to the block diagrams shown in Figs. 2.1, 2.3, 2.4 is proved (very often, this can be easily done), the achievable performance of these control schemes, whatever the name, is no better than that of the standard control system configurations. (The performance might be inferior if the block diagrams are chosen that inherently were to limit the order of the compensators, as in Example 1 in Chapter 8.) On the other hand, some of the potentially equivalent block diagrams may have certain advantages from the implementation point of view.

2.9 MIMO feedback systems

Multi-input multi-output (MIMO) systems have several command inputs, and several output variables are controlled simultaneously. For example, if the number of commands is 2, and the number of outputs is 3, this is a 2×3 system. The controlled variables could be, for example, angles of different bodies or angles in different dimensions of the same body.

The number of feedback *loops* does not necessarily correlate with the number of inputs and outputs. Very often, a multiloop system is employed to improve the performance of a single-input, single-output (SISO) system. For example, the systems shown in Figs. 2.11 and 2.12 are multiloop SISO systems.

An example of a MIMO system is shown in Fig. 2.20 where different plant variables are regulated by separate loops. The transfer function of the plant for each loop is the plant transmission from the actuator output to the sensor input. The transfer function from the ith actuator output to the jth sensor input ($i \neq j$) shown by the dashed line is called a ***coupling transfer function***. If the coupling transfer functions are all zero, then the multi-loop system is just a set of individual single-loop systems. In many cases, coupling exists but is small.

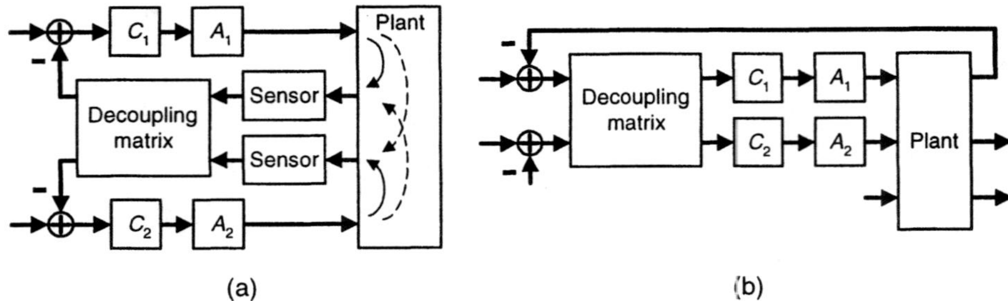

Fig. 2.20 2×2 MIMO system with loops to control nearly independent variables. The decoupling matrix can be placed in (a) the feedback path or (b) in the forward path.

Most often, actuators are relatively expensive and their number in engineering systems needs to be reduced to a minimum. Therefore, as a rule, only one actuator is assigned to do a specific job (Example 3 below offers exceptions): one actuator moves the plant in one direction, the second in another, etc. Or, in the case of an electrical signal generator, one actuator varies the signal frequency, the second one the signal amplitude, the third one, the temperature of the quartz resonator, etc. Because of this, the actuator loops in the block diagram in Fig. 2.20 are already to a large extent decoupled, i.e., the terms on the main diagonal of the plant matrix (from plant actuators to plant sensors) are substantially bigger than the off-diagonal terms.

Coupling between loops can be compensated for by using a ***decoupling matrix***, whose outputs only reflect the action of an appropriate actuator. The decoupling matrix makes the feedback loops independent of each other, simplifying the design and improving the system performance. The decoupling can be done in the feedback path by decoupling the sensor readings, or in the forward path by decoupling the signals going to actuators. Either method can make the loops independent of each other, but there is substantial difference between the methods: the matrix needs to be precise when placed in the feedback path, and can be less precise when placed in the forward path.

A decoupling matrix for linear plants can be found by inverting the matrix of known coupling transfer functions. If the coupling transfer functions do not contain pure delay, the decoupling matrix is causal and can be implemented with a digital or an analog computer. However, since the plant parameters are not known exactly, decoupling is never perfect.

The following types of multi-loop systems are most often encountered in practice: local actuator feedback, vernier type control with actuators differing in speed and in power, and nearly decoupled control where each of the actuators dominantly affects a specified output variable.

When fast action is of utmost importance, complex engineering and biological systems are typically arranged as an aggregation of several SISO mechanisms with large and relatively wideband feedback in each loop and a complex precision ***commander*** producing commands to the mechanisms. When the action need not be very fast but the accuracy is of prime concern, additional slower common feedback loops are added to precisely control the output variables – as shown in Example 1 in Section 2.1.

Example 1. The azimuth angle of the antenna in Fig. 1.1(b) might be regulated as well as the elevation, and the result would be a 2-input, 2-output system. The coupling between the elevation and the azimuth loops is typically small, and can be calculated and compensated which results in practically decoupling the loops.

Example 2. Spacecraft attitude controllers are commonly arranged as three separate loops for rotating the spacecraft about the x-, y-, and z-axes. The spacecraft inertia matrix is not symmetrical about all the axes. Therefore, the transfer function about one axis depends on the rotation angle and velocities about the other axes, and the three controllers are coupled and cannot be considered as three separate SISO systems. Good decoupling can be achieved over most of the frequency bandwidth of interest where the spacecraft parameters are well known and the decoupling matrix transfer functions can be accurately calculated. However, over some frequency ranges, for example, at the slosh modes of the propellant in the fuel tanks, the spacecraft parameters have much larger uncertainty and the calculated decoupling matrix is not very accurate. The uncertain coupling necessitates a reduction in the feedback in the control loops as will be discussed later, in Section 4.4.

Example 3. Multiple actuators of the same type can be used to achieve the appropriate power and/or balance. An example is the use of multiple power plants on jet transports is shown in Fig. 2.21. The output is one of the variables defining the airplane attitude and velocity (i.e., this block diagram shows only a part of the entire control system).

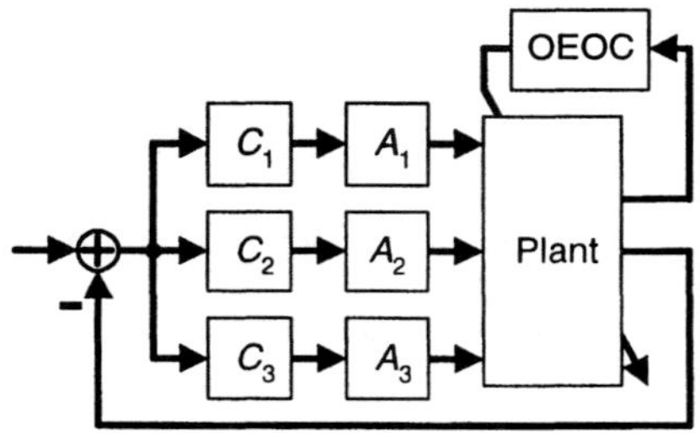

Fig. 2.21 Several parallel power plants system

The arrangement provides redundancy, i.e., one-engine-out capability (OEOC). Special control modes (auto or manual) may be necessary to support this sort of operation. For this purpose, additional feedback loops using aerodynamic control surfaces are applied so that a single actuator can power the plant independently in the event that the other actuators fail. The system is a multiloop MIMO system.

Example 4. In a TV set or in a VCR, there are several hundred feedback loops. More than 90% of the loops control electrical variables (currents, voltages), and some of the loops control image color and brightness, speed of the motors, and tension of the tape. The majority of the loops are analog, but some are digital, particularly those for tuning the receiver and for controlling the display. This, say, 300×300 MIMO feedback system is conventionally designed with frequency domain methods, as if the loops were independent, i.e., as if the system were merely a combination of 300 SISO systems. The variables to be controlled are to a large extent independent, i.e., the diagonal terms are dominant in the 300×300 matrix. Only seldom is some primitive decoupling used in the forward path. The decoupling matrix is sometimes included in the feedback path to calculate the variables fed back from the sensors' readings.

The design of a MIMO controller as a combination of several independent loops has the important advantages of structural design. It simplifies the system testing and troubleshooting, improves reliability, and simplifies the work of modification and redesign. To meet these goals, most engineering devices are designed structurally, in spite of the mathematically attractive idea of combined optimization of the entire 300×300 multivariable system which, ideally, must produce at least as good or better performance – but at the price of losing the advantages of the structural approach.

2.10 Problems

1 For a tracking system ($B = 1$) with T equal to (a) 5; (b) 20; (c) –80; (d) 120; (e) 2.72 find the value of the prefilter R that makes the closed-loop transfer function equal to 1.

2 Find the compensator and feedback path transfer coefficients for a system without a prefilter so that the system is equivalent to the system of Problem 1, with $AP = 10$ in both systems.

3 Find the compensator C and feedforward path gain coefficient FF for a system without a prefilter, with $B = 1$, and with $AP = 10$, so that the system is equivalent to the system of Problem 1.

4 Include link B in the feedback path in the block diagram depicted in Fig. 2.1. Derive an expression for the input-output transfer function.

5 $C = 2$, $A = 1$. Plant gain coefficient P is uncertain within the 10 to 20 range, and nominal plant gain coefficient $P_o = 15$.

Calculate the input-output gain coefficient ranges without and with a feedforward path. Does the feedforward affect the ratios of the maximum to the minimum input-output gain coefficient?

6 The loop gain coefficient is inversely proportional to the frequency. At what frequency ranges is the benefit of using feedforward most important?

7 Plant $P(s) = 10/(s + 10)$, the actuator model includes a linear gain block with gain coefficient $A = 10$ followed by a saturation link with threshold 1, and $C(s) = 0.3(s + 0.35)/(s + 3)$. The feedforward path transfer function is $0.1a(s + 10)/(s + a)$. The command is sinusoidal, with possible frequencies from 0 to 10 Hz.

Plot the frequency responses with MATLAB. Choose coefficient a such that the signal amplitude at the input to the saturation block will not exceed the threshold. What is the bandwidth of the feedforward?

8 Same problem as the previous one, but the input signal is a step 1 V. Make simulations with MATLAB, find a by trial and error.

9 Find the Bode sensitivity of transfer function $W_1 + W_2$ to W_1 if:
(a) $W_1 = 100$ and $W_2 = 2$;
(b) $W_1 = 50$ and $W_2 = 50$;
(c) $W_1 = -9$ and $W_2 = 10$;
(d) $W_1 = 10/s$ and $W_2 = 100/s^2$.

10 Find the Bode sensitivity to the transfer functions of the links P, A_V, and A_M in Fig. 2.13. (*Hint*: Use the chain rule. First, employ Bode sensitivity for a single-loop system for the composite link including the main and vernier channels; then, multiply the composite link Bode sensitivity by the sensitivity of the composite channel transfer function to variations in only one channel.) Give a numerical example.

11 Derive the expression for the sensitivity to variations in A_M of Black's feedforward system shown in Fig. 2.6. Compare two cases: (a) when $A_M = 10$ and the values of the rest of the links' gain coefficients are nominal, i.e., $A_E = 1$ and $B = 0.1$, and (b) when A_E deviates by 3 dB from the nominal value of 10.

12 In Fig. 2.6, $A_E = 95$, $B = 0.01$. Find the sensitivity of the output to variations in A_M when A_M is (a) 100; (b) 105; (c) 150.

13 For the conditions described in the previous problem, and considering the maximum output signal (saturation threshold) in the main amplifier to be $A_{MS} = 10$, find the maximum output signal in the error amplifier. What is the conclusion?

14 In Black's feedforward system shown in Fig. 2.6, the error amplifier A_E has internal feedback F_E. Using the chain rule, find the sensitivity of the system's input-output transfer function to the error amplifier gain variations.

15 How many loops, according to Bode's definition, are in the systems diagrammed in Fig. 2.22?

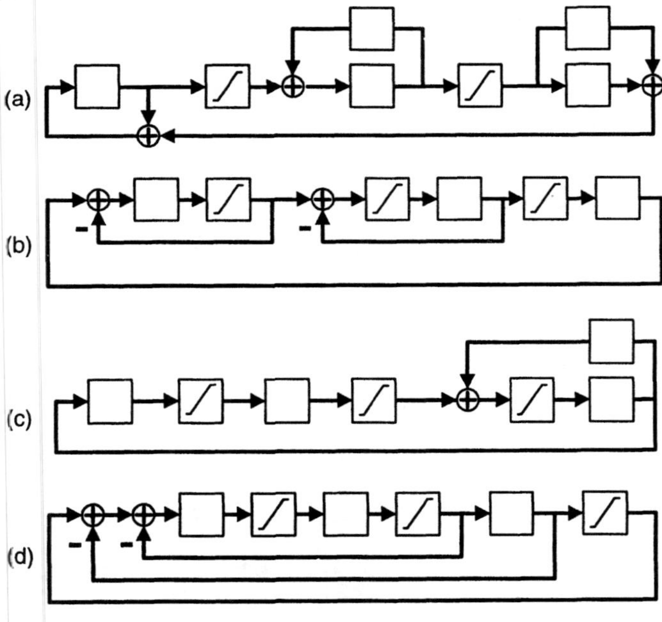

Fig. 2.22 Feedback systems

16 (a) In Fig. 2.11, the driver gain coefficient changes with temperature by ±30% from the nominal, the actuator changes by ±15% from the nominal, and plant transfer function is uncertain within 2 dB. The loop gain in the driver loop is 30 dB, and in the actuator loop (with the driver loop closed) is 10 dB. Find the total uncertainty in the plant loop gain.

(b) Same problem but the driver gain coefficient uncertainty is ±3 dB, actuator uncertainty is ±2 dB, and plant uncertainty is ±2 dB.

(c) Same problem as in (b) but with loop gain in driver loop 40 dB and 20 dB in the actuator loop.

(d) Same problem as in (b) but with loop gain in driver loop 20 dB and 20 dB in the actuator loop.

17 Explain why a pair of actuators, one high-power and sluggish, and one low-power and very fast, should typically cost less than a single powerful and fast actuator.

18 Find the input-output transfer function for the system shown in Fig. 2.10.

19 Find the input-output function of the system shown in Fig. 2.15.

20 For the multi-loop feedback system described in Fig. 2.11, find the input-output transfer function and the sensitivities to variations in driver, actuator, and plant.

21 Derive expressions for input-output transfer functions for the systems shown in Fig. 2.23.

22 (a) Calculate the decoupling matrix for the system where the sensor readings x', y', z' are related to the actuator outputs x, y, z by:

$$\begin{aligned} x' &= 2x + 0.2y + 0.3z, \\ y' &= 0.1x + 2.1y + 0.1z, \\ z' &= 0.04x + 0.1y + 1.9z. \end{aligned}$$

Since the diagonal terms are dominant,

$$x \approx 0.5x', \ y \approx 0.5y', \ z \approx 0.5z'.$$

A better approximation to x is found by substituting these first approximations to y and z into the first equation:

$$x \approx 0.5x' - 0.05y' - 0.075\,z'.$$

Proceed with better approximations to y, z. Compare these expressions with the exact solution found with MATLAB script:

```
A = [2 0.2 0.3; 0.1 2.1 0.1; 0.04 0.1 1.9]; inv(A)

ans =      0.5038    -0.0443    -0.0772
          -0.0235     0.4795    -0.0215
          -0.0094    -0.0243     0.5291
```

Draw the flowchart for the solution in the form shown in Fig. 2.24, and put the numerical values in. (See also Problem 6.10 for an analog computer implementation of the decoupling matrix using six op-amps (two quad op-amp IC) and 18 resistors.)

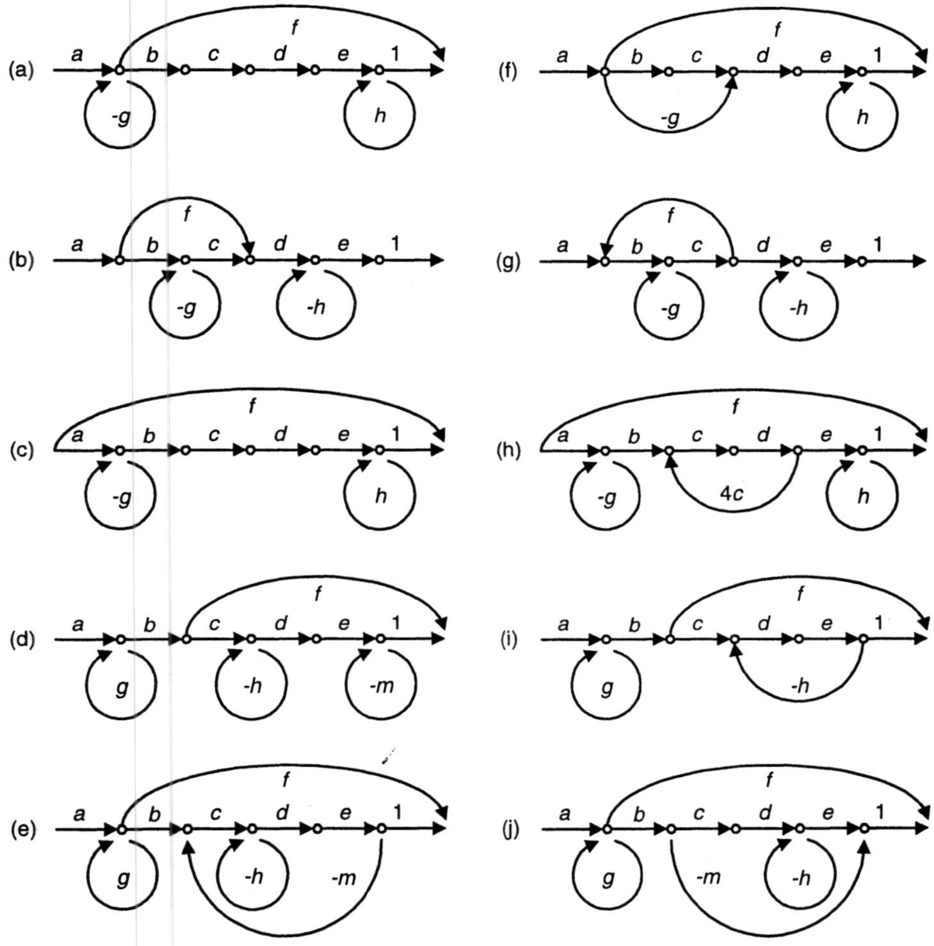

Fig. 2.23 Flowcharts

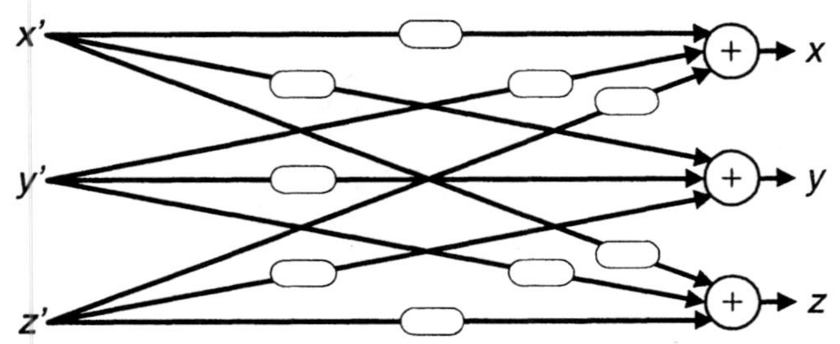

Fig. 2.24 Decoupling matrix flowchart

(b) Calculate the decoupling matrix for the system where the sensor readings x', y', z' are related to the actuator outputs x, y, z by:

$$x' = 2x + y + 0.3z,$$
$$y' = 0.1x + 2y + 0.5z,$$
$$z' = 0.4x + 0.5y + 1.9z.$$

by inverting the coefficient matrix with MATLAB.

(c) Same as (b) for:

$$x' = 3x + 0.4y + 0.3z,$$
$$y' = 0.3x + 2.1y + 0.2z,$$
$$z' = 0.04x + 0.1y + 1.9z.$$

(d) Same as (b) for:

$$x' = 2x + 0.1y + 0.1z,$$
$$y' = 0.1x + 3.1y + 0.1z,$$
$$z' = 0.04x + 0.4y + 1.9z.$$

(e) Same as (b) for:

$$x' = x + y - z,$$
$$y' = x - y + z,$$
$$z' = -x + y + z.$$

(This arrangement of three piezoactuators and three load cells has been used in the spacecraft vibration isolation system described in Section 6.4.2.)

23 The frequency of a quartz oscillator depends on the crystal temperature and on the power supply voltage (the voltage changes the capacitances of *pn*-junctions of the transistor that participate in the resonance contour). The temperature of the environment changes from 10° C to 70° C. The power supply voltage uncertainty range is from 5 V to 6 V.

The oscillator elements are placed in a small compartment ("oven") equipped with an electrical heater and a temperature sensor. The temperature and the dc voltage are regulated by control loops. The thermal loop return ratio is 600. The dc voltage stabilizing loop return ratio is 200. The references are 70° and 5 V, and the loops maintain the quartz temperature close to 70° and the power supply voltage, close to 5 V.

For the employed quartz crystal and the transistor, the dependencies of the frequency of oscillation on the crystal temperature and on the power supply voltage are well approximated in the neighborhood of the references by linear dependencies with coefficients 10^{-4} Hz/°C and -10^{-3} Hz/V. The maximum disturbances in temperature and voltage are 60° (when the environment temperature is 10°) and 1 V (when the power supply voltage is 6 V). Fig. 2.25 shows the flow-chart for calculations of the effects of the disturbances.

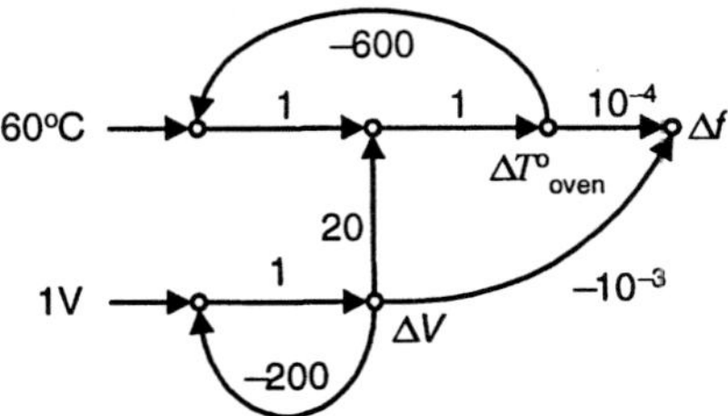

Fig. 2.25 Flowcharts representing the effects of the voltage and temperature variations on variations of the oscillator frequency

The loops are coupled since the dc voltage also affects the power dissipated in the transistor and, consequently, the oven temperature with the rate 20°C/V.

The flowchart represents a double-input single-output system. No decoupling between the control loops is required since the coupling is small and one-directional, from the voltage to the temperature loop. (The effect of temperature on the voltage loop is negligibly small.)

Calculate the total range of the frequency variations Δf due to the instability of the environment temperature and the power supply voltage.

24 Prove that, generally, when sensitivity is 0, redundancy is always provided. (*Hint*: Use bilinear relation $W = (aw + b)/(cw + d)$ for the general dependence of a linear system transfer function W on a link transfer function w.)

Chapter 3

FREQUENCY RESPONSE METHODS

Some requirements to control systems are typically expressed in frequency domain (such as disturbance rejection), while some others are most often formulated in time domain (such as rise time and overshoot). The latter need to be converted into the frequency-domain specifications in order to use frequency-domain design methods. Formulations of the time-domain requirements are commonly very simple, and the equivalent frequency-domain formulations are also simple. The requirements can be translated between the domains with simple approximate relations.

Since most control systems are of the low-pass type, responses of standard low-pass filters are reviewed for future references.

Typical closed-loop frequency responses for homing and tracking systems are considered.

The Nyquist stability criterion is derived and its applications reviewed. Stability margins are introduced and the Nyquist stability and the absolute stability discussed.

The Nyquist-Bode criterion is developed for multiloop systems' stability analysis with successive loop closure. Feedback systems with unstable plants are analyzed with the Nyquist criterion and with the Nyquist-Bode criterion. The effect of saturation on the system stability is briefly discussed.

Static error reduction is considered for systems of the first, the second, and the servo types.

The notion of minimum phase (m.p.) function is introduced.

The theorem is considered of equality to zero of the integral of the feedback in dB over the frequency axis.

The Bode integral of the real part of a function is applied to evaluation of impedances. The Bode integral of the imaginary part of a function is applied to estimation of feasible changes in the loop gain response.

The meaning and the significance of the Bode general phase-gain relationship are clarified, and the procedure for calculating the phase from a given gain response is explained.

The problem of finding the Bode diagram from a given Nyquist diagram is considered.

Non-minimum phase lag is studied. A criterion is derived for the transfer function of two m.p. parallel paths to be m.p.

The use of MATLAB and SPICE is illustrated for feedback system modeling and analysis.

When the book is used for a single-semester introductory control course, Section 3.9.3 can be bypassed.

3.1 Conversion of time-domain requirements to frequency domain

3.1.1 Approximate relations

Since signals can be substituted by the sums of their sinusoidal components, and in linear links, the signal components do not interfere (i.e., the superposition principle applies), linear links are fully characterized by their frequency responses. The formulas for the Fourier method are often derived with the Laplace transform using complex variable $s = \sigma + j\omega$. The Laplace transform is also used to make conversions between the time domain and the frequency domain responses. For brevity, we will write $W(s)$

even when we only mean the frequency response $W(j\omega)$. We assume the reader is already familiar with using frequency responses. If not, Appendix 2 can be of help.

Frequency responses are widely employed for characterizing links and design specifications. The feedback response required for best disturbance rejection is commonly specified in frequency domain since the disturbances are most often characterized by their spectral density, i.e., in frequency domain. High-order compensators and plants are also most often characterized by their frequency responses.

The time-domain characterization, on the other hand, is commonly applied to systems which are required to transfer signals without distortions. A step-function or a series of step-functions is usually employed as the input test-signal, and the output is specified in time domain.

Given a mathematical description of a linear system, conversion between the frequency and time responses and specifications is easily performed by computer. Analytical transformation between the time-domain function and the Laplace transform expression can be obtained in MATLAB by functions `laplace` and `invlaplace`. The time and frequency responses can be plotted with standard MATLAB plotting commands (or with SPICE simulation). Yet, it is important to be able to make the approximate conversion mentally for the purposes of creation and analysis of specifications to systems and subsystems, resolution of the trade-offs, and comparison of available versions of conceptual design. This can be done using the simple rules described below.

The **3 dB bandwidth** *is the bandwidth of a low-pass system up to the frequency where the gain coefficient decreases* $\sqrt{2}$ *times, i.e., by 3 dB*. For the first-order low-pass transfer function $a/(s + a)$, the 3 dB bandwidth is the pole frequency

$$f_p = a/(2\pi)$$

as shown in Fig. 3.1(a). The time response of such a link to step function input is

$$1 - \exp(-at)$$

(see Section A2.2 in Appendix 2). It is shown in Fig. 3.1(b).

The line tangent to the time response at $t = 0$ is at. The time it takes the signal to rise to 0.9 is found from the equation

$$1 - \exp(-at_r) = 0.9$$

to be

$$t_r \approx (1/f_p)/3. \qquad (3.1)$$

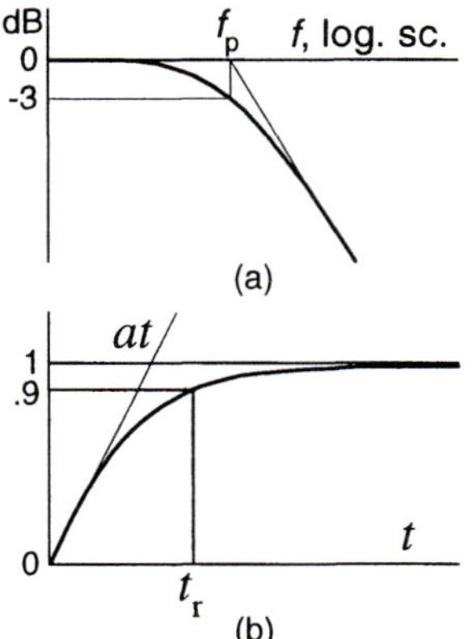

Fig. 3.1 (a) Frequency response and (b) time response for first-order link $a/(s + a)$

In other words, *rise time is approximately one-third of the period* $1/f_p$ *related to 3 dB bandwidth*. This rule is employed for calculating the bandwidth required for the rise time not to be longer than prescribed.

Example 1. A telecommunication antenna, a 10" diameter dish, to be placed on a balloon flying in the Venus atmosphere, needs to be pointed to Earth with 0.5° accuracy. The rate of the attitude variations of the balloon can reach 5°/sec. Therefore, the rise time of the antenna attitude control system must be smaller than 0.1 sec which translates into the 3 dB closed-loop bandwidth of at least 3 Hz, or, approximately, the crossover frequency $f_b > 1.5$ Hz. These calculations of the required feedback bandwidth are sufficiently accurate for the conceptual design, even though the closed-loop transfer function will be not first-order for which (3.1) was derived but higher-order.

For higher-order low-pass transfer functions, the rise time is still roughly approximated by (3.1) where under f_p the cut-off frequency of the frequency response is understood. However, the transient response is more complicated, and the deviation of the output from the desired step-function is commonly characterized by the five parameters shown in Fig. 3.2: ***delay time*** t_d, ***rise time*** t_r, ***settling time*** t_s of settling within the dynamic error envelope, ***overshoot***, and ***steady state*** (***static***) ***error***, all of them required to be small.

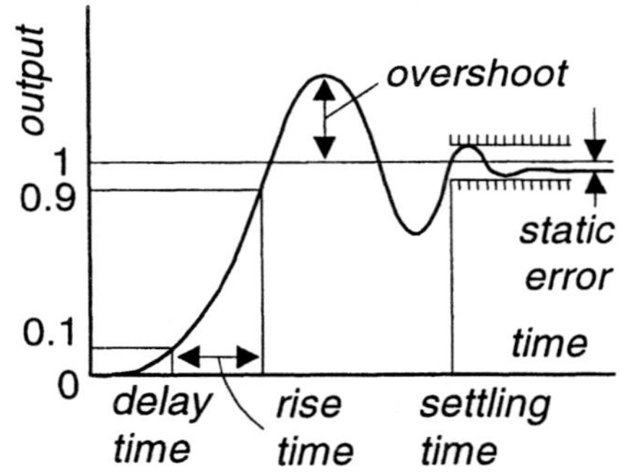

Fig. 3.2 Time-response to step-function input

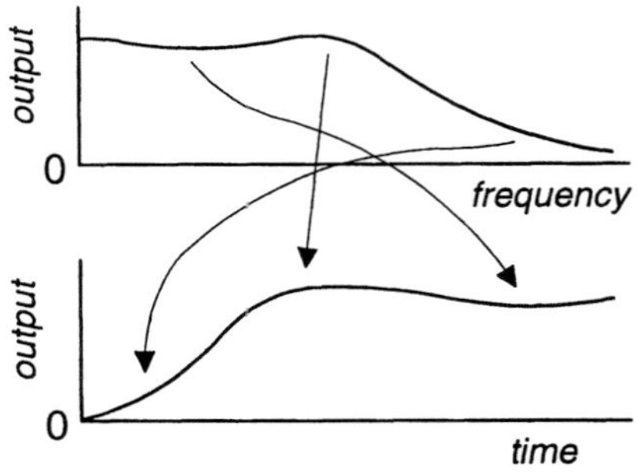

Fig. 3.3 Relations between the frequency domain and time domain regions

At zero time, the first n time-derivatives vanish for the systems with an nth-order pole at high frequencies as follows from the initial value Laplace transform theorem (see Section A2.3 in Appendix 2). Therefore, increasing n flattens the time-response at small times and increases the delay time.

The initial and the final value Laplace transform theorems relate the gain coefficient at lower frequencies to the time-response at longer times, and the gain coefficient at higher frequencies, to the time-response at smaller times, as is indicated in Fig. 3.3. For a low-pass system with relatively smooth responses, we cam assume that the transient response at specific times is predominantly affected by the gain coefficient over specific frequency intervals. Numerically, according to (3.1), the time-response at the time of 1 second is mostly affected by the gain coefficient at and around 0.3 Hz, i.e., by the gain over the 0.1 to 1 Hz frequency interval; the output at the time of 1 ms, by the gain over the 1 kHz to 10 kHz interval, etc.

Therefore, in Fig. 3.2, the rise time corresponds to the operational bandwidth and the settling time corresponds to the lower-frequency gain. The static error corresponds to the dc gain. It is zero when the dc gain is one.

An important correspondence also exists between the *slope* of the logarithmic gain-frequency response (Bode diagram) and the time-response *curvature*. For the gain responses with constant slope shown in Fig. 3.4(a), the Laplace transform gives the time responses shown in Fig. 3.4(b). Particularly, the delay time increases with the high-frequency asymptotic slope of the gain Bode diagram.

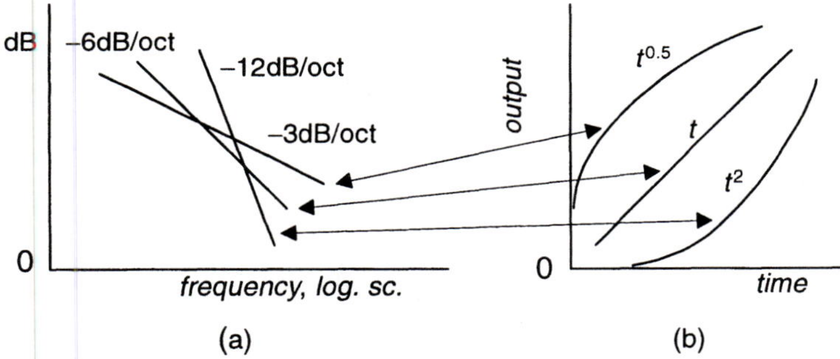

Fig. 3.4 Correlation between the slope of gain-frequency response and the curvature of time-domain step-response

From the gain and the slope of the gain response at specific frequencies, we can roughly reconstruct the time-response at specific times. Notwithstanding the relations' imprecision, they render very useful leads for system analysis and computer-aided iterating and tuning.

Example 2. The plant in the PLL in Fig. 1.3 is a VCO. It is an integrator since the *frequency* ω of the VCO is proportional to its input signal but the output variable applied to the phase detector is the *phase*. Therefore, the VCO transfer function is k/s where k is some coefficient that characterizes the VCO gain coefficient k/ω. Thus, when ω increases twice (by an octave), the gain coefficient decreases twice (by 6 dB), i.e., the slope of the gain response is – 6 dB/oct.

There exist plants which are double integrators k/s^2 and triple integrators k/s^3. The slopes of their gain responses are, respectively, –12 dB/oct and –18 dB/oct.

Example 3. When the plant and the loop gain responses have asymptotic slope –18 dB/oct, the closed-loop response also has this slope since at higher frequencies the loop gain vanishes. Then, the closed-loop transient response at small times will be proportional to the third power of time.

Example 4. The frequency response for the transfer function

$$T(s) = \frac{9000}{(s+30)(s+300)}$$

is plotted with MATLAB in Fig. 3.5. The output time-response to the step input for the same link is shown in Fig. 3.5(b). We can trace the correspondences shown in Figs. 3.3 and 3.4 on these responses.

At high frequencies, the transfer function degenerates into $9000/s^2$, i.e., into a double integrator. The slope of the gain response becomes –12 dB/oct, or, which is the same, – 40 dB per decade. (Note that each decade contains $\log_2 10 \approx 2.3$ octaves. The octaves in Fig. 3.5 are, for example, from 10 to 20, from 20 to 40, from 30 to 60; each octave has the same width on the logarithmic frequency scale.)

We see some correlation between the slope of the gain response and the phase response: when the slope is zero, the phase is zero; when the slope of the gain response approaches – 40 dB/dec, the phase approaches –180°.

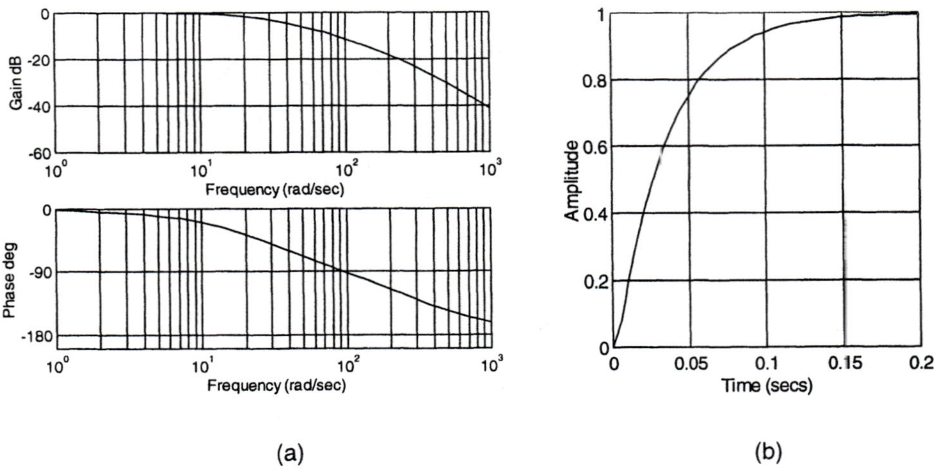

Fig. 3.5 (a) Frequency-domain and (b) time-domain responses for $T(s) = 9000/[(s + 30)(s + 300)]$

3.1.2 Filters

Since most feedback control systems are of the low-pass type, their responses can be better understood from their similarity to the responses of the standard low-pass filters.

Low-pass filters are most often employed for attenuating high-frequency noise and disturbances outside of the filter pass-band. When the shape of the signals must be preserved, then:

(a) the filter gain must be nearly the same for all important frequency components of the signal, and

(b) the dependence of the filter phase shift on frequency must be close to linear, i.e., the slope of this dependence, which is the group time delay, must be the same for all these components.

The curvature of the phase response causes different delays for sinusoidal components of different frequencies and has a profound effect on the overshoot. At some moment, various signal components which are not in phase at the input, come all nearly in phase at the output and cause the overshoot.

Fig. 3.6(a) and (b) show the gain and phase responses of several low-pass filters. The phase response of an ideal filter with 40 dB selectivity is extensively curved (it follows the response in Fig. 3.40 which is the weight function of the Bode integral, to be studied in Section 3.9.6).

The gain response of the ***Chebyshev*** (equiripple) filter bends sharply at the corner frequency, and its phase response is also significantly curved.

The ***Butterworth*** filter has ***maximum flat*** gain-frequency response, i.e., the first n derivatives of the gain response of the nth-order filter are equal to zero at zero frequency. The filter has a lesser selectivity than the Chebyshev filter, and its phase response is less curved.

The higher the order of a Chebyshev or a Butterworth filter, the sharper is the gain responses selectivity, and the phase shift response is more curved.

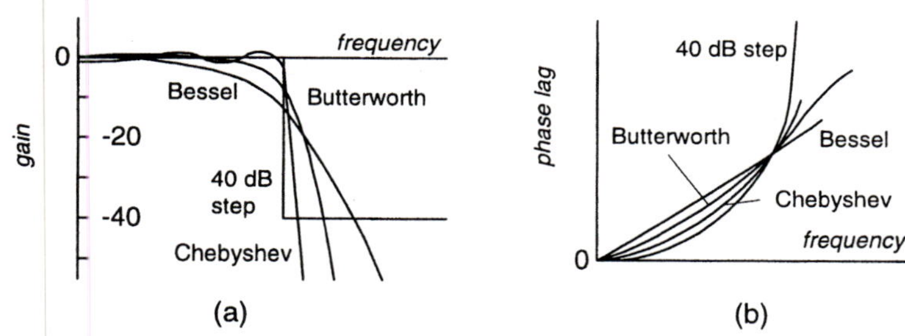

Fig. 3.6 Frequency responses of the (a) gain and (b) phase for the step 40 dB low-pass filter and for a Chebyshev, a Butterworth, and a Bessel filters

The curvature of the phase responses manifests itself in the overshoots shown in Fig. 3.7. The higher the order of a Chebyshev or a Butterworth filter, the higher is the overshoot and the longer is the settling time.

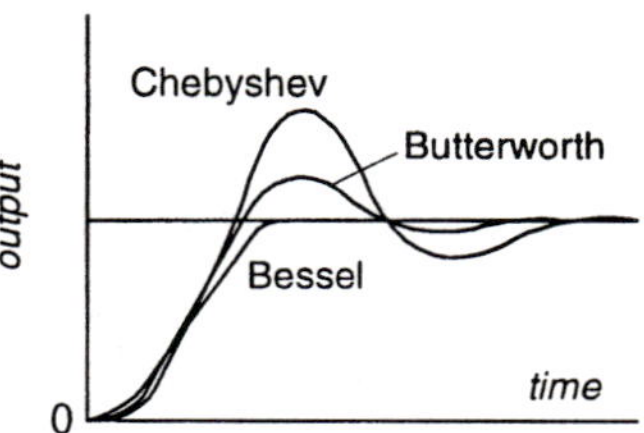

Fig. 3.7 Time step-responses for the filters

Example 1. For the 3rd-order low-pass Butterworth filter with normalized bandwidth 1 rad/sec, the overshoot is 8% and the settling time to 10^{-5} accuracy is 25 sec; for the 8th-order filter, the overshoot is 16% and the settling time is 60 sec.

The phase shift of the ***Bessel filter*** (or ***Thompson***, or ***linear phase filter***) is approximately proportional to frequency. The higher the order of a Bessel filter, the better is the phase response linearity, and the smaller are the overshoot, the rise time, and the settling time. The filter transfer function is

$$B(s) = \frac{b_o}{b_n s^n + b_{n-1} s^{n-1} + \ldots + b_k s^k + \ldots + b_o}$$

where $b_k = (2n - k)!/[2^{n-k}(n-k)!\,k!)]$.

Example 2. For the settling error of 10^{-5}, the settling time for the first-order normalized Bessel filter is 11.5 sec, for the 3rd-order, 8 sec, and for the 8th-order, 4 sec. The overshoot of the 8th-order filter is 0.35%.

±1% random variations of the denominator coefficients of the 8th-order Bessel filter transfer function increase the settling time up to 5 sec (i.e., by 1 sec).

Example 3. The transfer functions of the second- to fourth-orders Bessel filters are the following:

$$\frac{3}{s^2 + 3s + 3}, \qquad \frac{15}{s^3 + 6s^2 + 15s + 15}, \qquad \frac{105}{s^4 + 10s^3 + 45s^2 + 105s + 105}.$$

The gain- and step-responses for the three Bessel filters are shown in Fig. 3.8. It is easy to recognize which of the three unmarked responses corresponds to the highest-order filter: the one with steepest slope of the gain high-frequency asymptote, the one with largest negative phase at higher frequencies, and the one with largest time delay at small times. The phase responses do not look linear since the phase shift is plotted against the frequency axis with the logarithmic scale.

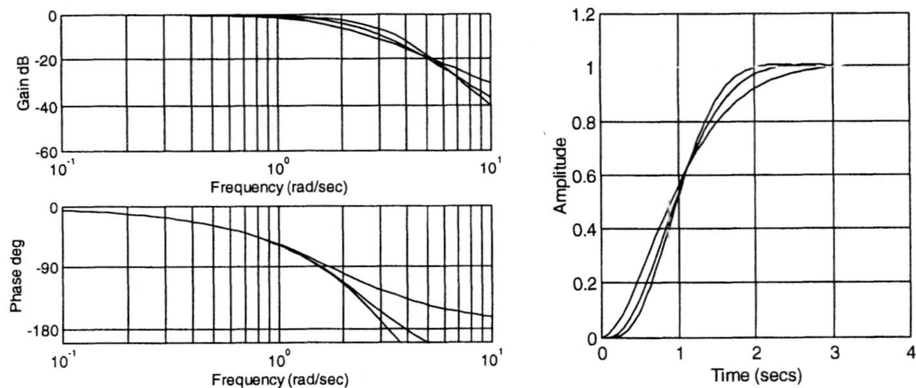

Fig. 3.8 Frequency- and step-responses of Bessel filters of second to fourth orders

3.2 Closed-loop transient response

The Nyquist diagram in the neighborhood of the critical point −1 is shown in Fig. 3.9(a). At crossover frequency f_b, $|T| = 1$. From the isosceles triangle shown in this picture,

$$|F(f_b)| = 2\sin|(-180^\circ - \arg T(f_b))/2| \tag{3.2}$$

(here, arg indicates the angle in degrees). Commonly, the angle $|-180^\circ - \arg T(f_b)|$ is less than 60°, and, as a result, $|F(f_b)| < 1$, i.e., the feedback becomes positive. When a tracking system has no prefilter, its closed-loop gain $|M| = |T/F|$ becomes greater than 1 at f_b, and the closed-loop gain response $20\log|M|$ has a hump as shown in Fig. 3.9(b).

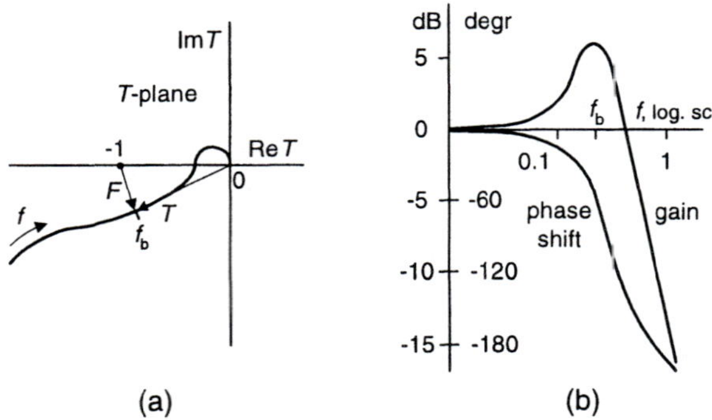

Fig. 3.9 (a) Nyquist diagram, (b) closed-loop gain and phase responses

The maximum of $|M|$ is commonly at a frequency somewhat smaller than f_b, and the value of this maximum

$$\max|M| \approx |M(f_b)| = 1/\{2\sin[(180^\circ - \arg T)/2]\}. \tag{3.3}$$

The hump is large when arg $T(f_b)$ is close to 180°, i.e., when the Nyquist diagram approaches the point −1.

Example 1. When the angle of $T(f_b)$ approaches $-180°$, then $|F|$ decreases to 0 and $|M|$ grows infinitely. When the angle is $-150°$, then $|M| \approx 2$, and the hump is 6 dB high as shown in Fig. 3.9(b). The resulting overshoot is about 50% as shown in Fig. 3.10.

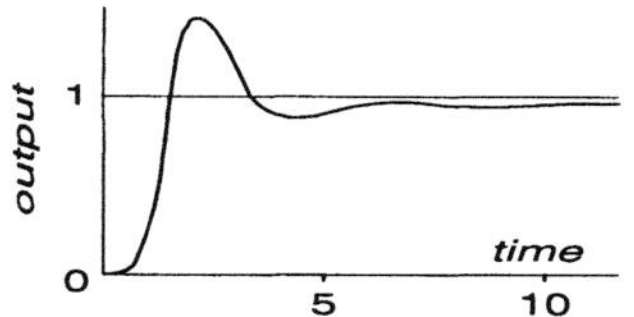

Fig. 3.10 Closed-loop transient response to step input

In a feedback system with a prefilter (or with a non-unity feedback path, or with a command feedforward path), the open-loop and closed-loop responses can be optimized independently of each other (as was shown in Sections 2.1 and 2.2), and there is no need to compromise the loop response in order to reduce the closed-loop overshoot.

For the overshoot in a closed-loop feedback system to be small, and the output to settle with high accuracy in a rather short time, the closed-loop gain response (together with the prefilter) must approximate a Bessel filter response. The prefilter must therefore equalize the hump of the closed-loop response from the summer to the output, i.e., must incorporate a broad notch. An example of such a prefilter is given in Section 4.2.3.

If the prefilter (or the feedback path, or the command feedforward link) cannot be implemented to be exactly optimal, it should at least in average make the closed-system phase response linear. This will prevent most of the harmonical components of the signal from reaching the output in phase at any time.

Homing systems do not have command summers and, therefore, do not have prefilters. The homing system open-loop response must be made such that the closed-loop transient response be as desired.

Example 2. For a homing missile, the response of interest is that of the missile direction to the disturbance which is the changing direction to the target caused by the target motion. Commonly, the disturbance is not measured or observed and only the error is measured, the difference between the missile direction and the direction to the target. The closed-loop transfer function from the disturbance to the missile direction is $1/F$. If the phase stability margin at f_b is small, the frequency response of $-20\log|F|$ has a large hump, the transient response of the output becomes too oscillatory, and there exists an effective maneuver for the target to avoid being hit. Typically, $|\arg T(f_b)|$ does not exceed 135° in such systems.

3.3 Root locus

The transfer function $T(s)/F(s)$ of a closed-loop system is infinitely big for the signal components corresponding to the poles of the function. As long as the pole is in the left half-plane of s, this doesn't create a problem. However, when one of the poles has a positive real part, i.e., the pole s_p is in the right half-plane of Laplace variable s, the transmission is infinitely big for the signal which is growing in time. Then, some components of the random input noise will be exponentially magnified causing the system's output to grow exponentially. Such a linear system is considered unstable and cannot perform as a control system. Verification of the system stability is one of the major tasks in control system design.

When the links in the loop are each inherently stable, i.e., all poles of their transfer functions are in the left half-plane of s, then, certainly, all poles of the transfer function

$T(s)$ are in the left half-plane of s, and the open-loop system is stable. When the loop is closed, some of the poles of the transfer function $T(s)/F(s)$ can appear in the right half-plane of s, and in this case, the system is unstable. It is interesting to trace what happens when the loop is "gradually" closed: instead of switching the loop open or closed, we place a link with the gain coefficient k in the loop, and we will continuously increase k from 0 to 1. Correspondingly, the poles of the transfer function will change continuously from the poles of the open-loop system to the poles of the closed-loop system. Their trajectories on the s-plane are called ***root loci***. The root loci start at the poles of T. Invasion of a root locus into the right half-plane of s indicates that the system becomes unstable.

There exist rules for drawing the root loci manually and for using the loci for the feedback system design. The rules provide for simple low-order system analysis but the method becomes cumbersome when applied to high-performance systems which are high-order systems. The root locus will be further discussed in Section 8.2.

Example 1. $T(s) = 10(s + 2)/(s^3 + s^2 + s)$. The open-loop and closed-loop poles can be calculated with the MATLAB commands

```
n = [0 0 10 20]; d = [1 1 1 0];
roots(d)          %open-loop poles
  ans = 0           -0.5000 ± 0.8660i
roots(n + d)      %closed-loop poles
  ans = -1.6551   0.3275 ± 3.4608i
```

The root loci in Fig. 3.11 are plotted with:

```
n = [10 20]; d = [1 1 1 0];
rlocus(n,d)
hold on
k = [0.05 0.1 0.2 0.5 1];
rlocus(n, d, k)
title('k = [0.05 0.1 0.2 0.5 1]');
hold off
```

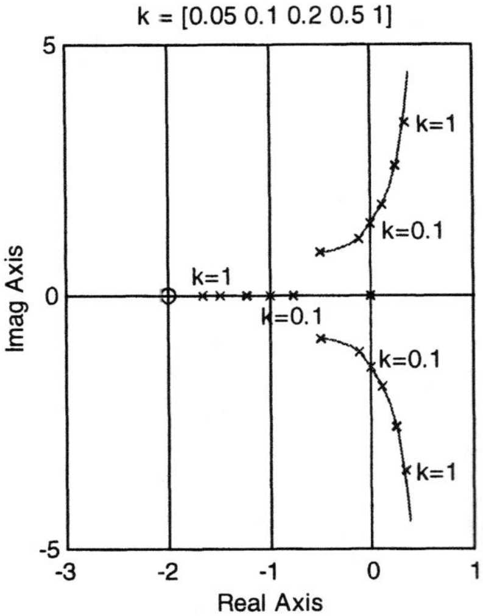

Fig. 3.11 Root loci for $T(s) = 10(s + 2)/(s^3 + s^2 + s)$.

As k increases and approaches infinity, the pole at the origin moves to the left and approaches −1.6551. The loci of the two complex poles end at $0.3275 \pm j3.4608$. The system is stable with the coefficient k up to 0.1, when the complex poles become purely imaginary, $\pm j1.42$. After that, the complex poles migrate to the right half-plane of s and the system becomes unstable.

When a pole of $T(s)/F(s) = T(s)/[T(s) + 1]$ crosses the $j\omega$-axis at a certain frequency, this function at this frequency becomes ∞, which can only happen if $T(j\omega)$ at this frequency becomes −1, i.e., $|T| = 1$ and $\arg T = \pm n\pi$, where n is an odd integer.

From here, a simple stability criterion follows: *if an open-loop system is stable, the closed-loop system is stable if* $|T| < 1$ *at all frequencies at which* $\arg T = \pm n\pi$ *where n is odd*, i.e., in practice, when $\arg T = \pm\pi$. This criterion is convenient but it is not a necessary one. A necessary and sufficient stability criterion based on the open-loop frequency response will be derived in the next section.

3.4 Nyquist stability criterion

Stability criteria allow verification of system stability without pole position calculations or experiments. The most convenient among them is the ***Nyquist criterion***. It allows passing a judgment on whether the system is stable by observing the plot of the open-loop frequency response, measured or calculated.

We will consider a single-loop feedback system which consists of linear time-invariable links whose transfer functions are rational with real coefficients. We assume that the system is stable when the loop is disconnected, i.e., the transfer function $T(s)$ does not have poles in the right half-plane of s.

Therefore, *the closed-loop transfer function T/F can only have poles in the right half-plane of s if some of the F(s) zeros are in the right half-plane of s*.

The zeros of the function

$$F(s) = \frac{(s - s_i)(s - s_j)(s - s_k)\ \dots}{(s - s_p)(s - s_q)(s - s_r)\ \dots} \tag{3.4}$$

are s_i, s_j, s_k, ... Let us derive a condition for one or several of the zeros to appear in the right half-plane of s. This is the condition of the closed-loop system instability.

Consider a simple closed contour $\boldsymbol{c}_i$ in the s-plane crossing neither poles nor zeros of $F(s)$ and encompassing no poles and one zero s_i as shown in Fig. 3.12(b). The rest of the zeros of $F(s)$ are outside the contour. While s makes a clockwise round trip about the contour $\boldsymbol{c}_i$, the vector $s - s_i$ shown in Fig. 3.12(b) completes a clockwise revolution. It is easy to notice that the vectors $s - s_j$, $s - s_k$, ... related to zeros outside the contour $\boldsymbol{c}_i$ exercise no such revolutions.

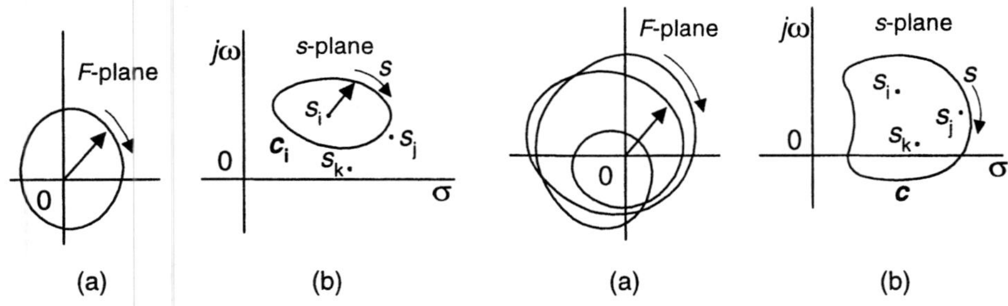

Fig. 3.12 (a) Revolution of a rational function $F(s)$ caused by (b) the trip of s about a closed contour c_i on the s-plane

Fig. 3.13 (a) Revolutions of a rational function $F(s)$ caused by (b) s having moved about a closed contour $\boldsymbol{c}$

The phase of the vector $F(s)$ is the sum of the phases of its multipliers. Therefore, a revolution of a multiplier about s_i changes the phase of $F(s)$ by 2π, i.e., it makes the vector $F(s)$ complete a clockwise revolution about the origin as shown in Fig. 3.12(a).

Consider next a contour $\boldsymbol{c}$ encompassing no poles and several zeros s_i, s_j, s_k as shown in Fig. 3.13(b). When s makes a full trip about $\boldsymbol{c}$, the argument of each multiplier of the kind $s - s_i$ changes by 2π. Therefore, *the number of clockwise revolutions of the locus F(jω) about the origin indicates the numbers of zeros within the contour* $\boldsymbol{c}$, as shown in Fig. 3.13(a).

To find the number of zeros of $F(s)$ in the right half-plane of s, the contour $\boldsymbol{c}$ should radius arc as shown in Fig. 3.14(b). The contour encompasses no poles of F since the poles of $F = T + 1$ are the poles of T. The number of revolutions of the locus of $F(j\omega)$ shown in Fig. 3.14(a) about the origin gives the number of zeros of $F(s)$ in the right half-plane of s. For this particular diagram, the number is 2, reflecting two complex conjugate zeros, s_i and s_j, in the right half-plane of s. Therefore, this particular closed-loop system is unstable.

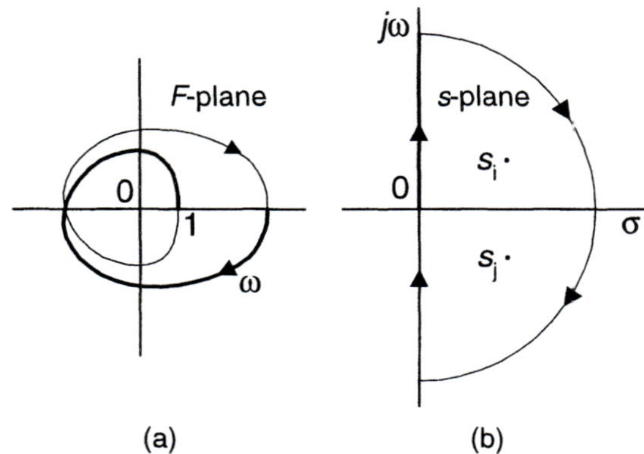

Fig. 3.14 (a) Nyquist diagram for F and (b) the contour surrounding the right half-plane of s

As mentioned before, the return ratio disappears in physical systems when s is big. Therefore, $F(s)$ becomes 1 while s moves along the infinite-radius arc. Thus, the locus in Fig. 3.14(a) is the mapping of the $j\omega$-axis, and the right half-plane of s maps into the inside of the contour on the F-plane.

The function $F(s)$ is rational with real coefficients. Complex conjugation of s makes F complex conjugate. Therefore, the locus of F consists of two image-symmetrical halves relating respectively to positive and negative frequencies.

The part of the locus of F drawn for positive ω is shown by the thicker curve and is called the ***Nyquist diagram***. The diagram makes half the number of revolutions of the whole locus which reflects the existence of zero s_i in the first quadrant of the s-plane.

The ***Nyquist criterion*** follows: *If a linear system is stable with the feedback loop open, it is stable with the loop closed if and only if the Nyquist diagram for F does not encircle the origin of the F-plane*.]

Although the Nyquist criterion has been proven here for only rational transfer functions, it is valid as well for transcendental transfer functions since transcendental transfer functions can be closely approximated by rational functions. For instance, the transcendental pure delay function in a medium with distributed parameters can be approximated by a rational transfer function describing a system with many small-value lumped elements. Therefore, the Nyquist criterion can be applied to all practical systems described by calculated or measured gain and phase frequency responses.

The Nyquist diagram is commonly drawn for $T = F - 1$ as shown in Fig. 3.15 (compare this plot with the locus in Fig. 3.14(a)). The ***critical point*** in this case is -1, instead of the origin.

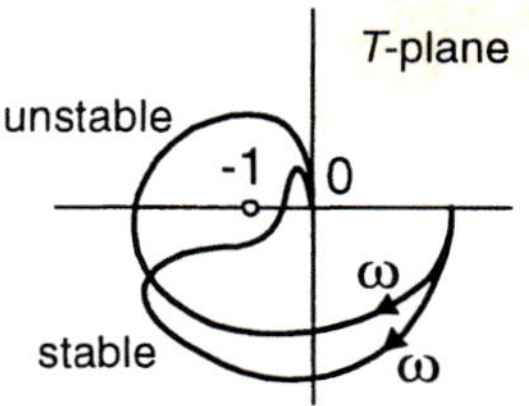

Fig. 3.15 Nyquist diagrams on *T*-plane

Most importantly, the Nyquist diagram tells us not only whether the system is stable or not, but also how to make the system stable by reshaping the loop response (by changing the compensator): by reducing the loop gain over a specific frequency range, and/or reducing the loop phase lag at certain frequencies. ***Loop shaping*** will be described in detail in Chapter 4.

As mentioned in Chapter 1, Nyquist diagrams can be plotted also on the *L*-plane, where the critical points are $(180^\circ \pm n360^\circ,\ 0\ \text{dB})$. Notice that when a Nyquist diagram on the *T*-plane encircles the critical point $(-1, 0)$ in the clockwise direction, the diagram on the *L*-plane typically passes to the left of the critical point $(180^\circ,\ 0\,\text{dB})$ in the counterclockwise direction.

3.5 Robustness and stability margins

Practical systems are required to be not only stable but also ***robust***, i.e., remaining stable when the plant parameters, and consequently the return ratio, deviate from the nominal values. The ***stability margins*** guard the critical point. They are often formed as shown in Fig. 3.16(a), by a segment on the *T*-plane, or, which means the same, by rectangles on the *L*-plane in Fig. 3.16(b). If the Nyquist diagram for the nominal plant does not penetrate the boundary of the stability margin, the system will remain stable for a certain range of variations of the plant parameters.

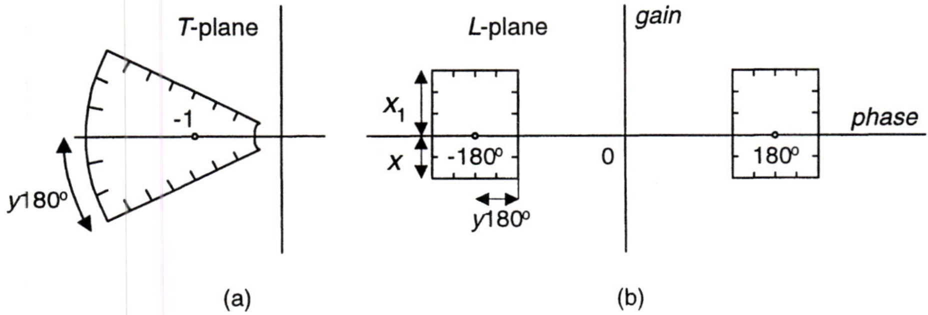

Fig. 3.16 Amplitude and phase stability margins on (a) the *T*- and (b) *L*-planes

The shape of stability margins shown in Fig. 3.16 assumes that plant parameter variations in gain and in phase are not correlated. This is typical for many practical plants. Consider, for example, a force actuator driving a rigid body plant. Variations in the force and in the plant's mass change the plant gain but not the plant phase shift. On the other hand, variations of small flexibilities in the plant change the phase shift but not much the gain in the frequency region of the crossover. Another example is the volume control depicted in Fig. 1.19, where the loop gain changes up to 10,000 times without any change in the loop phase shift. The disk stability margin, as in Fig. 3.20(c), is less suitable for practice.

The values of the lower and upper ***amplitude stability margins*** x, x_1 shown in Fig. 3.16 are typically 6 to 10 dB, and the values of the ***phase stability margin*** $y180^\circ$ are typically 30° to 45°. (These values are also sufficient to guard from some nonlinear phenomena in the control loop, studied in Chapter 12.)

As illustrated in Fig. 3.17, over the frequency range where the loop gain is within the interval $[-x, x_1]$, the phase stability margin is preserved and the system is ***phase-stabilized***, and over the frequency range where the angle of T is within the interval $[-180^\circ(1 + y), -180^\circ(1 - y)]$, the gain stability margin is preserved and the system is ***gain-stabilized***. Fig. 3.17(b) shows the Nyquist diagram of a system which is both phase- and gain-stabilized at frequencies below f_1 and above f_2, and either gain-stabilized or phase-stabilized between these frequencies.

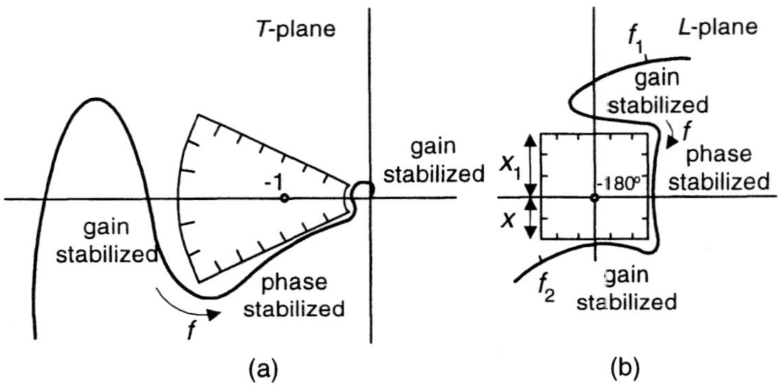

Fig. 3.17 Gain and phase stabilizing in (a) the *T*-plane and (b) the *L*-plane

Nyquist stability relates to a stable system whose Nyquist diagram crosses the negative axis to the left of the point −1. Examples are shown in Fig. 3.17(a) and 3.18(a). Such systems are only gain-stabilized (not phase-stabilized) at some frequencies (such as f_1 and f_2 in Fig. 3.18(a)) at which the loop gain is larger than 1.

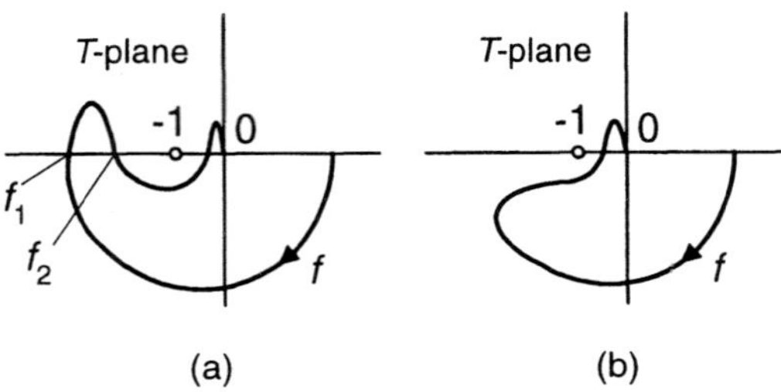

Fig. 3.18 Nyquist diagrams for (a) Nyquist-stable and (b) absolutely stable systems

Practical systems all include nonlinear links, at least, the saturation of the actuator, and are required to be ***globally stable***, i.e., to remain stable after any set of initial conditions. No ***limit cycle*** conditions, i.e., conditions of periodic oscillation, are allowed. (These issues will be discussed in more detail in Chapters 9 – 11.)

When the only nonlinear link in the loop is the actuator, then, loosely speaking, its saturation may reduce the equivalent loop gain while retaining the loop phase shift. This causes the equivalent Nyquist diagram to "shrink." While shrinking, the diagram of a Nyquist-stable system crosses the point −1. At these specific frequency and signal level, the equivalent return ratio becomes −1 which is the condition of self-oscillation. Therefore, *using Nyquist stability should be avoided in those feedback systems which have no nonlinear links other than the actuator saturation*.

The ***absolute stability*** notion relates to systems with a saturation link and whose Nyquist diagram is like that in Fig. 3.18(b), i.e., not crossing the critical point while shrinking. For the absolute stability, the stability margins are typically chosen as shown in Fig. 3.19.

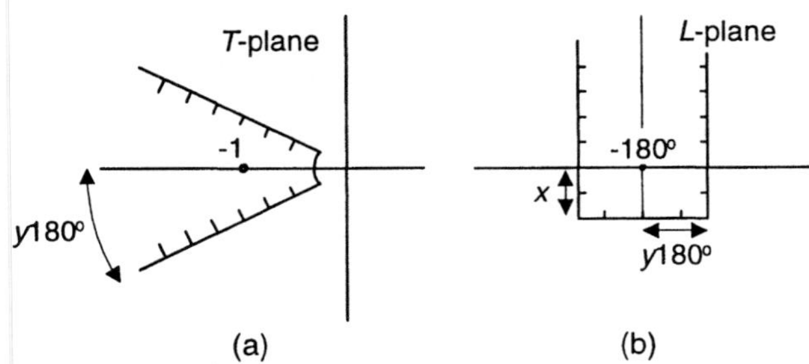

Fig. 3.19 Stability margins for a single-loop system with saturation on the *T*- and *L*-planes

Absolute stability will be studied in Chapter 10 in a more precise manner.

Example 1. According to the Nyquist criterion, the systems with the Nyquist diagrams shown in Fig. 3.20(a) – (e) are stable, while system (f) is unstable. The feedback is positive (i.e., $|T + 1| < 1$) if and only if T is on the unit radius disk centered at the point −1.

In Fig. 3.20(c), T is real and less than −1 at two frequencies, f_1 and f_2. This system is Nyquist-stable. In other words, the return signal at these frequencies is in phase and of a larger amplitude than the signal applied to the loop input. That a feedback system with this sort of T is stable was first found experimentally during development of feedback amplifiers at The Bell Laboratories, and was later proven theoretically by H. Nyquist.

(Using Nyquist stability is acceptable and even beneficial if special nonlinear dynamic links are introduced in the loop to exclude the possibility of self-oscillation. Designing such nonlinear links is described in Chapters 10 and 11.)

One of the Nyquist criterion's advantages is the simplicity of estimating the effects of multiplicative variations of the loop gain coefficient. For example, it is seen from the diagram in Fig. 3.20(a) that reducing the loop gain without changing the loop phase shift will not make the system oscillate, that increasing the gain by a factor of 2 will cause oscillation, and that increasing the loop phase delay by 60° at the frequency where $|T| = 1$ (the crossover frequency f_b) will also cause oscillation.

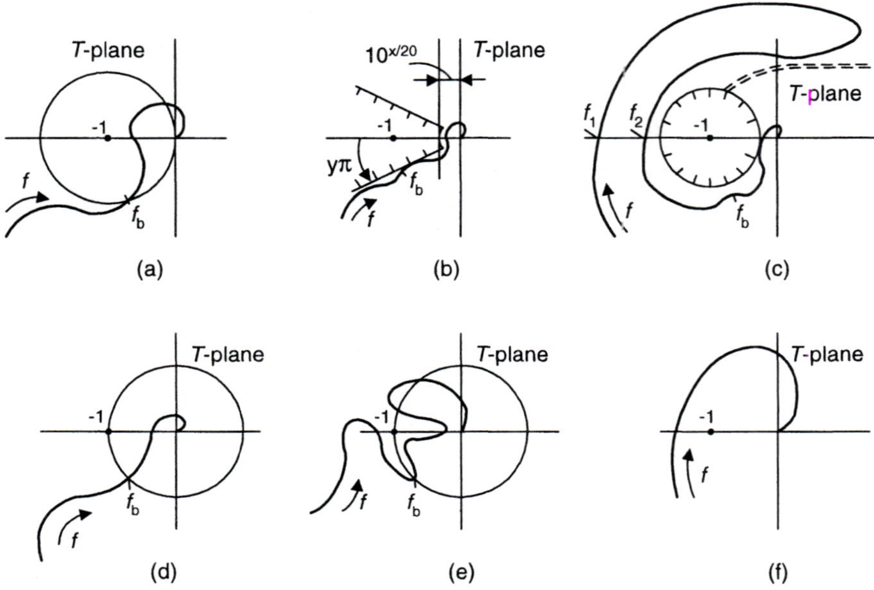

Fig. 3.20 Nyquist diagrams: (a) feedback is positive at frequencies at which the Nyquist diagram is on the disk, (b) Bode stability margins, (c) disk stability margin, (d) robust system, (e) non-robust system, (f) unstable system

Example 2. The linear scale is inconvenient for drawing Nyquist diagrams for practical systems, where the loop gain coefficient changes by several orders of magnitude between the lowest frequency of interest and the crossover frequency. The inventor of feedback amplifiers H. Black employed a circular coordinate system with gain in dB, and phase in degrees, as illustrated in Fig. 3.21. Another example of using these coordinates is given in Appendix 13, Fig. A13.26.

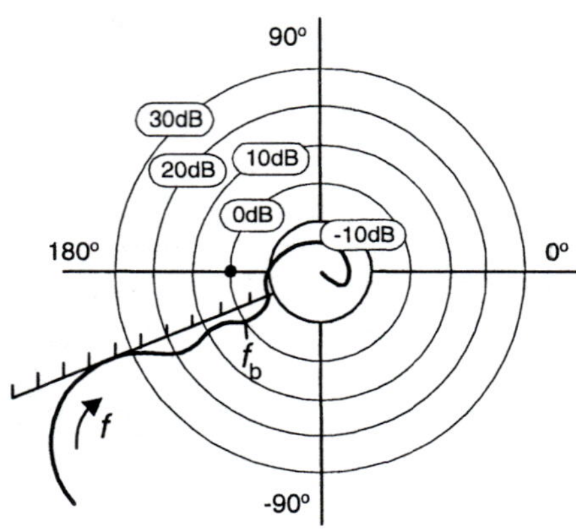

Fig. 3.21 Nyquist diagram with logarithmic magnitude scale

Example 3. The *L*-plane diagrams equivalent to the corresponding *T*-plane diagrams of Fig. 3.20 are shown in Fig. 3.22. The vertical axis corresponds to –180°.

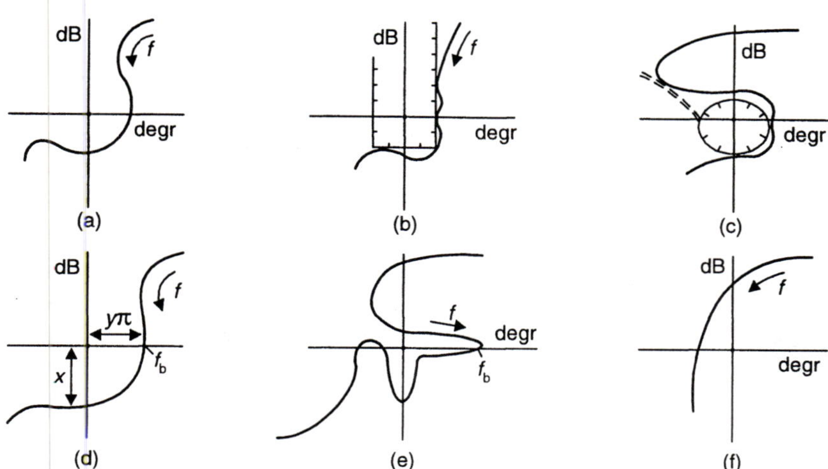

Fig. 3.22 Nyquist diagrams on the *L*-plane: (a) negative and positive feedback areas, (b) Bode stability margins, (c) disk stability margin, (d) robust system, (e) non-robust system, (f) unstable system

Example 4. In some of the contemporary literature, stability margins are defined as if they need only apply at discrete points instead of a solid boundary, and this concept is referred to as "guard-point" stability margins.

The ***guard-point phase margin*** *is the phase margin at the frequency where the gain margin is zero*, i.e., at f_b.

The ***guard-point gain margin*** *is the gain margin at the frequency where the phase margin is zero*.

This interpretation is acceptable and convenient when the order of *T*(*s*) is low, and even when the order of *T*(*s*) is high but the response is smooth, but it does not suffice in general. Because of the inescapable trade-off between stability margins and the available feedback, if a high-order compensator is optimized for closed-loop performance while only the guard-point stability margins are enforced, the Nyquist diagram might end up looking like Figs. 3.20(e) and 3.22(e) – these loop responses approach the critical point much too closely. This sort of misuse of the guard-point stability margins may be responsible for the misconception that high-order compensators "generally" lead to non-robust systems.

3.6 Nyquist criterion for a system with an unstable plant

The systems discussed until now were considered stable with the feedback loop disconnected. However, some physical systems are unstable without the feedback, i.e., *T*(*s*) has poles in the right half-plane of *s*. We would like to know what feedback will make such systems stable.

Now, consider the function *F*(*s*) = *T*(*s*) + 1 which may have poles and zeros within the contour in Fig. 3.14(b). Having a *pole* within the contour causes the plot in (a) to rotate in the direction opposite to the revolution produced by a *zero* within the contour (b). Consequently, when s completes a trip about the contour, the number of revolutions of the function locus equals the difference between the numbers of zeros and poles of $F(s)$ within the contour. The corresponding Nyquist diagram encircles the critical point a number of times equal to the difference between the number of poles and the number of zeros in the first quadrant of the s-plane. The rule follows:

In order for the system to become stable when the feedback is applied, the Nyquist diagram must encircle the critical point the number of times equal to the number of open-loop poles in the first quadrant of the s-plane, and these encirclements must be in the counterclockwise direction.

It is common to describe the plant's instability as the result of an internal feedback loop, where the links characterize physical processes and relations between the variables in the plant. This method is routinely applied to the analysis of various unstable plants: plants with aerodynamic instability, wind flutter, thermal flutter, gas turbulence, and the inverted pendulum. Consider the following two examples.

Example 1. In the system diagrammed in Fig. 3.23(a), the plant is unstable since the internal feedback path transmits the signal back to the input of the gain block in phase at the resonance frequency $\omega = 2$ of the feedback path. The Nyquist diagram for the internal loop encloses the critical point clockwise as shown in Fig. 3.23(b).

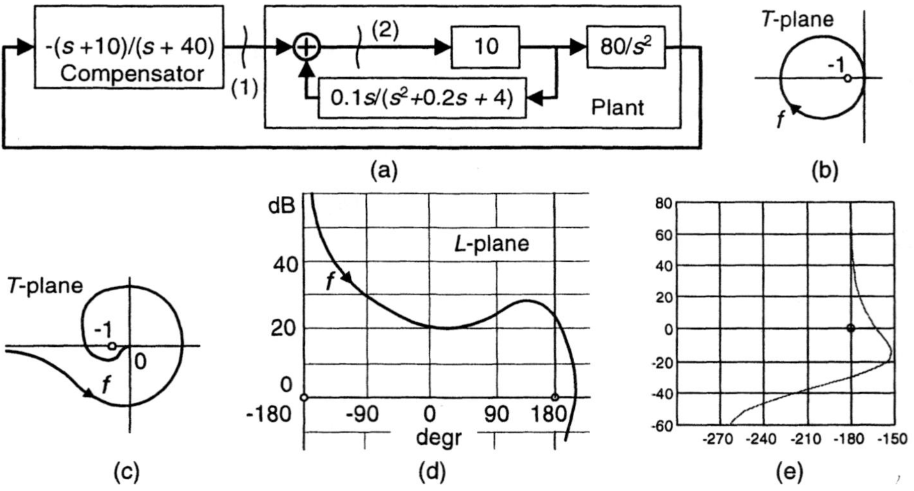

Fig. 3.23 (a) Feedback system with an unstable plant, (b) Nyquist diagram for the internal plant loop, (c) Nyquist diagram for T_1, not to scale, (d) Nyquist diagram for T_1 on the L-plane, (e) L-plane Nyquist diagram for T_2

Because of the internal feedback, the plant transfer function

$$P(s) = \frac{800(s^2 + 0.2s + 4)}{(s^2 - 0.8s + 4)s^2}$$

possesses a pair of complex conjugate poles with positive real parts, as can be judged by the negative sign of the damping coefficient in the denominator polynomial. With the compensator transfer function shown in Fig. 3.23(a), the main loop transfer function at cros-section (1) is

$$T_1(s) = \frac{800(s+10)(s^2+0.2s+4)}{(s+40)(s^2-0.8s+4)s^2}.$$

The locus of $T_1(j\omega)$ shown in Fig. 3.23(c) encircles the critical point -1 in the counterclockwise direction. This indicates that the closed-loop system is stable. The locus on the L-plane shown in Fig. 3.23(d) correspondingly encircles the critical point (0 dB, 180°) in the clockwise direction.

Another way of this system stability analysis is to look at the cros-section (2), where the loop transfer function is found to be

$$T_2(s) = \frac{-0.1s^4+76s^3+960s^2+1920s+3200}{s^5+42s^4+84s^3+160s^2}.$$

The Nyquist diagram for this cros-section is shown in Fig. 3.23(e). The system is stable.

Example 2. The high-gain amplifier shown in Fig. 3.24(a) would be stable by itself, but small parasitic input-output capacitance C causes self-oscillation in the absence of the standard feedback loop via R_2. The transfer function for the parasitic feedback path is shown in Fig. 3.24(b). The Nyquist diagram for the parasitic loop is shown in Fig. 3.24(c). When the amplifier is operated with the conventional feedback path $B = -R_1/(R_1 + R_2)$ closed, the amplifier gain is much smaller (say, only 10 when $B = 0.1$ as indicated in Fig. 3.24(b)), the parasitic feedback loop gain is small, and the system is stable. Another point of view on the problem is, that the parasitic feedback path transfer function $B_{par} = s/[s + 1/(RC)]$ is negligible when compared with the normal feedback path transfer function B, with which the amplifier is stable. Hence, the system can be analyzed without taking into account the parasitic feedback.

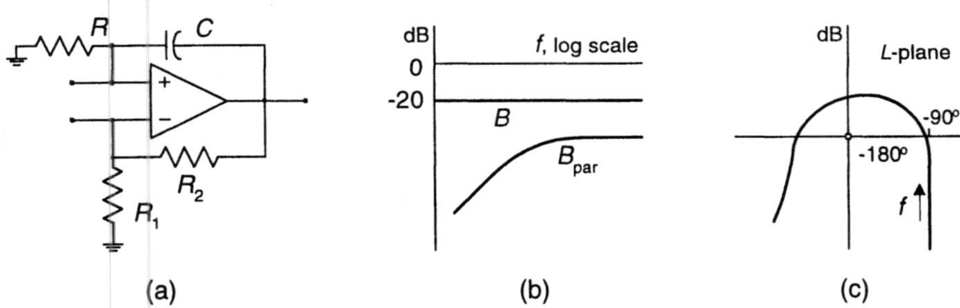

Fig. 3.24 System with parasitic feedback via capacitor C: (a) schematic diagram, (b) Bode diagrams for the feedback paths, and (c) Nyquist diagram for the parasitic feedback loop

3.7 Successive loop closure stability criterion

Example 1. Consider the multiloop system shown in Fig. 3.25.

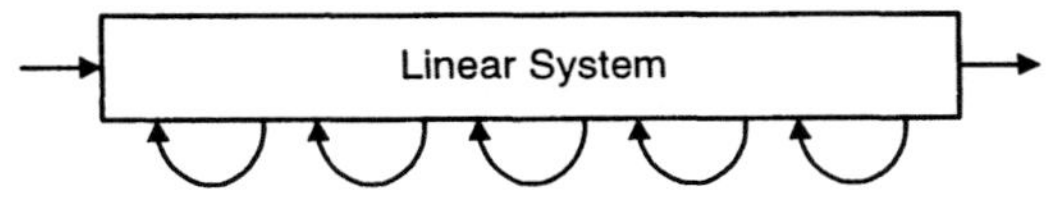

Fig. 3.25 Block diagram of a multiloop system

Assume the system is stable when all loops are

disconnected, and start closing the feedback loops successively which will eventually lead to the system with all the loops closed. A series of five such Nyquist diagrams is exemplified in Fig. 3.26.

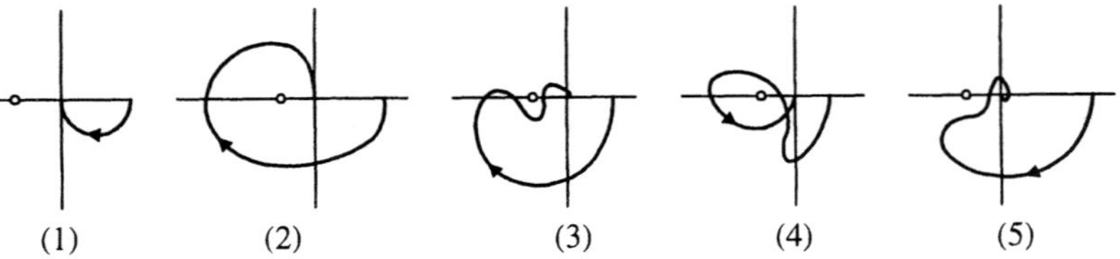

Fig. 3.26 Nyquist diagrams for the successive loop closure stability analysis

It is seen in Fig. 3.26 that after the first loop is closed, the system remains stable; after the second loop is closed, the system becomes unstable and the system transfer function possesses one pole in the first quadrant of the s-plane; after the third loop is closed, the system remains unstable since the Nyquist diagram indicates no change in the difference between the numbers of poles and zeros in the right half-plane of s; after the fourth loop is closed, the system becomes stable since the diagram encircles the critical point once in the counterclockwise direction; and the system remains stable after the fifth loop is closed since the fifth Nyquist diagram does not encircle the critical point.

By generalizing the procedure given in the Example 1, Bode formulated the ***Nyquist criterion for multiloop systems*** as follows:

When a linear system is stable with certain loops disconnected, it is stable with these loops closed if and only if the total numbers of clockwise and counterclockwise encirclements of the point (–1,0) are equal to each other in a series of Nyquist diagrams drawn for each loop and obtained by beginning with all loops open and closing the loops successively in any order leading to the system normal configuration.

The order in which the loops are closed can be chosen at the convenience of the designer.

Example 2. The system diagrammed in Fig. 3.27 contains two local loops and a common loop. The system is viewed as a three-loop system in Fig. 3.27(a). In this case, drawing three Nyquist diagrams is required for the stability analysis. The order of closing the loops and making the Nyquist diagrams can be the following: first, drawing the Nyquist diagram for the common loop (with local loops open), second, closing the common loop and drawing the first local loop diagram, and third, drawing the Nyquist diagram for the remaining local loop. Thus, we would have to draw and to analyze all three diagrams since it is not evident than any of the three diagrams avoids encircling the critical point.

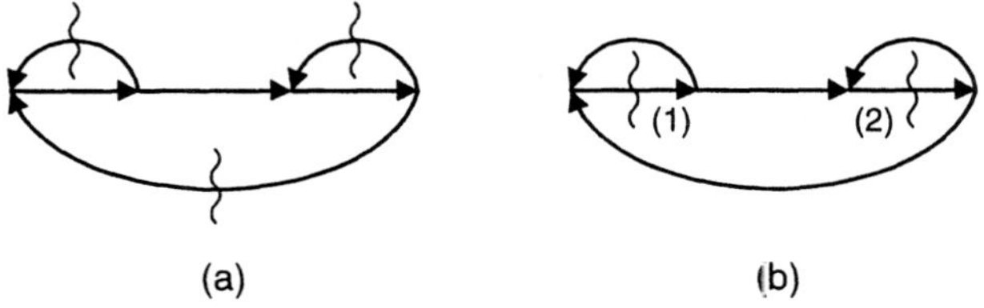

Fig. 3.27 Stability analysis with (a) three and (b) two cross-sections of the feedback loops

It is often evident, however, that the local loops are inherently stable because the phase lag in the loops is small. When this is the case, it is worth starting with these loops and not drawing the related diagrams since it is already known that the diagrams will not enclose the critical point. It remains, therefore, only to draw the diagram for the common loop, with the local loops closed.

Alternatively, the system can be analyzed using two cross sections in the forward paths as shown Fig. 3.27(b). These cross sections eliminate all feedback loops, and the analysis can be performed using only two Nyquist diagrams. Further, when the first local loop is stable by itself, it is convenient to start by closing the cross section (1) without drawing the related Nyquist diagram since it is known that the diagram should not encircle the critical point. Then, only the Nyquist diagram for the cross section (2), with the cross section (1) closed, needs to be drawn and analyzed.

Nevertheless, the simplest approach might be inconsistent with the convenience of verifying global stability. For this purpose, the order of closing the loops can be chosen such that it reflects the order in which the loops desaturate when the signal level gradually decreases [9].

3.8 Nyquist diagrams for the loop transfer functions with poles at the origin

Fig. 3.28 diagrams a mechanical rigid body plant driven by a force actuator. Depending on the type of the sensor employed – an accelerometer, a rate sensor, or a position sensor – the plant transfer function is correspondingly a constant $1/M$, a single integrator $1/(Ms)$, or a double integrator $1/(Ms)^2$.

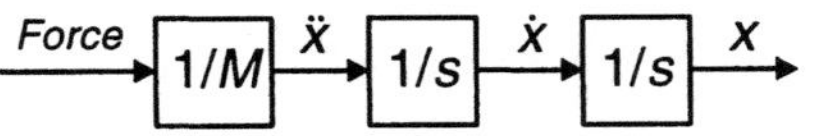

Fig. 3.28 Mechanical rigid body plant

The frequency responses for these plant functions are shown in Fig. 3.29(a). The gain for the single and double integrators is infinite at zero frequency. The poles (the singularities) for the single- and double-integrator plants are at the origin of the s-plane as shown in Fig. 3.29(b).

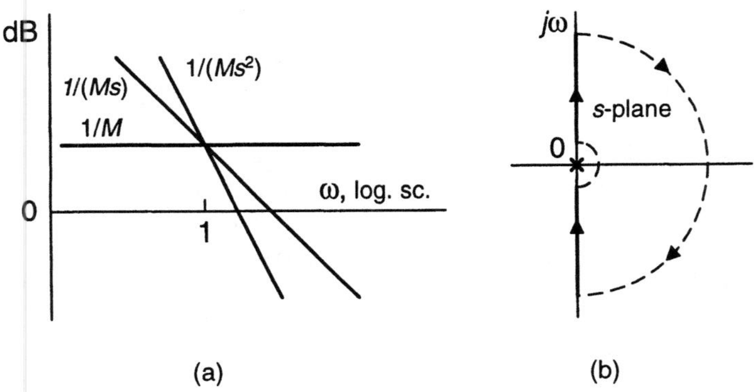

Fig. 3.29 (a) Bode diagrams for the plant as a constant, a single integrator, and a double integrator and (b) the related contour on the s-plane

The Nyquist diagram is the mapping of the contour encompassing the right half-plane of s. The contour should be chosen such that the function on the contour does not turn into infinity. Therefore, if the transfer function $T(s)$ possesses poles on the $j\omega$-axis, the contour should avoid the poles. Particularly, when n poles are at the origin, i.e.,

$$T(s) = \frac{(s - s_{z1})(s - s_{z2})(s - s_{z3})(s - s_{z4}) \ldots}{s^n (s - s_{p1})(s - s_{p2})(s - s_{p3}) \ldots}$$

the origin can be avoided by an infinitesimal-radius arc as shown in Fig. 3.29(b). On this arc, in the close vicinity of the pole, $|T|$ is infinitely large, and its phase changes by $n\pi$ as the phase of s changes by π along the arc. Therefore, the small arc in the s-plane maps onto the T-pane as an $n\pi$ arc of infinite radius. Half of this arc becomes a part of the Nyquist diagram.

The Nyquist diagram for a single-integrator plant is shown in Fig. 3.30(a). For a double-integrator plant, i.e., for a loop transfer function with a double pole at the origin, the arc is twice longer as shown in Fig. 3.30(b).

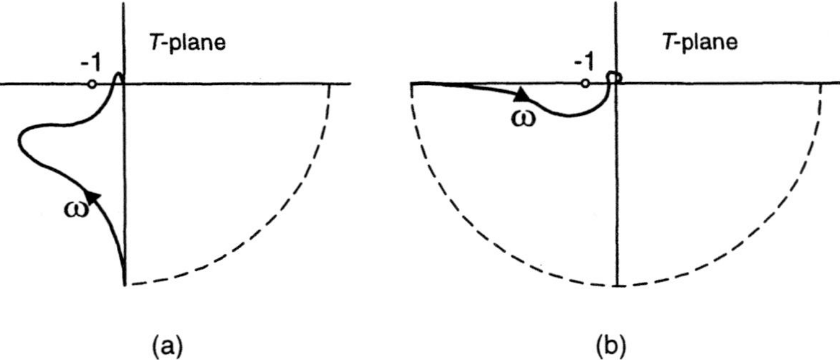

Fig. 3.30 Nyquist diagram for a stable feedback system with (a) a single integrator in the loop and (b) a double integrator in the loop

The feedback control loops are often classified as servomechanisms of ***Type 0*** (or ***servo type***), ***Type 1***, and ***Type 2***. This number is the number of poles of $T(s)$ at the origin. The Nyquist diagram shown in Fig. 3.30(a) is of Type 1, and the diagram shown in Fig. 3.29(b) is of Type 2. Nyquist diagrams of Type 0 are shown in Fig. 3.18(a) and (b). The shape of the Nyquist diagram and the related loop frequency response at lower frequencies define the steady-state response to commands and the steady-state error reduction.

The properties of these three types of servomechanisms can be understood by considering the feedback system shown in Fig. 3.31. The plant output variable is the angle θ. Two differentiators are added after the plant for the sake of analysis, to calculate the angular velocity Ω and the acceleration α. A double integrator is added to keep the return ratio unchanged.

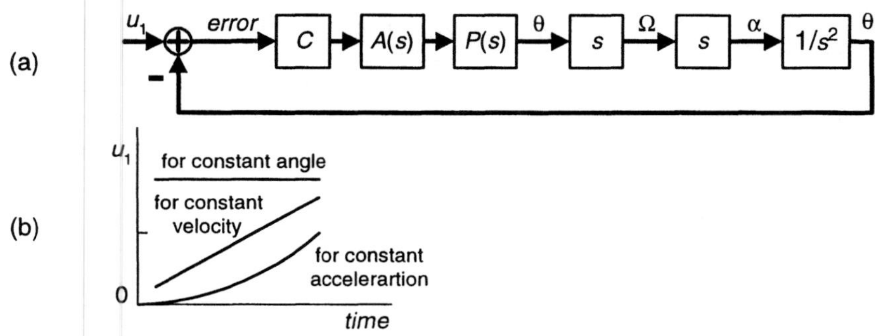

Fig. 3.31 (a) Feedback system block diagram and (b) the profiles of the command

Assuming the error is small, i.e., the return signal nearly equals the command, let us consider the problem of keeping constant one of the variables: θ, or Ω, or α. In order to do this, specific commands $u_1(t)$ need to be applied to the input. In Fig. 3.31(b), the time-functions are shown for, respectively, constant angle, constant angular velocity, and constant acceleration commands.

To reduce the static error of the controlled variable, the dc gain coefficient (i.e., the gain coefficient at $s \to 0$) *from the error to the controlled variable must be large.*

The Type 0 system has finite loop gain at zero frequency, and the dc gain from the summer output to the plant output (from *error* to θ) is finite. Therefore, the steady-state error of the angle θ is small but finite. On the other hand, the forward path gain coefficients from the *error* to the velocity and to the acceleration at $s \to 0$ are infinitely small. If in the commanded $u_1(t)$, the velocity or the acceleration is constant, the output will not track the command.

The Type 1 system has an infinite loop gain coefficient at zero frequency. The gain from the error to θ is also infinite, but the gain to velocity Ω is finite, and the gain coefficient to acceleration α is zero. Therefore, when constant angle θ is commanded, the error is the angle error and it is zero; when constant velocity is commanded, the velocity error is finite; and when constant acceleration is commanded, the acceleration error is not corrected at all.

Next, consider the effects of disturbances entering the feedback system at different points: disturbance in position, disturbance in velocity, and disturbance in acceleration as shown in Fig. 3.32. Disturbance in position is commonly caused by misalignment of mechanical parts; disturbance in velocity, by an extra velocity component of moving parts of the plant or by drift in time of the values of some of the plant's parameters; and the disturbance in acceleration, by disturbance torque due to wind, magnetic forces, etc.

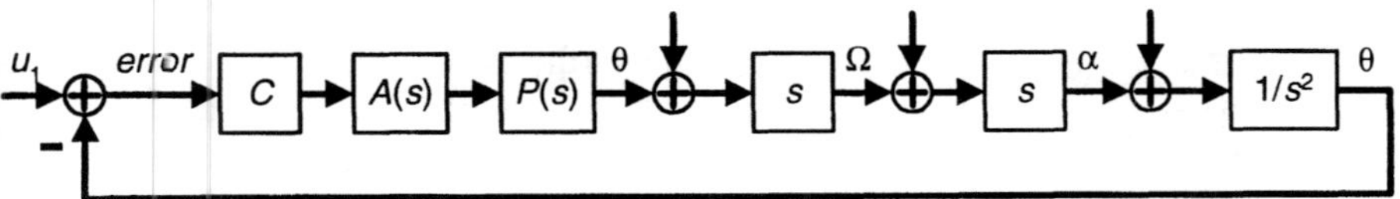

Fig. 3.32 Disturbances in angle, velocity, and acceleration in a feedback system

In a Type 1 system, a disturbance entering at the point of velocity is reduced infinitely since the feedback is infinite, but this disturbance causes a finite change in the angle since the gain coefficient at dc from the angle to the velocity is infinitely small. Thus, the constant velocity disturbance causes a "hang-up" error in position (i.e., an error that does not decay in time). To eliminate this error, a Type 2 system should be employed. The Type 2 servomechanism is also referred to as a "zero-velocity error" system. In this system, the steady-state errors in position and in velocity are zero, and there is finite reduction in the steady-state error in acceleration.

In some systems, the return ratio has a triple pole at zero frequency. These systems have larger loop gain and better accuracy at low frequencies, but the low-frequency phase lag in such systems approaches 270° and the system is not absolutely stable. For such a system to remain stable after the actuator becomes overloaded, the compensator must be made nonlinear, as will be discussed in Chapters 9–13.

3.9 Bode integrals

3.9.1 Minimum phase functions

Synthesis (design) of a stable feedback system using the Nyquist criterion is not quite straightforward. For example, if, to correct the shape of the Nyquist diagram, one decided to reduce the gain at some frequencies and did so, he might find out that this gain change affected the phase shift at other frequencies, and the system is still unstable, although with a quite different shape of the Nyquist diagram.

The Nyquist criterion uses three variables: frequency, loop gain, and phase shift. These variables are interdependent. H. W. Bode showed that in most practical cases using only two of them (the frequency and the gain) suffices for feedback system design. This greatly simplifies the search for the optimal design solution.

A logarithmic transfer function can be presented in the form

$$\theta(s) = A(s) + B(s),$$

where $A(s)$ is the even part of the function and $B(s)$ is the odd part of the function. When s is replaced by $j\omega$, then A becomes the real part of the function, i.e., the gain, and B becomes the imaginary part, i.e., the phase shift.

The real and imaginary parts of $\theta(j\omega)$ are related, although not in a unique way. It is always possible to add a constant to the gain without affecting the phase, and to add extra phase lag (but not phase lead!) without affecting the gain. This extra phase lag, called ***nonminimal phase*** (***n.p.***) lag, will be considered in more detail in Section 3.12. Functions without such phase lag are called ***minimum phase*** (***m.p.***).

It will be demonstrated later in this chapter that m.p. transfer functions have no zeros in the right half-plane of s, so that m.p. transfer functions of stable systems have neither zeros nor poles in the right half-plane of s. Therefore, m.p. logarithmic transfer function of stable systems are analytical in the right half-plane of s.

As will be shown further, the phase delay in the feedback loop limits the available feedback. Therefore, it is desirable for the transfer functions of the feedback loop links to be m.p. Designers of control loops use m.p. functions in compensators and, if possible, employ actuators and plants with m.p. transfer functions. The phase-gain relations in m.p. functions are of special interest for feedback system designers.

Example 1. A passive two-pole is stable in the conditions of being open or shorted. Therefore, its impedance and admittance have no zeros or poles in the right half-plane of s. Therefore, these functions are m.p.

exactly on $j\omega$-axis do not have place if we assume some mimimum dissipation.) The logarithms of these functions are also m.p. functions.

Example 2. Fig. 3.33 shows a ladder passive electrical network. The output-to-input ratios of the network can be expressed as voltage- and current-transfer functions, the transimpedance (ratio of the output voltage to input current) and the transadmittance (ratio of the output current to the input voltage).

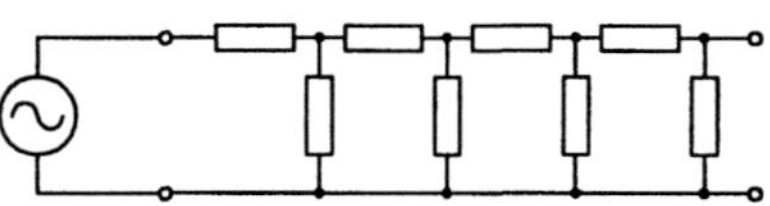

Fig. 3.33 Ladder network

It is evident that a signal applied to the input of a ladder network arrives at the output unless at least one of the series branches is open, or one of the parallel branches is shorted. Therefore, the output-to-input ratio zeros are produced by poles of the impedances of the series two-poles and by zeros of the impedances of the shunting two-poles. When the two-poles are passive with some dissipation, their poles and zeros are not in the right-half plane of s or on its boundary, so the network transfer function does not have zeros in the right half-plane of s and on the $j\omega$--axis. Logarithmic transfer functions also have these properties. Bode named such functions minimum-phase functions (meaning minimum phase lag). He proved that the phase delay of the minimum-phase function is the smallest among the transfer functions with the same gain frequency response and is uniquely defined by the gain frequency response.

3.9.2 Integral of feedback

Consider the function $\theta(s)$ which has no singularities on the border and inside the right half-plane of s and is limited at high frequencies and therefore can be approximated for large s by the series

$$\theta(s) = A_\infty - \frac{B_1}{s} - \frac{A_2}{s^2} - \frac{B_3}{s^3} - \frac{A_4}{s^4} - \ \dots \tag{3.5}$$

By integrating $\theta(s)$ about a contour enclosing the right half-plane of s (see Appendix A4.1), Bode proved that

$$\int_0^\infty (A - A_\infty)\, d\omega = -\frac{B_1}{2\pi}. \tag{3.6}$$

Now, assume that return ratio $T(s)$ of a system that is open-loop stable turns into a/s^n for large s, where a is a coefficient, and let $\theta = \ln F$. Then, when s is large,

$$\theta(s) = \ln(1+T) \approx \ln\left(1 + \frac{a}{s^n}\right) \approx \frac{a}{s^n}.$$

In the most common case of $n \geq 2$, by comparing this expression with (3.6) we see that $B_1 = 0$, and the ***integral of feedback*** is zero:

$$\int_{-\infty}^{\infty} \ln|F|\, d\omega = 0 \tag{3.7}$$

When the feedback is negative, $|F| > 1$ and therefore $\ln|F| > 0$, and vice versa. Hence, as illustrated in Fig. 3.34, <u>*the feedback integral over the frequency region where*</u>

the feedback is negative is equal to the negative of the integral over the range of positive feedback. The larger the negative feedback and its frequency range, the larger must be the area of positive feedback. Typically, as illustrated in Fig. 1.7, positive feedback concentrates near the crossover frequency f_b. The positive feedback area in Fig. 1.7 looks much smaller because this picture was drawn with a logarithmic frequency scale.

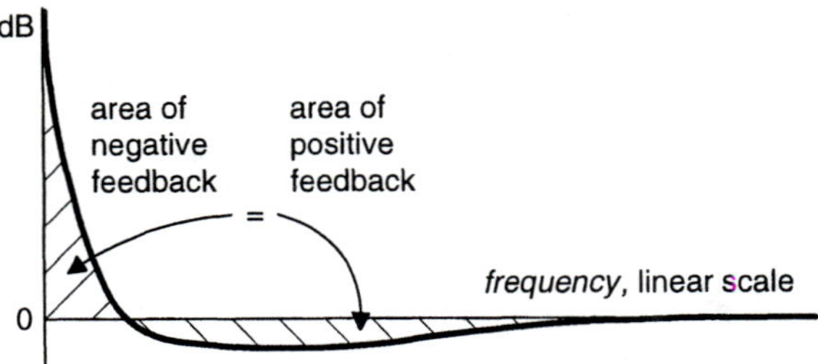

Fig. 3.34 Negative and positive feedback areas

Therefore, *if negative feedback reduces the effect of disturbances in certain frequency regions, there must exist a frequency region of positive feedback where these effects are increased*. In practice, the feedback decreases the output mean-square error since the feedback becomes positive only at higher frequencies where the error components are already reduced by the plant, which is typically a kind of low-pass filter.

Corollary 1. *If the area of negative feedback over the functional frequency bandwidth needs to be maximized, the area of positive feedback must be maximized as well*. The value of the positive feedback depends on the distance from the Nyquist diagram to the critical point. Therefore, this distance should be kept minimal over the bandwidth of positive feedback, or, in other words, *the Nyquist diagram should follow the stability margins' boundary as closely as possible, and the chosen stability margins should not be excessive*.

Corollary 2. Since the positive feedback is concentrated within a few octaves near the crossover frequency, *the accuracy of loop shaping in the crossover area is of extreme importance* for achieving maximum negative feedback over the functional frequency bandwidth.

Corollary 3. Shifting the loop gain response toward higher frequencies doesn't change the area of positive feedback on the logarithmic frequency scale and therefore increases it on linear frequency scale. Negative feedback increases correspondingly. Therefore, *feedback bandwidth must be increased until some limiting factors (plant uncertainty, noise effects) will prevent further increase of it*.

3.9.3 Integral of resistance

Next, let $\theta(j\omega) = Z(j\omega) = R(j\omega) + jX(j\omega)$ stand for the impedance of the parallel connection of a capacitance C and a two-pole with impedance Z', as shown in Fig. 3.35, where Z' is assumed not to reduce to zero at infinite frequency and to be limited at all frequencies. Then, at higher frequencies $Z \approx 1/(j\omega C)$. Comparing this formula with (3.5) gives $B_1 = 1/C$, and from (3.6) follows an equation called the ***resistance integral***:

$$\int_0^\infty R \, d\omega = \frac{\pi}{2C} \qquad (3.8)$$

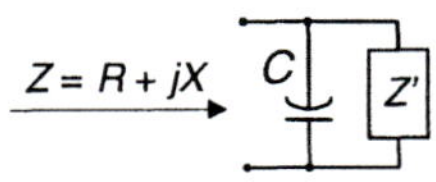

Fig. 3.35 Two-pole Z' shunted by capacitance C

It is seen that *the area under the frequency response of the resistance R is exclusively determined by the parallel capacitance C*. The frequency responses of R in Fig. 3.36(d) relate to the two-poles of Fig. 3.36(a), (b), and (c) with different Z' but the same C. The area under the curves is the same. It is also seen that the maximum value of R over the desired frequency band can be achieved if R equals zero outside the operational band, which can be achieved by using as Z' a filter loaded at a matched resistor.

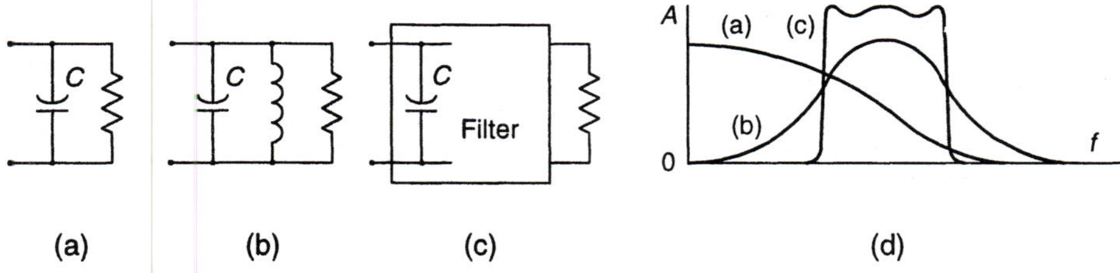

Fig. 3.36 Two-poles made of reactive two-ports loaded at resistors: (a) low-pass, (b) resonance, (c) Chebyshev band-pass filter, and (d) their resistive components

The similarly derived integral of the real part of the admittance Y,

$$\int_0^\infty \operatorname{Re} Y(\omega)\, d\omega = \frac{\pi}{2L}, \tag{3.9}$$

is valid for the dual circuit shown in Fig. 3.37, where the admittance Y' of the remaining part of the circuit does not turn to 0 at infinite frequency, i.e., does not contain a series inductance. The relations (3.8) and (3.9) are widely applied in radio frequency and microwave engineering for the evaluation of the available bandwidth-performance product in systems where the stray reactive element, C or L, becomes critical (in particular, in the input and output circuits of wide-band high-frequency amplifiers, or in the parallel or series feedback paths of such amplifiers).

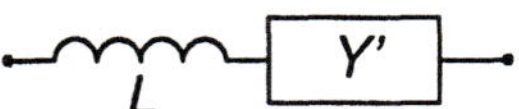

Fig. 3.37 Stray inductance limiting the real part of admittance

The resistance integral is also useful for estimating the available performance of control and active damping of mechanical flexible structures. Important classes of flexible plants include active suspension systems, micromachined mechanical systems, and large, relatively lightweight actively controlled and damped structures in zero gravity environment.

In mechanical flexible structures where some flexible modes need to be damped, sometimes the damper can be connected only to the port where a mass or a spring limits the bandwidth of a disturbance isolation system. In this case, diagrammed in Fig. 3.38, to achieve maximum performance over a specified bandwidth, Z' can be implemented electrically and connected to the mechanical structure via an electromechanical transducer.

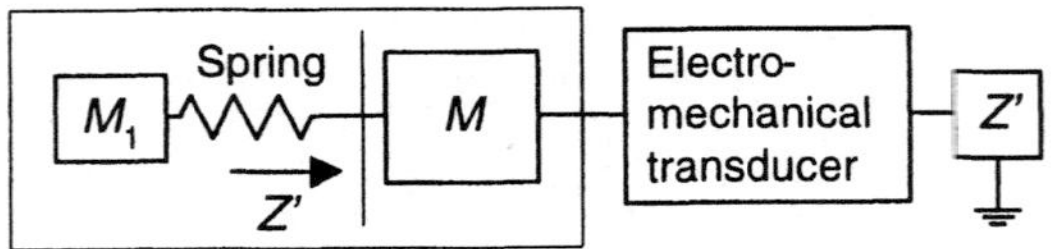

Fig. 3.38 Active damping of a mechanical structural mode by connecting an active damper with impedance Z'

3.9.4 Integral of the imaginary part

The relation known as the ***phase integral*** is

$$\frac{2}{\pi}\int_{-\infty}^{\infty} B(u)\, du = A_\infty - A_o \tag{3.10}$$

where B is the phase shift of an m.p. function, A_o and A_∞ are the values of the gain at respectively zero and infinite frequency, $u = \ln(\omega/\omega_c)$, and ω_c can be arbitrary. In other words, the integral is taken along the frequency axis with logarithmic scale (the equation is derived in Appendix 4).

The integral can be conveniently applied to the difference between the two gain-frequency responses A'' and A' joining at higher frequencies as shown in Fig. 3.39, and therefore having the same value of A_∞. By (3.10), this difference is

$$\Delta A_o = (A_o'' - A_o') = -\frac{2}{\pi}\int_{-\infty}^{\infty} (B'' - B')\, du.$$

In (3.10), the phase, the gain, and the frequency units are related to the natural logarithm. When the units are converted to degrees, dB, and decades, the low-frequency gain difference is

$$\Delta A_o,\ \mathrm{dB} = 0.56a\,(\mathrm{dec} \times \mathrm{degr}) \tag{3.11}$$

where a is the difference in the phase integrals, i.e. the difference in the areas under the phase responses. (The use of this formula will be illustrated in Section 5.5.)

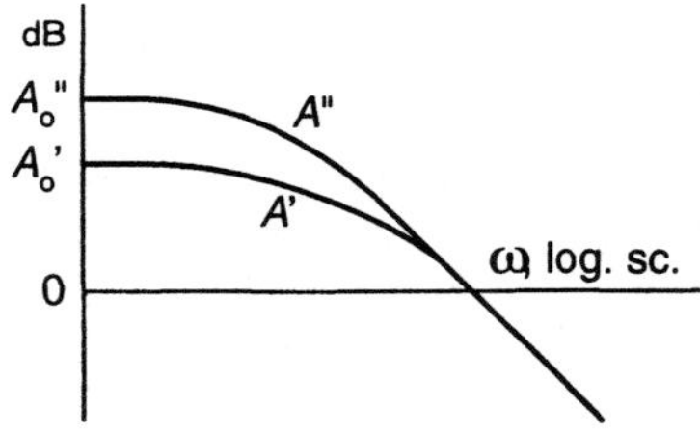

Fig. 3.39 Two gain responses having a common high-frequency asymptote

It follows that an increase in the loop gain in the band of operation is accompanied by an increase of the area under the frequency response of the loop phase lag. Hence, the larger the feedback, the larger must be the area of the phase lag. In particular, the available feedback is larger in Nyquist-stable systems because of their larger loop phase lag.

3.9.5 Gain integral over finite bandwidth

Still another important relation is

$$\int_{\omega=0}^{\omega=1} (A - A_\infty)\, d \arcsin\omega = \int_{1}^{\infty} \frac{B}{\sqrt{\omega^2 - 1}}\, d\omega. \tag{3.12}$$

With application to feedback systems, (3.12) means that *if the phase lag at frequencies $\omega \geq 1$ needs to be preserved, then the loop gain response can be reshaped in the functional band $\omega \leq 1$ as long as the area of the gain plotted against arcsinω is not changed*. Applications of this rule will be studied in Chapter 4.

3.9.6 Phase-gain relation

Given the gain frequency response $A(\omega)$ of a minimum phase transfer function, it is possible to find the phase shift B at a specified frequency ω_c using the famous Bode formula (see Appendix 4 for the proof):

$$B(\omega_c) = \frac{1}{\pi} \int_{-\infty}^{\infty} \frac{dA}{du} \ln \coth \frac{|u|}{2} du \tag{3.13}$$

where $u = \ln(\omega/\omega_c)$. This formula uses natural logarithm units for the angle, attenuation, and frequency.

As seen, *the phase shift is proportional to the slope of the gain response versus the frequency with logarithmic scale*.

For example, for a single integrator with transfer function $1/s$, the slope is -6 dB/oct and the phase shift is $-90°$. For a double integrator, the slope is -12 dB/oct and the phase shift is $-180°$. For the gain response having constant slope -10 dB/oct, the phase shift, proportionally, is $-150°$. The convenience of relating the phase to the slope of the gain with this Bode formula is the reason why the gain responses drawn with logarithmic frequency scale became called Bode diagrams.

Since the integral in (3.13) is taken from $-\infty$ to ∞ along the u-axis, i.e. over the frequencies from 0 to ∞, *the phase shift at any specified frequency depends on the gain response slope at all frequencies*.

The extent of this dependence is determined by the weight function $\ln \coth |u/2|$, charted in Fig. 3.40 with the script

```
u = linspace(-3,3,200);
b = log(coth(abs(u/2)));
plot(u b,'w');
grid
```

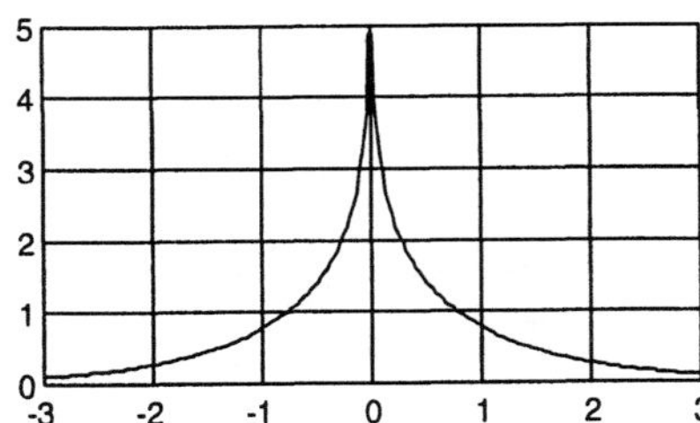

Fig. 3.40 Weight function ln coth$|u/2|$

Due to the selectiveness of the weight function, *the neighborhood of $u = 0$, i.e. the neighborhood of the frequency ω_c (at which the phase shift is being calculated), contributes much more to B_c than the remote parts of the Bode diagram do*.

The Bode phase-gain relationship is equivalent to the stability of m.p. systems, and to the ***causality principle***, i.e. the rule that the result (output) never precedes the cause (input) (see Appendix 3). The integral can be proved with countour integration (see Appendix 4) or with the Hilbert transform.

It is important that this and other Bode integrals are applicable to the transfer functions of physical systems which are not necessarily rational.

The Bode integral allows one to exclude the phase from the three variables – gain, phase, and frequency – employed in the Nyquist stability criterion. The relation between the remaining two variables, the gain and the frequency, expressed in the form of the Bode diagram, contains also the information about the phase shift. Therefore, the system design can be based on only the gain responses which is very convenient. Bode diagram methods are widely applied to practical links which are m.p. or whose n.p. components are small or can be additionally calculated and, therefore, accounted for separately.

The sequence of frequency responses in Fig. 3.41 illustrates the structure of the Bode formula:

in (a), the Bode diagram is shown;
in (b), the slope of the Bode diagram is plotted;
(c) shows the weight function centered at ω_c;
(d) shows the slope response multiplied by the weight function.

The area under response (d) gives the phase shift at ω_c. Notice that Fig. 3.41 is only illustrating the structure of the phase-gain relation. A practical method for the phase calculation will be described in the next section.

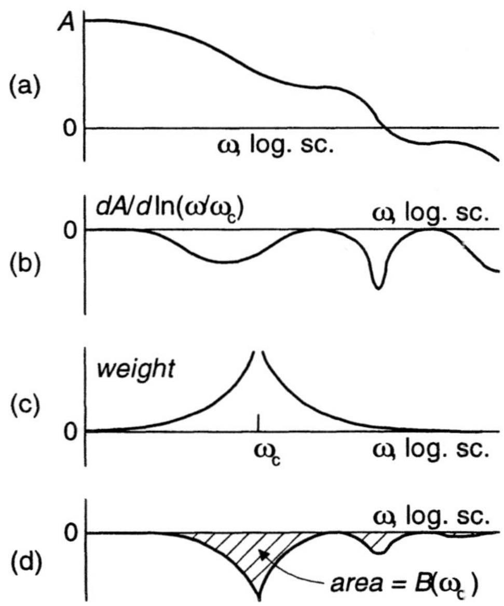

Fig. 3.41 Phase shift calculation at frequency ω_c: (a) Bode diagram, (b) Bode diagram slope, (c) weight function centered at ω_c, (d) product of the slope and the weight

Example 1. Calculate the phase at frequency $\omega_c < 1$ which relates to the following low-pass response: the gain is 0 dB up to the corner frequency $\omega = 1$, then decreasing with unit slope, $dA/du = -1$.

From (3.13), the phase is

$$B(\omega_c) = -\frac{1}{\pi} \int_{-\ln\omega_c}^{+\infty} \ln\coth\frac{|u|}{2} du$$

where the lower integration limit corresponds to the corner frequency $\omega = 1$, which is the frequency where the gain slope changes from zero to -1, i.e. $u = \ln(1/\omega_c)$. Since

$$\ln\coth\frac{|u|}{2} = \ln\left|\frac{1+\omega_c/\omega}{1-\omega_c/\omega}\right|,$$

and $du = d\ln(\omega/\omega_c) = \omega^{-1}d\omega$, we can rewrite the expression for $B(\omega_c)$ as

$$B(\omega_c) = -\frac{1}{\pi} \int_{1}^{\infty} \ln\left|\frac{1+\omega_c/\omega}{1-\omega_c/\omega}\right| \frac{d\omega}{\omega}.$$

At small ω_c, the logarithm equals $|2\omega_c/\omega|$. Then

$$B(\omega_c) \approx -\frac{2}{\pi}\omega_c \int_{1}^{\infty} \frac{d\omega}{\omega^2}$$

and finally,

$$B(\omega_c) \approx -\frac{2}{\pi}\omega_c. \tag{3.14}$$

The phase shift is negative and proportional to the frequency.

Corollary. Since the gain response of any low-pass filter can be approximated in a piece-linear manner, *the phase of a low-pass filter at low frequencies* is a sum of functions, each proportional to the frequency, so that this sum as well *is proportional to the frequency*.

3.10 Phase calculations

Accurate calculation of the phase lag from the gain response is rarely needed in engineering practice, and computer programs developed for calculation of the integral (3.13) are used rather infrequently. However, approximate calculation of the phase is quite often required during the conceptual stage of the design, and for small readjustments of the loop frequency responses. For these purposes, a modified version of a graphical procedure suggested by Bode is described below.

In Fig. 3.42, the phase responses are plotted for the gain ray that originates at f_c with the slope of $-6n$ dB/oct (dashed line), and for the segments (ramps) of the gain response with the slope of $-6n$ dB/oct over w octaves centered at f_c, i.e. at $u = 1$.

In general, if the segment's slope is a dB/oct, then the left scale of the phase should be multiplied by $a/6$, or the right scale multiplied by $a/10$.

Bode diagrams can be approximated piece-linearly by segments and rays, and the phase responses related to these can be added up. Bode proved that even a crude approximation of A renders a fairly accurate phase frequency response. For the responses typical for automatic control, the number of segments need not be large.

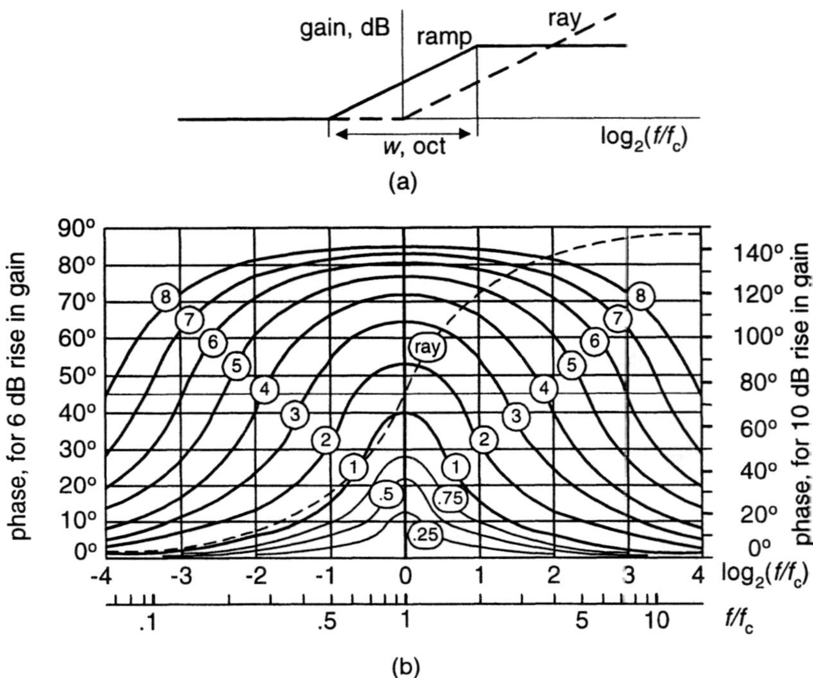

Fig. 3.42 (a) Ramp gain response with constant slope over w octaves centered at f_c, and the gain ray starting at f_c; (b) phase responses corresponding to this gain response, for different w, and (dashed line) phase response corresponding to 6 dB/oct ray

Example 1. The gain response is approximated by two segments with nonzero slope, three segments with zero slope (no phase is related to these segments), and a ray as illustrated in Fig. 3.43. The phase response is then obtained as the sum of the elementary phase responses, each related to a single segment or a ray of the gain frequency response. The total phase response is the sum of the three phase responses.

Example 2. A loop response crosses the 0 dB line at the frequency 800 Hz. In an attempt to increase the feedback at lower frequencies, it is contemplated to make the Bode diagram steeper by 6 dB/oct over an octave centered at 200 Hz. What will be the effect of this change on the guard-point phase margin? The effect can be calculated with the help of the chart in Fig. 3.42. From the curve marked "1," at the distance of 2 octaves from the center, the phase is 13°. That is, the guard-point phase stability margin will be reduced by 13°.

Piece-linear approximation of $A(\omega)$ is particularly useful for trial-and-error procedures of finding a physically realizable response for $\theta(j\omega)$ that maximizes a certain norm while complying with a set of heterogeneous constraints (such as weighted maximization of the real component over a given frequency range under the limitation in the form of a prescribed boundary for the frequency hodograph of the function).

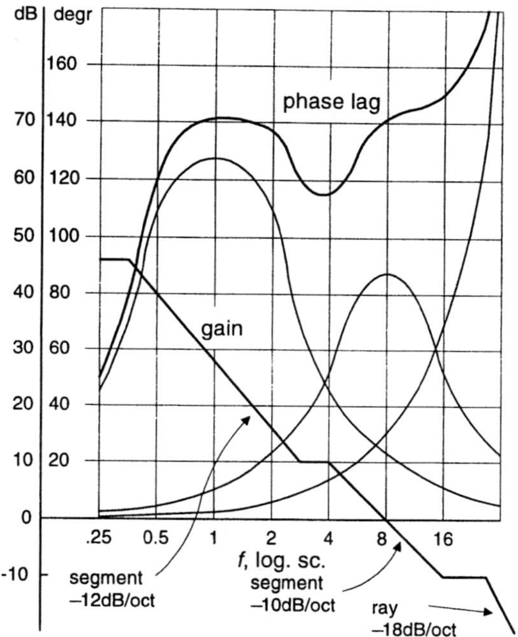

Fig. 3.43 Phase calculation for piece-linear approximation of a Bode diagram

Starting with some initial response for A, say, A', we could calculate the related response B' and get a physically realizable $\theta' = A' + jB'$. Next, changing the gain response as seems reasonable, we would find the related phase response, etc. As a rule, the process converges rapidly, and the accuracy of the graphical procedure is sufficient. (The appropriate scales are: 10 dB/cm and 1 oct/cm for sketches; 5 dB/cm and 0.5 oct/cm for more accurate calculations.)

Example 3. A piece-linear gain response can be viewed as a sum of several ray responses. A ray starting at ω_o with the slope $-12n$ dB/oct can be approximated by $(s^2 + \omega_o s + \omega_o^2)^{-n}$; here n is not necessarily an integer. MATLAB function BONYQAS described in Appendix 14 is based on this approximation. It calculates and plots the m.p. phase response and the Nyquist diagram related to a piece-linear gain response specified by the vector of corner frequencies, the vector of the gains at these frequencies, and the low-frequency and high-frequency asymptotic slopes.

3.11 From the Nyquist diagram to the Bode diagram

From a known Bode diagram, phase can be calculated and a Nyquist diagram can be plotted. The inverse problem is, given the function $B(A)$, i.e. the shape of the Nyquist diagram for an m.p. function, to find the Bode diagram. Although no analytical solution exists to this problem, the solution can be found numerically with a computer by approximating the Nyquist diagram with a high-order rational function. Alternatively, the responses important for practice can be found rather easily with an iterative procedure utilizing the Bode method of finding $B(\omega)$ from $A(\omega)$.

The iterative process consists of the following steps: (a) plotting a first-guess Bode diagram composed of segments and rays, (b) calculating the related phase lag, (c) plotting the Nyquist diagram, (d) correcting the first-guess Bode diagram, etc., and converges rapidly for smooth-shaped Nyquist diagrams. However, for Nyquist diagrams with some sharp angles (which are optimal for many systems), the convergence of this procedure is slow. The convergence is improved when a sharp corner response is included in the set of elementary functions. For this purpose we can use the function

$$\theta(jf) = \ln \frac{1}{\sqrt{1-f^2} + jf} \tag{3.15}$$

which is plotted in Fig. 3.44(b). This response is low-pass. It has the peculiar property of having the gain of 0 dB over the frequency band $\omega \le 1$, and the phase lag of $\pi/2$ for $\omega \ge 1$. The frequency locus of the ratio in (3.15) is plotted in Fig. 3.44(a).

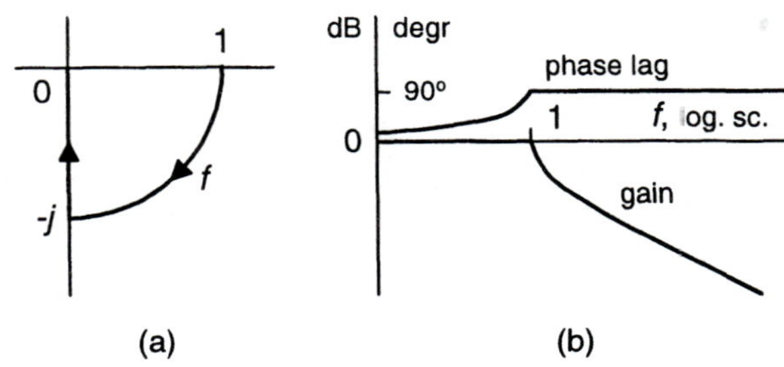

Fig. 3.44 (a) Locus of the ratio in (3.15) and (b) frequency responses for the gain and phase of (3.15)

The high-pass response shown in Fig. 3.45(a) can be calculated by substituting f in (3.15) by $-1/f$; band-pass response shown in Fig. 3.45(b) can be obtained by substituting f in (3.15) by $(f - 1/f)$ [2].

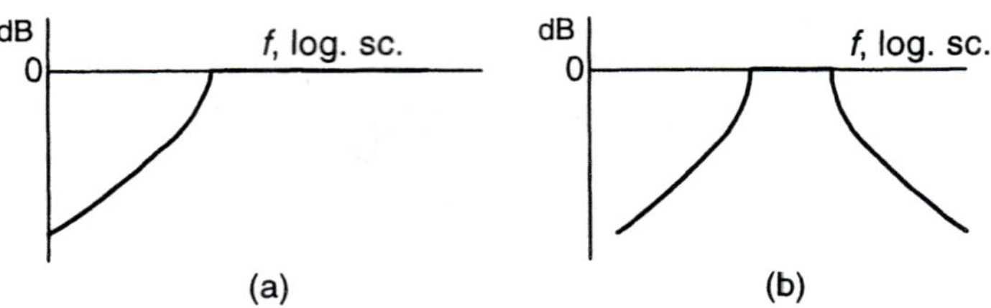

Fig. 3.45 Bode diagrams for high-pass and band-pass transforms of function (3.15)

Multiplication of the function θ by n, i.e. raising the expression under the sign of logarithm to the power n, preserves the function's m.p. property (although not the p.r. property). The asymptotic phase lag will be then $-n\pi/2$. For example, for the phase lag of $-150°$, the power coefficient is $n = 5/3$, and the asymptotic slope is -10 dB/oct as shown in Fig. 3.46(a) and (b).

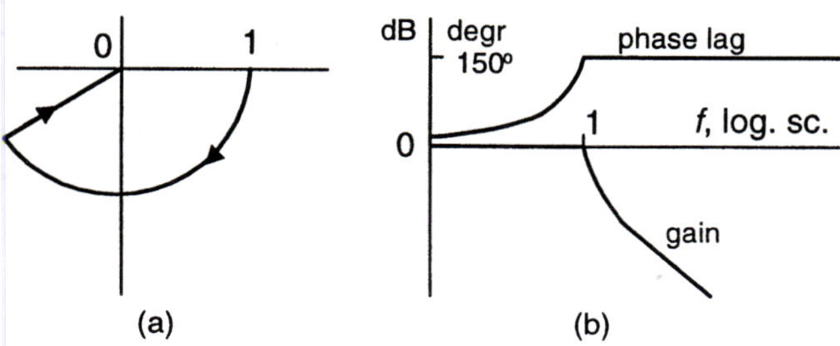

Fig. 3.46 (a) Nyquist diagram and (b) frequency responses for gain and phase of function (3.15) raised to the power 5/3

Combining these frequency responses with piece-linear responses, one can compose Nyquist diagrams with sharp angles. For example, the sum of the four responses: (1) low-pass, (2) high-pass, (3) constant slope and (4) a constant (not shown in the picture) produces the response shown in Fig. 3.47. This response can be used as a part of the Nyquist diagram for a Nyquist-stable system. This part can be connected to the rest of the diagram by the same procedure as that used in Fig. 3.47 to "glue" together the low-pass and high-pass responses.

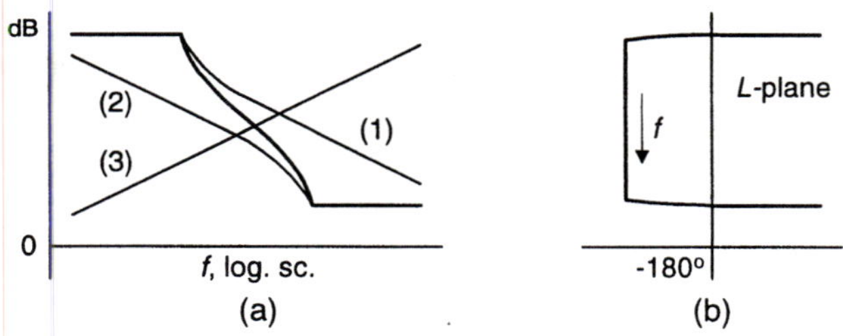

Fig. 3.47 (a) Example of a gain response composed of several elementary responses and (b) the response on the *L*-plane

3.12 Non-minimum phase lag

Non-minimum phase lag is the link's lag in excess of that given by the Bode formula. Notice that the Bode formula was derived for functions with a finite number of poles and zeros (i.e. rational functions), without zeros in the right half-plane of s. So, we might expect the non-minimum phase lag to appear (a) in systems with distributed parameters described by transcendental functions and (b) in systems whose transfer functions have zeros in the right half-plane.

In the systems with distributed parameters, the n.p. lag is caused by the time τ of the signal propagation over the media. This lag is often called ***transport lag*** and is proportional to the frequency:

$$|B_n| = \omega\tau .$$

The transport lag is substantial when the feedback loop is physically long and/or the speed of the signal propagation in the media is low as might happen in thermal, pneumatic, or acoustical systems. The transport lag of electrical signals in feedback amplifiers can be significant when the feedback bandwidth reaches hundreds of MHz.

The effect of the right-sided zeros s_i is exemplified in Fig. 3.48. In (a), poles and zeros of m.p. function θ_m are shown. In (b), poles and zeros are shown of an n.p. function θ which has some zeros in the right half plane of s, these zeros being mirror images of some of the zeros of θ_m. Notice that $|\theta_m(j\omega)| = |\theta(j\omega)|$ since the magnitude of each multiplier $(j\omega - s_i)$ is preserved.

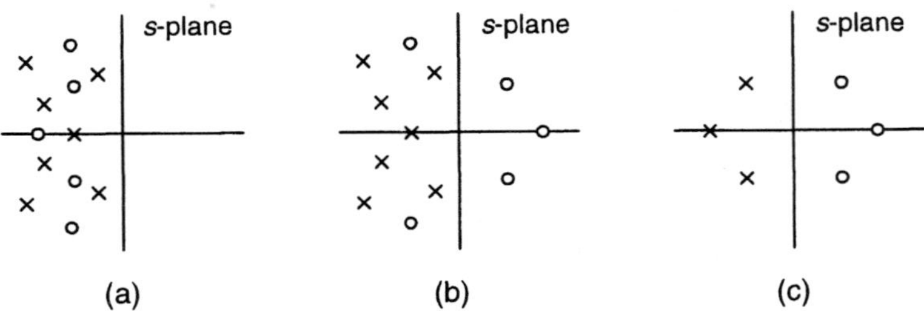

Fig. 3.48 Poles and zeros of (a) m.p.f. θ_m, of (b) n.p.f. θ, and of (c) the all pass θ/θ_m

The ratio $\theta_n = \theta/\theta_m$ is called ***pure n.p. lag*** because its phase

$$B_n = B - B_m = \arg \theta_n(j\omega)$$

represents the phase lag of θ in excess of the phase lag of θ_m. Since $|\theta_n| = 1$ at all frequencies, the function $\theta_n(s)$ is also called ***all pass***. As shown in Fig. 3.48(c), zeros s_i of θ_n are in the right half-plane and they are the mirror images of the poles which are in the left half-plane of s. Since the zeros are either real or come in complex conjugate pairs, θ_n can be expressed as

$$\theta_n = \prod_i \frac{s + s_i}{s - s_i}\,.$$

In particular, each *real* zero $s_i = \sigma_i$ contributes n.p. shift

$$B_{ni} = 2\arctan(\omega/\sigma_i)\,. \tag{3.16}$$

For $\omega < 0.4\sigma_i$, i.e. for $B_{ni} < 0.8$ rad,

$$B_{ni} \approx 2\omega/\sigma_i\,.$$

A sum of linear functions is a linear function. Therefore, generally, if the n.p. lag B_n is less than 0.8 rad, its frequency dependence can be approximated by the linear function

$$B_n \approx |B_n(f_c)|\frac{f}{f_c} \tag{3.17}$$

where f_c is some frequency.

3.13 Ladder networks and parallel connections of m.p. links

The ladder network diagrammed in Fig. 3.49 consists of parallel and series two-poles.

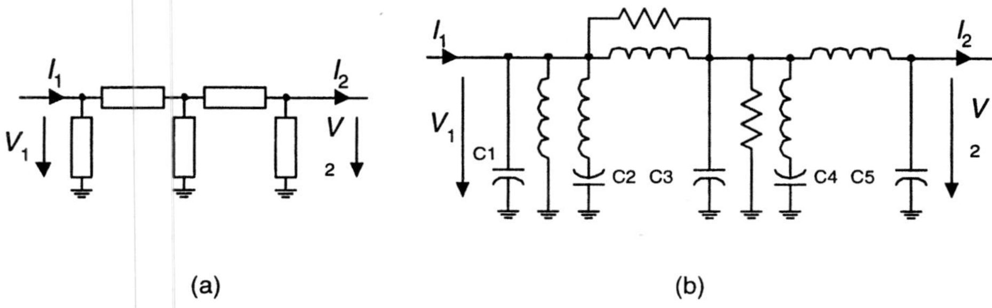

Fig. 3.49 Electrical ladder network: (a) general, (b) example

The transfer function for translational motion propagation in the x-direction via the mechanical system depicted in Fig. 3.50 is equivalent to that of the electrical ladder network shown in Fig. 3.49(b), if we use the force-to-current, velocity-to-voltage electromechanical analogy (described in detail in Section 7.1.1).

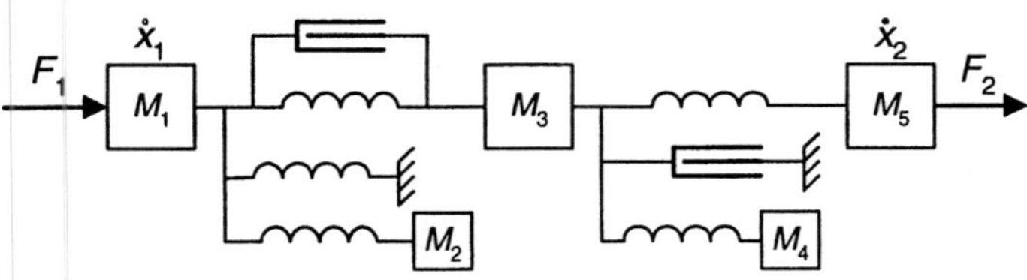

Fig. 3.50 Mechanical ladder network

In ladder networks, as previously mentioned in Section 3.9.1, transfer function zeros can only result either from infinite impedances in the series branches or from zero impedances in the shunting branches. Hence, as long as the branch impedances are positive real and do not have right-sided poles and zeros, the transfer function cannot possess zeros in the right half-plane. Therefore, the transfer function of a passive ladder network is always m.p.

A general network transfer function can be presented as a sum (resulting from parallel connection) of several m.p. transfer functions. Therefore, the right-sided zeros of a general network transfer function can only result from mutual cancellation of the output signals of two or several parallel paths, which is feasible even for s in the right half-plane. In other words, an n.p. transfer function can result from parallel connection of several ladder networks.

Similarly, a mechanical system transfer function can become n.p. when the signal propagates from the input to the output along different paths, or along the same path but in different motion modes, as, for example, when the incident signal excites translational and torsional modes which add up in the motion of the target node.

It is therefore of interest for control system designers to have a simple method of detecting when a parallel connection of several paths, each of them m.p., becomes n.p.

Fig 3.51 shows parallel connection of two links, W_1 and W_2. The composite link's transfer function is $W_1 + W_2$. *If both W_1 and W_2 are stable and m.p., then the function $W_1 + W_2$ is m.p. if and only if the Nyquist diagram for W_2/W_1 does not encompass the point -1.*

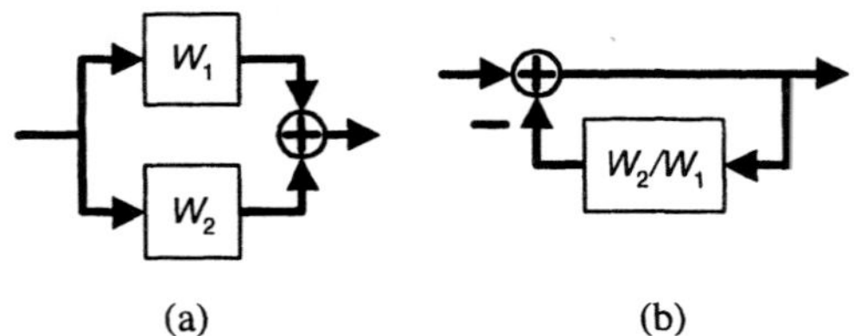

Fig. 3.51 (a) Parallel connection of two links, (b) closed-loop system

The proof is the following: The sum $W_1 + W_2 = W_1(1 + W_2/W_1)$. Here, W_1 possesses neither zeros nor poles in the right half-plane of s. Therefore, $W_1 + W_2$ has no right-handed zeros if $(1 + W_2/W_1)$ has no such zeros, i.e. if $1/(1 + W_2/W_1)$ has no such poles. The latter expression is the transfer function of the system (b), which stability can be verified with the Nyquist diagram for W_2/W_1.

When the Nyquist criterion is used, the index 1 should be assigned to the path with larger gain at higher frequencies in order for the ratio $|W_2/W_1|$ to roll off at higher frequencies. Applications for this criterion will be exemplified in Sections 5.9 and 6.3. The minimum phase property of a system including more than two parallel paths can be verified with the analog of the Bode-Nyquist criterion for successive loop closure.

3.14 Problems

Many problems on Laplace transform and frequency responses can be found at the end of Appendix 2.

1 Derive formula (3.1).

2 What frequency bandwidth is required for the rise-time to be less than (a) 5 sec (b) 0.5 sec (c) 2 msec (d) 1 nsec (e) 2.72 psec?

3 Calculate the rise-time for the system with the highest-order frequency response from those shown in Fig. 3.8.

4 The plant is a rigid body, the input signal is a force, the output is an (a) acceleration (b) velocity (c) position. What is the slope of the plant gain response, in dB/oct and dB/dec?

5 The plant is a capacitor, the input signal is a current, the output is a (a) voltage (b) charge. What is the slope of the plant gain response, in dB/oct and dB/dec?

6 Plot with MATLAB the Bessel filter responses using transfer functions given in Section 3.1.2, with linear and logarithmic frequency scales (use MATLAB commands `linspace`, `logspace`, `plot`).

7 Using some filter design software, print the plots for the gain and phase of a 3rd-order Butterworth filter.

8 The phase stability margin in a homing system is (a) 25°; (b) 35°; (c) 45°; (d) 55°. What is the hump in dB on the closed-loop response (calculate or use the Nichols plot)?

9 Use the relationship shown in Fig. 3.8 to explain why the Bessel filter has nearly no overshoot.

10 Why are Nyquist-stable systems stable? (The gain about the loop in such systems exceeds 1, the phase is 0, hence, the return signal comes back in phase and with increased amplitude. It is counter-intuitive to suggest that such a system is stable – but it is!)

11 Prove the Nyquist Criterion using the notion of continuous dependence of the roots of a polynomial on its coefficients. Hint: Modify the system under analysis to be definitely stable, and further change the coefficients gradually and continuously until they reach their true values, while observing the topology of the locus of $F(s)$.

12 If the $j\omega$-axis maps onto the locus of $F(j\omega)$, onto what area is the right half-plane of s mapped? What is the mapping of the zeros of $F(s)$?

13 The system is stable open-loop. The Nyquist diagram makes 2 revolutions about the critical point. Is the closed-loop system stable? How many poles in the right half-plane does the closed-loop transfer function have?

14 Plot the Bode diagram and the L-plane Nyquist diagram for the system with an unstable plant shown in Fig. 3.23. Change the value of damping in the local feedback path of the unstable plant, and check the system stability. The following SPICE program analyzes the equivalent electrical schematic diagram in Fig. 3.52.

```
** unst_pl.cir for feedback loop with unstable plant
* compensator, inverting, zero s=10, pole s=40,
* crossover s=20
GC 2 0 1 0 1
RC1 1 0 1MEG     ; for SPICE, to make node (1) non-floating
RC2 2 0 1
RC3 2 3 0.33333
LC 3 0 0.033333 ; Z2=(s+10)/(s+4)
 * plant loop, non inverting, closed loop 0.1s/(s^2+0.2s+4)
EP 5 0 2 4 10
GP1 4 0 5 0 1
RP 4 0 0.5
CP1 4 0 10
LP 4 0 0.025     ; Z5=0.1s/(s^2+0.05s+4)
 * plant integrators, (4) to (6) to (7)
GP2 6 0 0 5 80
GP3 7 0 0 6 1
RP2 5 0 1MEG     ; for SPICE
RP3 6 0 1MEG     ; for SPICE
RP4 7 0 1MEG     ; for SPICE
CP2 6 0 1
CP3 7 0 1
 * loop closing resistor
RL 7 1 1MEG      ; to close the loop, place semicolon
                 ; after M to reduce RL
 *
VIN 8 0 AC 1
RIN 8 1 1
 .AC DEC 100 0.1 20
 .PROBE          ; (if using PSPICE)
 .END
```

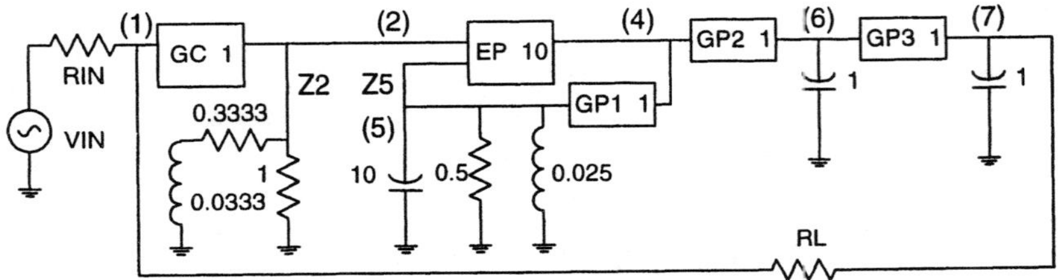

Fig. 3.52 SPICE model (de-floating 1 MEG resistors not shown) of the system with an unstable plant shown in Fig. 3.23

When using the PROBE postprocessor to plot open-loop responses with convenient scales, plot `vdb(7)` and `0.2*vp(7)`. To plot the Nyquist diagram, make the *x*-axis scale linear, the *x*-variable being `vp(7)`, and the range –50, 400; plot `vdb(7)`. To plot the plant loop Bode diagram, plot `vdb(4)-vdb(5)+20`. Alternatively, use MATLAB, SIMULINK, or some other analysis program.

15 In the plant of the system shown in Fig. 3.23, change the gain block gain coefficient to 5. Is the system stable?

16 Depending on the angle of attack of the horizontal stabilizer, the airplane in Fig. 3.53(a) can be statically unstable. The block diagram for the pitch autopilot feedback loop is shown in Fig. 3.53(b). Here, τ is the torque and J is the moment of inertia about the center of gravity CG. Consider $J = 2000$, the plant aerodynamics transfer function $-1000s + 2000$, the control surface gain coefficient of 1000, and the compensator transfer function $C = 400(s + 5)/(s + 20)$. Use SPICE, MATLAB, or SIMULINK. Is this airplane stable with the autopilot?

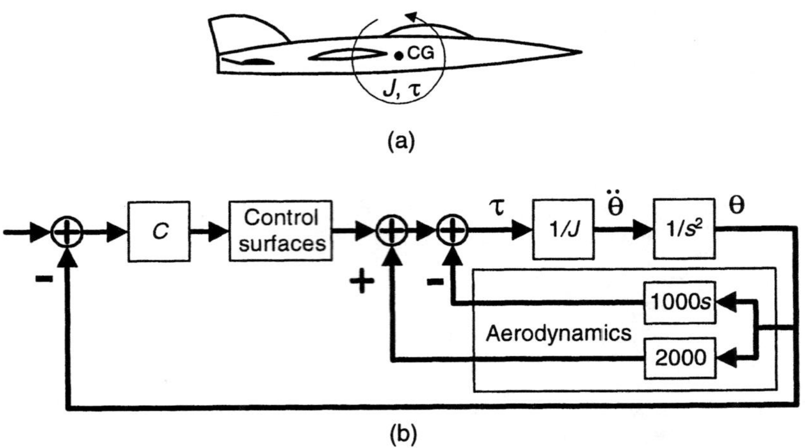

Fig. 3.53 System with aerodynamic instability in the plant

17 Show that the function (3.15) is p.r.

18 Which of the following functions can be used a s a model of a passive physical plant, and for plotting a Nyquist diagram for stability analysis?
(a) $T = 250(s + 5)/[(s + 5)(s + 5)(s + 5)]$;
(b) $T = 250(s + 5)/[(s + 5)(s + 5)]$;

(c) $T = 250(s + 5)(s + 50)(s + 50)/(s + 500)$;
(d) $T = 250(s + 5)(s + 5)(s + 50)(s + 500)$.

19 In literature, you can often read the sentence: "The feedback in this system is negative." Does this statement need qualification? Of what should you be aware?

20 What is the slope of the Bode diagram and the phase shift of the function: (a) $s^{-0.6}$; (b) $s^{-1.6}$; (c) $s^{-0.2}$; (d) $s^{-0.8}$; (e) $s^{-2.7}$.

21 What are the guard-point gain and phase stability margins if $T = 6200(s + 2.5)/[(s + 5)(s + 5)(s + 50)]$? Is the feedback negative at the crossover frequency?

22 If the slope of a Bode diagram at all frequencies is (a) –9 dB/oct; (b) –40 dB/dec; (c) –8 dB/oct; (d) –30 dB/dec, what is the phase shift?

23 If the phase shift at all frequencies is (a) 90° (b) –40° (c) –150° (d) –210°, what is the gain response?

24 The gain response can be approximated by a segment
(a) 3 octaves wide with the slope –10 dB/oct;
(b) 1.5 oct wide with the slope –6 dB/oct;
(c) 2 oct wide with the slope –12 dB/oct;
(d) 3 oct wide with the slope –20 dB/dec;
(e) 1 oct wide with the slope 5 dB/oct.
What is the phase shift at the frequencies at the ends and the center of the segment (use the plot in Fig. 3.42)?

25 Draw a Nyquist diagram for a loop transfer function (a) having a pole at the $j\omega$-axis; (b) with triple integration in the loop.

26 Using the Bode phase-gain relation, explain why the Bessel-type gain response has better phase linearity than other filters.

27 Find the expression for n.p. lag at lower frequencies for the transfer function having zeros (a) (3, –5, 0.1); (b) (–1, –3, –10); (c) (10, 5, –3); (d) (5 ± j12); (e) (4, –15, 0.1); (f) (–1, –4, 10); (g) (10, – 5, –3); (h) (6 ± j1.2).

28 Sketch a Nyquist diagram on the *L*-plane related to the Bode diagram in Fig. 3.18(a).

29 Plot phase responses for piece-linear approximations of certain Bode diagrams shown by the responses (1), (2), (3), (4) in Fig. 3.54. For convenience, copy the pages with Figs. 3.42 and 3.54 using the same magnification (either 1:1 or 1:1.5). To copy the plots from one sheet of paper to another, superimpose the two sheets on a light table (or against a window glass), or make and use a transparent paper copy of the page with the problem. For $w = 1.5$, or 2.4, interpolate.

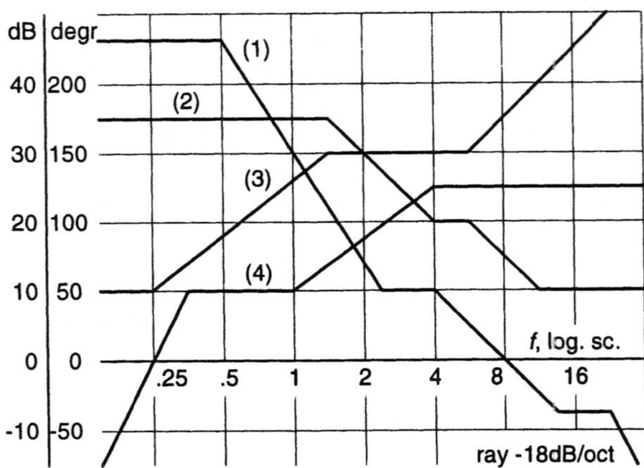

Fig. 3.54 Piece-linear Bode diagrams for phase calculation exercises

30 Fig. 3.55 shows the loop gain responses of the thruster attitude control about the *x*- and *z*-axes of a spacecraft. Approximate the responses by segments and rays. Calculate the phase shift.

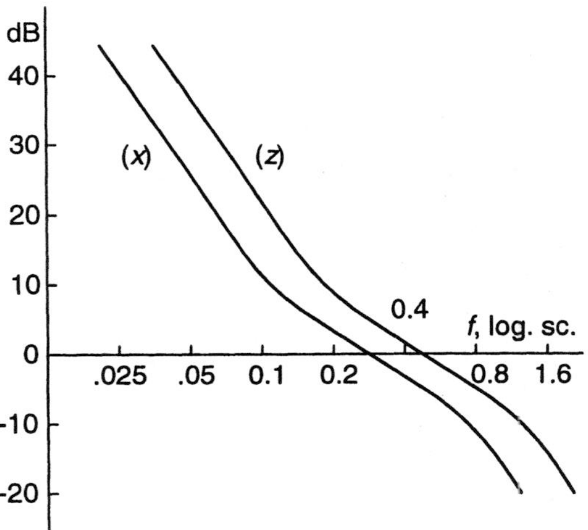

Fig. 3.55 Bode diagrams for attitude control loops of a spacecraft

31 Is $W_1 + W_2$ m.p. if:
(a) $W_1 = 1000/[s^2(s+10)]$, $W_2 = (s+2)(s+5)/[(s+200)(s+500)]$?
(b) $W_1 = 10000/[s^2(s+20)]$, $W_2 = (s+4)(s+10)/[(s+400)(s+1000)]$?
(c) $|W_1| < |W_2|$ at all frequencies, and W_1 and W_2 are m.p.?

32 In a homing system the closed-loop response is that of the 4th-order Bessel filter. Find the loop transfer function and plot the open-loop Bode diagram.

Answers to selected problems

3 The cut-off frequency for the highest order filter is approximately 2 rad/sec, and the period for this frequency is approximately 3 sec, and a third of it is the rise time

≈ 1 sec. From Fig. 3.8 where the transient response is plotted, we see that the rise time (from 0.1 to 0.9 of the output) is about 1 sec, i.e. formula (3.1) gives a good result.

9 Since the slope of the Bessel filter gain response increases gradually with frequency and does not reach large values before the gain is already very small, the curvature of the transient response also increases gradually and does not reach large values, which guarantees the absence of an overshoot.

11 Start with a simple system with $T(s) = 0.1$ that is definitely stable. Augment the transfer function of T to polynomials of high order but with infinitesimal coefficients. Start changing the coefficients while observing changes in the resulting Nyquist diagram. If a root of the characteristic polynomial migrates into the right half-plane of s, it must cross the $j\omega$-axis. At the crossing frequency the closed-loop gain must become infinite and F, therefore, must become 0. If this never happens, i.e. the locus of F during the modifications never crosses the origin, i.e. the locus of T never crosses the point −1, then the system must be stable.

24 (a) The phase shift is 110° at the center of the segment, 63° at its ends.

26 Bessel filter gain response can be approximated by a piece-linear response as shown in Fig. 3.56 (not to scale). It is seen that the phase response is rather close to linear because of the contribution to the phase lag at lower frequencies made by the gain response slope at lower frequencies. For higher selectivity filters (Butterworth, Chebyshev, Cauer), the gain response slope at lower frequencies is negligible, but the slope becomes very steep right after the passband ends, and within the passband, correspondingly, the phase lag is smaller at lower frequencies and higher at higher frequencies thus making the phase response curved.

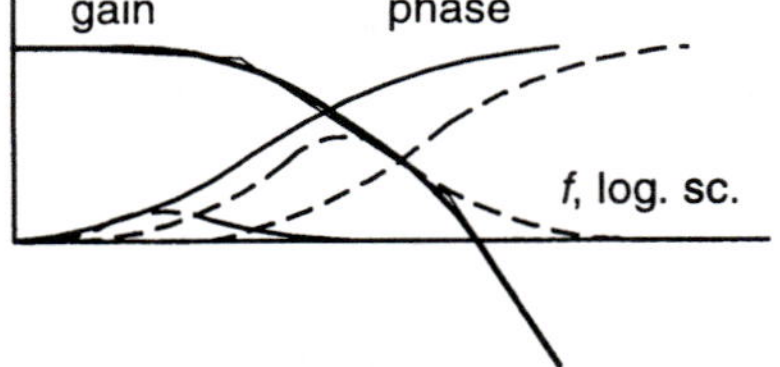

Fig. 3.56 Bessel filter gain response and phase response

Another explanation is that to make the phase lag proportional to frequency, the phase lag increase over an octave should double such increase over the preceding octave. Therefore, roughly speaking, the increase in the slope of the gain response should double each octave.

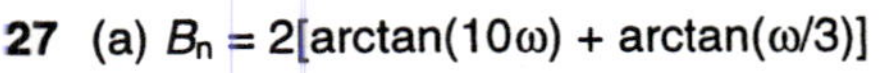

27 (a) $B_n = 2[\arctan(10\omega) + \arctan(\omega/3)]$

32 $105/(s^4 + 10s^3 + 45s^2 + 105s)$.
The response is plotted in Fig. 3.57.

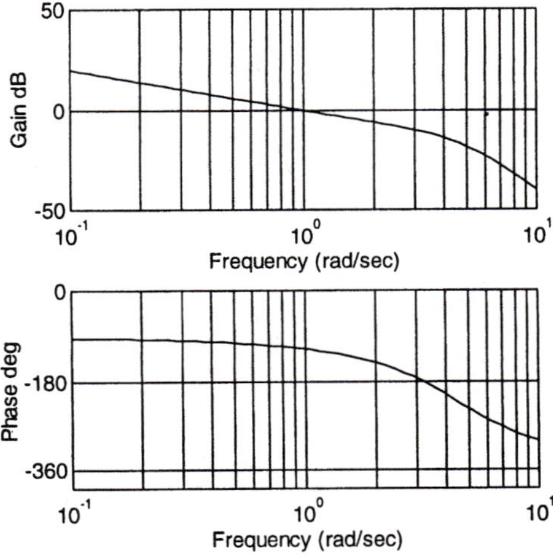

Fig. 3.57 Bode diagram

Chapter 4

SHAPING THE LOOP FREQUENCY RESPONSE

The problem of optimal loop shaping encompasses two fairly independent parts that can be solved sequentially (thus making the design structural):

The first part is *feedback bandwidth maximization* which is solved by appropriately *shaping the feedback loop response at higher frequencies* (in the region of crossover frequency and higher).

The second part is of *distribution of the available feedback over the functional feedback band.*

The feedback bandwidth is limited by the sensor noise effect at the system output, the sensor noise effect at the actuator input, plant tolerances (including structural modes), and nonminimum phase lag (analog and digital) in the feedback loop. The optimal shape of the loop gain response at higher frequencies, subject to all these limitations except the first, includes a Bode step.

The Bode step is presented in detail as a loop shaping tool for maximizing the feedback bandwidth. The problem of optimal loop shaping is further described and the formulas are presented for calculation of the maximum available feedback over the specified bandwidth.

The above solution is then generalized by application of a Bode integral to reshaping the loop gain response over the functional bandwidth (i.e.,, for solving the second part of the loop shaping problem). It is shown that the feedback is larger and the disturbance rejection improved in Nyquist-stable systems.

Loop shaping is described for plants with flexible modes, for collocated and non-collocated control, and for the loops where the plant is unstable. The effect of resonance mode coupling on the loop shaping in MIMO systems is considered.

It is described how to shape the responses of parallel feedback channels to avoid nonminimum phase lag while providing good frequency selection between the channels.

When the book is used for an introductory control course, Sections 4.2.5 – 4.2.7, 4.3.3, 4.3.6, 4.3.7, 4.4, and 4.5 can be omitted.

4.1 Optimality of the compensator design

Webster's Collegiate Dictionary defines the word "optimal" as "most desirable," and practical engineering views optimality as a provision for best customer satisfaction.

In application to practical control systems, the controller performance means some combination of accuracy, speed of action, repeatability, reliability, and disturbance rejection. The trade-offs between these requirements are commonly quite clear and/or can be resolved with Bode integrals. The feedback controller design is incomplete without estimation of the theoretically best available performance.

The second kind of the design trade-off is that of the controller performance versus complexity. As a rule, compensators and prefilters are many times less expensive than the plants. Hence, in order to improve system performance, it is worthwhile to make them very close to optimal, even at the price of making them complex. The order of the compensator to be reasonably close to the optimal is, typically, 8 to 15.

Example 1. Increasing the compensator order from 4 to 12 and including in the compensator several nonlinear links will only add 20 to 30 lines of code to the controller software, or several extra resistors, capacitors and operational amplifiers, if the compensator is analog – and may substantially improve the system's performance. For example, if the settling time of some expensive manufacturing machinery with a short repetition cycle of operation can be reduced by 20% while retaining the same accuracy, the resulting time per operation might be reduced by, say, 5%, and the number of pieces of the equipment at the factory can be correspondingly reduced by 5%, with additional savings on maintenance. Or, a fighter's maneuverability can be noticeably improved. Or, the yield of a chemical process can be raised by 2% – etc. This is why the compensators should be designed to provide close to optimum performance, and not only the performance specified by a customer representative who doesn't know in advance what kind of performance might be available.

Even when the accuracy of the system with a simple controller suffices, it still pays to improve the controller, since with larger margins in accuracy, the system will remain operational when some of the system parameters degrade to the point that without the better controller, the system fails.

Linear compensators are fully defined by their frequency responses. Therefore, the problem of optimal linear compensator design is the problem of optimal loop response shaping. The theory of optimal feedback loop shaping should be able to provide the answers to the questions: (a) what performance is feasible, and (b) what loop response achieves this performance limit.

Commonly, control systems are initially designed as linear with some idealized plant model. Still, the design must result in a sound solution when the idealized plant is replaced by the physical plant and actuator models. *Physical system models must reflect the uncertainty in the system parameters, the asymptotic behavior of the transfer functions at higher frequencies, the sensor noise, and the nonlinearities*.

Example 2. In the paper "When is a linear control system optimal?" (see [128] in the bibliography to Ref. [9]) which was considered by many a cornerstone of the so called modern control, the following definition is given: "*a feedback system is optimal if and only if the absolute value of the return difference is at least one at all frequencies*". Meeting this condition assures dynamic optimality which is rigorously defined in the paper. The problem with this theory is, however, that it cannot be applied to physical systems. In physical systems the loop gain drops faster at higher frequencies than that of an integrator, and according to Bode integral (3.7), the absolute value of the return difference cannot be "at least one at all frequencies."

The process of control law design that deserves to be called optimal must provide timely information to the system engineers about the achievable control system performance and the related possibility of relieving certain requirements to the system hardware, which may permit replacing some initially chosen components and subsystems by simpler and cheaper ones. That is, the optimality requirements must relate to the entire engineering system and not only to the controller in the narrow sense.

Example 3. In Chapter 1, we analyzed some design of antenna elevation angle control – without proving that this design is the best possible. (It is not. The real-life controller is multiloop, and includes signal feedforward and high-order nonlinear compensators.) Reasonable questions for the customer to ask are: – Is this design the best? If not, by how much can it be improved?

Normally, we should not even ask: – and, at what cost? – since, as we already mentioned, the cost of the compensator is small compared with the cost of the antenna dish (although better or additional sensors, bearings, and servomotors can add to the system cost).

Also, the loop responses in this example do not include the plant structural modes. Given the modes, the loop gain at higher frequency should be rolled off fast to avoid instability.

The system engineer asked the control designer to find the shape of the control loop response which is optimal for some specific task, and to estimate the available performance without completing the lengthy compensator design. Can this be done?

Example 4. A flexible mode of a structure may be anywhere between 30 and 40 Hz. By this resonance, the environmental vibration noise is accentuated and the accuracy of the device under control decreases. The feedback system therefore must be able to reject the noise over the entire bandwidth 30 to 40 Hz. How to implement the maximum feedback over this frequency range, and what is the value of this feedback? Can it be quickly calculated without designing the controller?

4.2 Feedback maximization

4.2.1 Structural design

We already mentioned the advantages of structural design in Section 2.9 (and will further discuss them in Section 7.2.1). The structural design flowchart for the basic feedback control system is shown in Fig. 4.1.

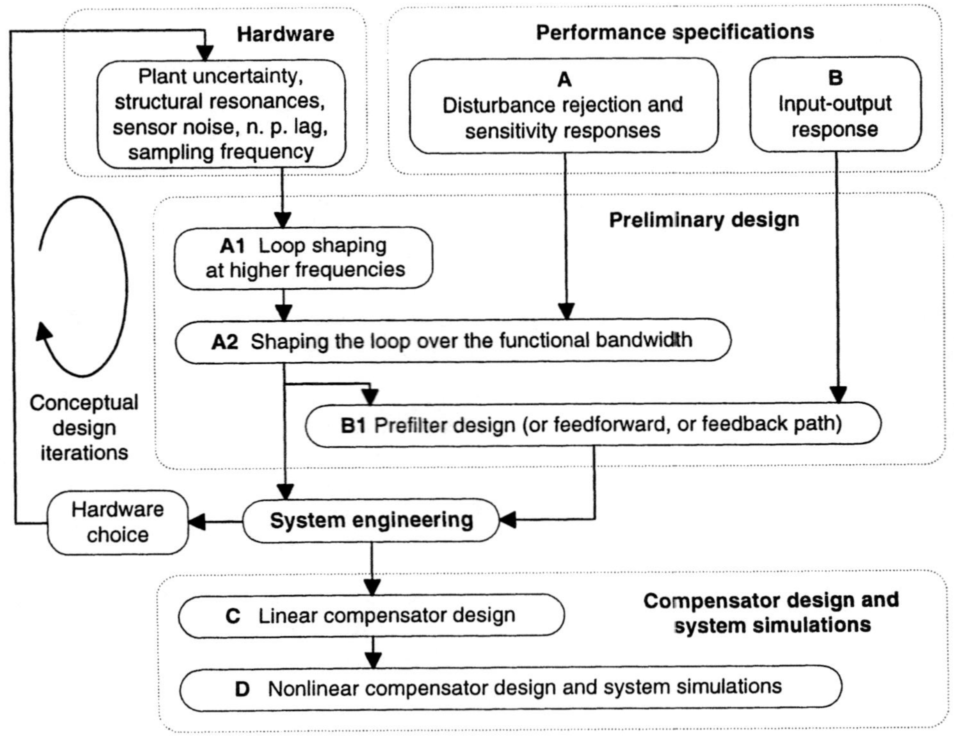

Fig. 4.1 Feedback system design flowchart

The performance specifications comprise the desired responses of

A the disturbance rejection and the sensitivity, i.e.,, the loop gain, and

B the nominal command-to-output transfer function.

The preliminary design addresses specification **A** first, by ***loop response shaping***. This can be subdivided into

A1 achieving maximum feedback bandwidth by appropriate loop shaping at higher frequencies, and

A2 distributing of the achieved feedback over the functional frequency range so as to exceed the worst-case specifications.

Since the first subproblem is to a large extent independent from the second one, it makes sense to solve them sequentially. At this stage, the responses need not be expressed by rational functions. They can be expressed by rational or transcendental functions, by plots, or by tables.

After the loop is shaped, the command-to-output response is modified (if required) by adding an appropriate prefilter or command feedforward to meet specifications **B**.

During the conceptual design of complex engineering systems, only the steps **A1** and **A2** (and sometimes **B1**) need to be performed to provide the system engineers with the accurate data on the available control performance. The system engineers evaluate different versions of the hardware/software configurations iteratively, using the preliminary design results on the *available* control performance. Inclusion of only the stages **A1, A2, B** in the iteration loop makes the loop fast.

After the system engineers finally decide which hardware configuration is the best, the compensator and prefilter are designed. The compensator response is found by subtracting the plant response from the loop response. Then

C the compensator and prefilter responses are approximated by m.p. rational functions and implemented as algorithms or as analog circuits, and

D the compensator is augmented with nonlinear elements and the system is simulated.

4.2.2 Bode step

In physical plants, the uncertainty of the plant parameters increases and the gain drops at higher frequencies. In electrical circuits, this happens due to stray capacitances and inductances, in thermal systems, due to thermal resistances and thermal capacitances, and in mechanical systems, the same happens due to the plant flexibility. Because of this and because of the sensor noise which will be studied in Sections 4.3.2 and 4.3.3, the loop gain must sharply decrease at higher frequencies. For the purpose of analysis, the loop gain Bode diagram can be sufficiently well approximated at these frequencies by a line with a constant ***asymptotic slope*** of $-6n$ dB/oct. The slope is rather steep in physical feedback systems with $n \geq 2$. In other words, $|T(j\omega)|$ *decreases at higher frequencies at least as* ω^{-2}*, and the integral of feedback* (3.7) *is therefore zero*.

The crossover region studied in this section is the region of transition between the functional frequency band and the higher-frequency band where the feedback becomes negligible. As was stated in corollaries in Section 3.9.2, *shaping the Bode diagram in the crossover region* (step A1 from Section 4.2.1) *is critical in achieving a maximum area of positive feedback near the crossover and, therefore, achieving a maximum area of negative feedback in the functional feedback band*.

Physical systems include actuators with saturation. If a system contains no other nonlinear links, absolute stability is required and the stability margin boundary must be as shown in Fig. 4.1(a). From the Bode integral of phase it follows that to maximize the feedback, the phase lag must be the maximum allowable, i.e., the Nyquist diagram should follow the boundary curve as closely as possible. Such a diagram is shown by the thin line. The gain monotonically decreases with increasing frequency and eventually degenerates into the high-frequency asymptote with the associated phase shift $-n90^{\circ}$.

It will be shown below that this Nyquist diagram corresponds to the Bode diagram shown in Fig. 4.1(b). The loop gain response is piece-linear with corner frequencies f_d and f_c. The related loop phase lag is less than $(1-y)180^{\circ}$ until the loop gain becomes smaller than $-x$ dB.

The system is phase-stabilized with the margin not less than $y180^{\circ}$ up to the frequency f_d where the loop gain drops to $-x$. Because of the Bode phase-gain relation, the slope of the Bode diagram at these frequencies is approximately $-12(1-y)$ dB/oct.

The high-frequency asymptotic loop response is considered known. It is defined by:
(a) the asymptotic slope $-6n$ dB/oct,
(b) the point on this asymptote with coordinates $(f_c, -x)$ as shown in Fig. 4.2(b), and
(c) the nonminimum phase lag at this frequency $B_n(f_c)$.

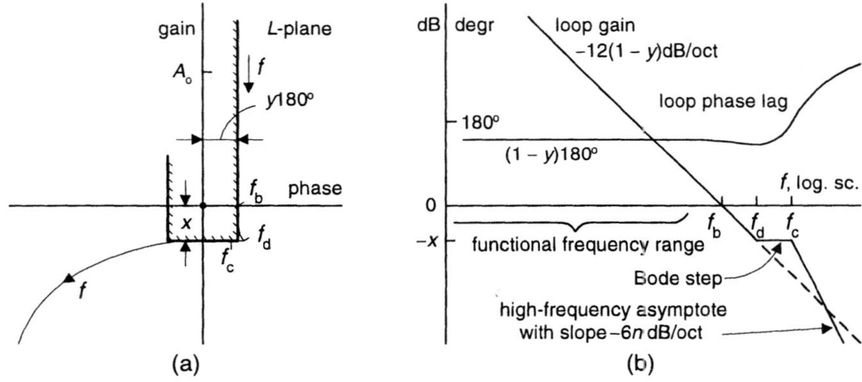

Fig. 4.2 Bode step: (a) absolute stability boundary on the L-plane, (b) piece-linear gain response with related phase lag response that produce the Nyquist diagram shown in (a) which approximates the boundary

The transition between the slope $-12(1-y)$ dB/oct and the high-frequency asymptotic slope must be as short as possible to increase the loop selectivity: to maximize the loop gain in the functional frequency range while reducing the loop gain at higher frequencies. The transition is provided by the ***Bode step*** made at the gain level of $-x$ dB as shown in Fig. 4.2(b).

Without the step, the phase lag in the crossover area would be too large due to the steep high-frequency asymptote and the non-minimum phase lag. The step reduces the phase lag at the crossover frequency – but also reduces the loop selectivity (i.e., given the high-frequency asymptote, reduces the feedback bandwidth). Therefore, *the length of the step must not be excessive*.

The nonminimum phase lag $B_n(f_c)$ is assumed to be less than 1 radian which is true in well-designed systems. With the linear approximation (3.17), the nonminimum phase lag at frequencies lower than f_c is

$$B_n \approx |B_n(f_c)|\frac{f}{f_c}.$$

The phase lag related to the asymptotic slope ray which starts at f_c can be expressed with (3.14) as, approximately,

$$\frac{2}{\pi}n\frac{f}{f_c}.$$

Consider next the "discarded" dashed-line ray which is the extension of the main slope line beyond the frequency of the beginning of the step f_d. The phase lag related to this ray is expressed with (3.14) as, approximately,

$$\frac{2}{\pi}2(1-y)\frac{f}{f_d}.$$

To make the loop phase lag at frequencies $f < f_d$ equal to $(1-y)180^\circ$, the sum of the phase contribution of the asymptotic slope and the nonminimum phase lag should equal the phase contribution of the "discarded" dashed-line ray. This consideration is expressed as

$$\frac{2}{\pi}n\frac{f}{f_c} + |B_n(f_c)|\frac{f}{f_c} \approx \frac{2}{\pi}2(1-y)\frac{f}{f_d}.$$

From this equation, the Bode step frequency ratio is

$$\frac{f_c}{f_d} \approx \frac{n+\frac{\pi}{2}|B_n(f_c)|}{2(1-y)}. \tag{4.1}$$

For the typical phase stability margin of 30°, i.e., $y = 1/6$,

$$f_c/f_d \approx 0.6n + B_n(f_c). \tag{4.2}$$

Example 1. When the specified stability margins are 30° and 10 dB, and the crossover frequency $f_b = 6.4$ kHz, then the slope is –10 dB/oct and $f_d = 2f_b = 12.8$ kHz. Further, when $n = 3$ and $B_n(f_c) = 0.5$ rad, then from (4.2) $f_c/f_d = 2.3$ so that $f_c \approx 30$ kHz.

The Nyquist diagram for the Bode step response closely follows the stability boundary in Fig. 4.2(a). In practice, this response is approximated by a rational transfer function, and the corners of the Nyquist diagram become rounded. Examples of loop responses with Bode steps will be given in Sections 5.6, 5.7, and 5.11, in Chapter 13, and in Appendix 13.

The loop response with a Bode step should be employed when the dominant requirement is maximizing the disturbance rejection, i.e., maximizing the feedback in the functional frequency range. This case is common but not ubiquitous. Noise reduction requirements and certain implementation issues may require Bode diagrams to be differently shaped in the crossover area.

In any case, the Bode diagram must be made shallow over some range in the crossover frequency region to ensure the desired phase stability margin. There are several options for where to do this: to the right of f_d by the Bode step, to the left of the crossover frequency as in systems where the sensor noise is critical, and over a frequency range nearly symmetrically situated about the crossover as in the so-called *PID* controller which will be discussed in Chapter 6. Loop responses without any Bode step typically provide 4 to 20 dB less feedback in the functional frequency range.

The output transient response to a step-disturbance in a homing system with a Bode step and 30° to 40° phase stability margin has substantial overshoot. If this overshoot exceeds the specifications, the loop gain response in the neighborhood of f_b should be made shallower. However, this will reduce the available feedback and the disturbance rejection.

For a system which has an explicit command input, a prefilter or one of its equivalents can be introduced to ensure good step-responses without reducing the available feedback.

4.2.3 Example of a system having a loop response with a Bode step

In this section an example of the implementation of a loop response with a Bode step is considered. This response has been used as a prototype for several practical control systems.

Example 1. The plant

$$P(s) = \frac{1}{s^2}\frac{10-s}{10+s}$$

is a double integrator with an n.p. lag. The compensator function $C(s)$ is a ratio of a 3rd-order to a 4th-order polynomials (the compensator design methods are described in the next chapter). The loop transfer function

$$T(s) = C(s)\frac{1}{s^2}\frac{10-s}{10+s} = \frac{11s^3+55s^2+110s+36}{s^4+7.7s^3+34s^2+97s+83}\frac{1}{s^2}\frac{-s+10}{s+10} = \frac{n(s)}{d(s)}$$

is plotted in Fig. 4.3 with

```
n = conv([11 55 110 36], [-1 10]);
d = conv([1 7.7 34 97 83 0 0], [1 10]);
w = logspace(-1,1,200);
bode(n,d,w)
```

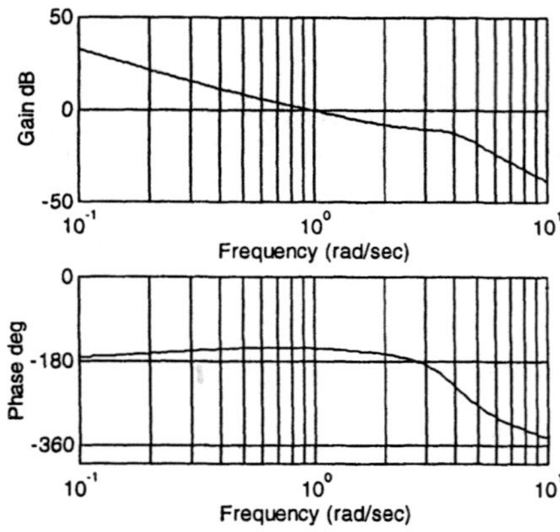

Fig. 4.3 Open-loop frequency response

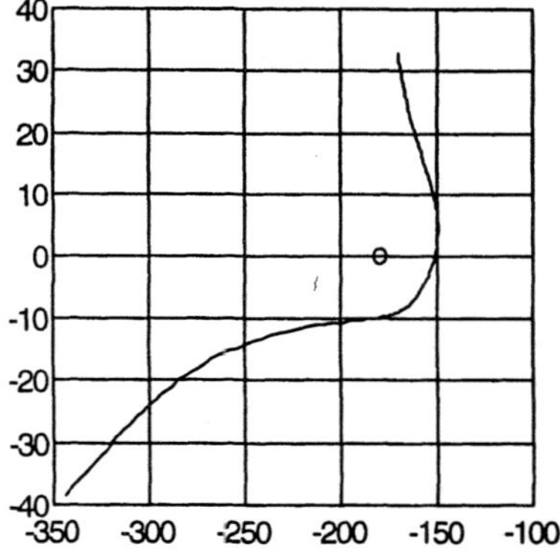

Fig. 4.4 *L*-plane Nyquist diagram

The crossover frequency $\omega_b = 1$ rad/sec. The *L*-plane Nyquist diagram is plotted in Fig. 4.4 with

```
[mag, phase] = bode(n,d,w);
plot(phase,20*log10(mag),'w', -180,0,'wo'); grid
```

The slope of the loop gain at lower frequencies approaches -12 dB/oct and the loop phase shift approaches $-180°$. This response provides larger feedback at lower frequencies than the response in Fig. 4.2(b) following the stability margin boundary in Fig. 4.2(a) but violates the stability margin at these frequencies. To ensure stability and good transient responses to large commands and disturbances (which saturate the actuator), the loop compensation must be nonlinear. Such compensation is described in the end of this example and explained at length in Chapters 10, 11, and 13.

The closed-loop transfer function $M = n/g$ where $g = n + d$ is plotted in Fig. 4.5 with

```
n = conv([0 0 0 1],n);
g = n + d;
w = logspace(-1,1,200);
bode(n,g,w)
```

(The zeros are added to n to make the dimensions of vectors n and d equal so that MATLAB can calculate g.)

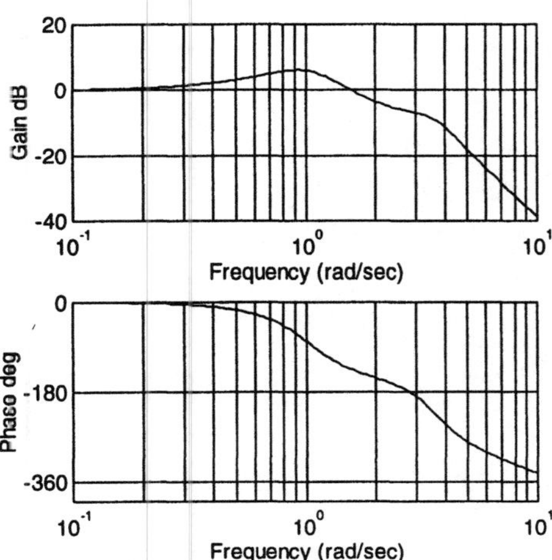

Fig. 4.5 Closed-loop frequency response

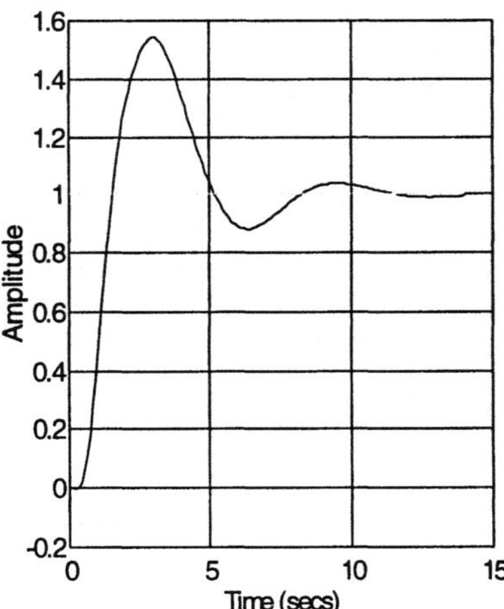

Fig. 4.6 Closed-loop step-response

As should be expected, the closed-loop step-response plotted by `step(n,g)` and shown in Fig. 4.6 has a large, 55% overshoot. The overshoot can be reduced by introduction of a prefilter $R(s)$.

The command-to-output response of the system with the prefilter should be close to the response of a linear phase (Bessel, or Gaussian) filter. This can be achieved by using a notch prefilter

$$R(s) = \frac{s^2 + \omega_b s + 0.81\omega_b^2}{s^2 + 2\omega_b s + 0.81\omega_b^2}.$$

The notch is tuned to $0.9\omega_b$ with the gain at this frequency equal to $20\log(1/2) = -6$ dB. In the case considered, $\omega_b = 1$. The prefilter response is plotted in Fig. 4.7 with

```
nr = [1 1 0.81]; dr = [1 2 0.81]; bode(nr, dr, w)
```

and the closed-loop response with the prefilter is plotted in Fig. 4.8 with

```
nc = conv(nr, n); gc = conv(dr, g);
bode(nc, gc, w)
```

The phase response in Fig. 4.8 has a lesser curvature than that in Fig. 4.5. We might therefore expect a better transient response. Indeed, the overshoot is only 7% on the step-response for the closed-loop system with the prefilter plotted by `step(nc,gc)` and shown in Fig. 4.8; the rise time and the settling time remain approximately the same.

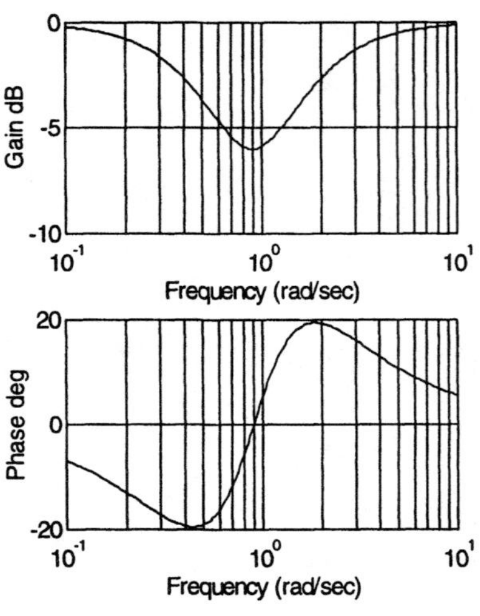

Fig. 4.7 Prefilter frequency response

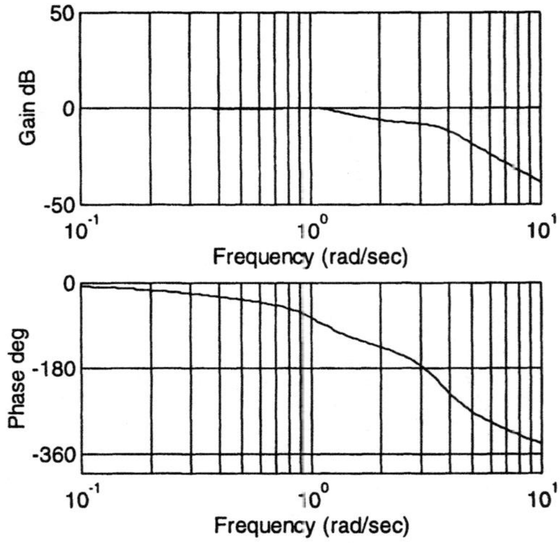

Fig. 4.8 Closed-loop response of the system with a notch prefilter

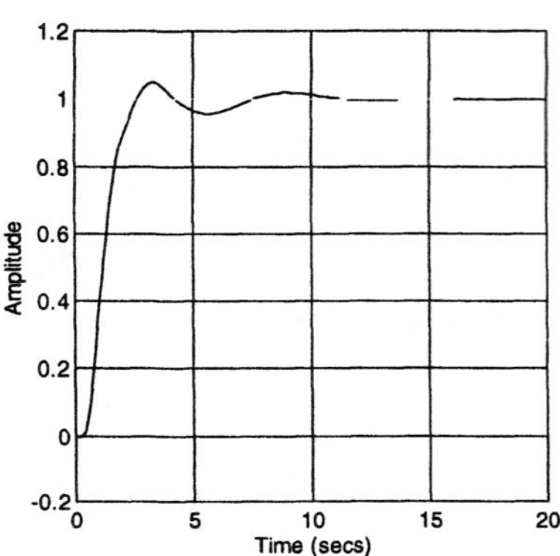

Fig. 4.9 Closed-loop step-response of the system with a notch prefilter

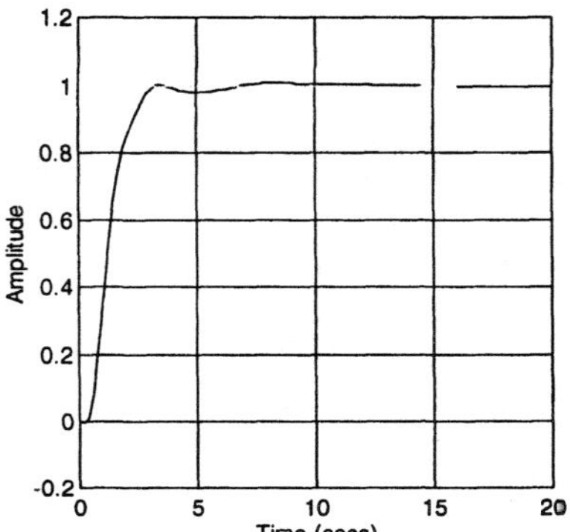

Fig. 4.10 Closed-loop step-response of the system with the prefilter composed of two notches

The gain response in Fig. 4.8 still has a hump at $\omega = 1$ which contributes to the phase nonlinearity and to the overshoot. By compensating for the hump with a small, 1.6 dB additional notch with

```
nr2 = [1 0.5 1]; dr2 = [1 0.6 1];
nc2 = conv(nr2, nc); gc2 = conv(dr2, gc);
bode(nc2, gc2, w)
step(nc2, gc2)
```

one can further reduce the overshoot as shown in Fig. 4.10. A still better and more economical equalization can be devised and implemented.

The equalization becomes less accurate when the plant parameters deviate from the nominal. The plant gain variations can be specified by a multiplier k. The effects of variations in k can be calculated as follows:

```
n = conv([11 55 110 36], [-1,10]);
d = conv([1 7.7 34 97 83 0], [1 10]);
k = 1; n = conv(n, k);                    % specify k
w = logspace(-1,1,200);
bode(n, d, w)                             % open-loop response
[mag, phase] = bode(n, d, w);
plot(phase, 20*log10(mag))                % Nyquist diagram
grid
n = conv([0 0 1], n);
g = n + d;
bode(n, g, w)                             % closed-loop response
nr = [1 1 0.81]; dr = [1 2 0.81];
bode(nr, dr, w)                      % prefilter notch response
nc = conv(nr, n); gc = conv(dr, g);
bode(nc, gc, w)              % closed-loop response with notch
nr2 = [1 0.5 1]; dr2 = [1 0.6 1];
bode(nr2 ,dr2, w)                    % second notch response
nc2 = conv(nr2, nc); gc2 = conv(dr2, gc);
bode(nc2, gc2, w)            % closed-loop response with notches
step(nc2, gc2)                       % step response with notches
```

For ±20% variations in the plant gain coefficient, the step-responses are shown in Fig. 4.11. The overshoot remains less than 10%.

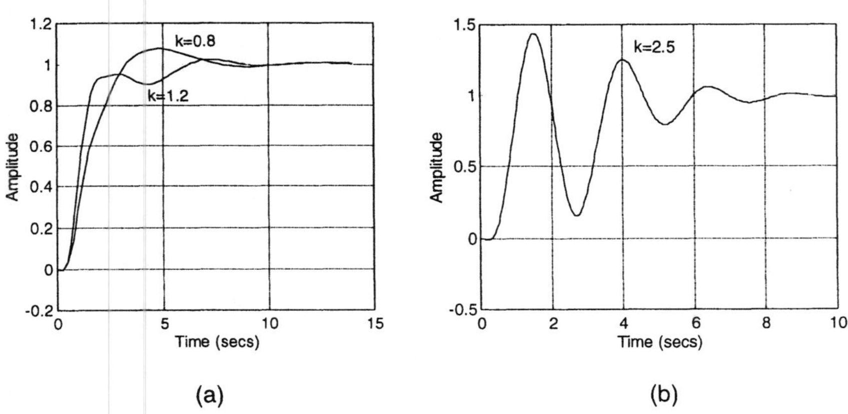

(a) (b)

Fig. 4.11 Closed-loop step-responses of the system with notches, (a) for $k = 1.2$ and $k = 0.8$ and (b) for $k = 2.5$

When the plant's gain coefficient increases to 2.5, i.e., the plant's gain increases by 8 dB from the nominal thus reducing the gain stability margin to only 2 dB, the transient response becomes oscillatory as shown in Fig. 4.11(b); still, this is not a catastrophic failure of the controller.

With $\pm 40\%$ variations in k the overshoot remains under 20%, and with $k \geq 3.2$ (i.e., with $20 \log k = 10.1 > x$) the output starts growing exponentially; the proof (simulation) is left as a recommended student exercise. It is also recommended to make simulations with small (say, 5%) variations in the coefficients of the polynomials in the compensator transfer function and to observe that these changes do not critically affect the system performance, and therefore, the coefficients can be appropriately rounded.

To ensure that large-amplitude commands which overload the actuator will not trigger self-oscillation, and to improve the transient responses for large-amplitude commands, the compensation can be made nonlinear. This is done by, first, splitting the compensator transfer function $C(s)$ into the sum of two functions $C_1(s)$ and $C_2(s)$ which represent transfer functions of two parallel paths, the first path being dominant at lower frequencies; and second, by placing a saturation link with an appropriate threshold in front of the first path's linear link. The related theory and design methods are described in Chapters 10, 11, and 13. The transfer functions of the paths can be found as

$$C_1 = a_o/(s + p_1) \quad \text{and} \quad C_2 = [C(s) - C_1(s)]$$

where p_1 is the lowest pole of $C(s)$ (p_1 can be 0), and a_o is its residue. For finding p_1 and a_o, the MATLAB function `residue` or the function `bointegr` from the Bode Step toolbox in Appendix 14 can be used.

Example 2. A dc motor rotates a spacecraft radiometer antenna whose moment of inertia $J = 0.027\,\text{Nm}^2$. θ is the antenna angle, $s\theta$ is the antenna angular velocity, and U is the voltage applied to the motor. The motor winding resistance $R = 2\,\Omega$. The motor constant (the torque-to-current ratio) $k = 0.7$ Nm/A. The torque is $k(U - E_B)/R$ where the back electromotive force $E_B = ks\theta$. The angle $\theta = [k(U - E_B)/R]/(s^2J)$.

From the latter two equations, the voltage-to-angle transfer function

$$\frac{\theta}{U} = \frac{k}{RJ}\frac{1}{s[s + k^2/(RJ)]} = \frac{13}{s(s + 9.07)}.$$

If the loop n.p. lag equals that in Example 1, and if the loop response needs to be the same as in Example 1, the compensator and driver transfer function should be $C(s)$ from Example 1 multiplied by

$$C_1 = \frac{U}{\theta}\frac{1}{s^2} = \frac{RJ}{k}\frac{s + k^2/(RJ)}{s} = \frac{0.077(s + 9.07)}{s}.$$

The MATLAB code for this transfer function's numerator and denominator is

```
k = 0.7; res = 2; ja = 0.027; a = k*k/(res*ja);
nc1 = conv(res*ja/k,[1 a]); dc1 = [1 0];
```

Example 3. To use the response in Example 1 as a prototype for a control system with crossover frequency $f_b = 12$ Hz, s must be replaced by s/ω_b (where $\omega_b = 12 \times 2\pi \approx 75.4$) in the transfer functions for the loop and for the prefilter. The numerator `ns` and the denominator `ds` of the loop transfer function can be found with

```
n = conv([11 55 110 36], [-1,10]); wb = 75.4;
d = conv([1 7.7 34 97 83 0 0], [1 10]);
format short e;  [ns, ds] = lp2lp(n, d, wb)
```

and the numerators and denominators of the two notch transfer functions of the prefilter with

```
nr = [1 1 0.81]; dr = [1 2 0.81];
[nrs, drs] = lp2lp(nr, dr, wb)
nr2 = [1 0.5 1]; dr2 = [1 0.6 1];
[nr2s, dr2s] = lp2lp(nr2, dr2, wb)
```

The results will be displayed with single precision (`format short e`), although the calculations will be performed and the numbers stored with double precision.

(Notice that if we apply the `lp2lp` transform not to the entire loop but only to the compensator and retain $1/s^2$ as the plant, the new compensator transfer function must be multiplied by ω_b^2.)

Example 4. We can now write an example of a design specification for a time-invariable controller for a plant with a prescribed nominal transfer function P_n, with high-frequency flexible modes, and with hard saturation in the actuator:

- At $\omega > 8$, the nominal loop gain coefficient should not exceed $0.1/\omega^3$ in order to gain-stabilize uncertain flexible modes.
- The output effect of disturbances must be minimized.
- With the plant $P = kP_n$ where $0.8 < k < 1.2$, in the linear state of operation (for small commands), the rise time must be less than 2 sec.
- The overshoot/undershoot for small-amplitude step-commands must be:
  for the nominal plant, under 5%;
  for $0.8 < k < 1.2$, under 10%;
  for $0.6 < k < 1.4$, under 20%;
  for $0.4 < k < 2.5$, the overshoot must be under 50%.
- These norms on the overshoot are also applied to large-amplitude step-commands (up to overloading the actuator 10 times).
- No step command may trigger self-oscillation, for $0.4 < k < 2.5$.

The specifications on the overshoot guarantee minimum performance for the worst case of maximum plant parameter variations, and also assure that most systems in production (with the plant parameters close to the nominal) have much better performance than the worst case.

From Example 1 it is seen that these specifications can be met when the stability margins are 30° and 10 dB. If smaller stability margins are chosen, the sensitivity in the frequency neighborhood of f_b increases and the variations in k cause larger deviations of the transient response from the nominal.

4.2.4 Reshaping the feedback response

We proceed next with step A2 from Section 4.2.1, i.e., with shaping the loop response over the functional feedback bandwidth.

In many practical cases the disturbances' amplitude decreases with frequency, and the loop gain should also decrease with frequency. If, further, the system is phase-stabilized with a constant stability margin, the loop gain response is similar to that shown in Fig. 4.2 and to response (3) in Fig. 4.12(a).

To match a specified disturbance rejection response, the constant-slope gain response must be modified. Fig. 4.12(a) gives several examples of feasibly reshaped loop gain responses. The responses redrawn in Fig. 4.12(b) on the arcsin f scale have the same area under them over the frequency interval [0, 1]. Therefore, the phase response at frequencies $f > 1$ and the stability margins can be the same for all shown gain responses (recall (3.12) from Section 3.9.5).

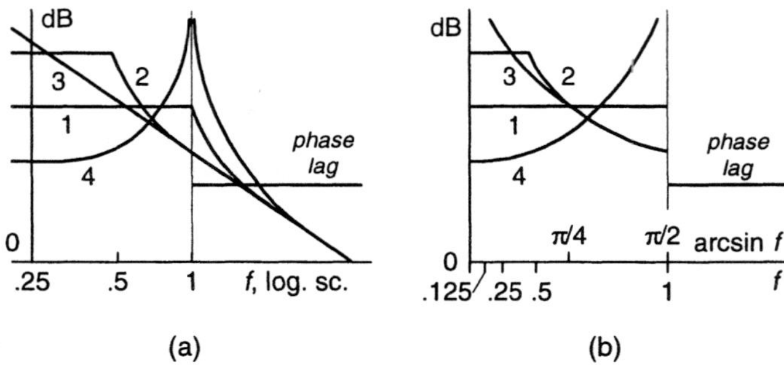

Fig. 4.12 Reshaped loop gain responses on (a) the logarithmic frequency scale and (b) the arcsin f scale

Any frequency can be chosen to be the normalized frequency $f = 1$ (or $\omega = 1$). For control system design it is commonly convenient to use as the normalized frequency the frequency at which the constant-slope response gain is approximately 10 dB.

Example 1. Propellant sloshing in Cassini spacecraft's tanks causes structural modes with gain magnitude up to 8 dB. To provide gain stabilization with 8 dB upper stability margin over the range where the modes can be, the nominal loop gain needs to be at least 16 dB. Fig. 4.13 shows the loop gain response to suit the problem, obtained by reshaping the constant slope response. Numerically, the loss in feedback at lower frequencies can be estimated by application of the rule (3.12) (preservation of the area of the loop gain) when the plot is redrawn on arcsin f scale, with the normalized frequency $\omega = 1$ at the upper end of the slosh mode range.

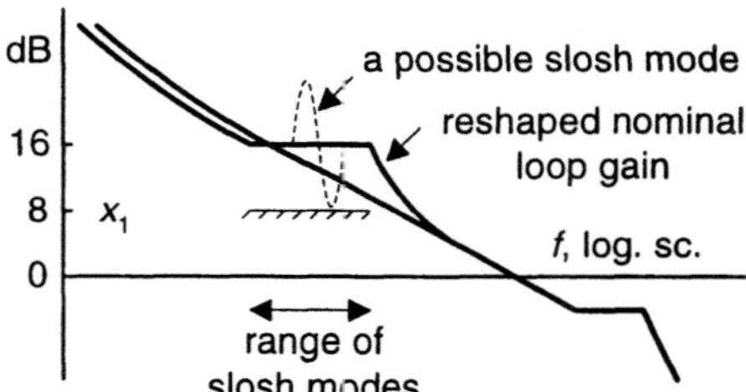

Fig. 4.13 Reshaping the loop gain response

4.2.5 Bode cutoff

When the frequency components of the expected disturbances within the functional band $f < 1$ have the same amplitudes, the loop gain within the functional band should be constant as shown in Fig. 4.14. The value $A_o = 20 \log |T|$ must be maximized.

To find this response, Bode made use of the function $\theta(jf)$ defined by (3.15). The function

$$A_o + 2(1 - y)\theta(jf) \tag{4.3}$$

has the high-frequency asymptote with the slope $2(1 - y)n$ dB/oct. It replaces the constant-slope response in Fig. 4.2, as shown in Fig. 4.14(a). It is seen from the picture (and from the formulas) that this loop gain at $f = 1$ equals the value A_o that the constant-slope response has at $f = 0.5$. In other words, the functional bandwidth of A_o dB feedback in the ***Bode optimal cutoff*** becomes extended by one octave.

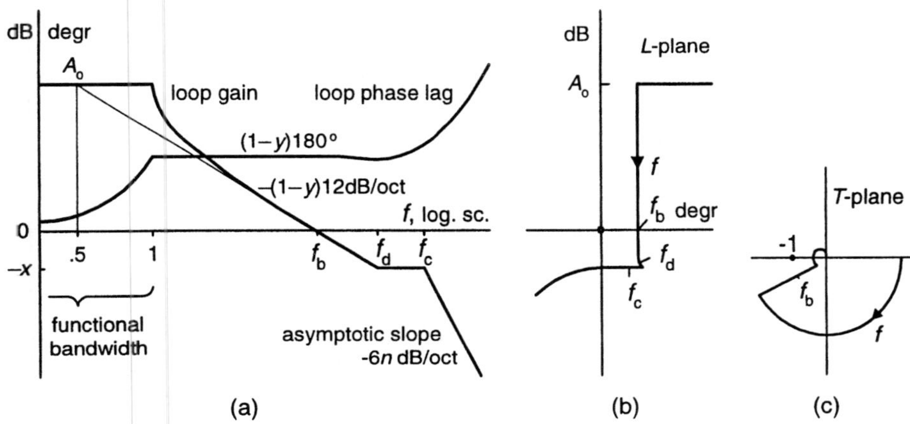

Fig. 4.14 Bode optimal cutoff, (a) Bode diagram, (b) Nyquist diagram on L-plane, (c) Nyquist diagram on T-plane

From the triangle shown in Fig. 4.14(a), the available loop gain A_o is the product of the slope $12(1 - y)$ dB/oct and the feedback bandwidth in octaves plus 1 (the extra octave is that from 0.5 to 1):

$$A_o \approx 12(1 - y)(\log_2 f_b + 1)\,. \tag{4.4}$$

In the common case of 30° stability margin, i.e., $y = 1/6$,

$$A_o \approx 10\,(\log_2 f_b + 1)\,. \tag{4.5}$$

This simple formula is quite useful for rapid estimation of the available feedback.

Example 2. The prescribed stability margins are 30° and 10 dB, the feedback is required to be constant over the bandwidth of [0,200] Hz, and the crossover frequency f_b is limited by the system dynamics to 6.4 kHz. From (4.5) the available feedback is 60 dB.

4.2.6 Band-pass systems

In some systems, for example in vibration suppression systems, the frequency band of functional feedback does not include dc. The band can be viewed as centered at some finite frequency f_{center}. Generally, the physically realizable band-pass transfer function can be found by substituting

$$s + (2\pi f_{center})^2/s \tag{4.6}$$

for s in a low-pass prototype transfer function [2]. The loop response obtained with (4.6) from the low-pass Bode optimal cutoff is shown in Fig. 4.15. Notice that in Fig. 4.15(b), two critical points for the Nyquist diagram to avoid are shown: $-180°$ and $180°$, each of the points being a mapping of the T-plane point -1 onto the L-plane.

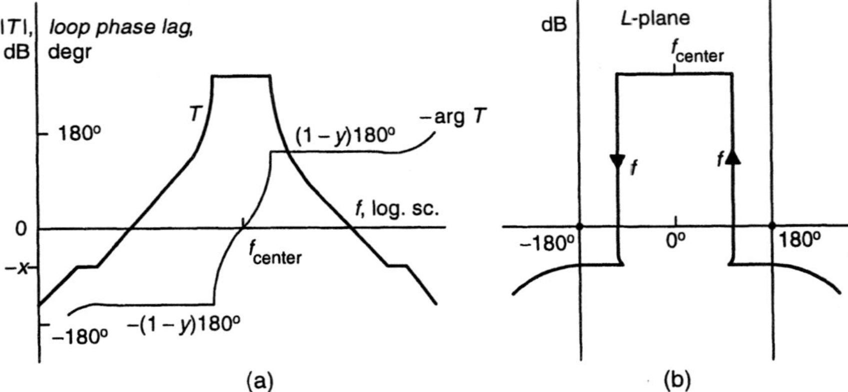

Fig. 4.15 Band-pass optimal Bode cutoff: Bode diagram (a) and Nyquist diagram (b)

The absolute bandwidth Δf of the available feedback is an invariant of the transform (4.6) as is illustrated in Fig. 4.16, and it equals the bandwidth of the low-frequency prototype f_o. (The bandwidths of the three responses in the picture do not look equal because the frequency scale is logarithmic.) It is seen that a higher f_{center} corresponds to a smaller relative bandwidth and steeper slopes of the band-pass cutoff.

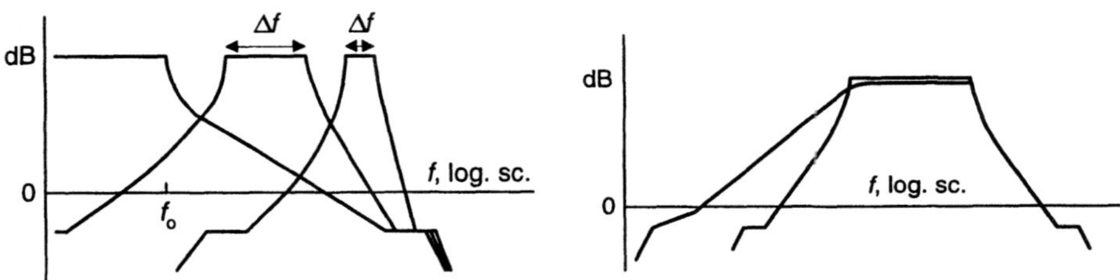

Fig. 4.16 Preservation of operational bandwidth of the band-pass transform

Fig. 4.17 Bode diagrams for a wide-band band-pass system

When the relative functional bandwidth is fairly wide, more than 2 octaves, the steepness of the low-frequency slope has only a small effect on the available feedback since the absolute bandwidth of the entire low-frequency roll-off is rather small. This case is shown in Fig. 4.17.

4.2.7 Nyquist-stable systems

As mentioned in Section 4.2.2, and as will be detailed in Chapters 10 and 11, the responses shown in Figs. 4.2, 4.12, 4.16, 4.17 are tailored to guarantee stability of a system with a saturation link in the loop. If, however, the system is furnished with an extra dynamic nonlinear link of special design (described in Chapters 10, 11, and 13), the loop phase lag, the slope of the Bode diagram, and the available feedback can be increased by using the Nyquist-stable system loop response shown in Fig. 4.18 instead of the phase-stabilizing response shown by the thin line. Here, x_1 and x represent the upper and lower amplitude stability margins. At frequencies where $A > x_1$, the system is only gain-stabilized. The integral of the phase lag in this system is larger than in the absolutely stable system, and as a result, the feedback over the functional bandwidth is larger. This response can be generated by pasting together several elementary responses [9].

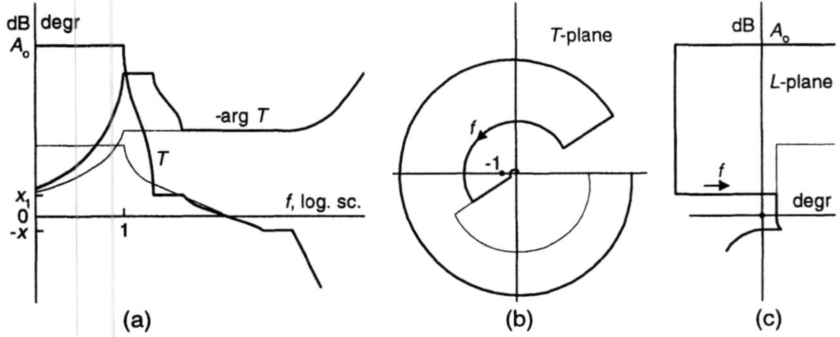

Fig. 4.18 Comparison of a Nyquist-stable system with an absolutely stable system (thin lines): (a) Bode diagrams, (b) Nyquist diagrams (not to scale), (c) Nyquist diagrams on the *L*-plane

Fig. 4.19(a) shows a simplified response which is easier to implement (although it provides somewhat less feedback). The essential features of the response are the steep slope of $-6n_1$ dB/oct before the upper Bode step and the presence of two Bode steps, the width of the lower step calculated with (4.1), and of the upper step from f_g to f_h with the similarly derived formula

$$\frac{f_h}{f_g} = 0.6n_1 \,. \tag{4.7}$$

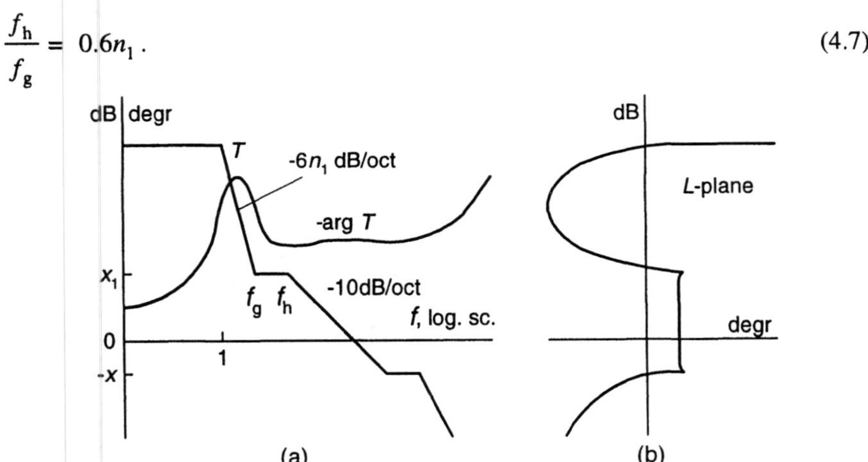

Fig. 4.19 Simplified Nyquist-stable loop response: (a) Bode diagram and (b) Nyquist diagram on the *L*-plane

The larger the integral of phase, the larger is the available feedback. The phase lag, however, cannot be arbitrarily big. A certain boundary curve $A(B)$ exemplified in Fig. 4.20 is specified by the features of nonlinear links in the loop (nonlinear compensators will be discussed in Chapters 10–13).

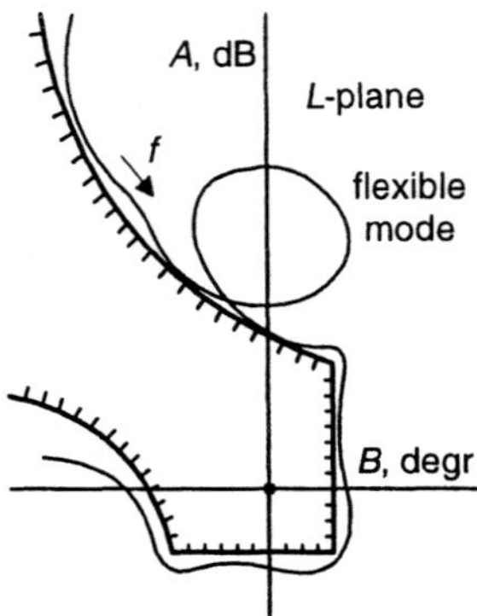

Fig. 4.20 The Nyquist diagram should not penetrate the boundary curve specified by the properties of the nonlinear dynamic compensator.

In Fig. 4.20, the Nyquist diagram is shown with a loop on it caused by a flexible mode of the plant. At frequencies of this mode, the phase stability margin is excessive. In accordance with the phase integral this reduces the achieved feedback, but the feedback deficit due to the loop is rather small since the mode resonance is narrow and the excess in the integral of phase is small.

Type 1 and Type 2 systems (recall Section 3.7) are Nyquist-stable. The stability in such systems can be achieved with upper and lower Bode steps. In practice, the transition between the steep low-frequency asymptote and the crossover area is most often made gradual to simplify the compensator transfer functions, thus reducing somewhat the available feedback at lower frequencies.

4.3 Feedback bandwidth limitations

4.3.1 Feedback bandwidth

In Section 4.3 we will discuss physical constrains on the high-frequency loop gain. However, first we need to clarify the definitions of the term ***feedback bandwidth***. In the literature and in the professional language of control engineers, this term may have any of the following three interpretations indicated in Fig. 4.21:

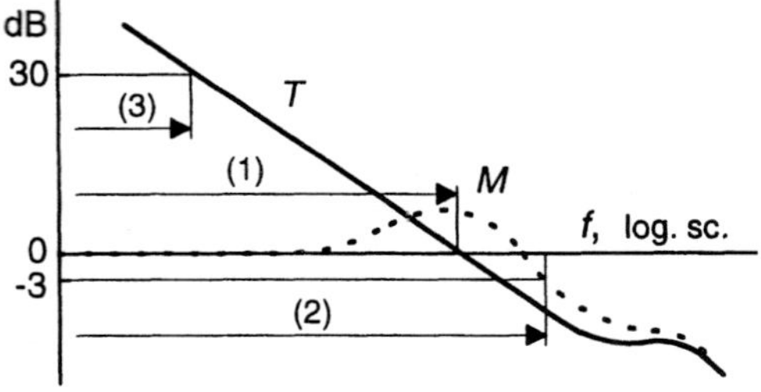

Fig. 4.21 Feedback bandwidth definitions

(1) The ***crossover frequency*** f_b, i.e., the bandwidth of the loop gain exceeding 0 dB. In this book, this definition

for feedback bandwidth is accepted.

(2) The frequency f_M where $|M| = 1/\sqrt{2}$, i.e., $20\log|M| \approx -3$ dB. This frequency is the ***tracking system 3 dB bandwidth***. The frequency is typically from $1.3f_b$ to $1.7f_b$.

(3) The frequency up to which the loop gain retains a specified value (e.g., the bandwidth of 30 dB feedback). This bandwidth is also called the ***bandwidth of functional feedback***.

4.3.2 Sensor noise at the system output

Next, consider the system shown in Fig. 4.22. The source N represents the sensor noise (here, N is understood to be the mean square amplitude of the noise). Since from the noise input to the system output, the system can be viewed as a tracking system with unity feedback, the mean square amplitude of noise at the system output is

$$N_{out} = NT/F.$$

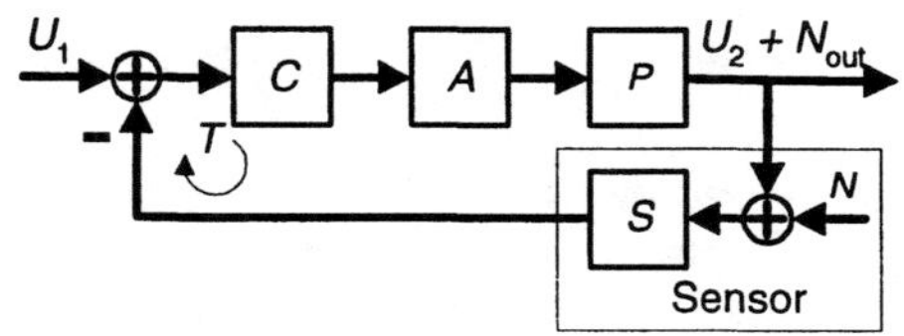

Fig. 4.22 Sensor noise effect at system's output

In shaping the loop gain response, the trade-off is between output noise reduction and disturbance rejection. Larger feedback bandwidth leads to larger output noise, but smaller disturbances. The output mean square error caused by the noise can be found by computer simulation of the output time-responses. Another way to do this is to find the output noise power by integration (on a computer or even by graphical integration) of the frequency-domain noise responses.

Example 1. Consider the Bode diagram shown in Fig. 4.23. This loop gain response has a rather steep cutoff after f_b to reduce the output noise effect, but shallower gain response and smaller feedback at lower frequencies. The phase stability margin is large. The hump on the response of $|M|$ is small; therefore, the overshoot in the transient response is also small.

This response is employed when the plant is already fairly accurate and there is no need for large feedback at lower frequencies, and positive feedback near the crossover frequency should be reduced to reduce the output effect of the sensor noise. In such a system, command feedforward is commonly used to improve the closed-loop input-output response.

When the loop response is steep as shown in Fig. 4.24, the output noise increases because of the positive feedback at the crossover frequency and beyond. This causes a substantial increase in the output noise since the contribution of the noise spectral density to the mean square error is proportional to the noise bandwidth. On the other hand, this response provides better disturbance rejection. The loop response should be therefore shaped in each specific case differently to reduce the total error.

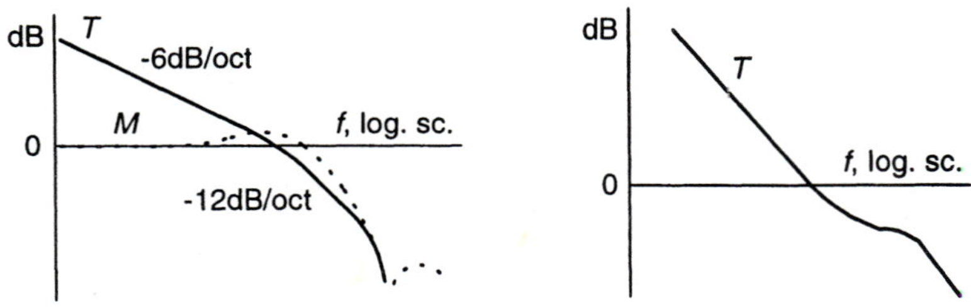

Fig. 4.23 Shallow slope response **Fig. 4.24** Steeper slope response

Example 2. Consider the spacecraft attitude control system in Fig. 4.25 which uses a gyro as a sensor. The system is accurate except at the lowest frequencies where the gyro drift causes attitude error. The drift is eliminated by a low-frequency feedback employing a second sensor, a star tracker. The optimal frequency response for the star tracker loopis that which reduces the total noise from the two sensors, i.e., which reduces the mean square error of the system output variable. The calculations can be performed in the frequency domain or by using the LQG method described in Chapter 8. Since the star tracker noise varies with time, depending on whether bright stars are available in its field of view, the feedback path responses need to be varied to maintain the minimum of the error. Such an adaptive system is illustrated in Chapter 9.

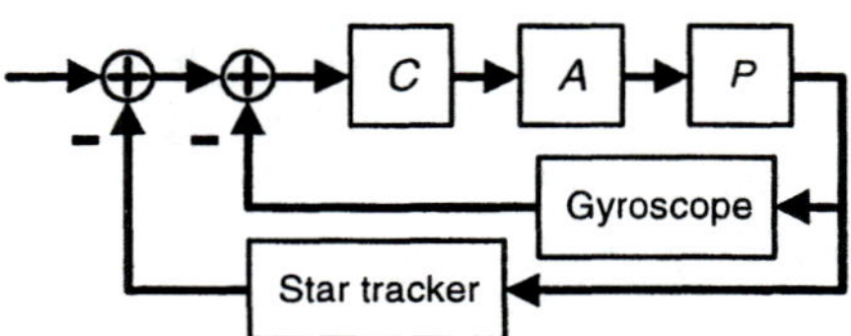

Fig. 4.25 Spacecraft attitude control system using two sensors

4.3.3 Sensor noise at the actuator input

In the control system diagram in Fig. 4.26, the noise source N reflects the noise from the sensor and the noise from the pre-amplifier in the compensator. When the signal at the input of the saturation link is below the saturation threshold, the noise effect at the input of the saturation link is

$$N_{\mathrm{A}} = \frac{NCA}{CAP+1}.$$

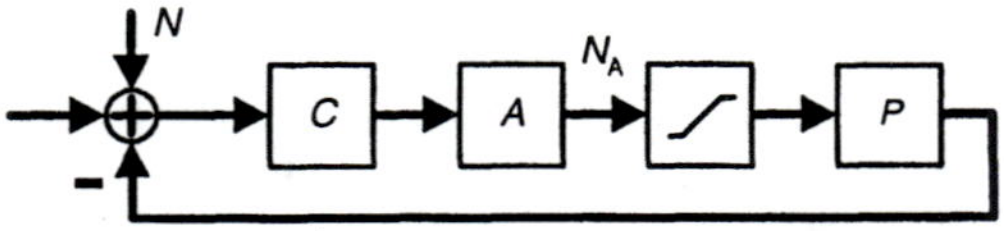

Fig. 4.26 Noise source in a feedback system

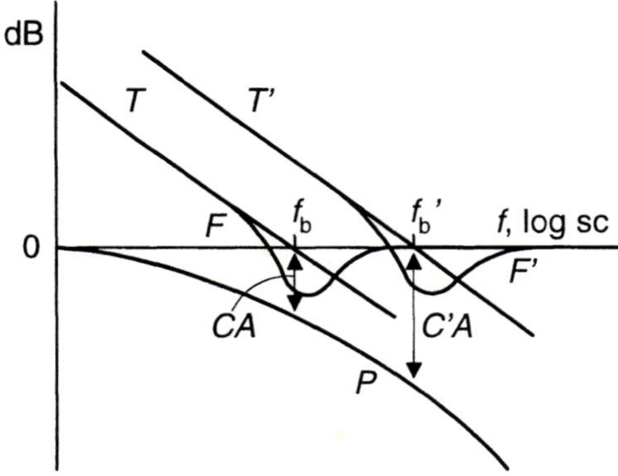

Fig. 4.27 Noise level at saturation link input

With the typical responses of P and T shown in Fig. 4.26, $|CAP| >> 1$ at lower frequencies. At these frequencies the noise $N_A \approx N/P$ does not depend on C. On the other hand, at frequencies higher than f_b, $N_A \approx NCA$, and reducing $|C|$ decreases the noise. With proper loop shaping, the noise N_A is most prominent at the frequencies within 2 to 4 octaves above f_b.

It is seen from Fig. 4.27 that the change in the return ratio from T to T' is attained at the price of increasing $|CA|$ to $|C'A|$, i.e., at the price of a bigger noise effect at the input to the nonlinear link in Fig. 4.26. The noise amplitude and power increase not only because $C' > |C|$, but also because the noise power is proportional to the frequency bandwidth.

When the noise overloads the actuator, the actuator cannot transfer the signal. As a result, the effective gain of the actuator drops, the distortions of the signal increase, and the control system accuracy decreases. Hence, the noise effect at the input to the actuator must be bounded. This restricts the available feedback in the operational band by constraining $|C|$.

The optimal shape of the Bode diagram which provides maximum feedback bandwidth while limiting the noise effect can be found by experimenting with computer simulation. Typically, the responses which are best in this sense contain Bode steps.

Example 1. In an existing system, the bandwidth is limited by the noise at the input to the actuator. If a better amplifier and better sensors become available with half the noise mean square amplitude, the feedback bandwidth can be increased. Maintaining the same mean square noise amplitude at the actuator input, the feedback bandwidth can be increased 1.4 times (since the mean square amplitude of the white noise is proportional to the square root of the noise bandwidth).

4.3.4 Non-minimum-phase shift

The n.p. lag in the feedback loop typically should be less than 1 rad at f_c, or else to compensate for it, the Bode step would have to be very long. This limitation on the feedback feasibility can be critical for loops including a substantial delay. Transport delay, which causes a phase lag proportional to frequency, can be particularly large.

Let us consider two examples of audio systems with large transport delay.

Example 1. Since speaker systems are expensive and their frequency responses are difficult to equalize, since the sound wave reflections from the room walls change the frequency responses on the way to listener's ears, and since good quality inexpensive microphones are easily available, it would be commercially advantageous to make acoustical feedback from a microphone placed in the vicinity of one's ears. Is it possible to maintain good sound quality over the entire range of audio signals up to 15,000 Hz using a feedback system like that shown in Fig. 4.28? Probably not, since nobody does this. There must be a good reason. We might suspect that the reason is the excessive time of the signal propagation about the feedback loop. Let us check it out.

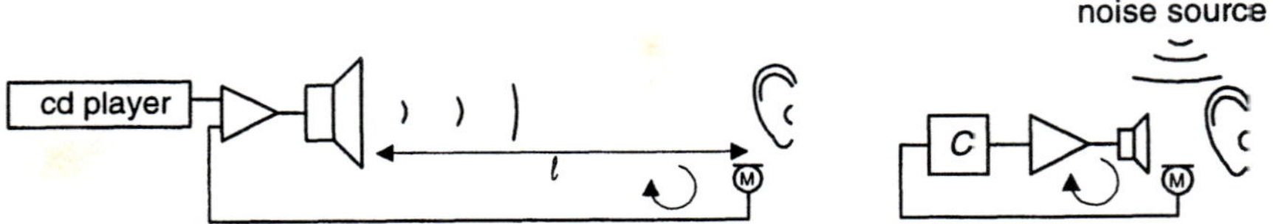

Fig. 4.28 This type of real time acoustical feedback system is not feasible

Fig. 4.29 Sound suppression feedback system

The speed of sound being 330 m/sec, and the distance l between the speakers and the microphone being 2 m, the transport delay is 6.6 msec. For the frequency $f_c = 15 \times 4 = 60$ kHz, the phase lag $B_n(f_c) = 2\pi \times 60{,}000 \times 0.0066 \approx 2500$ rad which is 2500 times the allowable limit. Thus, real-time feedback in this system is not possible. (The response can be equalized by an adaptive system using plant identification.)

Example 2. A system for noise rejection is diagrammed in Fig. 4.29. The microphone is placed at the point where it is desired to keep the noise minimum. The sound from the speaker cancels the noise. The acoustic signal propagation between the speaker and the microphone introduces nonminimum phase lag into the feedback loop. To reduce the phase lag, the distance between the microphone and the compensating speaker should be short. The assembly of the microphone and the speaker is commonly mounted on a helmet.

When 30 dB of feedback is required up to 5 kHz, then frequency $f_d = 20$ kHz. At this frequency, B_n should not exceed 1 rad. Therefore, the pure delay is $(1/40{,}000)/2\pi \approx 0.000008$ sec and the distance between the speaker and the microphone should be shorter than 2.6 mm.

Another source of non-minimum phase lag is the delay in the analog-to-digital conversion when the control is digital. The effects of this delay will be analyzed in the next chapter.

4.3.5 Plant tolerances

The plant gain typically decreases with frequency, and the tolerances of the plant transmission function increase with frequency as illustrated by the limiting curves in Fig. 4.30. In plants whose transfer functions have only real poles and zeros, the plant responses and their variations are commonly smooth and monotonic. Such responses are typical for temperature control and rigid-body position control. Since the minimum necessary stability margins must be satisfied for the worst case, which is typically the case of the largest plant gain, the feedback will be smaller in the case of the minimum plant gain. This way the plant response tolerances reduce the minimum guaranteed feedback.

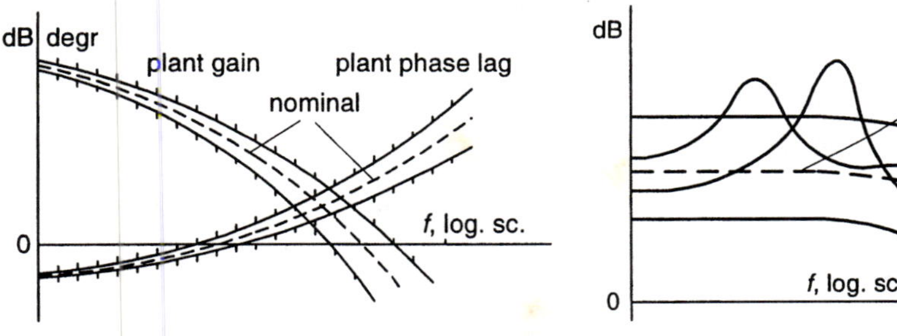

Fig. 4.30 Boundaries of monotonic plant transfer functions

Fig. 4.31 Plant gain frequency responses

For plants with monotonic responses, it is convenient to consider some ***nominal*** plant response, as shown in Fig. 4.30. The feedback loop design is then performed for the nominal plant. Larger plant tolerances – larger deviations from the nominal – require increased stability margins for the nominal response, and thereby limit the nominal available feedback.

For computer design and simulation, the plant uncertainty is most often modeled as ***multiplicative uncertainty***, that is multiplication of the loop transfer function by some error response (i.e., addition of some uncertainty to the gain and phase responses). The multiplicative uncertainty is typically either a constant as in Example 1 in Section 4.2.3 (see Fig. 4.11) or a function of frequency, as those shown in Fig. 4.30.

In general, the dependence of the plant transfer function on its varying parameters can be complicated. For some plants, parameter uncertainty causes deviations from the nominal plant responses which are neither symmetrical nor monotonic, like those shown in Fig. 4.31. Uncertainties are also sometimes modeled by vector addition of some error response to the transfer function (***additive uncertainty***).

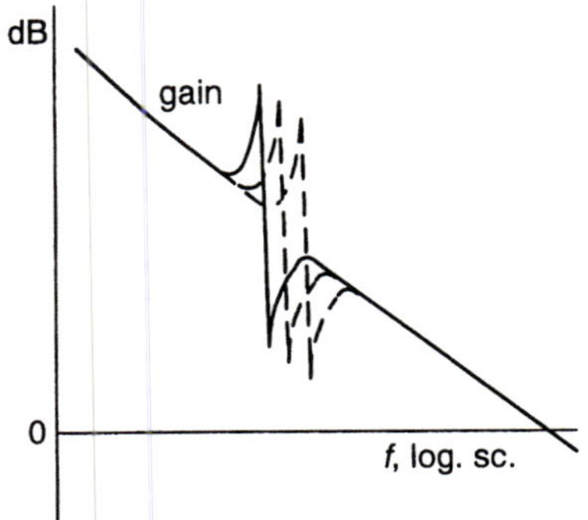

Fig. 4.32 Plant structural resonance on a Bode diagram

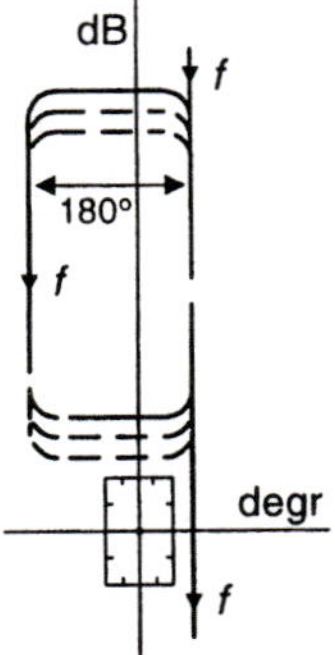

Fig. 4.33 Plant structural resonance on an *L*-plane Nyquist diagram

Flexible plants, i.e., plants composed of rigid bodies connected with springs and dampers, have structural resonances corresponding to specific modes of vibration. Stiffness and mass variations in flexible plants change the pole and zero frequencies as shown in Fig. 4.32. Similar responses are obtained in low-loss electrical systems such as transformers and filters. The resonances typically produce loops nearly 180° wide on the L-plane Nyquist diagram as shown in Fig. 4.33. Neither multiplicative nor additive uncertainty conveniently characterizes these effects. It is better to describe such plants by specified uncertainties in the transfer function poles and zeros.

4.3.6 Lightly damped flexible plants; collocated and non-collocated control

Some of the poles and zeros of flexible plants' transfer functions are only lightly damped, with the damping coefficients as small as 1% and even 0.1%. The loop gain responses of the loops with such plants exhibit sharp peaks and notches as in the examples in Figs. 4.32 and 4.34.

For the closed-loop system to be stable, the modes should be gain- or phase-stabilized. The modes which need attention are the modes that are not already gain-stabilized, i.e., those resulting in the loop gain falling within the interval from $-x$ to x_1, as the modes 2 and 3 shown in Fig. 4.34(a). Increasing the modal damping can reduce the value of the modal peak and notch and gain-stabilize the mode. Otherwise, the mode needs to be phase-stabilized as shown in Fig. 4.34(b).

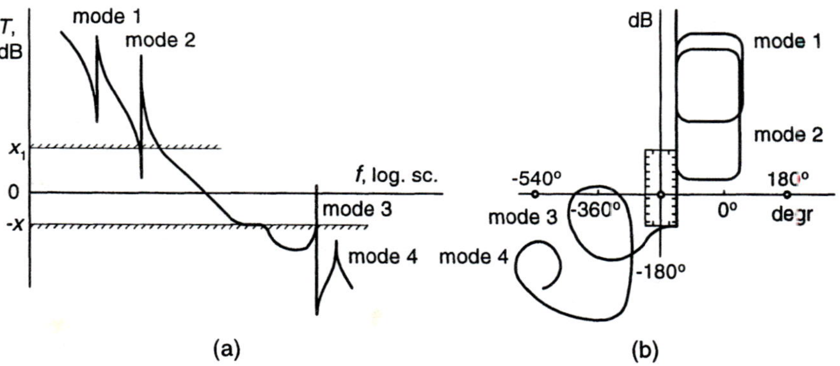

Fig. 4.34 Modes (a) on the loop gain response and (b) on the L-plane Nyquist diagram

To phase-stabilize mode 3, a phase lag might be added to the loop to center the mode at the phase lag of –360°, so the resonance loop will be kept away from the critical points –180° and –540° as illustrated in Fig. 4.34. The required phase lag can be obtained by introducing a low-pass filter in the loop (this is a better solution than adding n.p. lag since the filter will provide the additional benefit of attenuating modes of higher frequencies).

If the plant's phase uncertainty at the frequency of the flexible mode is large, phase-stabilization is not feasible and the mode must be gain-stabilized. A typical case of the gain stabilization of a structural mode is shown in Fig. 4.35. The Bode step allows a steep roll-off at frequencies beyond the step. Gain-stabilization of the mode reduces the feedback in the functional band. The average loop gain at the frequency of the mode must be no higher than $(20 \log Q + x)$ dB. Damping of the high-frequency mode would allow increased feedback bandwidth.

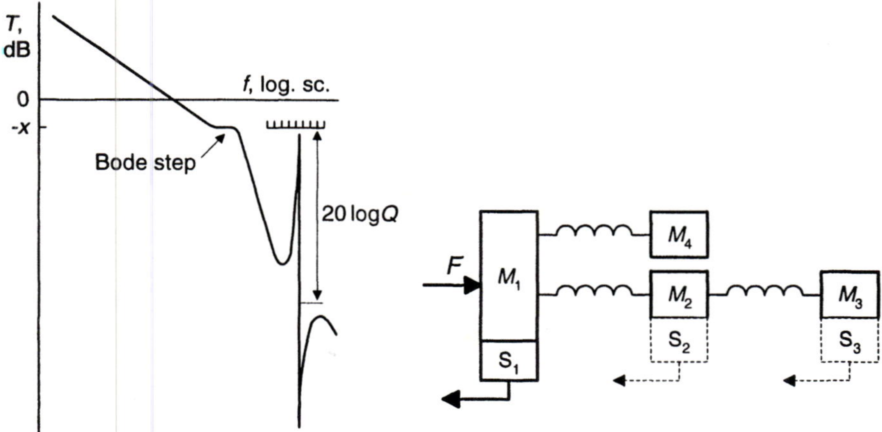

Fig. 4.35 Gain-stabilization of a high-frequency mode

Fig. 4.36 Mechanical plant with flexible appendages and the sensor collocated with the actuator

Next, consider flexible plant model for translational motion shown in Fig. 4.36, consisting of rigid bodies with masses M_1, M_2, ... connected with springs. The actuator applies a force to the first body. The motion sensor S_1 is collocated with the actuator and senses the motion of the first body, so the control is called ***collocated***.

If the sensors are velocity sensors, the transfer function from the actuator force to the sensor S_1 is, in fact, the plant driving point impedance (or mobility). The driving point impedance of a passive system is positive real (see Appendix 3), and its phase belongs to the interval [–90°, 90°]. The impedance function of a lossless plant has purely imaginary poles and zeros which alternate along the frequency axis, and the phase of the plant transfer function alternates between 90° and –90°. Flexible plants are discussed in more detail in Chapter 7.

Example 1. The plant of the control system having the loop response shown in Fig. 4.35 is non-collocated since the plant has a pole at zero frequency, and then the mode's pole and zero follow, i.e., a pole follows a pole. If the mode pole-zero order were reversed, the control would be collocated.

Example 2. A collocated force-to-velocity translational transfer function of a body with two flexible appendages resonating respectively at 3.32 and 7.35 rad/sec and negligible damping is

$$P(s) = \frac{1}{s}\,\frac{s^2+10}{s^2+11}\,\frac{s^2+50}{s^2+54}.$$

The masses of the appendages in this example are approximately 10 times smaller than the mass of the main body which explains why the poles are rather close to the zeros. The gain and phase responses are plotted in Fig. 4.36 with

```
n = conv([1 0 10],[1 0 50]);
d = conv([1 0],[1 0 11]); d = conv(d,[1 0 54]);
w = logspace(0, 1, 1000); bode(n,d,w)
```

When the sensor is a position sensor or an accelerometer, an extra integrator or differentiator should be added to this transfer function which changes the slope respectively by – 6 dB/oct or 6 dB/oct.

When the control is collocated, a flexible appendage adds a zero-pole pair to the loop response as shown in Figs. 4.34 (two lower-frequency modes) and 4.37. The modes do not destabilize the system since the phase lag only *decreases* by 180° between the added zero-pole pair. (However, the mode reduces the integral of phase, and therefore the average gain slope and the available feedback somewhat decrease.)

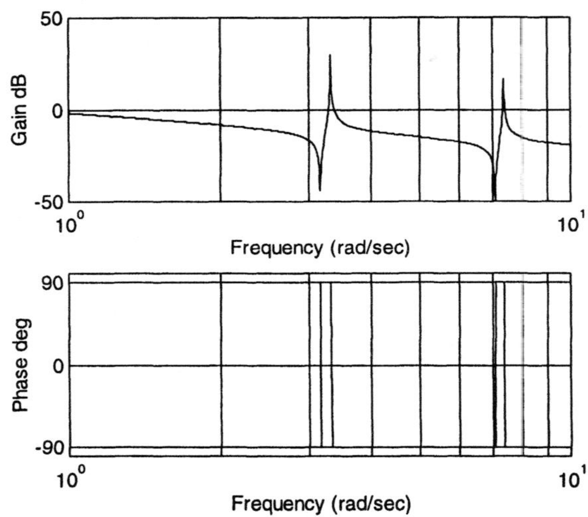

Fig. 4.37 Bode diagram for a collocated mechanical plant with two flexible appendages

Placing the sensor on any other body makes the control ***non-collocated***. In this case the spring connecting the bodies introduces an extra unwelcome phase lag into the loop.

Thus, the sensor location defines whether the control is collocated or non-collocated. The trade-off associated with where to place the sensor, on M_1, on M_2, or on M_3 in Fig. 4.36, is very often encountered in practice. The sensor must be placed within the power train someplace from the actuator to the tip of the tool or other object of control. When the sensor is placed closer to the actuator, i.e., on M_1, the feedback bandwidth can be widened but it is the position of the first body that is controlled. The flexibility between the bodies will introduce the error in the position of the tip. However, when the sensor is placed on the tool, i.e., on M_2, or on the tip of the tool, on M_3, then we are controlling exactly the variable that needs to be controlled, but the feedback bandwidth must be reduced as shown in Fig. 4.35. The best results can be obtained by combining these sensors.

The collocated and non-collocated control will be further discussed in Sections 7.8.3 and 7.8.4.

Example 3. A Nyquist diagram for a flexible plant (Saturn V controller) is given in Appendix 13, Fig. A13.26. Many flexible modes are seen on this diagram. While the controller was being designed, serious discussions were going on about where to place the gyros: closer to the location that needed to be better controlled (in this case the control would be non-collocated), or closer to the engines where the control would be collocated and would be easier to implement. It was eventually decided to play it safe and place the gyros closer to the engines.

Most of the modes are gain-stabilized, only the large mode seen in the upper right sector of the diagram is phase-stabilized. The loop phase lag at the mode's central frequency is 315°. The plant parameter variation must not reduce this phase lag by more than 45° or else the stability margins will be violated. The control at these frequencies is analog. A digital controller with insufficiently high sampling frequency would cause large phase uncertainty (as will be discussed in Section 5.10.7) and would make the system unstable.

Example 4. In pneumatic systems, compressibility of the air in the cylinder of an actuator creates a series "spring" between the actuator and the plant. This makes the control non-collocated and reduces the available feedback bandwidth.

Example 5. In Example 1 of Section 4.2.5, the slosh modes are non-collocated. Because of this, gain-stabilization of the modes is required and is implemented as shown in Fig. 4.14.

4.3.7 Unstable plants

Unstable plants are quite common. For example, the Saturn V launch vehicle and some airplanes are aerodynamically unstable; a slug formed in the combustion chamber or a turbulence in the chamber can make a rocket unstable; rotation of a prolate spacecraft is unstable; a large-gain electronic amplifier without external feedback circuitry is often unstable. For the purposes of analysis and design, an unstable plant can be equivalently presented as a combination of a stable forward path link P with internal feedback path B_{int} that makes the plant unstable, as shown in Fig. 4.38(a).

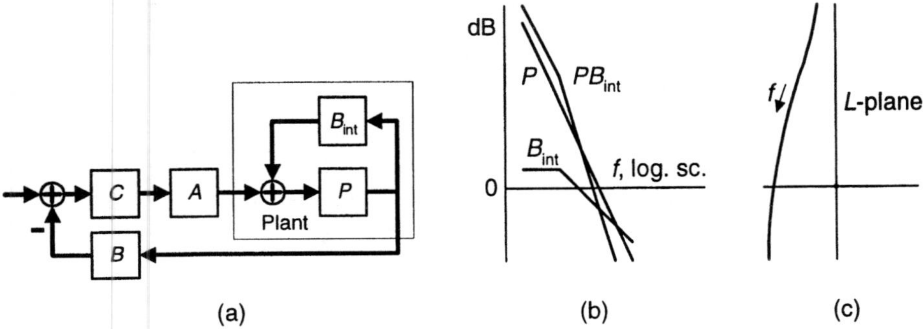

Fig. 4.38 (a) Plant with internal feedback, and the diagrams of (b) Bode and (c) Nyquist for the internal feedback

Example 1. Consider the system diagrammed in Fig. 4.38(a). Assume that the plant is a double integrator with an internal feedback path having a low-pass transfer function $B_{int} = b/(s + a)$. The Bode diagrams are shown in Fig. 4.38(b). The internal loop phase lag exceeds 180° at all frequencies, and the plant becomes unstable as seen from the Nyquist diagram in Fig. 4.38(c).

There are two convenient ways of analyzing and designing such systems.

- When the loop is disconnected at the input to the link P, the loop transfer function is $T = (B_{int} + BCA)P$. After the desired frequency response for T is specified, the required transfer function for the compensator is $C = (T/P - B_{int})/(AB)$. The method is especially convenient when B_{int} is small compared with BCA.
- The compensator can be directly designed for the unstable plant. In this case, the main-loop Nyquist diagram must encompass the critical point in the counterclockwise direction, as required by the Nyquist-Bode multiloop stability criterion.

4.4 Coupling in MIMO systems

As stated in examples in Section 2.9, coupling is typically negligible in well designed control systems where the number of the actuators is kept small. However, at some frequencies the coupling can be large, uncertain, and create stability problems.

In mechanical structures with multi-dimensional control, the actuators are typically applied in mutually orthogonal directions so that the coupling between the corresponding feedback loops is relatively small. However, the plant might include some flexible attachment, like an antenna, a solar panel, or a magnetometer boom on a spacecraft, as shown in Fig. 4.39. The attachment's flexible mode can be excited by any of the actuators (reaction wheels, thrusters), and will provide signals to all the sensors. This coupling may occur at any frequency within a certain frequency range defined by uncertainty in the mass and stiffness of the appendage.

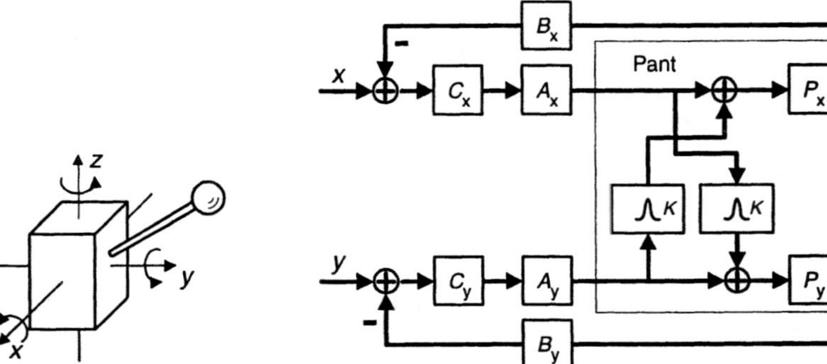

Fig. 4.39 Mechanical plant with a flexible appendage

Fig. 4.40 Block diagram for the coupled attitude control loops

Because of the coupling, the block diagram for the coupled loops looks like that shown in Fig. 4.40. Here, $K(s)$ is the coupling transfer function. The return ratios for the controllers in x and y calculated without taking the coupling into account are $T_x = C_x A_x P_x B_x$ and $T_y = C_y A_y P_y B_y$.

The system can be designed with the Bode-Nyquist multiloop system stability criterion as follows. First, the y-actuator is disabled and T_x is shaped so that the x-loop is stable and robust. Then, A_y is switched on (while the x-actuator is kept on). The transfer function in the y-channel between A_y and P_y is $1 + (T_x/F_x)K^2$. The compensator C_y is then shaped properly to make the system stable (sometimes, this is not possible). The gain-stabilization in the range of the flexible mode is the best choice since phase-stabilization is difficult because the transfer function $K^2 = a^2M/(s^2 + 2\zeta\omega + \omega_v^2)^2$ contains double complex poles with large associated phase uncertainty; here, a is some coefficient. If gain-stabilization cannot be used and the system needs to be phase-stabilized, C_x should be modified to make the response in the x-loop shallower over the frequency range of coupling. The associated reduction of available feedback is unavoidable. Bode diagrams for stability analysis with the successive loop closure criterion, for the x- and y-loops, may look like those shown in Fig. 4.41.

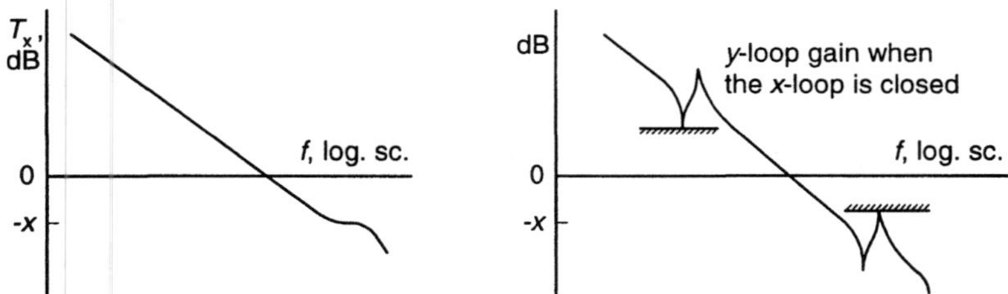

Fig. 4.41 Gain responses of the attitude control loops: (a) x-loop while y-loop open, (b) y-loop while x-loop closed

Coupling between x- and y-controllers can be also caused by the effects of rotation about the z-axis (even in a spacecraft without a flexible appendage). In this case, x-actuators produce rotation about the y-axis, and y-actuators produce rotation about the x-axis. The effect is only profound at frequencies close to the frequency of rotation about the z-axis. The system analysis and design are similar to those in the case of flexible mode coupling.

An x-y positioning table is shown in Fig. 4.42. The translational motions may become coupled via rotational motion due to the load asymmetry and due to structural flexibility, especially at the frequency of the structural mode of the rotation.

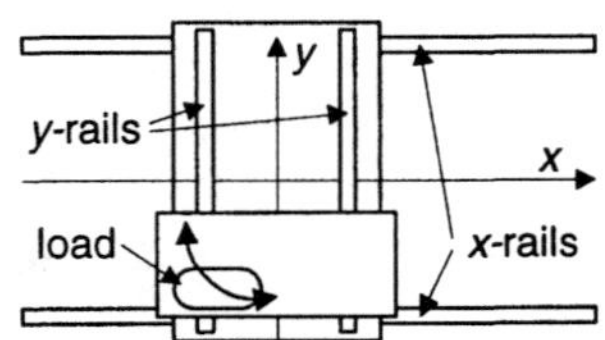

Fig. 4.42 x-y positioner

When the number of the actuators is large (there are many thousand separate muscles in the trunk of an elephant), each control loop should use position (or velocity) and force sensors. This compound feedback makes the loop transfer function less sensitive to plant parameter variations, and makes the output mobility of the actuator dissipative and damping the plant. This also reduces the variations of the loop coupling that are caused by variations of the load and the plant parameters. Loop decoupling algorithms can then be used effectively. Design of a loop with prescribed actuator mobility will be discussed in Chapter 7.

4.5 Shaping parallel channel responses

In MIMO and even in SISO control systems, several paths are often connected in parallel, especially when several actuators or sensors are employed. As was demonstrated in Chapter 3, if two stable m.p. links W_1 and W_2 are connected in parallel as shown in Fig. 4.43(a), then the total transfer function $W_1 + W_2$ can become n.p. The frequency responses of the parallel channels should be shaped properly for the combined channel to be m.p.

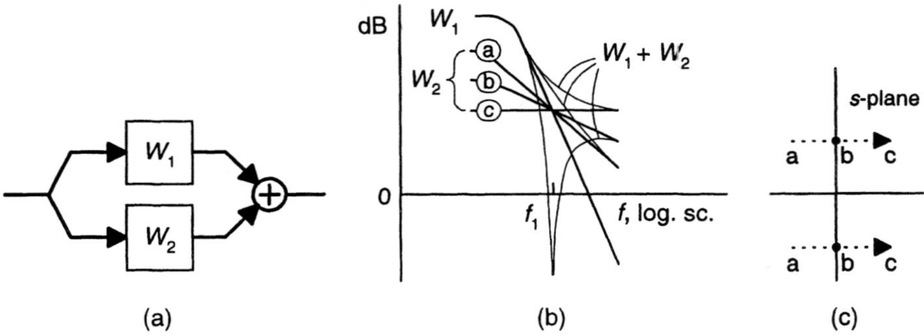

Fig. 4.43 (a) Parallel channels, (b) Bode diagrams for W_1, W_2, and $W_2 + W_1$, and (c) s-plane root loci for $W_1 + W_2$

Example 1. The low-pass links W_1 and W_2 are connected in parallel as shown in Fig. 4.43(a). The steep roll-off W_1 and the three versions of shallower roll-off W_2 are shown in Fig. 4.43(b). At frequency f_1 where the gains are equal, the phase difference between the two channels is, respectively, less than π, equal to π, and more than π. The thin lines show the logarithmic responses of $|W_1 + W_2|$ obtained by vector addition of the links' output signals. When the phase difference between the channels at f_1 is π, the outputs of the links cancel each other and therefore, the composite link transfer function has a pair of purely imaginary zeros $\pm j2\pi f_1$. If the slope of W_2 is gradually changed, the root loci for the zeros of the transfer function cross the $j\omega$-axis as shown in Fig. 4.43(c) and the total transfer function becomes n.p.

As has been proven, the sum $W_1 + W_2$ is m.p. if and only if the Nyquist diagram for W_1/W_2 does not encompass the point -1. Since the ratio W_1/W_2 is also stable and m.p., one can determine whether the Nyquist diagram encloses the critical point by examining the Bode diagram for W_1/W_2.

When the tolerances of the parallel channel transfer functions are not negligible, they can produce large variations in $W_1 + W_2$. The sensitivities of the sum to the components,

$$\frac{\dfrac{d(W_1 + W_2)}{W_1 + W_2}}{\dfrac{dW_1}{W_1}} = \frac{W_1}{W_1 + W_2}\,\frac{d(W_1 + W_2)}{dW_1} = \frac{1}{1 + W_2/W_1}$$

and

$$\frac{\dfrac{d(W_1 + W_2)}{W_1 + W_2}}{\dfrac{dW_2}{W_2}} = \frac{1}{1 + W_1/W_2},$$

become unlimited as the ratio W_1/W_2 approaches -1. To constrain the sensitivities, *the hodograph of W_1/W_2 should be required not to penetrate the* **safety margin** *around* *the point −1*.

Analogous to the stability margins, the ***phase safety margin*** is defined as $y\pi$, and the ***amplitude safety margins***, as x and x_1.

A common practical reason for using two parallel links is that one of the links (actuators, or sensors) works better at lower frequencies, and the second link works better at higher frequencies. Combining them with frequency selection filters generates a link (actuator, or sensor) that is good over a wide frequency range. The composed link transfer function must be m.p. so that it can be included in the feedback loop. However, excessive selectivity of the filters can make the composite link n.p.

Fig. 4.44(a) shows the responses of the low-pass link W_1 and the high-pass link W_2, with different selectivities. Fig. 4.44(b) shows the Bode diagrams for the ratio W_1/W_2. It is seen that when the difference in the slope between W_1 and W_2 responses increases, the Bode diagram for the ratio steepens, the related phase lag increases, and the critical point becomes enclosed by the related Nyquist diagram.

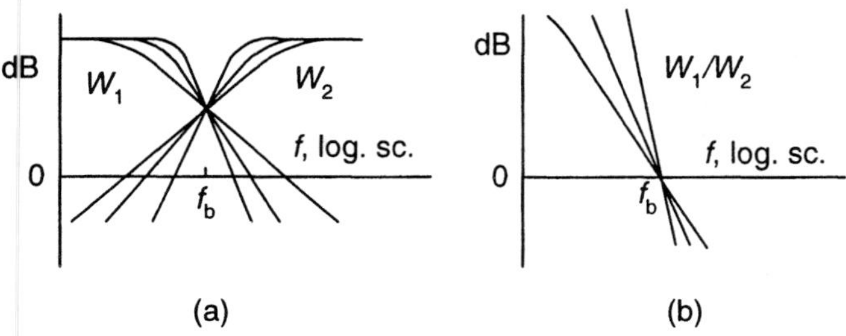

Fig. 4.44 Bode diagrams for (a) W_2, W_1 and (b) W_1/W_2

In order to preserve sufficient safety margins while keeping the slope of W_1/W_2 steep, the Bode diagram for W_1/W_2 can be shaped as in a Nyquist-stable system, Fig. 4.45(b). Then, W_1 and W_2 can be as illustrated as in Fig. 4.45(a). Responses of this kind are particularly useful for systems with the main-vernier actuator arrangement described in Section 2.7.

An alternative to shaping the responses and then approximating them with rational functions is direct calculation of the channel transfer functions. Given the transfer function of the first link, the transfer function of the second link can be found directly as $W_2 = 1 - W_1$. This method works well if the links are precise (as when sensors' readings are combined). If the links are imprecise (like actuators and different signal paths through the plant), and the selectivity is high, then the link parameter variations should be accounted for and sufficient safety margins introduced.

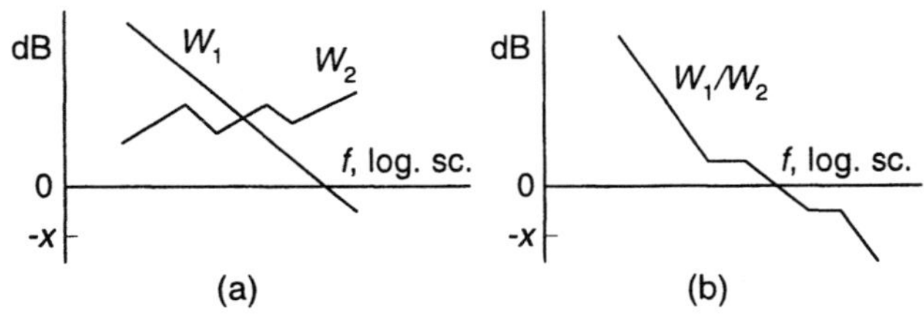

Fig. 4.45 (a) Frequency-selective responses for W_1 and W_2 and (b) Nyquist-stable shape of Bode diagram for W_1/W_2 which preserve m.p. character of $W_1 + W_2$

Example 2. If W_1 is a low-pass,

$$W_1 = \frac{a}{s+a},$$

then the second channel transfer function

$$W_2 = 1 - W_1 = \frac{s}{s+a}$$

is a high-pass, and the ratio

$$\frac{W_1}{W_2} = \frac{a}{s}.$$

The phase safety margin is 90° at all frequencies. The margin is large, but the selectivity between the channels is not high.

Example 3. To improve the selectivity, the first link is chosen to be a second-order low-pass Butterworth filter:

$$W_1 = \frac{1}{s^2 + s\sqrt{2} + 1}. \tag{4.9}$$

The second link transfer function is then

$$W_2 = 1 - W_1 = \frac{s(s+\sqrt{2})}{s^2 + s\sqrt{2} + 1}, \tag{4.10}$$

and the ratio

$$\frac{W_1}{W_2} = \frac{1}{s(s+\sqrt{2})}.$$

The plots shown in Fig. 4.46 indicate that the safety margins are less than in the previous example but still quite wide. With higher-order filters, safety margins become smaller.

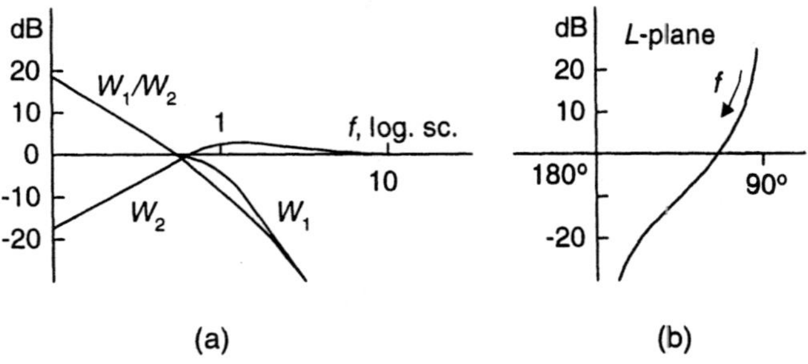

Fig. 4.46 Bode (a) and Nyquist (b) diagrams for W_1, $W_2 - W_1$, W_1/W_2

4.6 Problems

1 The slope at the crossover frequency 40 Hz is (a) –6 dB/oct (b) –9 dB/oct (c) –10 dB/oct (d) –12 dB/oct. What is the feedback in dB at 1.25 Hz if the slope remains the same down to this frequency? What would be the loop phase lag if the slope is constant at all frequencies?

2 What is the length of the Bode step and f_b if
(a) the asymptotic slope is –18 dB/oct, $x = 10$ dB, $y = 1/6$, the asymptote passes the –10 dB level at 1 kHz, and the n.p. lag at 1 kHz is 0.5 rad;
(b) the asymptotic slope is –12 dB/oct, the asymptote passes the 0 dB level at 5 kHz, and the n.p. lag at 5 kHz is 1 rad;
(c) the asymptotic slope is –12 dB/oct, the asymptote passes the –10 dB level at 100 Hz, and the n.p. lag at 100 Hz is 0.5 rad;
(d) the asymptotic slope is –24 dB/oct, the asymptote passes the –10 dB level at 50 kHz, and the n.p. lag at 100 kHz is 0.6 rad.

3 Sketch the phase lag response and the Nyquist diagram for the Bode cutoff in which the Bode step is omitted, with low-frequency slope –10 dB/oct, $x = 10$ dB, and asymptotic slope –18 dB/oct. Is the system stable?

4 At the frequency of structural resonance $f_{st} = 120$ Hz or higher there is a narrow resonance peak in the plant gain response and the plant phase at this frequency is completely uncertain. (The phase uncertainty is the result of using a digital controller with sampling frequency 100 Hz. Digital controllers will be studied in detail in Chapter 5.) The system must be therefore gain-stabilized at the frequency of the structural resonance f_{st}.

Using the asymptotic slope –18 dB/oct, 30° and 10 dB stability margins, and assuming that $f_c = 2f_b$, and $f_d = 4f_d$ (which are typical numbers and should be used for initial estimates), express f_b as a function of f_{st} and Q. Make a sketch of the loop gain response similar to that shown in Fig. 4.47, but with numbers on it, for the peak value $20 \log Q$ equal to: (a) 20 dB; (b) 25 dB; (c) 30 dB; (d) 40 dB; (e) 50 dB.

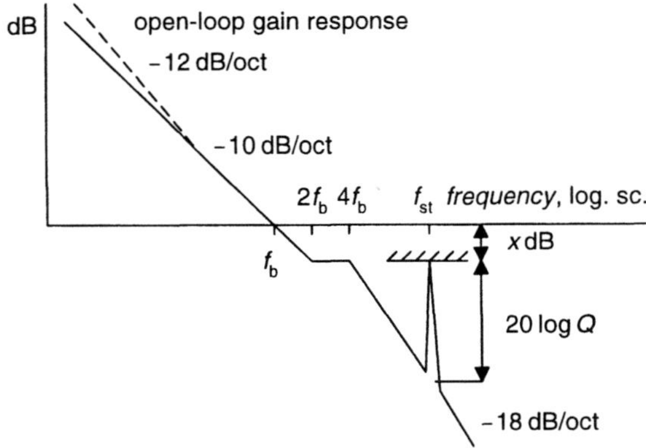

Fig. 4.47 Feedback bandwidth limitation due to a structural resonance

5 The same as in Problem 4, but the resonance uncertainty range starts at 170 Hz.

6 The same as in Problem 4, but the resonance uncertainty range starts at 85 Hz.

7 The crossover frequency f_b is 1 kHz, $y = 1/6$. The system is phase-stabilized at all frequencies up to f_b, with Bode optimal loop response. What is the available feedback over the bandwidth (a) [0,50], (b) [0,30], (c) [30,60]?

8 (a) What is the feedback in dB at 1.5 Hz if the crossover frequency is 300 Hz and the main slope is –10 dB/oct?
(b) What is the maximum available feedback in dB at 1.6 Hz when the feedback is kept constant at frequencies below 1.6 Hz, the system is phase-stabilized with 30° stability margins, and the crossover frequency is 300 Hz?

9 The required feedback at 10 Hz is 40 dB, and the feedback should increase at lower frequencies with the slope –10 dB/oct. What is the crossover frequency? What are the frequencies at the beginning and the end of the Bode step if the step's length is 0.8 oct and $x = 10$ dB?

10 In a system with feedback bandwidth 100 Hz, amplitude stability margin 10 dB, phase stability margin 30°, and no n.p. lag, the attenuation in the feedback loop is required to be large over 1.5 kHz where there might be flexible modes in the plant. Calculate what attenuation at 1.5 kHz is available in the loop response with a Bode step if the asymptotic slope is chosen to be (a) –12 dB/oct; (b) –18 dB/oct. What is the conclusion? Explain the result by referring to the shape of the weight function in the Bode phase gain relation.

11 In a GaAs microwave feedback amplifier, there are two gain stages, and the length of the feedback loop is 1 mm. The speed of signal propagation is 150,000 km/sec. At what frequency is $B_n = 1$ rad? Considering this frequency as f_c, what is the length of the Bode step? What is the available feedback over the bandwidth (a) 0 to 3 GHz, (b) 1.5 to 3 GHz, (c) 0 to 2 GHz, (d) 2 to 3 GHz?

12 Find the loop transfer function and the prefilter for the system with crossover frequency: (a) 0.2 Hz; (b) 6 Hz; (c) 2 kHz; (d) 6 kHz; (e) 2 kHz; (f) 6 MHz; (g) 2 MHz; (h) 4 rad/sec; (I) 100 rad/sec. Use the 1 rad/sec crossover prototype described in Section 4.2.3.

13 The actuator and plant transfer function AP is:
(a) $4(10 - s)(s + 2)/[s^2(10 + s)(s + 7)]$;
(b) $2(10 - s)(s + 2)/[s^2(10 + s)(s + 8)]$;
(c) $3(10 - s)(s + 4)/[s^2(10 + s)(s + 9)]$;
(d) $0.6(10 - s)(s + 5)/[s^2(10 + s)(s + 10)]$;
(e) $2.72(10 - s)(s + 6)/[s^2(10 + s)(s + 11)]$;
Find the compensator that makes the loop transfer function the same as in the example studied in Section 4.2.3, where

$$T(s) = C(s)\frac{1}{s^2}\frac{10-s}{10+s} = \frac{11s^3 + 55s^2 + 110s + 36}{s^4 + 7.7s^3 + 34s^2 + 97s + 83}\frac{1}{s^2}\frac{-s+10}{s+10}.$$

14 Determine the band-pass transform from the low-pass optimal cutoff with frequency range
(a) [0, 1] rad/sec to the bandwidth [30, 70] Hz;
(b) [0, 2] rad/sec to the bandwidth [50, 70] Hz;
(c) [0, 3] rad/sec to the bandwidth [60, 70] Hz;
(d) [0, 4] rad/sec to the bandwidth [40, 70] Hz;
(e) [0, 5] rad/sec to the bandwidth [30, 120] Hz;
(f) [0, 1] rad/sec to the bandwidth [30, 100] Hz.

15 Draw a Nyquist diagram on the *T*-plane (not necessarily to scale, only to show the shape) for the Nyquist-stable system in Fig. 4.19.

16 Initial analysis with a low-order compensator has shown that in the plant hardware configuration A, larger feedback is available than in configuration B. The system engineer assumed (wrongly) that feedback will be larger in configuration A even when, later, a better controller will be developed. Therefore, he decided that configuration A should be chosen. Devise a counterexample to prove that optimal shaping for the Bode diagram must be used for initial analysis as well. (Hint: Use a plant with a flexible mode.)

17 Using the phase-gain chart in Fig. 3.42 (or the program from Appendix 5), calculate the phase response for a Nyquist-stable system whose loop gain response is: from 0 to 10 Hz, 60 dB; from 10 to 20 Hz, –50 dB/oct; from 20 to 80 Hz, 10 dB; from 80 to 320 Hz, –10 dB/oct; from 320 to 640 Hz, –10 dB; from 640 Hz, –18 dB/oct.

18 An extra management level was added to a four-level management system. How will it affect the speed of accessing the market and adjusting the product quantities and features (make a rough estimate)?

19 In a Nyquist-stable system with a response like that shown in Fig. 4.19, f_b = 100 Hz, the phase stability margin is 30°, and the upper and lower gain stability margins are 10 dB. Calculate the frequencies at the ends of the upper and lower Bode steps if the slope at lower frequencies is (a) –12 dB/oct, (b) –18 dB/oct, and the asymptotic slope is (a) –18 dB/oct, (b) –24 dB/oct.

20 The loop gain plot crosses the 0 dB level at 200 kHz, and the rest is as in Problem 19, versions (a) and (b).

21 The unstable plant can be equivalently represented as a stable plant with a feedback path B_1. The path from the plant output to the plant input via the regular feedback loop links is 40 dB larger than via path B_1 at most frequencies, but only by 20 dB larger over some narrow frequency range. How to approach the design of the feedback system?

22 Two parallel links have the following transfer functions respectively:

(a) $$W_1(s) = \frac{(s+5)(s+8)}{(s+40)(s+100)} \quad \text{and} \quad W_2(s) = \frac{1000}{(s+3)(s+10)(s+20)};$$

(b) $$W_1(s) = \frac{(s+6)(s+10)}{(s+50)(s+125)} \quad \text{and} \quad W_2(s) = \frac{1800}{(s+4)(s+13)(s+27)};$$

(c) $$W_1(s) = \frac{(s+2.5)(s+4)}{(s+20)(s+50)} \quad \text{and} \quad W_2(s) = \frac{125}{(s+1.5)(s+5)(s+10)};$$

(d) $$W_1(s) = \frac{(s+10)(s+16)}{(s+80)(s+200)} \quad \text{and} \quad W_2(s) = \frac{8000}{(s+6)(s+20)(s+40)}.$$

Is the composite link $W_1(s) + W_2(s)$ m.p.?

23 Two minimum phase links in parallel made a nonminimum phase link. A third link with constant gain coefficient *k* has been added in parallel to the two links. How would you find the minimum k for the total transfer function to be m.p.?

24 Find transfer function $G(s)$ of a high-pass in parallel to low-pass $12/[(s+3)(s+4)]$ so that the total transfer function is 1. Is this $G(s)$ realizable?

25 Prove that if a transfer function of a linear passive two-port is m.p. with some passive impedances of the signal source and the load, the transfer function is m.p. with any other passive source and load impedances.

Answers to selected problems

1 (a) 1.25 Hz is 5 octaves below 40 Hz. Therefore, the feedback is $6\times5 = 30$ dB.

2 (a) The frequency $f_c = 1$ kHz; then, from (4.2) $f_c/f_d = 0.6\times3 + 0.5 = 2.3$ (i.e., the step length is $3.32 \log 2.3 = 1.2$ oct), $f_d = 0.435$, and $f_b = f_d/2 = 0.22$ kHz.

3 (a) Since $y = 1/6$, the slope is -10 dB/oct. There are $3.32 \log(1000/50) = 4.3$ oct down to 50 Hz from the crossover. Then, according to (4.5), the available feedback is $(4.3 + 1)\times10 = 53$ dB.

4 (a) $18 \log_2(f_c/f_{st}) = 20 \log_{10} Q$, and $f_b = f_c/4$, then $f_b = f_{st}/2^{(20 \log Q)/18+2}$, i.e.,

```
fst = 120; Q = 10; fb = fst/2^[20*log10(Q)/18 + 2]
fb = 13.8881
```

The diagram is shown in Fig. 4.48.

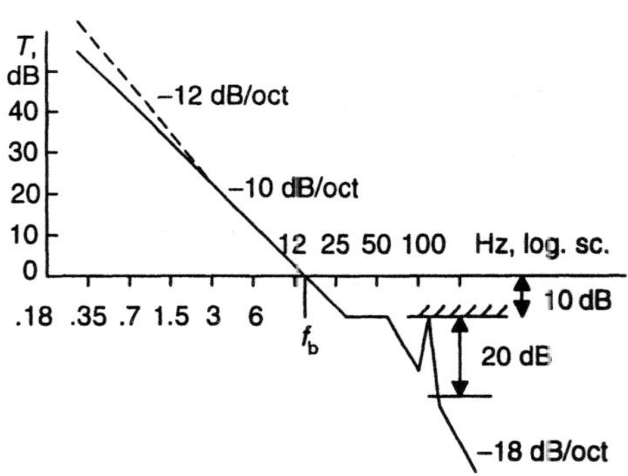

Fig. 4.48 Loop response Bode diagram

With the slope of the Bode diagram −10 dB/oct, disturbance rejection in dB at frequency $f < f_b/2$ depends on the feedback bandwidth f_b approximately as $10 \log_2(f_b/f)$. At lower frequencies $f < f_b/6$, the slope can be increased to −12 dB/oct as shown by the dashed line. (In this case, a nonlinear dynamic compensator must be employed to assure global stability as described in Chapters 10 and 11.)

When discussing the tradeoff between the resonance frequency and the resonance quality (Q) of the object of control and the available disturbance rejection at a meeting with mechanical designers, it is helpful for control engineers to have prepared plots like those shown in Fig. 4.49, exemplifying the available disturbance rejection for two structural resonance frequencies, $f_{st} = 50$ Hz and 100 Hz.

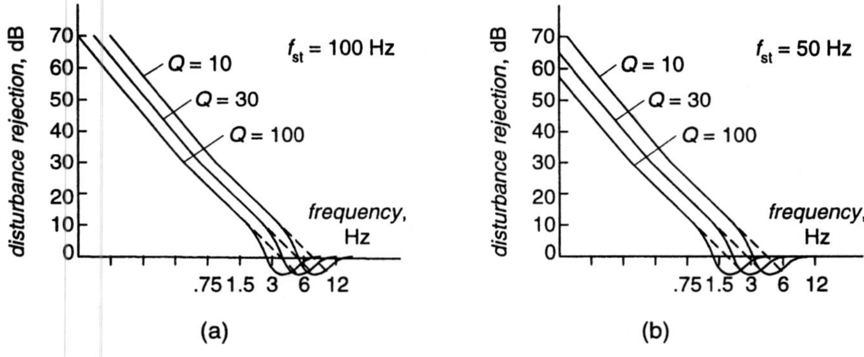

Fig. 4.49 Dependence of disturbance rejection on the structural mode frequency and damping

Chapter 5

COMPENSATOR DESIGN

Compensators, in software or hardware, are commonly built having rational transfer functions.

An asymptotic Bode diagram is piece-linear. The slope of the segments of the diagram is $6n$ dB/oct, where n is an integer. The asymptotic Bode diagram is used to approximate the conceptual Bode diagram for the compensator (which may include segments of any slope). From the asymptotic Bode diagram the poles and zeros of the compensator transfer function are immediately evident.

Approximation of an arbitrary constant-slope gain response based on asymptotic Bode diagrams is described.

Lead and lag links are defined, their asymptotic responses shown, and it is shown how to use them to increase or reduce the slope of the compensator Bode diagram.

A set of normalized plots is presented for a second-order low-pass functions having complex poles with different damping coefficients.

The allocation of transfer function poles and zeros to cascaded links affects the compensator dynamic range. This is demonstrated and recommendations are formulated about how to construct a compensator by cascading the links of the first- and second-order.

Compensators consisting of a parallel connection of links can be made in such a way that each of the links dominates over a certain frequency range. This is a convenient way to make steps on the loop response.

A digital compensator can be viewed as a modification of an analog one in which the analog integrators are replaced by discrete trapezoidal integrators. The trapezoidal integrator is analyzed, the z-transform introduced, and the conversions of the polynomial coefficients tabulated.

The Laplace and Tustin transforms are compared, and it is shown that the Tustin transform is adequately accurate for practice.

A design sequence is recommended for digital controllers. It is explained how to generate block diagrams, equations, and computer code for the first- and second-order digital compensator links. A complete compensator design example is presented with the derivation of a prototype analog compensator including linear and nonlinear links, Tustin transform z-functions, equations of compensator links, and computer code.

The effect of aliasing is described and the loop response is considered for the reduction of the aliasing errors with reduced penalty to the available feedback.

5.1 Accuracy of the loop shaping

The desired frequency response of the compensator has been presented in previous chapters in terms of curves of logarithmic magnitude and phase versus logarithmic frequency. In general these curves imply transcendental functions of s. For implementation, a rational approximation to the ideal transcendental response is almost always required.

The accuracy of the constant-slope implementation directly affects the value of the available feedback. Fig. 5.1(a) shows the ripples on the responses of the gain and phase caused by the Chebyshev (equiripple) approximation of a desired constant-slope response by a rational transfer function. The higher the order of the rational function, the better can be the approximation and the smaller will be the ripples. Fig. 5.1(b) shows the relation between the ripple amplitudes of the gain and phase responses. In order for the *maximum* phase lag not to exceed $(1 - y)\pi$, the *average* phase lag must be reduced by the value of the phase ripple as seen in Fig. 5.1(c), and the average slope of the gain response must be reduced correspondingly.

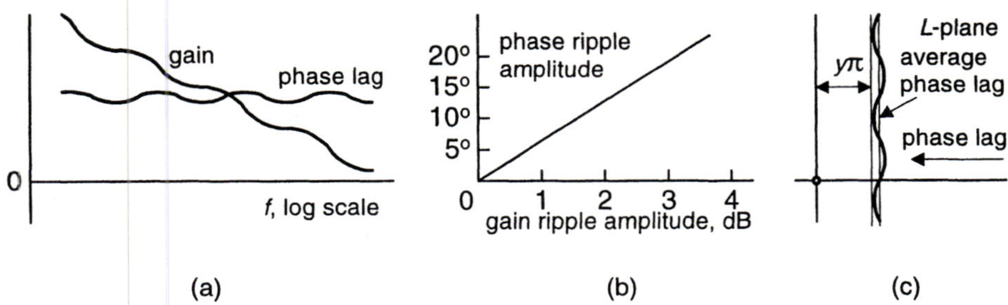

Fig. 5.1 Gain and phase lag diagrams (a), relations between the ripple amplitudes (b), and the ripples on the Nyquist diagram (c)

Example 1. A 5° phase ripple amplitude, i.e., 10° peak-to-peak phase ripples, will force an increase in the average phase margin by 5°. The average slope of the gain response will therefore be reduced by

$$(5°/180°) \times 12 \text{ dB/oct} = 1/3 \text{ dB/oct}.$$

For the typical 4 octaves length of the cut-off, the loss in the feedback will be 1.2 dB – which is marginally acceptable. The corresponding gain ripple amplitude is 0.75 dB as seen in Fig. 5.1(b), and the peak-to-peak ripples in the gain response are 1.5 dB.

Thus, *the required accuracy in the loop gain response is typically not better than ±0.5 dB and the compensator need not be precise*.

The higher the order of the compensator, the smaller the ripples and the better the accuracy of the loop response. Typically, the designer can achieve feedback within 2 dB to 5 dB of the theoretically available using a compensator of order 8 to 15.

5.2 Asymptotic Bode diagram

The gain for transfer function s^{-n} is expressed as $-20n \log \omega$, and the Bode plot is a straight line with the slope $-20n$ dB/dec, or $-6.021n$ dB/oct $\approx -6n$ dB/oct as shown in Fig. 5.2. The use of octaves is preferred since decades do not provide the necessary resolution. The corresponding phase shift is constant at $-n90°$.

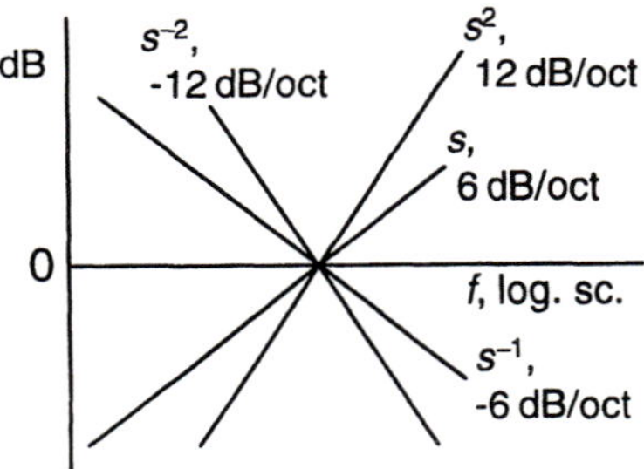

Fig. 5.2 Constant-slope Bode diagrams

Generally, a rational transfer function of the Laplace variable s can be expressed as

$$L(s) = k\prod\frac{(s - s_{zi})}{(s - s_{pj})}$$

where k is a real coefficient and the zeros s_{zi} and the poles s_{pj} can be real or make complex conjugate pairs. The frequency response is calculated after replacing s by $j\omega$. The gain frequency response is

$$20\log|L(j\omega)| = 20\log k + \Sigma\, 20\log|j\omega - s_{zi}| - \Sigma\, 20\log|j\omega - s_{pj}|.$$

The gain plots (Bode diagrams) for transfer functions with respectively the single real negative zero $20\log|(j\omega + 2\pi f_z)/2\pi f_z|$ and with the single real negative pole $20\log|2\pi f_p/(j\omega + 2\pi f_p)|$ are shown in Fig. 5.3(a),(b) by thin lines.

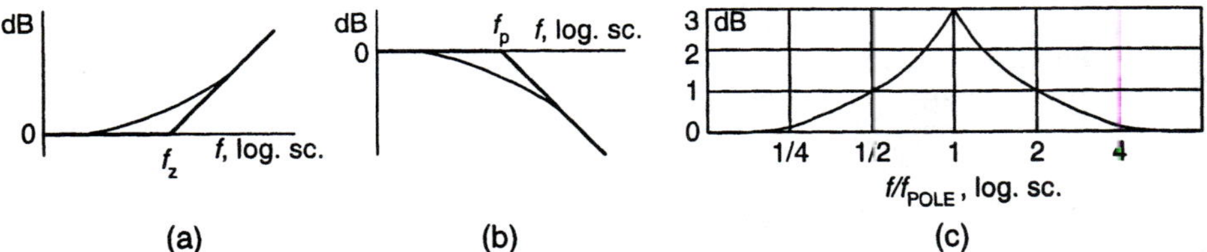

Fig. 5.3 Bode diagrams for a single zero (a), single pole (b), and the error of the asymptotic gain approximation (c)

The responses can be approximated by their high- and low-frequency asymptotes, shown by the thicker lines. These are two straight lines which have slopes 0 dB/oct and ±6 dB/oct, and which intersect at the ***corner frequency*** equal to the frequency of the zero f_z or of the pole f_p.

For the calculation of the asymptotic gain, the expression $(j\omega + 2\pi f_z)$ *is replaced by its larger component, whether real or imaginary*. The error of such approximation is shown in Fig. 5.3(c). The error is 3 dB at the corner, 1 dB one octave away from the corner, and less than 0.1 dB at two octaves.

The diagrams are drawn against the f or ω frequency axis. A convenient scale for drawing the diagram is 10 dB/cm, 1 oct/cm.

For a function having several real poles and zeros, *the piece-linear* ***asymptotic Bode diagram*** *bends upword at each zero corner frequency by 6 dB/oct, and bends downward, by −6 dB/oct, at each pole frequency*.

Example 1. The asymptotic Bode diagram for the function

$$L(j\omega) = \frac{(j\omega + 0.5)(j\omega + 2)(j\omega + 20)}{(j\omega + 0.2)(j\omega + 0.3)(j\omega + 10)}$$

can be drawn as follows. First, the value of the asymptotic Bode diagram is calculated at some frequency, for instance, at the frequency $\omega = 1$. It is

$$L(j\omega) = \frac{j\omega \times 2 \times 20}{j\omega \times j\omega \times 10} = \frac{4}{j\omega} = \frac{4}{j},$$

i.e., the gain is 12 dB as shown in Fig. 5.4.

The slope of the asymptotic diagram at this frequency corresponds to the resulting power of $j\omega$, i.e., it is −6 dB/oct. Then, the asymptotic diagram can be drawn by "bending" the response up at the zeros, and down at the poles, by 6 dB/oct. The actual Bode diagram shown by the thin line is obtained by adding the asymptotic error responses.

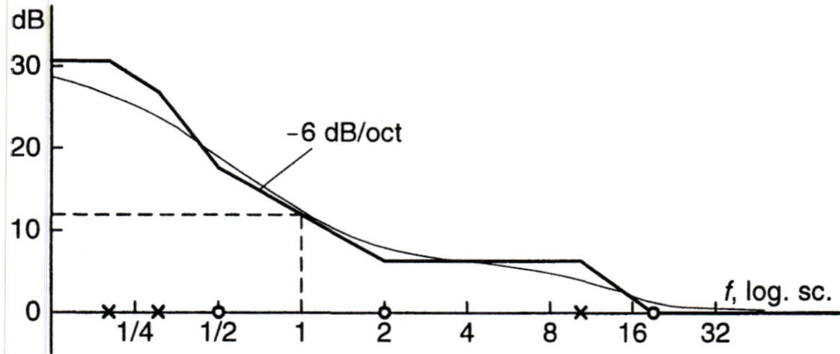

Fig. 5.4 Asymptotic and actual Bode diagrams of a rational transfer function

As seen the asymptotic diagram is fairly close to the actual Bode diagram. Asymptotic diagrams are widely used for conceptual design, and for the purpose of finding a rational function which gain approximates a gain response defined by a plot.

The shape of the actual Bode diagrams is commonly verified by computer simulation. For the above example, the polynomial coefficients of the numerator and denominator of the transfer function can be found from the known roots of the numerator and denominator with

```
rn = [-0.5 -2 -20];
rd = [-0.2 -0.3 -10];
num = poly(rn);
denum = poly(rd);
```

and the Bode plot can be obtained with

```
bode(num, denum)
```

or by using the command `freqs`.

5.3 Approximation of constant-slope gain response

As mentioned in Chapter 4, constant-slope segments are important components in shaping the loop frequency response.

Any constant-slope function can be decomposed into a product of a rational and an irrational functions of s: $s^{-p} = s^{-m} s^{-q}$ where p and q are real, m is an integer, and $0 < q < 1$. The irrational function s^{-q} can be approximated by a rational function whose poles and zeros are real and alternate along the frequency axis as illustrated in Fig. 5.5(a). The Bode plot of a rational approximation to s^{-q} is shown in Fig. 5.5(b). The pole-zero spacings a and b result in an average slope of $-6b/(a+b)$ dB/oct, which should equal $-q$ dB/oct.

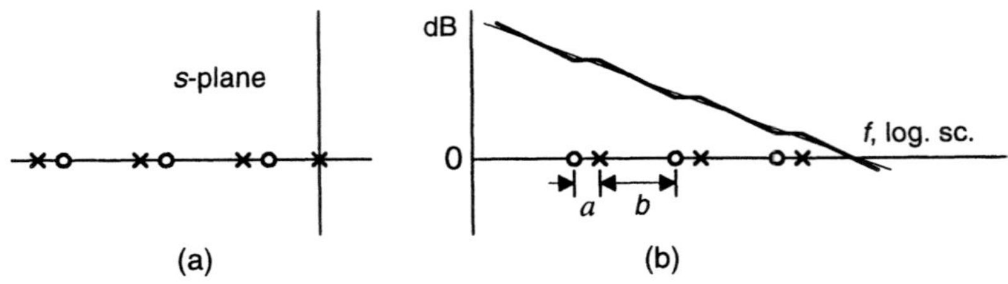

Fig. 5.5 Alternating pole-zero approximation to s^{-q}

Example 1. Fig. 5.6 shows the gain and phase responses for a simple rational approximation to $s^{-1/2}$ using three zero-pole pairs spaced evenly, an octave apart:

$$T(s)=\frac{(s+1/16)(s+1/4)(s+1)}{(s+1/32)(s+1/8)(s+1/2)}. \tag{5.1}$$

The gain slope is nearly constant at -3 dB/oct over a wide frequency range. At the corner frequencies, as the gain transitions to the flat asymptotes, the phase lag is nearly 45°/2, in accordance with the Bode phase-gain relation. The −38° phase at the center differs from the -45° phase of the half-integrator due to the effects of the gain asymptotes.

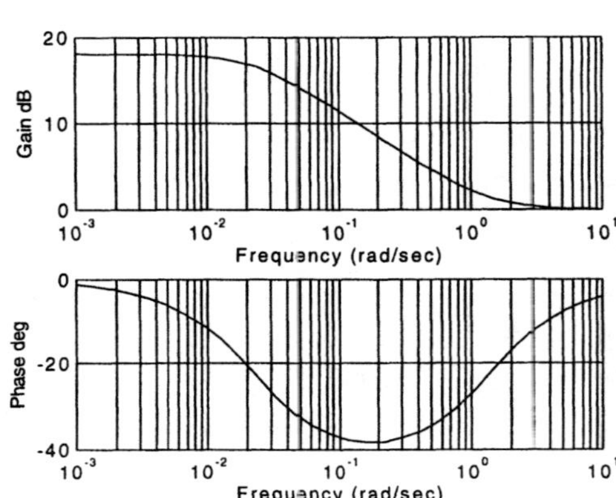

Fig. 5.6 Gain and phase responses of a function with alternating real poles and zeros

Example 2. A rather extreme example is the following approximation to $s^{-1/3}$ which was generated by a curve-fitting program:

$$C(s)=\frac{0.4415s^6+2.234s^5+1.861s^4+0.4276s^3+0.02954s^2+0.0005682s+0.000002178}{s^6+2.462s^5+1.3037s^4+0.2007s^3+0.009201s^2+0.00010989s+0.0000001979}. \tag{5.2}$$

Over the frequency range 0.01 to 10 Hz, the gain response deviates from the ideal −2.007 dB/oct by less than 0.05 dB, while the phase remains within 0.05° of the ideal −30°. This high accuracy is rarely necessary since the required implementation accuracy for compensators in control systems is typically only 0.5 dB or so.

Curve-fitting computer programs can be used for approximation of desired compensator responses. In most cases, this works out well. Sometimes, however, the programs produce transfer functions which are too sensitive to the polynomial coefficient variations. Therefore, the transfer functions must be checked over by changing the coefficients by increments which reflect the software round-off errors and the tolerances of the hardware implementation. On the other hand, approximation of the desired curve by the placement of the poles and zeros as described in Sections 5.4–5.7, and then implementation of the compensator by cascading low-order links as described in Section 5.6, always results in a robust design.

5.4 Lead and lag links

The ideal compensator response should be determined by subtracting the known actuator/plant frequency response from the desired loop response, as discussed in Chapter 4. Computer generation of frequency responses makes the iterative design of the compensator quick and effective. When linear models of the actuator and plant are available, the iteration can be carried out using a trial compensator in series with the actuator and plant, until the desired loop frequency response is achieved. With some experience, convergence will require no more than 5 – 10 iterations.

For the purpose of iterative structural design, it is best to regard the compensator as being composed of elementary building blocks. The simplest of these has the transfer function with a pole-zero pair

$$\frac{s + 2\pi f_z}{s + 2\pi f_p} \tag{5.3}$$

where f_z and f_p are the frequencies of the zero and the pole, respectively. This transfer function is called ***lead*** when the zero precedes the pole, i.e., the zero is at a lower frequency than the pole ($f_z < f_p$). The transfer function is called ***lag*** when the pole comes first ($f_p < f_z$). Fig. 5.7 shows asymptotic Bode diagrams and Bode diagrams for lead and lag transfer functions.

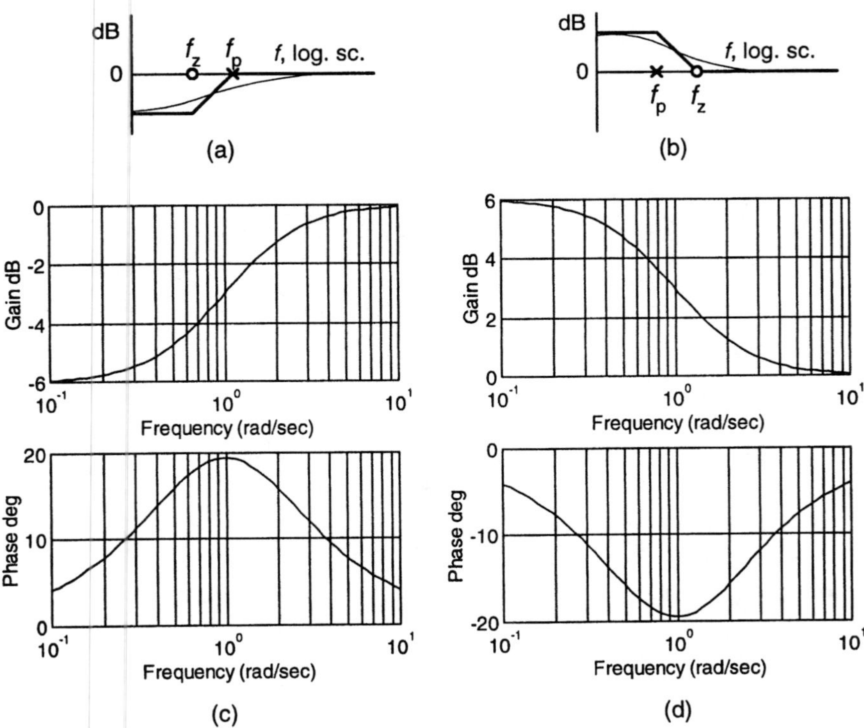

Fig. 5.7 Lead (a) and lag (b) Bode asymptotic diagrams and (thin lines) Bode diagrams; frequency responses for (c) lead $(s + 0.7)/(s + 1.4)$ and (d) lag $(s + 1.4)/(s + 0.7)$

The larger the pole-zero separation, the larger is the gain change from the lower to the higher frequencies, and the larger is the phase lead (or phase lag).

Adding a lead link makes the loop gain Bode diagram locally shallower thus reducing the gain at low frequencies and locally reducing the phase lag.

Fig. 5.8 shows the use of a compensator to change the slope –12 dB/oct of a double integrator to the desired slope of –10 dB/oct with a single lead and with two leads. The use of two lead links provides a closer approximation to the desired Bode diagram than could be achieved with just one lead link with a larger pole-zero separation.

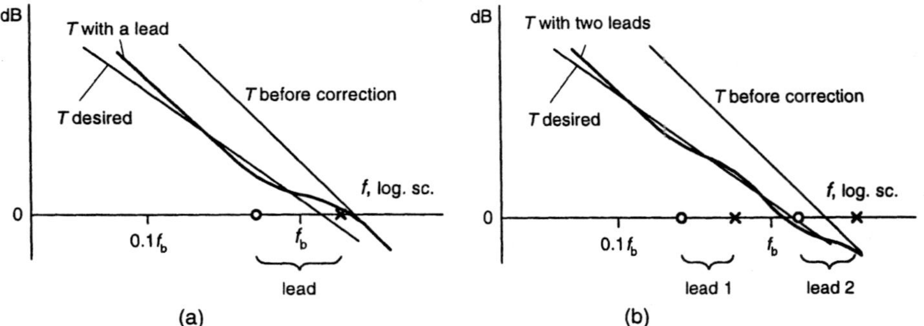

Fig. 5.8 Compensation of a double integrator with (a) one and (b) two lead links

Lag compensation steepens locally the Bode diagram, thus increasing the loop gain at lower frequencies as shown in Fig. 5.9. As in the case of lead compensation, several lag links are sometimes needed for better approximation accuracy.

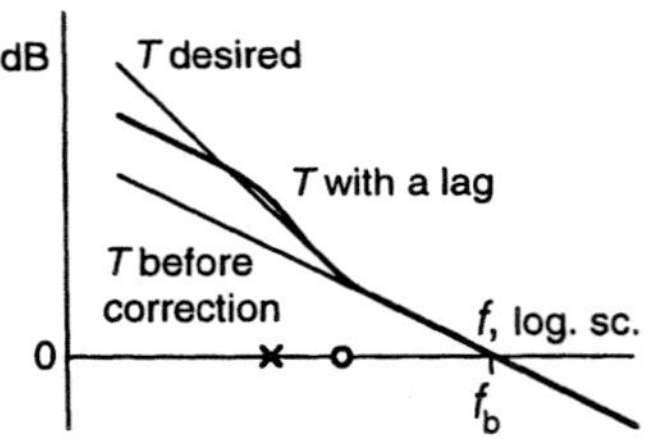

Fig. 5.9 Lag compensation

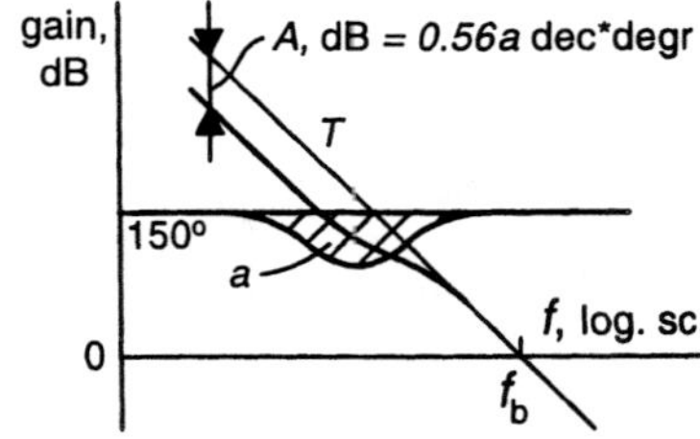

Fig. 5.10 Use of the phase integral

Before deciding to introduce further compensation to improve the approximation to the desired response, one can use Bode's phase integral (3.10), (3.11) to estimate the available improvement in the feedback.

Example 1. In a system, the desired loop gain response is a straight line with slope –10 dB/oct which corresponds to constant 150° phase lag as shown in Fig. 5.10. The current gain response is somewhat shallower, and there is an excessive phase stability margin with the area a decade$\times$degr. The phase integral indicates that elimination of this excess will yield an additional 0.56a dB of feedback at lower frequencies. The trade of the excessive phase margin for larger low-frequency loop gain can be made by introducing lag links.

5.5 Complex poles

Upper and lower Bode steps need to be reasonably sharp. To implement these sharp angles on the Bode diagram, complex poles and zeros are required. Complex poles can be also required to compensate for the plant response, and to shape the loop gain over the functional frequency band.

The normalized gain and phase lag frequency responses for the complex pole pair function

$$\frac{\omega_o^2}{s^2 + Q^{-1}\omega_o s + \omega_o^2} = \frac{\omega_o^2}{s^2 + 2\zeta\omega_o s + \omega_o^2} \tag{5.4}$$

are presented in Fig. 5.11. The magnitude of the function (5.4) at the resonance where $s = j\omega_o$ and the first and the last terms cancel each other is the ***quality factor*** $Q = 1/(2\zeta)$ where ζ is the ***damping coefficient***.

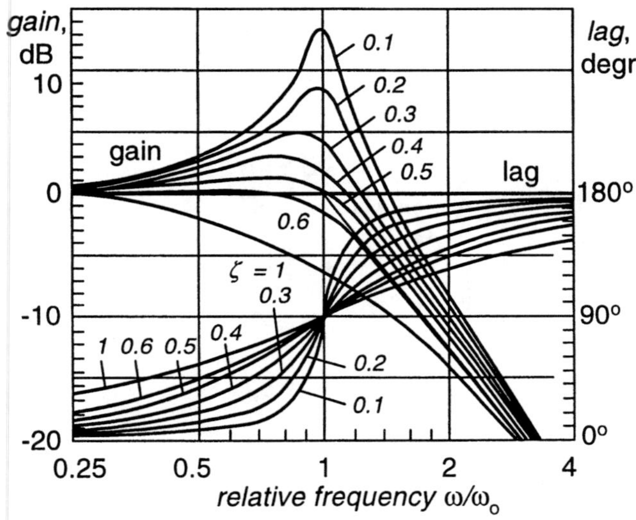

Fig. 5.11 Gain and phase lag responses for a complex pole pair

The asymptotic Bode diagram is calculated by retaining only the last term in the denominator of (5.4) at $\omega < \omega_o$, and only the first term at $\omega > \omega_o$.

Example 1. A simple low-pass filter can be obtained by using transfer function (5.4) with $\zeta = 0.5$.

Example 2. A third-order low-pass filter can be obtained by cascading a single-pole link and a link with transfer function (5.4). The frequency of the complex pole must be higher than the frequency of the real pole, and ζ should be chosen such that the peak of the complex poles compensates the roll-off of the real pole. This method will be used to form the Bode step in Fig. 5.14.

Example 3. The difference between the two logarithmic responses in Fig. 5.11 is a notch or a peak response. Its transfer function is the ratio of functions (5.4):

$$\frac{s^2+2\zeta_n\omega_o s+\omega_o^2}{s^2+2\zeta_d\omega_o s+\omega_o^2}.$$

This function equals to 1 at zero and infinite frequencies, and to ζ_n/ζ_d at the frequency of the resonance where $s = j\omega_o$. When $\zeta_n < \zeta_d$, a notch response results. When $\zeta_n > \zeta_d$, a peak response follows. The width of the notch or the peak depends on the chosen damping. Such notches have been used in the prefilter described in Section 4.2.3.

5.6 Cascaded links

When the elementary links of the compensator are cascaded, attention should be paid to the signal level at the link junctions so as not to impair the compensator's ***dynamic range***. This is the amplitude range of the signals the largest of which is not yet distorted by saturation in the nonlinear links, and the smallest is still substantially larger than the disturbance and noise mean-square amplitudes.

Example 1. Consider the implementation of the transfer function

$$W(s) = W_1(s)W_2(s) = \frac{s+2}{s+1000}\frac{s+500}{s+10} \tag{5.5}$$

as a cascade connection of the two links. The asymptotic gain responses for W_1 and W_2 are shown in Fig. 5.12(a). It is seen that the signals at lower frequencies are attenuated in the first link by 54 dB, and then amplified in the second link by 34 dB. This way of making the compensator is certainly not the best since after the attenuation, the signal drops dangerously close to the noise level, and after the amplification the noise floor will be raised.

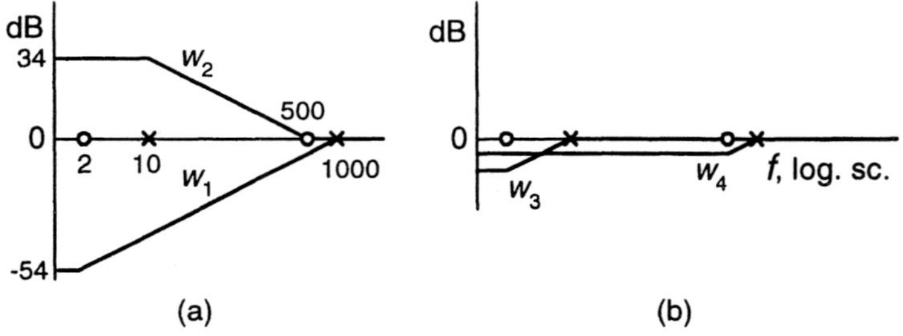

Fig. 5.12 Gain responses of two different implementations of the same transfer function

Assume that a 1 mV signal with various frequencies is applied to the input as shown in Fig. 5.13(a). At the junction between the links, the signal levels at 1 Hz and at 1 kHz differ, as shown in Fig. 5.13(a). Assume also that there is a 5 μV disturbance source at the junction of the links. Such disturbances may be caused by noise or interference in analog systems, and by round-off errors in digital systems. At 1 Hz, the signal amplitude is only 2 μV so that the signal will be heavily corrupted with the noise.

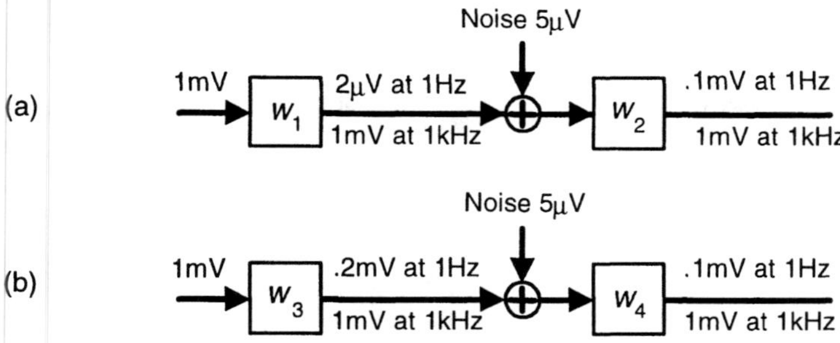

Fig. 5.13 Signal levels at link junctions for the responses in Fig. 5.12

It is better to implement the same transfer function by cascading the following links:

$$W(s) = W_3(s)W_4(s) = \frac{s+2}{s+10}\frac{s+500}{s+1000} \ . \tag{5.6}$$

The frequency responses for W_3 and W_4 are shown in Fig. 5.12(b). The signal levels have a much smaller dynamic range as indicated in Fig. 5.13(b), and even the smallest signal amplitude (0.2 mV at 1 Hz) remains much larger than the 5 μV noise.

The general rule is to avoid creating links with excessive attenuation or gain at any frequency, i.e., to keep in the same link the poles and the zeros which are close to each other. When this rule is followed, the link affects the slope of the total Bode diagram over a relatively small frequency range – which also simplifies iterative adjustments of the frequency responses.

Example 2. The plant is a single integrator $1/s$. The loop transfer function must behave as a single integrator at zero frequency. The gain and phase stability margins must be not less than 10 dB and 30°. The crossover frequency must be not less than 0.9 rad/sec, but the loop gain at frequency 10 rad/sec and higher must not exceed –35 dB. For this, the roll-off at higher frequencies must be –18 dB/oct or steeper, and the loop response must include a Bode step.

Let us design the compensator as a cascade connection of several links.

When $\omega_b = 1$ and $x = 10$ dB, then $\omega_d = 2$. The desired width of the Bode step (4.2) is $0.6 \times 3 = 1.8$, i.e., $\omega_c \approx 3.6$.

The asymptotic Bode diagram shown in Fig. 5.14 is composed of pieces with slopes – 6 dB/oct, –12 dB/oct, 0 dB/oct, and –18 dB/oct. At lower frequencies, it makes the slope – 6 dB/oct. The average slope is –10dB/oct at frequencies 0.15 to 2 rad/sec.

A pair of complex zeros at ω_c make the corner at the beginning of the Bode step.

A real pole at $\omega = 2.8$ and a pair of complex poles at a frequency somewhat smaller than ω_c form a third-order low-pass filter (as explained in Example 2 in Section 5.5) at the end of the Bode step and effect the desired asymptotic slope, –18 dB/oct.

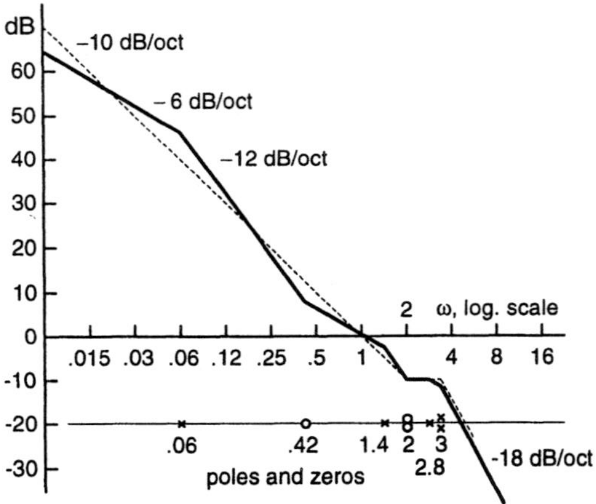

Fig. 5.14 Asymptotic Bode diagram

Damping coefficient of 0.5 is chosen for the complex zeros in order for the gain response to pass through the corner point as seen in Fig. 5.11. Damping coefficient of 0.4 is chosen for the complex poles to compensate for the rounding effect of the real pole at $\omega = 2.8$. The resulting loop transfer function is

$$T(s) = k\frac{1}{s+0.06}\ \frac{s+0.42}{s+1.4}\ \frac{2.8}{s+2.8}\ \frac{s^2+2s+4}{s^2+2.4s+9}\ \frac{1}{s}.$$

The compensator is composed of four first- or second-order links. This example shows that *even when the order of the plant transfer function is low, the order of the compensator transfer function must be reasonably high in high-performance controllers*.

From the condition that at $\omega = \omega_b = 1$ the asymptotic loop gain coefficient

$k\,(1\times1\times1\times2.8\times4\times1)/(1\times1.4\times2.8\times9\times1) = 1,$

the coefficient $k \approx 3.15$. By using this value of k initially, it was found that for the Bode diagram to pass close to the 0 dB level at $\omega = 1$, k must be increased to 4.

After multiplication of the polynomials in the numerator and denominator:

```
n = 4 * 2.8 * conv([1 0.42],[1 2 4])
   n = 11.2000    27.1040    54.2080    18.8160

d = conv(conv([1 0.06],[1 1.4]), conv([1 2.8],[1 2.4 9 0]))
   d = 1.0000  6.6600  23.3960  48.5880 38.1125  2.1168  0
```

the loop transfer function becomes

$$T(s) = \frac{11.2s^3+27.1s^2+54.2s+18.8}{s^6+6.66s^5+23.4s^4+48.6s^3+38.1s^2+2.12s}.$$

The loop responses are shown in Fig. 5.15(a). The Nyquist diagram on the L-plane plotted by

```
w = logspace(-1, 1);    [mag, phase] = bode(n, d, w);
```

```
plot(phase,20*log10(mag),'r', -180, 0,'wo')
title('L-plane Nyquist diagram')
set(gca,'XTick',[-270 -240 -210 -180 -150 -120 -90])
grid
```

is shown in Fig. 5.15(b).

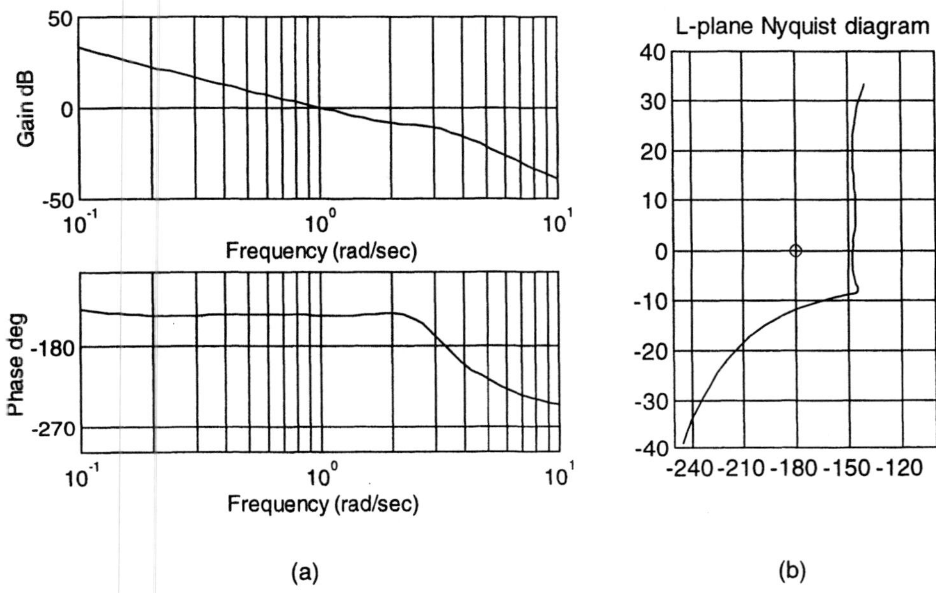

Fig. 5.15 Loop frequency response (a) and the Nyquist diagram (b)

It is recommended for students to play with this response, to modify it by changing the poles and zeros (or the coefficients of the polynomials) in order to get the feeling for the sensitivity of the response to the poles, the zeros, and the polynomial coefficients. For example, by increasing the complex pole damping coefficient the Nyquist diagram can be made more rounded; an n.p. lag can be added to the plant and the Bode step made, correspondingly, wider; the high-frequency asymptotic slope can be made steeper by adding complex poles or a notch, which will also require lengthening the step; the gain response at lower frequencies can be made steeper. The response can be shifted along the frequency axis by replacing s by as (as described in Section 4.2.3, Example 2), and along the gain axis, by multiplying the function by a constant.

5.7 Parallel connection of links

The compensator may be implemented also by connecting several links in parallel as shown in Fig. 5.16. The compensator transfer function is equal to the sum of the transfer functions of the elementary links $W_1 + W_2 + W_3$. The poles of the compensator are just the poles of the elementary links. The zeros of the compensator result from the interactions between the elementary links: the output is 0 at that value of s at which the sum of all the links' outputs is 0.

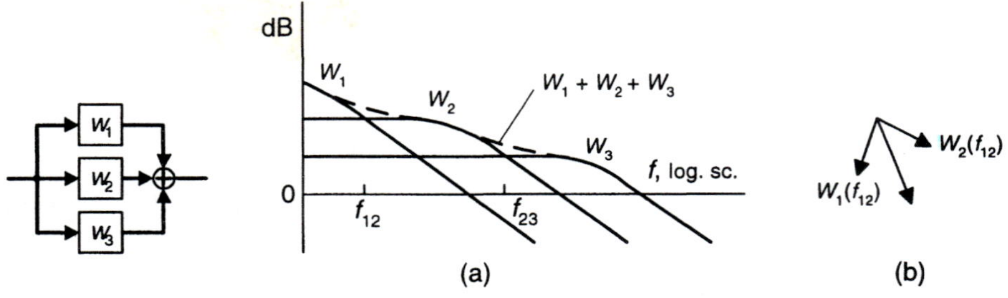

Fig. 5.16 Parallel connection

Fig. 5.17 (a) Parallel links' gain responses (b) output signals' addition at f_{12}

It is convenient to design the compensator such that each one of the parallel links dominates the response over a certain frequency band, as shown in Fig. 5.17(a). This way the links can be designed and adjusted one at a time. (This configuration provides also for an option of placing separate nonlinear links in the parallel paths, if required, as described in Chapters 11 and 13.) At the frequency at which the link responses cross, (i.e., at f_{12}, f_{23}, etc.,) the output can be found by vector addition of the links' transfer functions. Depending on the phase difference between the output signals of frequency-adjacent channels, the summed signal amplitude may be larger or smaller than the amplitude of the components.

Parallel links provide a convenient way of implementing Bode steps.

Example 1. In the system shown in Fig. 5.18(a), the plant is an integrator $1/s$. The crossover frequency must be 1 rad/sec.

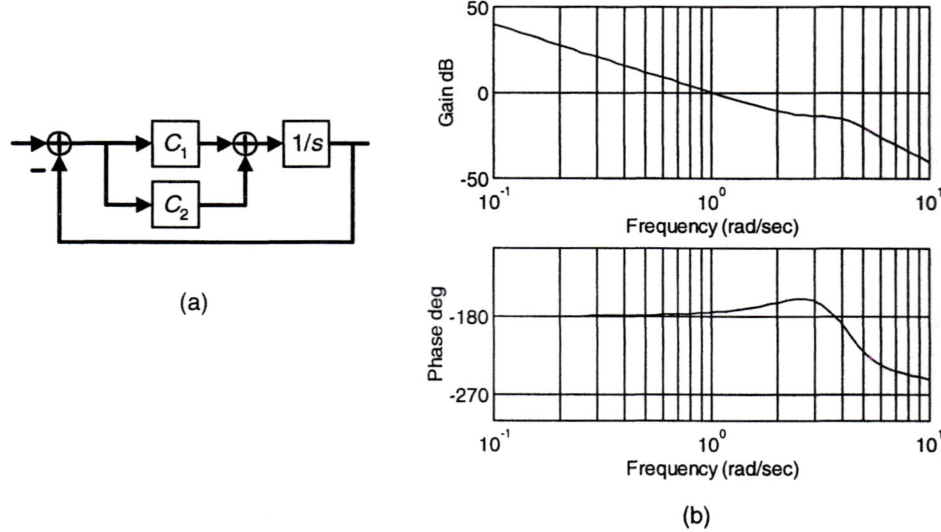

Fig. 5.18 (a) Bode step implementation and (b) the Bode diagram

The compensator is the parallel connection of the link $C_1 = 4/(s^2 + 4s)$ and the low-pass filter $C_2 = 5/(s^2 + 2.4s + 16)$ with $\zeta = 0.3$. In the filter, the resonance is at four times the crossover frequency, and the low-frequency gain is –10.1 dB. The loop response

$$T(s) = \frac{C_1(s) + C_2(s)}{s} = \frac{9s^2 + 29.6s + 64}{s^5 + 6.4s^4 + 25.6s^3 + 64s^2}$$

is shown in Fig. 5.18(b). It is seen that the Bode step is well implemented, but the phase lag at lower frequencies is too large. This can be remedied with lead links placed in C_1 or in the common path.

5.8 Simulation of a PID controller

The ***PID controller*** consists of three parallel branches: *I/s*, *P*, and *Ds*. The coefficients *I*, *P*, *D* define the transfer function of the controller. A saturation link is commonly placed in front of the *I*-channel to improve the controller performance in the nonlinear mode of operation, for large-level signals (this issue will be considered in Chapters 10 – 13). This controller does not implement a Bode step, is not optimal in most applications, but is simple and as such, quite popular.

We will consider an example with a double integrator plant with a flexible mode. The system block diagram is shown in Fig. 5.19. The plant includes two masses, *CP1* and *CP2*, a spring, and a dashpot (a device providing viscous friction).

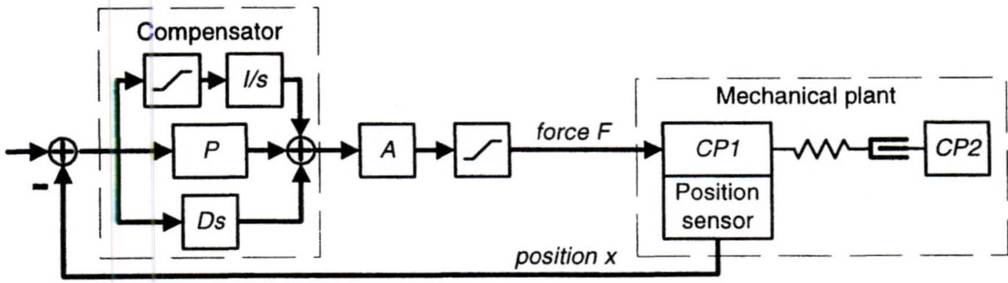

Fig. 5.19 Block diagram for a control loop a *PID* compensator and a double integrator plant having a flexible mode

The simulation can be performed in MATLAB, SIMULINK, or SPICE. When MATLAB is used, the transfer function for the plant should be first found. SIMULINK and SPICE allow the transfer function derivation to be skipped (using SIMULINK block diagrams for ladder network analysis is described in Section 7.6.1).

Example 1. In this example, we will use SPICE. In spite of a somewhat longer input file and the necessity to draw an equivalent schematic diagram, using SPICE has certain advantages: there is no need to generate a mathematical description of the plant if we use the common electromechanical analogies (force to current, velocity to voltage, mass to capacitor, spring to an inductor with the inductance equal to the inverse of the spring stiffness coefficient, and the dashpot to a resistor with the resistance equal to the inverse of the damping coefficient; the analogy is explained in detail in Section 7.1.1). The schematic diagram for the simulation is shown in Fig. 5.20. The current to node 13 represents the force *F*. The position *x* (voltage at node 2) is calculated from the plant velocity (voltage at node 13) by integration.

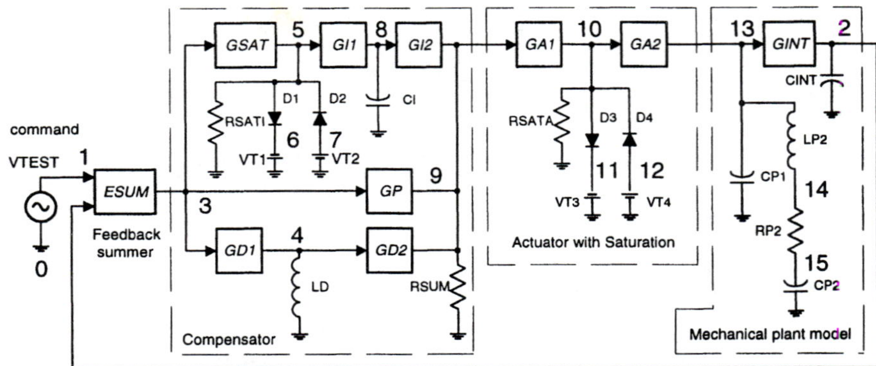

Fig. 5.20 SPICE model for the system shown in Fig. 5.19

The output currents of the three parallel paths in the compensator pass through the 1 Ω summing resistor RSUM, and the voltage on the resistor represents the sum of the outputs of the parallel paths. The integrators are imitated by ideal controlled current sources loaded into capacitors. Saturation links are implemented using opposite-biased diodes shunting a resistive load.

This diagram represents a simple way of simulating the performance of the block diagram in Fig. 5.19 in SPICE, and it does *not* describe the implementation of an analog compensator (compensator implementation will be considered in the next chapter).

The SPICE input file is shown below. Included in the file but not shown on the picture are the high-resistance leakage resistors connecting nodes 1, 2, 8, and 13 to ground, as required by SPICE.

```
*** PID example Figs. 5.19, 5.20 ***
ES    3 0 1 2 1              ; input signal summer
RSR1  1 0 1MEG               ; leakage resistor
RSP2  2 0 1MEG               ; leakage resistor
***
GSAT  5 0 0 3 0.001          ; saturation in I-path
RSAT  5 0 1K                 ; threshold = (0.7+VT1)*GI1/GSAT
D1    5 6 DIODE
D2    7 5 DIODE
 .MODEL DIODE D
VT1   6 0 1V
VT2   0 7 1V
***
GI1   8 0 0 5 1               ; I-path
CI2   9 0 0 8 10              ; integral coefficient
RSP8  8 0 1MEG                ; leakage resistor
***
GP    9 0 0 3 2               ; proportional coefficient
***
GD1   4 0 0 3 3               ; differential coefficient
LD    4 0 1
GD2   9 0 0 4 1
***
RS    9 0 1                   ; summing resistor
***
```

```
GA1   10 0 9 0 1             ; actuator gain coefficient
***
RSATA 10 0 1K                ; saturation in actuator
D3    10 11 DIODE
D4    12 10 DIODE
VT3   11 0 1V
VT4   0 12 1V
GA2   13 0 0 10 1            ; force source
***
CP1   13 0 5                 ; mass of the main body
RSP13 13 0 1MEG              ; leakage resistor
LP2   13 14 0.1              ; spring of flexible mode
CP2   12 0 0.5               ; mass of second body
RP    14 15 0.02             ; losses in the flexible mode
GINT  2 0 0 13 1
CINT  2 0 1                  ; integrator to generate position
***   V10 is force, V13 is velocity, V2 is position
***
VTEST 1 0 AC 1               ; use only when frequency responses
                             ; are tested
 .AC DEC 20 0.01 10 ; use only when frequency responses tested
** Pulse            (Vmin Vmax delay rise fall width period)
* VPULSE 10 PULSE ( 0V  10V   0S    0S    0S   500    500 )
                             ; when transient responses tested
* .TRAN 0.1 10               ; when transient responses tested
 .PROBE                      ; or other graphical postprocessor
 .END
```

Since the number of the nodes is only 15, this system can be simulated using a student version of SPICE which presently allows up to 25 nodes and available free of charge from Intusoft, or PSPICE® from Microsim. (In SPICE, the summers and nonlinearities can also be specified by algebraic expressions; in most versions of SPICE, the transfer functions can also be specified by their poles and zeros.)

To plot the closed-loop response, plot `vdb(2), vp(2)`.

To plot the loop response, connect the `ESUM` second input to 0 and then plot `vdb(2),vp(2)`.

To plot transient response, disable with an asterisk lines `VTEST` and `.AC` and enable `VPULSE` and `.TRAN`.

To plot the Nyquist diagram with logarithmic scales, change the abscissa scale to linear and make it `vp(2)`.

Running the program is recommended as a student exercise. The plots should be made for:

(a) frequency response of loop gain and phase so that stability margins can be checked;

(b) closed-loop frequency response; responses of the linear links (as the difference between the input and output signal levels and phases),

(c) transient closed loop response for different coefficients *P*, *I*, *D* and saturation thresholds.

The loop gain response should be similar to that shown in Fig. 5.21.

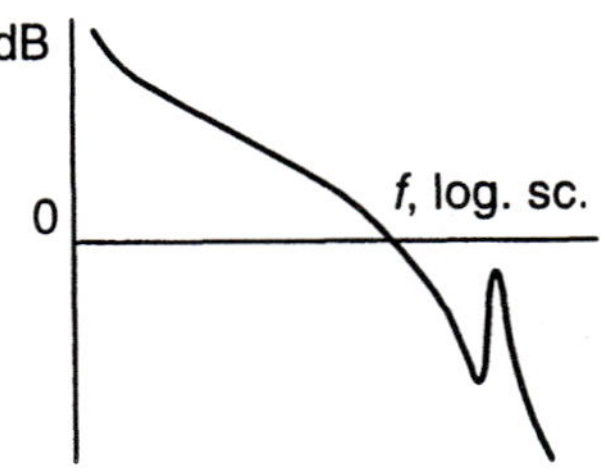

Fig. 5.21 Loop gain response

The loop response for *PID* controllers can be augmented by additional low-pass filters and notches, to reduce the loop gain at the frequencies of the plant structural resonance and provide gain stabilization over the frequency range of the resonance.

Example 2. A triple-pole low-pass filter (two complex poles and one real) with the cut-off frequency, approximately, 4 times larger than the crossover frequency, is placed in front of a single integrator plant. The coefficient D is chosen such that with only this path of the compensator connected, the loop gain at lower frequencies is $-x$ dB. The P coefficient is chosen such that with only this path on, the crossover frequency is where it must be. The integral term can be chosen such that with only this path on, the loop gain at the crossover frequency is $-x$ dB. The resulting response has the high-frequency asymptotic slope −18 dB/oct, and the lower-frequencies loop gain is lower than that of the response with the crisp Bode step by about 5 dB.

Drawing the asymptotic Bode diagrams and making MATLAB simulation is left as a recommended student exercise.

5.9 Analog and digital controllers

While many factors may affect the choice between an analog and a digital system, three important considerations are accuracy, bandwidth, and price.

When considering accuracy, it should be remembered that only the accuracies of the prefilter, summer, and feedback path link directly affect the output accuracy. The accuracy required of the compensator and also of command feedforward links is much lower, so generally these can be implemented with less accurate elements.

Analog circuitry can be made very accurate and stable in time. For example, the analog reference voltage source, resistors, and chopper-stabilized amplifiers inside bench-type digital voltmeters are accurate to 7 or 8 digits. The dynamic range of a common op-amp is 1 μV to 10 V. This dynamic range is equivalent to 23 bits.

In digital systems, the accuracy and the dynamic range are frequently limited by the dynamic range of the employed A/D converter which is, typically, 12 to 23 bits. An additional principal drawback of digital controllers is that the sample rate (typically, up to 1 MHz) and computational delay limit the bandwidth and the available feedback. Because of this, feedback loops with $f_b > 100$ kHz should be analog.

Other considerations are: the type of sensors used, the prices of a microcontroller and of D/A and A/D converters, power consumption, and ease of troubleshooting. Another important issue is the compensator's flexibility in being adjusted for different plants.

For the great majority of applications, either type of controller, analog or digital, can be designed such that it will perform well, and the choice between the two is determined mostly by the price of development and fabrication.

5.10 Digital compensator design

5.10.1 Discrete trapezoidal integrator

As is explained in Section 5.6, for the purpose of reducing the required dynamic range, it is advantageous to break the compensator into several cascaded links, each link related to different time-constants, i.e., different frequency regions. For a similar reason, to reduce the rounding errors, it is common in digital signal processing (DSP) to implement a high-order transfer function as a cascade connection of first-order and second-order functions, each such function having poles reasonably close to zeros.

The second-order transfer function

$$\frac{p_2 s^2 + p_1 s + p_o}{s^2 + q_1 s + q_o}, \tag{5.7}$$

or **biquad**, can be implemented using analog integrators $1/s$ with the feedback block diagram shown in Fig. 5.22 (verification of the correspondence of the equation to the block diagram is left as a student exercise).

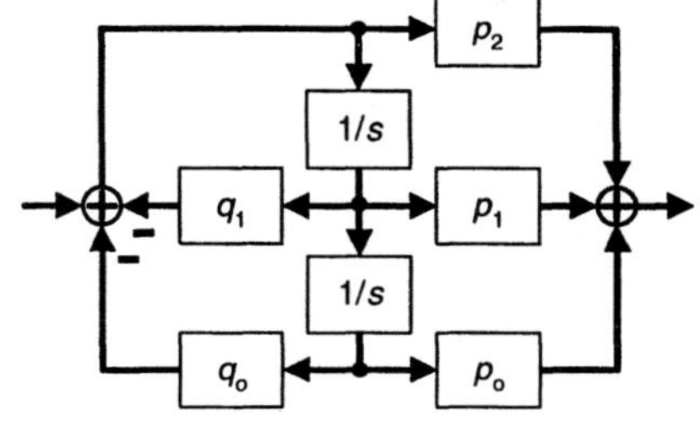

Fig. 5.22 Feedback implementation of a biquad transfer function

A similar block diagram can be used for digital implementation of the biquad operator, only the integration must be implemented in discrete steps performed at sampling instants.

Discrete trapezoidal integration is shown in Fig. 5.23. The input analog signal $u(t)$ is sampled at time intervals T_S, $2T_S$, $3T_S$ ... . The output $v(t)$ represents the shadowed area which approximates the integral of the input. The increment in $v(t)$ for the time interval from the sampling $\# n-1$ to the sampling $\# n$ is

$$v_n - v_{n-1} = \frac{u_{n-1} + u_n}{2} T_S . \tag{5.8}$$

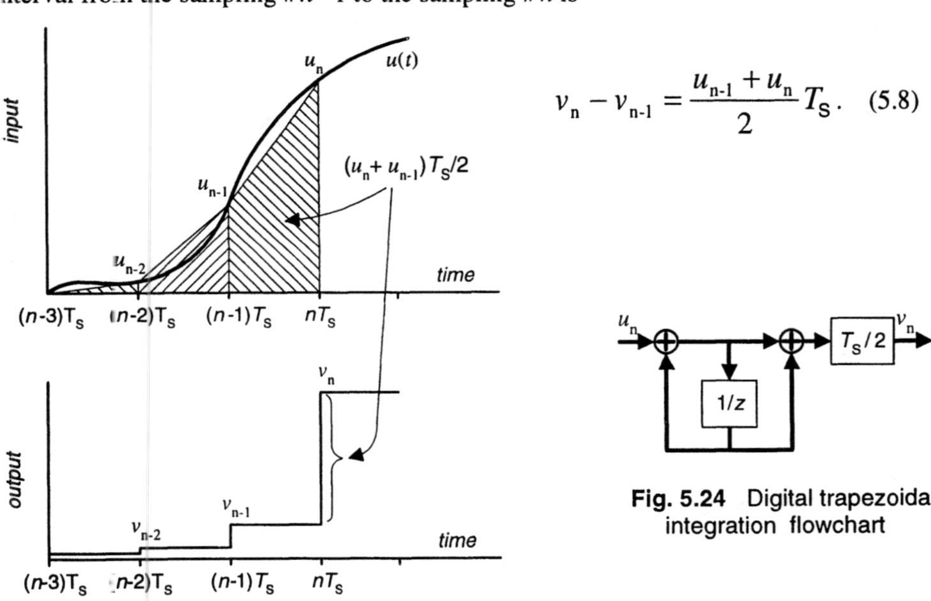

Fig. 5.23 Trapezoidal digital integration

Fig. 5.24 Digital trapezoidal integration flowchart

Next, let multiplication by z signify the time increase by one sample period so that

$$z v_{n-1} = v_n ,$$
$$z u_{n-1} = u_n .$$

With this symbolism, (5.8) can be rewritten as

$$v_n - z^{-1}v_n = \frac{z^{-1}u_n + u_n}{2}T_S$$

and the transfer function of the discrete trapezoidal integrator is

$$\frac{v_n}{u_n} = \frac{1+z^{-1}}{1-z^{-1}}\frac{T_S}{2}. \tag{5.10}$$

This formula is presented as a flowchart in Fig. 5.24. Operator z^{-1} can be implemented by storing a sampled value and recalling it after the sampling time T_S.

The biquad (5.7) can be implemented digitally by replacing integrators $1/s$ in the flowchart in Fig. 5.22 by digital integrators. An equivalent but simpler flowchart can be obtained by substituting (5.10) into (5.7) to replace $1/s$, and simplifying the resulting expression to obtain a function of z as a ratio of two second-order polynomials. The resulting flowchart looks like the block diagram in Fig. 5.22 but with different coefficients and with $1/z$ replacing $1/s$.

Next, three important comments need to be made.

- As seen in Fig. 5.23, the input signal value averaged over the interval of sampling is calculated by discrete integration only at the end of the sampling interval, while analog integrator output reaches the average value in the middle of the sampling interval. In other words, compared with analog integration, digital integration delays the signal by $T_S/2$.
- While a digital feedback system is being designed, the delay $T_S/2$ can be treated as nonminimum phase lag. As mentioned in Sections 4.2.2 and 4.3.4, this lag should not exceed 1 rad at the crossover frequency. Therefore, the ***sampling frequency*** $f_S = 1/T_S$ should be at least $2\pi/1 \approx 6$ times larger than f_b. With such a high sampling frequency, trapezoidal integration is quite accurate.
- Since it needs at least 2 samples per period to identify a sinusoid, digital signal processing is only feasible for sinusoidal signals with frequencies less than the ***Nyquist frequency*** $f_S/2$.

5.10.2 Laplace and Tustin transforms

To get an additional insight into the problem, we might duplicate the results of the previous section using the Laplace transform. Since $1/z$ signifies a delay by T_S, we can view $1/z$ as the Laplace transform of the delay which is

$$z = \exp\frac{s}{f_S}. \tag{5.11}$$

From here,

$$s = f_S \ln z. \tag{5.12}$$

Function (5.11) maps the strip of width $2\pi f_S$ of the left-hand plane of s, bounded by the dash-dotted lines in Fig. 5.25(b), onto the unit radius disk of the z-plane shown in Fig. 5.25(a). The origin of the s-plane maps onto the point (1,0) of the z-plane, and the points $-j\pi f_S$ and $j\pi f_S$, onto the point (–1,0) of the z-plane.

The Nyquist frequencies $\pm\, i\pi f_S/2$ of the s-plane map onto points $\pm\, j$ in the z-plane.

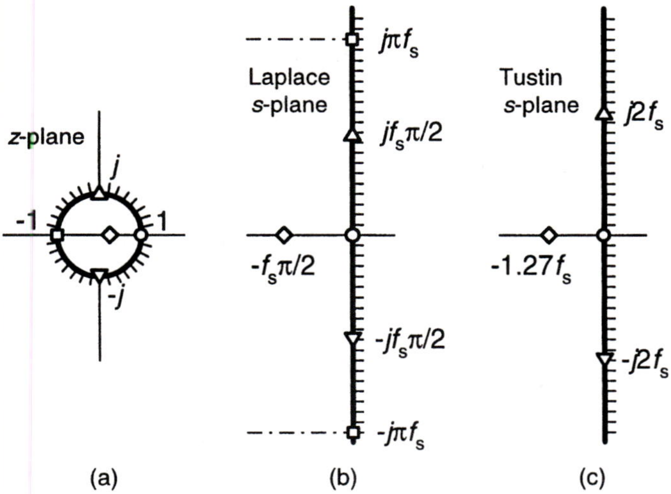

Fig. 5.25 Mapping of (a) the z-plane onto the s-plane with (b) function $s = f_S \ln z$ and with (c) Tustin transform $s = 2f_S(z-1)/(z+1)$

Near the origin, an exponent can be approximated well by a bilinear function, i.e., when x is small, $\exp(2x) \approx (1+x)/(1-x)$. In applications to DSP, this approximation is known as the ***Tustin transform***,

$$s = 2f_S \frac{z-1}{z+1} \tag{5.13}$$

i.e.,

$$z = \frac{2f_S + s}{2f_S - s}. \tag{5.14}$$

From (5.13), the expression for the integrator $1/s$ is

$$\frac{1}{s} = \frac{1+z^{-1}}{1-z^{-1}} \frac{T_S}{2}.$$

This expression is the same as (5.10) signifying that the Tustin transform uses trapezoidal integration. The Tustin transform shown in Fig. 5.25(c) maps the entire left half-plane of s onto the unit radius disk in the z-plane as shown in Fig. 5.25(a).

In practical control systems $f_b < 0.1f_S$ for the reasons cited in the ends of Sections 5.10.1 and 5.10.6. Therefore, at all frequencies below f_b the design uses only a small part of the first quadrant of the mapping in Fig. 5.25(a). Within this part of the quadrant, the Tustin transform is quite accurate. At frequencies of the Bode step, from $2f_b$ to $4f_b$, the accuracy of the Tustin transform, although decreased, still remains adequate since the required accuracy of the loop gain response implementation over this frequency range also decreases.

In fact, in application to compensator design, there is not much difference between the Laplace and Tustin transforms, and either one can be used. The Tustin transform is used most frequently. The Tustin transform can be matched to the Laplace transform at a specified frequency by ***pre-warping***. The advantages of using pre-warping are, however, insignificant and the method will not be discussed here.

It is seen from (5.14) that *z remains the same when s and f_S are similarly scaled up or down*; for example, when f_S is expressed in kHz, s must be expressed in krad/sec.

For the purpose of verifying the equations, the following evident properties of the z-transform can be used:

- a transfer function of s for $s = 0$, is equal to the transfer function of z for $z = 1$;
- a transfer function of s for $s = \infty$, is equal to the Tustin transform of z for $z = -1$;
- multiplying a function of s by a constant is equivalent to multiplying the function z by the same constant;
- the z-transform can be also found by mapping the poles and zeros of the function of s onto the z-plane, and then appropriately scaling the gain coefficient.

Example 1. The Tustin transform of the lead

$$\frac{s+5}{s+10}$$

with sampling frequency 100 Hz can be found by substituting s with expression (5.13). Table 5.1 in Section 5.10.4 gives the algebraic expressions for the coefficients of the function of z. Numerically, $m = 210$, $a_1 = 0.9762$, $a_o = -0.9286$, and $b_o = -0.9048$. The function of z is

$$\frac{0.9762 - 0.9286 / z}{1.0 - 0.9048 / z}.$$

Example 2. The same problem is solved with MATLAB by

```
n = [1 5];
d = [1 10];
fs = 100;                          % sampling frequency in Hz
[nd,dd] = bilinear(n,d,fs)

nd =   0.9762   -0.9286
dd =   1.0000   -0.9048
```

Example 3. The Tustin transform properties for the numeric values of the previous example are verified in the following.

For $s = 0$, the function of s is 0.5; for $z = 1$, the Tustin transform is also 0.5.

For $s = \infty$, the function of s is 1; for $z = -1$, the Tustin transform is also 1.

The zero of the function of s is -5, and the pole is -10. With Tustin transform (5.14), the zero and the pole of the function of z are found to be, respectively,

$$z_z = \frac{2f_S + s}{2f_S - s} = \frac{200-5}{200+5} = 0.95122\ ,$$

$$z_p = \frac{2f_S + s}{2f_S - s} = \frac{200 - 10}{200 + 10} = 0.904762\,.$$

Due to the finite accuracy of the calculations, the calculated values of the pole and the zero differ somewhat from the exact values. The errors caused by the calculation errors and by the fact that the Tustin transform is only an approximation to the Laplace transform, do not matter much for the digital functions in the forward path of the feedback loop. However, the error might not be acceptable for the links in the feedback path or in the prefilter. In this case, a good option is to use instead the command feedforward.

5.10.3 Design sequence

Digital control systems can be designed as follows:

- Given the required feedback bandwidth f_b, the sampling frequency f_S is chosen; commonly, $f_S \geq 10f_b$, and for better performance, $f_S \geq 50f_b$.
- By approximation of the optimal response, a rational transfer function of the analog compensator is determined and broken appropriately into second-order rational functions (as described in Section 5.5).
- For simulation and tuning in analog form, MATLAB, SIMULINK, or SPICE can be used. The sampling delay $T_S/2$ is imitated by introducing in the loop an all-pass with pole-zero pair $(s - f_S)/(s + f_S)$ or by placing a pole at frequency $f_S/3$.
- The Tustin transform is used to find functions of z which correspond to the second-order functions of s.
- The system is simulated using a digital control software package (for example, SIMULINK), or directly in C or in another implementation language.
- The functions of z are coded in, and the system is tested or simulated on a computer (including the nonlinearities).
- The stability margins are verified by increasing the loop gain and loop phase lag up to the moments when self-oscillation starts.

5.10.4 Block diagrams, equations, and computer code

The functions of z can be specified by block diagrams, by equations, or by computer code. The block diagrams for the first-order and the second-order functions are shown in Fig. 5.26. These block diagrams have forward and feedback paths.

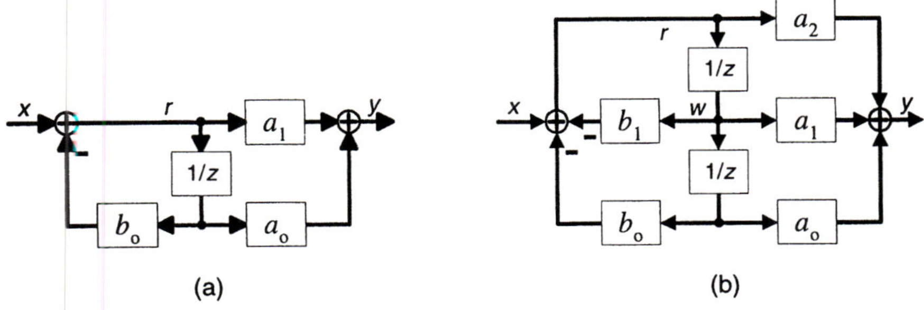

Fig. 5.26 Block diagrams for (a) the first-order link, (b) the second-order link

Using Mason's rule, the transfer function for the block diagram (a) is found to be

$$\frac{a_1 + a_o z^{-1}}{1 + b_o z^{-1}} \qquad \text{or} \qquad \frac{a_1 z + a_o}{z + b_o} \tag{5.15}$$

and for the block diagram (b),

$$\frac{a_2 + a_1 z^{-1} + a_o z^{-2}}{1 + b_1 z^{-1} + b_o z^{-2}} \qquad \text{or} \qquad \frac{a_2 z^2 + a_1 z + a_o}{z^2 + b_1 z + b_o}. \tag{5.16}$$

The coefficients a_i, b_i can be obtained by substituting (5.13) into the expressions for the transfer functions of s. For the function of s

$$\frac{p_1 s + p_o}{s + q_o},$$

which is bilinear, the coefficients are given in Table 5.1, where $m = 2f_S + q_o$.

Table 5.1 Coefficients for bilinear function of z

a_1	a_o	b_o
$(p_1 2f_S + p_o)/m$	$(-p_1 2f_S + p_o)/m$	$(q_o - 2f_S)/m$

For the biquad of s

$$\frac{p_2 s^2 + p_1 s + p_o}{s^2 + q_1 s + q_o}$$

the coefficients for the biquad of z are given in Table 5.2.

Table 5.2 Coefficients for biquad of z

a_2	a_1	a_o
$(4f_S^2 p_2 + 2f_S p_1 + p_o)/n$	$(-8f_S^2 p_2 + 2p_o)/n$	$(4f_S^2 p_2 - 2f_S p_1 + p_o)/n$
b_1	b_o	n
$(-8f_S^2 + 2q_o)/n$	$(4f_S^2 - 2f_S q_1 + q_o)/n$	$4f_S^2 + q_1 2f_S + q_o$

As already mentioned, the coefficients of the rational function of z can be calculated from the rational function of s with given numerator `num` and denominator `den` by the MATLAB function

```
[numd, dend] = bilinear(num, den, fs)
```

where `fs` is the sampling frequency. To verify the transform, frequency responses of digital filters can be found with the MATLAB command `freqz`.

For the first-order link in Fig. 5.26(a), the following C code equations can be used, with a_o, a_1, b_o renamed `A0`, `A1`, `B0`:

```
y = A0 * r;          /* using previous value of r to find **
                     ** first component of the output */
r = x - B0 * r;      /* updating r */
```

```
y += A1 * r;              /* adding second component to **
                          ** the output using updated r */        (5.17)
```

The variables are recalculated each cycle. The cycle starts with the new sample value of the input. First the value of the variable `r` that is stored in the previous cycle is used, then this variable is updated. The cycle repeats f_S times per second, controlled by some loop. The variables are commonly initialized to zeros, and they must be either static or global to keep the values stored to be used in the next cycle.

For the second-order link, similarly, the following code can be used:

```
y = A0 * w + A1 * r;          /* using previous values of r,w */
r1 = x - B0 * w - B1 * r;       /* updating r */
w = r;                          /* updating w */           (5.18)
r = r1;                         /* updating r */
y += A2 * r;                /* adding component to the output */
```

5.10.5 Compensator design example

A small parabolic telecommunication link antenna tracking the Earth has been placed on the Mars Pathfinder Lander. Two identical brushless motors with internal analog rate feedback loops articulate the antenna in two orthogonal directions. The motors are controlled by two independent identical SISO controllers.

The sampling frequency is 8 Hz. Were the delay caused only by the sampling, the crossover frequency would be $f_S/5 \approx 1.6$ Hz. However, since the computer must handle not only the motor control loops but also other higher priority tasks, there is an additional 500 msec delay caused by four real time interrupt (RTI) delays, 125 msec each. Also, due to limited bandwidth of the analog rate controllers for the motors (already designed), the motors have 50 msec delay. Since the total delay is not only 62.5 msec (of sampling) but $62.5 + 500 + 50 \approx 600$ msec, the realizable crossover frequency is lower in proportion to this delay, i.e.,

$$f_b < 1.6\times 62.5/600 \approx 0.17\,\text{Hz}.$$

The design has been done initially in the s-domain. The controller is nonlinear, and includes two cascaded linear links, $C_1 + 1$ and C_2. A saturation link placed in front of C_1 makes the transfer function of the compensator dependent on the signal level. When the signal level is below the saturation threshold, the compensator transfer function is $(C_1 + 1)C_2$. When the signal is high, the compensator transfer function is reduced to C_2. The operation of such nonlinear dynamic compensators (NDCs) will be further described in Chapters 10 and 13.

For small-signal amplitudes, the compensator function is

$$C = (C_1 + 1)\, C_2$$

where C_1 is a single-pole low-pass filter,

$$C_1 = 2.5/(0.0833 + s),$$

and C_2 is a lead link,

$$C_2 = (0.106 + s)/(2.23 + s).$$

The asymptotic gain frequency responses of the compensators are shown in Fig. 5.27.

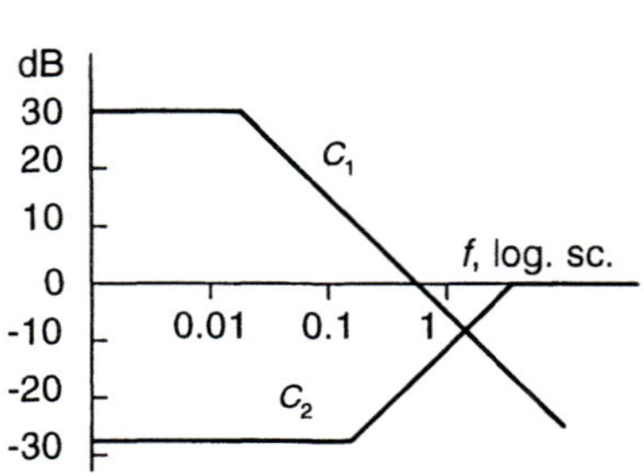

Fig. 5.27 Asymptotic Bode diagrams of compensators

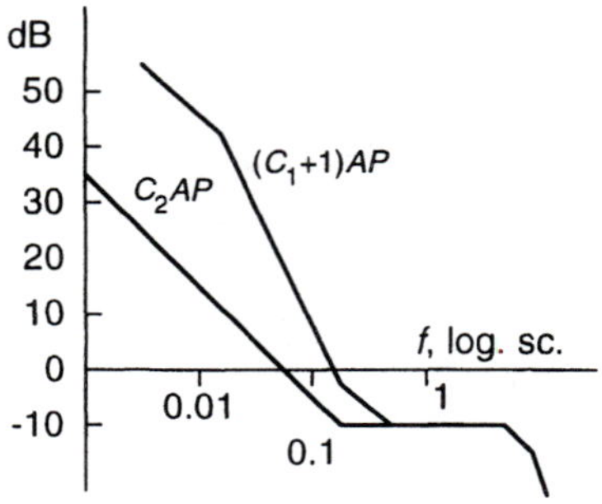

Fig. 5.28 Open-loop asymptotic Bode diagrams for small error (upper curve) and large error (lower curve)

The lead C_2 provides phase advance and partially compensates the following lags: the phase lag of up to 7 RTI (for extra robustness), i.e., up to 1.875 sec delay, and the delay of 0.05 sec of the closed analog rate loop.

The path C_1 is parallel to the path with unity gain. At zero frequency, C_1 becomes 30.

The asymptotic loop gain frequency responses are shown in Fig. 5.28 for the case of both C_1 and C_2 operational, and for the case of $C_1 = 0$ (lower curve). The Bode step is very long because of the necessity to compensate for large time delay of up to 7 RTI, and to reduce or eliminate the overshoot.

The system with these analog compensators was simulated in SPICE and in MATLAB. The phase delay of the sampling was imitated by an extra pole placed at frequency $f_S/3$. After several small adjustments to the initial response were made to obtain the desired stability margins, the design proceeded to conversion into digital compensation.

The following digital compensator equations were obtained from the analog controller functions with the Tustin transform:

$$C_1 = (0.15 + 0.15/z)/(1 - 0.99/z) \tag{5.19}$$
$$C_2 = (0.9 - 0.8883/z)/(1 - 0.75/z) \tag{5.20}$$

The coefficients in the equations have been rounded to the required accuracy. The second expression can be rewritten as

$$C_2 = 0.9(1 - 0.987z)/(1 - 0.75/z).$$

The accuracy of the coefficient 0.987 should be rather high since this value is subtracted from 1 at lower frequencies where z approaches 1. Thus, for the accuracy of the low-frequency gain coefficient to be better than 6% (i.e., 0.5 dB), the difference $1 - 0.987 = 0.013$ must be accurate to 6%, i.e., to 0.0008, so that the number 0.8883 in (5.20) should not be further rounded.

Equations (5.17) and (5.18) correspond to the flowchart shown in Fig. 5.29 (a),(b).

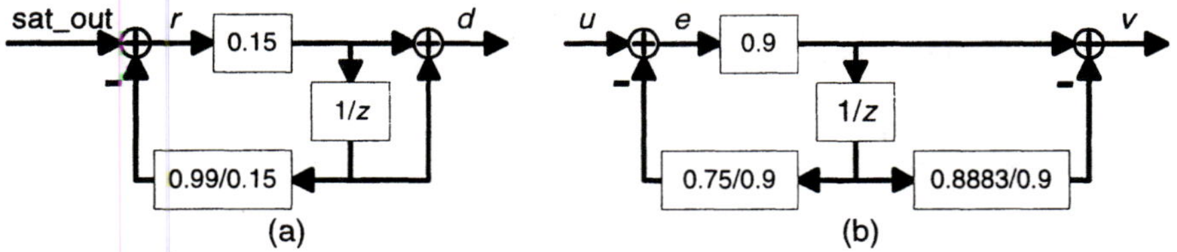

Fig. 5.29 Flowcharts corresponding to equations (5.19), (5.20)

The simplified feedback loop block diagram is shown in Fig. 5.30. The block diagram includes a saturation link in the higher-gain, low-frequency path; linear links C_1 and C_2; a scaling block that has saturation and a dead zone; a delay block; and a model of the plant (of the motor with its analog control electronics).

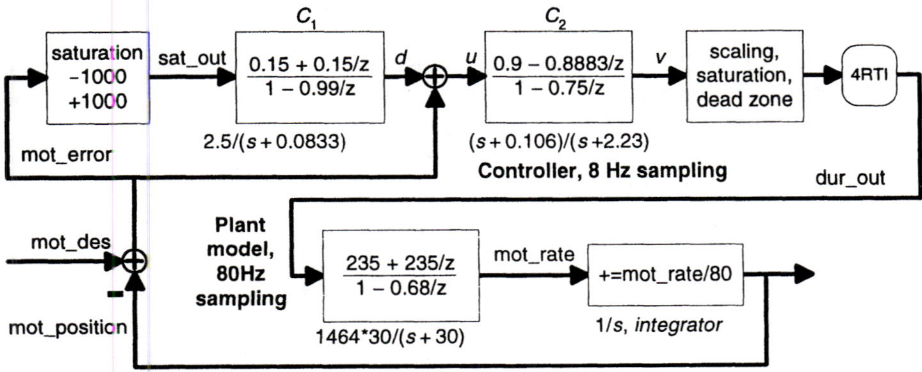

Fig. 5.30 Motor controller flowchart

The C code for the compensator follows:

```
#define PAR1       .15
#define PAR2       .99
#define PAR3       .9
#define PAR4       .75
#define PAR5       .987
#define THRESHP    1000
#define THRESHN   -THRESHP

global double r = 0.0;
global double d = 0.0;
global double e = 0.0;
global double u = 0.0;
global double v = 0.0;
global double sat_out = 0.0;

sat_out = mot_error;                 /* saturation */
if (mot_error > THRESHP)
 sat_out = THRESHP;
if (mot_error < THRESHN)
 sat_out = THRESHN;

d = PAR1 * r;                        /* compensator C1 */
r = sat_out + PAR2 * r;
```

```
d += PAR1 * r;
u = d + mot_error;

v = -PAR5 * e;                        /* lead C2 */
e = u + PAR4 * e;
v += PAR3 * e;
```

The variable `dur_out` is the duration of time that the motor is on during the sampling period of 125 msec. The motor is rate-stabilized by an analog loop with 30 msec rise-time. The motor transfer function is therefore that of an integrator (the angle of rotation is proportional to the time the motor is on) with an extra pole caused by the limited bandwidth of the analog rate loop. The motor (plant) transfer function of s is shown in Fig. 5.30, under the block.

For computer simulations in C, a digital motor model was employed. The transfer function of z is shown in Fig. 5.30 in the block. The rectangular sample-and-hold integrator was used for simplicity, and for better accuracy the sampling frequency for the model was set to 80 Hz, 10 times higher than that of the compensator. The data in the motor model is updated 10 times after each update in the controller. This system shown in Fig. 5.30 is an example of a multirate system (although the controller itself is single-rate). In multivariable controllers, different rates are frequently used, faster rates for processing rapidly changing variables, and lower rates for slowly varying variables.

5.10.6 Aliasing and noise

An A/D converter contains a ***sample-and-hold*** (S/H) link, i.e., a device which samples the signal and keeps this value at its output until the time of the next sampling. An example of such a link is shown in Fig. 5.31. The switch samples the input signal by closing for a short duration at the sampling times. The capacitor charges and holds the sampled value of the signal until the next sampling. The output of the S/H is processed digitally and then returned to the analog form by a D/A converter at the input to the actuator, or directly to the plant (when the actuator is also digital).

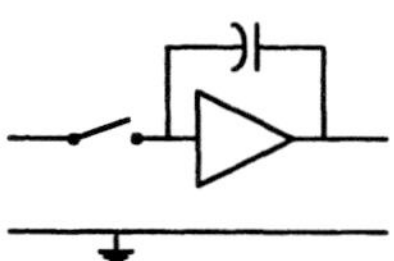

Fig. 5.31 Sample-and-hold link circuit diagram

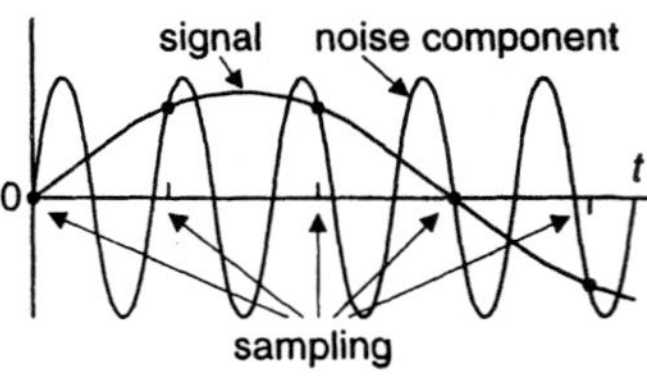

Fig. 5.32 Aliasing

The S/H link is a linear time-variable circuit, and as such, works as an amplitude modulator. Modulation of the high-frequency noise in the S/H link by the sampling frequency and its harmonics produces frequency-difference products that fall within the signal bandwidth. This effect called ***aliasing*** is illustrated in Fig. 5.32. It is seen that on the basis of the information sampled at discrete points, it is impossible to distinguish between the low-frequency signal with frequency f and the high-frequency signal with frequency nf. From here, two important implications for the control system design follow.

First, the effects of the high-frequency noise are added to the baseband signal at the output of the A/D converter.

Aliasing might introduce substantial error in the A/D conversion. To reduce this error by rejecting the high-frequency input noise, a high-order anti-aliasing low-pass filter is commonly installed at the input to the sample-and-hold link (or A/D converter) of DSP systems as shown in Fig. 5.33.

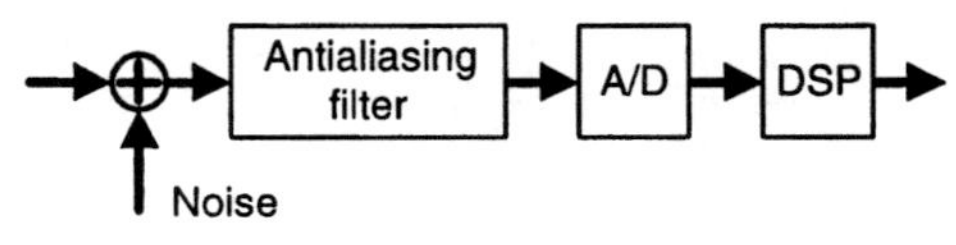

Fig. 5.33 Anti-aliasing filter

In closed-loop feedback systems like that shown in Fig. 5.34, the high-frequency sensor noise N causes the output noise N_{out} in the functional frequency band. The noise is reduced by the anti-aliasing filter.

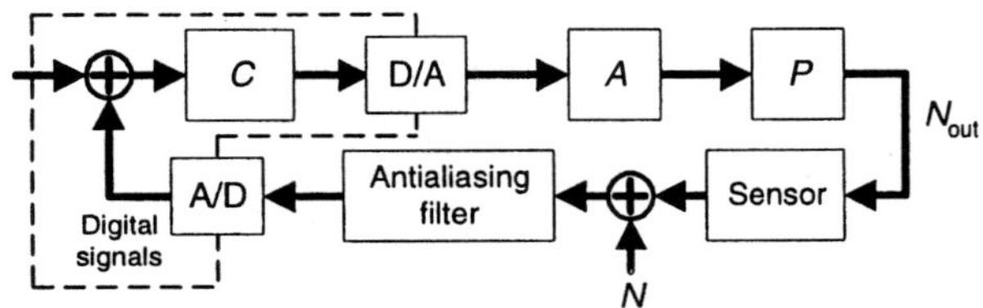

Fig. 5.34 Control system with antialiasing filter

The filter selectivity is limited by its the effect of the loop gain and input-output closed-loop response. To a certain extent, the attenuation of the antialiasing filter at higher frequencies can be equalized by an increase of the gain of the digital compensator (thus making the loop gain as desired) and the introduction of a prefilter in the command path to reduce the input-output closed-loop gain.

The sensor noise is transformed into the baseband largely from the frequencies close to f_S and its harmonics. Therefore, the attenuation of the anti-aliasing filter can be smaller between these frequency bands. The optimal response of the feedback antialiasing filter is therefore not monotonic, and the loop gain response which is close to the optimum looks like that shown in Fig. 5.35. Its feedback bandwidth is wider by 0.3 to 0.5 octaves than the monotonic response shown by the dotted line.

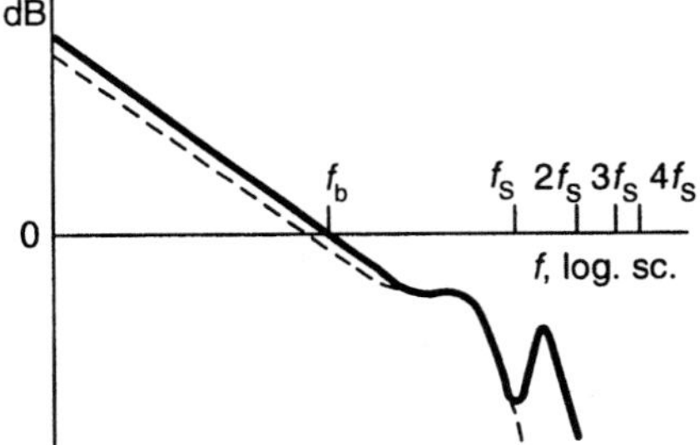

Fig. 5.35 Open-loop Bode diagrams for rejection of aliasing noise

Second, due to aliasing, the gain of the digital filter for a sinusoidal signal with Nyquist frequency is the same as the dc gain. Therefore, a combination of a large low-frequency gain of the digital compensator and a small attenuation in the analog plant near the Nyquist frequency may result in large loop gain and oscillation. In these situations, the sampling frequency must be substantially increased.

5.10.7 Transfer function for the fundamental

In this section, we present yet another view on the effect of digital compensation on the feedback loop. A linear digital compensator is a linear time-variable (LTV) link. As will be shown in Section 7.11, time-dependencies of linear systems can reduce the stability margins and can make the systems oscillate.

Consider the case when the signal is sinusoidal. The input and output signals of the S/H link with 12 samples per period are plotted in Fig. 5.36. Let's define the gain coefficient in fundamentals as the ratio of the amplitude of the output signal fundamental to the amplitude of the input signal. It is seen that the magnitude of this gain coefficient is approximately 1, and the phase lag is approximately 15°. It is clearly seen that the lag, as long as the sampling frequency is relatively high, is inversely proportional to the sampling frequency.

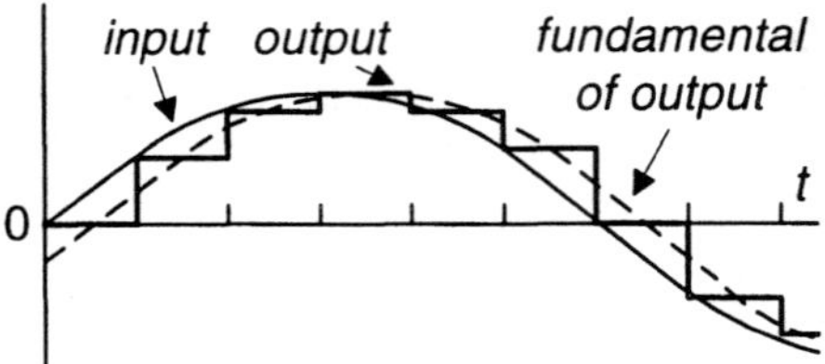

Fig. 5.36 Signal at the output of a sample-and-hold link

When the number of samples per signal period is only two (the Nyquist frequency case), the output of the S/H circuit is Π-shaped as shown in Fig. 5.37, and the output amplitude is $\sin\phi$, where ϕ is the phase shift between the sampling and the input signal.

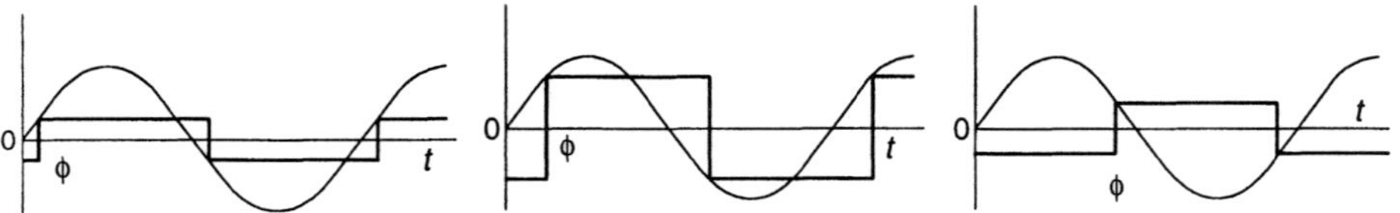

Fig. 5.37 Effect of the phase difference between the signal and the sampling

As follows from the Fourier analysis, the amplitude of the fundamental of the output is $(4/\pi)\sin\phi$.

When, in particular, $\phi = 90°$, then the phase lag is 90°, the output amplitude is 1, and the equivalent gain coefficient is $4/\pi$, i.e., the gain is 2.1 dB. When ϕ approaches 0 or 180°, the gain coefficient approaches 0. The uncertainty in the gain and phase (due to the uncertainty of ϕ) increases with the decrease of the sampling frequency, as can be seen by comparing Figs. 5.36 and 5.37. The system must be made stable with sufficient margins for all possible ϕ.

The caused by digitizing phase delay $2\pi f \times T_s/2 = \pi f/f_s$ can be modeled during simulation of a continuous system with a delay transfer function

$$\frac{-s + 4f_s}{s + 4f_s}$$

since the function

$$\frac{-j2\pi f + 4f_s}{j2\pi f + 4f_s} = \frac{1 - j(\pi/2)f/f_s}{1 + j(\pi/2)f/f_s}$$

can be approximated at lower frequencies by $1 - 2 \times j(\pi/2)f/f_s$, so that the phase shift approximately equals $\operatorname{atan} \pi f/f_s \approx \pi f/f_s$.

In Fig. 5.38 are shown the stability margin boundary and an example of a Nyquist plot on the logarithmic Nyquist plane of a well-designed LTI system. Consider next the effect of introducing an S/H link in this loop at the Nyquist frequency.

Oscillation in control systems, if it happens, is usually periodical, and the shape of the oscillation at the plant output is close to sinusoidal due to the plant low-pass filter properties (this will be discussed further in Chapter 11). Because of the relations between the phase and amplitude illustrated in Fig. 5.37, and because of the extra 2.1 dB of gain on the fundamental, the gain of the LTI loop at the Nyquist frequency must be below the boundary curves with maximum $-x - 2.1$ dB centered at $-270°$ and $-90°$, to rule out an oscillation at this frequency with any possible ϕ, with x dB margin. The resulting penalty in the available feedback is, typically, not large. On the other hand, if the Nyquist frequency should fall on the part of the Nyquist diagram which is phase stabilized, the penalty would be up to 90°, which would require reducing the slope of the Bode diagram and the feedback. This is why the sampling frequency must be kept sufficiently high.

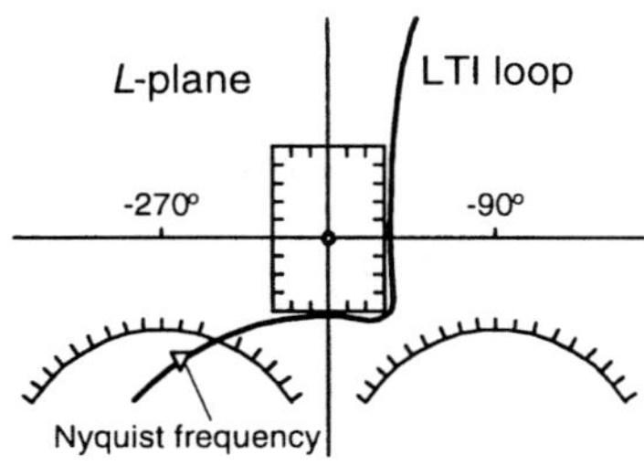

Fig. 5.38 Stability margin boundary and the loop transfer function uncertainty at Nyquist frequency

5.11 Command profiling

The actuator force (or torque) time-profile is often required not to include sharp peaks, and the settling time is required to be as short as possible. In these cases, the position or angle step-command needs to be smoothed out, either by passing it through a high-order Bessel prefilter or by replacing it with a smooth time-function.

The smooth rising of the command from 0 to q over the time interval $[0,\tau]$ can be expressed with $q[3(t/\tau)^2 - 2(t/\tau)^3]$, or with $q[\sin(\pi t/\tau - \pi/2) + 1]/2$.

Example 1. Assume $q = 2$, $\tau = 0.5$. MATLAB plots the first function with

```
ezplot('2*(3*(x/0.5)^2 - 2*(x/0.5)^3)',[0 0.5])
```

A smooth digital command with sampling frequency f_S can be expressed with function $(A/n)(1 + z^{-1} + z^{-2} + ... + z^{-(n-1)})^3$ where $n = \tau f_S$.

5.12 Problems

1 The crossover frequency is 100 Hz. The system must be phase-stabilized at all frequencies below the crossover with margin 30°. The loop gain approximates a constant-slope Bode diagram in the Chebyshev sense. By how much (approximately) can the feedback be increased at (a) 10 Hz, (b) 1 Hz, (c) 0.5 Hz, (d) 20 Hz, (e) 2.72 Hz, if, by using a higher-order compensator, the phase peak-to-peak ripples are reduced from 15° to 2°?

2 By addition of a real pole and a real zero, an asymptotic Bode diagram was made steeper by 6 dB/oct over the frequency interval from 10 Hz to 30 Hz. What are the pole and the zero frequencies? Will the new Bode diagram be more concave or more convex? What happens to the diagram if the pole and zero are interchanged?

3 Draw an asymptotic Bode diagram for the function having:
(a) gain coefficient 10 at $\omega = 0$; zeros (in ω, i.e., in rad/sec) -1, -3, -6, and poles (in ω) -0.5, -4, -8;

(b) gain 10 dB at $\omega = 2$, zeros (in ω) -2, -5, -5, and poles (in ω) -1, -3, -20;
(c) gain coefficient 10 at $f = \infty$, zeros (in Hz) -15, -30, -400, and poles (in Hz) -60, -100, -200;
(d) gain 20 dB at $f = 200$, zeros (in Hz) -100, -200, -1000, and poles (in Hz) 0, -10, -1600.
Use scales: 10 dB/1 cm, 1 oct/1 cm. Find the Bode diagrams from asymptotic responses using the rule for the error: 3 dB at pole, 1 dB one octave from the pole, $0.1 \approx 0$ dB two octaves from the pole.

4 Use MATLAB to make Bode plots for the function:
(a) $T(s) = 100/[s(s+15)(s+100)]$;
(b) $T(s) = 1000/[s(s+100)(s+500)]$;
(c) $T(s) = 5000/[s(s+200)(s+6000)]$;
(d) $T(s) = 200/[s^2(s+100)(s+1000)]$.

5 Find a rational function approximation of the constant slope function:
(a) slope 6 dB/oct, frequency range 1 to 10 Hz;
(b) slope 9 dB/oct, frequency range 1 to 10 Hz;
(c) slope 12 dB/oct, frequency range 1 to 10 Hz;
(d) slope 15 dB/oct, frequency range 1 to 10 Hz;
(e) slope 27 dB/oct, frequency range 1 to 10 rad/sec;
(f) slope 9 dB/oct, frequency range 1 to 10 rad/sec;
(g) slope 12 dB/oct, frequency range 1 to 10 rad/sec;
(h) slope 12 dB/oct, frequency range 1 to 10 rad/sec;
(i) slope 18 dB/oct, frequency range 1 to 10 rad/sec.
(j) slope 6 dB/oct, frequency range 1 to 10 rad/sec.

6 Draw asymptotic Bode diagrams and make the plots with MATLAB for the leads:
(a) $(s+2)/(s+15)$;
(b) $(s+0.1)/(s+0.2)$;
(c) $(s+0.5)/(s+2.5)$;
(d) $(s+2)/(s+4)$.
(e) $(s+2.72)/(s+21)$;
(f) $(s+1)/(s+16)$.

7 Draw asymptotic Bode diagrams and make the plots with MATLAB for the lags:
(a) $(s+15)/(s+2)$;
(b) $(s+1)/(s+0.2)$;
(c) $(s+5)/(s+2.5)$;
(d) $(s+8)/(s+4)$;
(e) $(s+7)/(s+2.72)$;
(f) $(s+16)/(s+2)$;
(g) $(s+8)/(s+4)$.

8 The phase stability margin is excessive by 10° over one decade. Find the lost feedback at lower frequencies.

9 If the peaking must be 8 dB, what is the damping coefficient ζ (use the plots in Fig. 5.10)? Find the polynomial corresponding to the peaking frequency 300 Hz.

10 Plot with MATLAB the normalized low-pass frequency response with a pair of complex poles, with the damping coefficient: (a) 0.0125; (b) 0.125; (c) 0.25; (d) 0.5; (e) 0.99.
Use the MATLAB function `lp2lp` to convert the transfer function to that having the

resonance frequency 5 Hz.

11 Plot with MATLAB the normalized band-pass frequency response with a pair of complex poles, with the damping coefficient: (a) 0.01; (b) 0.1; (c) 0.2; (d) 0.4; (e) 0.99 (obtain the response by multiplying the low-pass transfer function by s). Use the MATLAB function `lp2lp` to convert the transfer function to that having the resonance frequency 50 Hz.

12 Plot with MATLAB the normalized high-pass frequency response with a pair of complex poles, with the damping coefficient: (a) 0.02; (b) 0.2; (c) 0.3; (d) 0.5; (e) 0.99 (obtain the response by multiplying the low-pass transfer function by s^2). Use the MATLAB function `lp2lp` to convert the obtained response to that having the resonance frequency 15 Hz.

13 Plot a series of 5 notches with the notch amplitude 6 dB and various width, centered at (a) 1 rad/sec; (b) 10 rad/sec; (c) 10 Hz; (d) 1 kHz; (e) 2.72 kHz.

14 Break the compensator function into cascaded links:
(a) $5000(s+1)(s+2)(s+1000)/[s(s+20)(s+6000)]$;
(b) $100(s+0.1)(s+8)(s+200)/[s(s+20)(s+600)]$.
(c) $5000(s+1)(s+2)(s+1000)/[s(s+20)(s+6000)]$;
(d) $100(s+0.1)(s+8)(s+200)/[s(s+20)(s+600)]$.
(e) $5000(s+1)(s+2)(s+1000)/[s(s+20)(s+6000)]$;
(f) $100(s+0.1)(s+8)(s+200)/[s(s+20)(s+600)]$.

15 The m.p. component of the plant is $1/s$, and the n.p. lag of the plant is 1 rad at 2 kHz. The amplitude stability margin must be 10 dB. The asymptotic slope must be −18 dB/oct, the asymptote crossing the −10 dB level at 2 kHz. The loop must have a Bode step and −10 dB/oct constant slope down to 100 Hz. Design an analog compensator composed of cascaded links.

16 Using the same plant and requirements as in Problem 15, design an analog compensator composed of parallel links.

17 The feedback bandwidth is limited by the effect of the sensor noise. The loop gain response must be steep right after f_b, and to provide the stability margin, the slope of the loop gain must be only −6 dB/oct for two octaves below f_b. $B_n(f_b) = 1$ rad. Design the compensator for the following plant and feedback bandwidth:
(a) $1/[s(s+300)(s+1000)]$; $f_b = 3$ kHz;
(b) $10^{-3}/[s(s+30)(s+100)]$; $f_b = 300$ Hz;
(c) $10^{-6}/[s(s+3)(s+10)]$; $f_b = 30$ Hz.

18 Verify that equations (5.7) follow from the diagram in Fig. 5.22.

19 A high-order digital compensator was implemented in C without breaking the transfer function into second-order links. It was found that a single-precision simulation on different processors or using different compilers gave slightly different results, while double-precision simulation showed nearly identical results. After the DSP was modified by properly breaking the function of z into second-order multipliers, single-precision became sufficient to obtain the same results on all computers. Explain why this might have happened.

20 The poles of an analog compensator, in s, are:
(a) −3, −6, −8;

(b) –12, –60, –80;
(c) –13, –16, –85;
(d) –10, –600, –1500.
With sampling frequency f_S = 50 Hz, find the poles of the function of z using formula (5.14) and/or MATLAB command `bilinear`. (Hint: Each pole can be found by applying the function `bilinear` to the function $1/(s - s_{pole})$.)

21 Find the Tustin transforms from:
(a) $C(s) = (s + 3)/(s + 2)$,
(b) $C(s) = 5(s + 2)/(s + 3)$,
(c) $C(s) = 10(s + 5)/[s(s + 4)]$,
(d) $C(s) = 3(s + 7)/(s + 20)$,
(e) $C(s) = 15(s + 8)/[(s + 100)s]$,
(f) $C(s) = 2(s + 3)/s$.

22 For sampling frequency f_S = 10 Hz, convert to $C(s)$ from:
(a) $f(z) = (0.2174 + 2174/z)/(1 - 0.7391/z)$,
(b) $f(z) = (0.1200 + 0.1200/z)/(1 - 0.600/z)$,
(c) $f(z) = (1.33 - 0.4444/z)/(1 - 0.1111/z)$.
For sampling frequency f_S = 100 Hz, convert to $C(s)$ from:
(d) $f(z) = (0.22 + 22/z)/(1 - 0.74/z)$,
(e) $f(z) = 0.272 + 0.272/z)/(1 - 0.600/z)$,
(f) $f(z) = (1.1 - 0.4/z)/(1 - 0.1/z)$.

23 Design a digital compensator for the analog plant $P(s) = 50000(s + 200)/(s + 300)$, with the slope of the loop Bode diagram at frequencies below f_b approximately –10 dB/oct. Assume f_S = 10 kHz, the aliasing noise is of critical importance, and the gain and phase stability margins are, respectively, 10 dB and 30°. Consider:
(a) a version with f_b = 1 kHz, a Bode step, monotonic response, and asymptotic slope –12 dB/oct;
(b) a version with f_b = 1.4 kHz, a Bode step, and a notch at f_S as in Fig. 5.35.

24 Write a program in C for $f(z)$ (a),(b),(c) from Problem 22.

25 Consider Example 2 in Section 5.6. Remove the Bode step. In the function $T(s)$, remove the step-forming complex poles and zeros, and move the two real poles from $\omega = 2$ to the right until the guard-point phase stability margin becomes 30°. Where will these poles be? What will be the loop gain at $\omega = 10$? Are the technical specifications satisfied?

26 Make simulations of the system with *PID* controller shown in Fig. 5.19 in Section 5.8 with (a) MATLAB and (b) SIMULINK.

27 In a spacecraft scanning interferometer, a carriage with retroreflectors is being moved by a motor via a cable as shown in Fig. 5.39 to change the lengths of the optical paths. The carriage position range is 20 cm, the position must be accurate within 0.1 mm, and the velocity, within 3%. The lowest structural mode with the frequency in the 100 to 150 Hz range results from the cable flexibility.

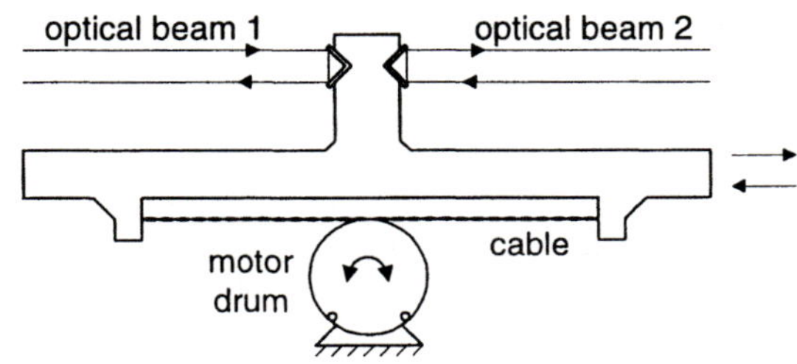

Fig. 5.39 Retroreflector carriage

In the block diagram in Fig. 5.40(a), the prefilter, the feedback summer, and the compensator are digital.

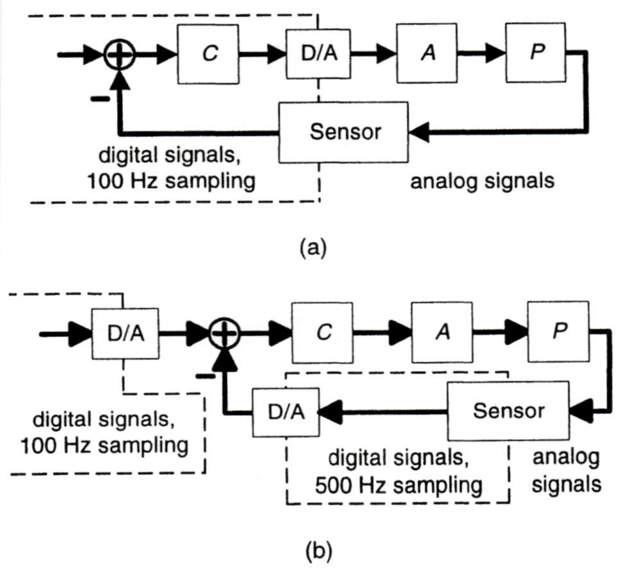

Fig. 5.40 Block diagrams of the carriage control options

The sampling frequency is limited to 100 Hz since the calculations are performed by the flight computer on a time-sharing basis with several other tasks. The position sensor (16-bit encoder) is connected directly to the motor shaft, so the control is collocated. (A more accurate position sensor, a laser interferometer, is used to measure the exact position of the carriage for taking the science data. This sensor is not used for closed-loop position control and is not shown in the pictures.) The D/A converter is placed at the input of the motor driver. The sensor output is digital. As discussed in Problem 4 in Chapter 4, the control bandwidth is limited to 6 Hz.

Estimate and compare the available feedback bandwidth and the control accuracy (a) in this case and (b) when: the sensor data is read with a rather high sampling rate; D/A converters are placed in the command and sensor paths; the prefilter, the command summer, and the compensator are analog. Is the accuracy of the analog circuitry sufficient?

Draw block diagrams. Consider the advantages and limitations of these two modes of the controller implementation.

Answers to selected problems

6 (a) The diagram is shown in Fig. 5.41.

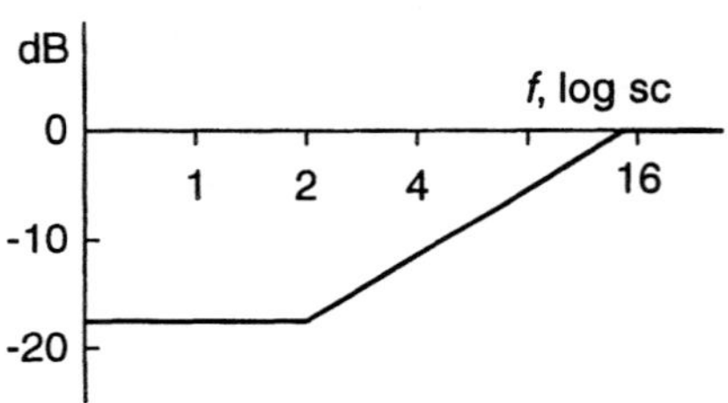

Fig. 5.41 Asymptotic Bode diagrams for the lead $(s + 2)/(s + 15)$

14 (a) $5000(s + 1)/s$; $(s + 2)/(s + 20)$; $(s + 1000)/(s + 6000)$

15 The frequency f_c must be 2 kHz since at this frequency n.p. lag is 1 rad. From (5.2), the Bode step ratio is 2.8 (i.e., 1.5 oct). Thus, $f_d = f_c/2.8 = 0.7$ kHz, and $f_b = f_d/2 = 0.35$ kHz, or $\omega_b = 2.2$ krad/sec. The required ideal loop response shown in Fig. 5.42 is similar to that shown in Figs. 5.15(a), 5.16 (Example 1 in Section 5.6) however with a wider Bode step. The asymptotic diagram shown by the solid line approximates the general shape of the ideal Bode diagram.

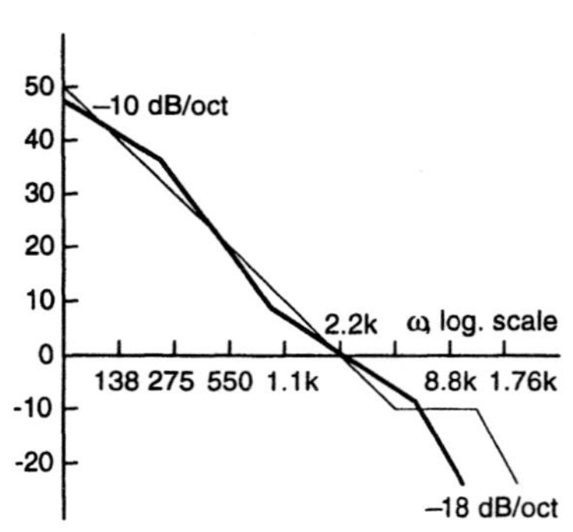

Fig. 5.42 Ideal and asymptotic Bode diagrams

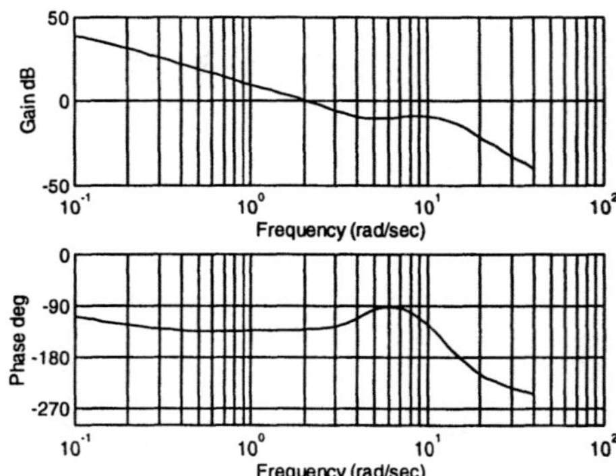

Fig. 5.43 Bode diagrams, ω in krad/sec

The expression in Example 1 for the return ratio,

$$T(s) = 10\,\frac{s + 0.4}{s + 0.1}\,\frac{1}{(s + 2)^2}\,\frac{s^2 + 1.6s + 4}{s^2 + 2.4s + 9}\,\frac{1}{s},$$

needs to be modified:

(a) It must be scaled to change ω_b from 1 to 2200. To avoid large numbers, we express ω in krad/sec and ω_b becomes 2.2. For the scaling, s should be replaced by $s/2.2$. The return ratio becomes

$$T_1(s) = 10\,\frac{s+0.88}{s+0.22}\;\frac{2.2^2}{(s+4.4)^2}\;\frac{s^2+3.52s+19.4}{s^2+5.28s+43.6}\;\frac{2.2}{s}\,.$$

(b) The Bode step frequency ratio must be increased from 1.5 to 2.8, and correspondingly, the poles at the end of the Bode step must be shifted up 2.8/1.5 = 1.9 times. Also, the two poles at 4.4 must be somewhat increased, say, to 5.5 (this is already shown in Fig. 5.40). The return ratio becomes

$$T_2(s) = 10\,\frac{s+0.88}{s+0.22}\;\frac{2.2^2\times(5.5/4.4)^2}{(s+5.5)^2}\;\frac{s^2+3.5s+19.4}{s^2+10.7s+157}\;\frac{1.9^2}{1}\;\frac{2.2}{s}\,.$$

(It can be seen that the corners in the asymptotic diagram in Fig. 5.40 correspond to the real poles and zeros of $T_2(s)$.) Or,

$$T_2(s) = 601\,\frac{(s+0.88)(s^2+3.5s+19.4)}{s(s+0.22)(s+5.5)^2(s^2+10.7s+157)}\,. \tag{5.21}$$

MATLAB function conv is used to multiply the polynomials in the numerator:

```
a = [601]; b = [1 0.88]; c = [1 3.5 19.4];
ab = conv(a,b); num = conv(ab,c)
```

and in the denominator:

```
d = [1 0]; e = [1 0.22]; f = [1 5.5]; g = [1 10.7 157];
de = conv(d,e); def = conv(de,f); deff = conv(def,f);
den = conv(deff,g)
```

The resulting return ratio is

$$T_2(s) = \frac{601s^3+2632s^2+13510s+10260}{s^6+21.9s^5+309.7s^4+2117.8s^3+5200.4s^2+1044.8s}\,.$$

The Bode diagram for this function is plotted in Fig. 5.42 with `w = logspace(-1,1.6)`, to properly scale the gain axis. The diagram is close to the desired. Notice that the frequency axis is erroneously labeled in rad/sec since we used, for simplicity, the `bode` command. The axis must be labeled in krad/sec.

The loop phase response in Fig. 5.42 does not yet include the n.p. lag. The lag can be modeled as described in Section 4.11, or instead, we can just add this phase lag (which is linearly proportional to the frequency) to the phase response in Fig. 5.42. If we do this, we will see that the system is stable with the desired stability margins.

The compensator transfer function is

$$\frac{T_2(s)}{P(s)} = 601\frac{(s+0.88)(s^2+3.5s+19.4)}{(s+0.22)(s+5.5)^2(s^2+10.7s+157)} \tag{5.22}$$

and can be presented as three cascaded links:

$$C_1(s) = 60.1\frac{(s+0.88)}{(s+0.22)(s+5.5)^2}\;, \qquad C_2(s) = \frac{5}{(s+5.5)^2}\;,$$

$$C_3(s) = 2\frac{s^2 + 3.5s + 19.4}{s^2 + 10.7s + 1570} .$$

In these expressions s is in krad/sec. To convert the compensator functions to functions of s in rad/sec (if desired), s should be replaced by $s/1000$.

16 Three solutions (among many possible) are given below.

(1) We might start with the compensator from Section 5.7 having two parallel paths:

$C_1 = 4/(s^2 + 4s)$ and $C_2 = 5/(s^2 + 2.4s + 16)$

for the plant $1/s$, with the loop response

$$T(s) = \frac{C_1(s) + C_2(s)}{s} = \frac{9s^2 + 29.6s + 64}{s^5 + 6.4s^4 + 25.6s^3 + 64s^2}$$

shown in Fig. 5.19(b). This response must be modified to widen the Bode step. This can be done by reducing C_1 approximately 1.2 times, and increasing C_2 1.5 times. The return ratio becomes

$$T_1(s) = \left(\frac{3.3}{s^2 + 4s} + \frac{7.5}{s^2 + 2.4s + 16}\right)\frac{1}{s} = \frac{10.8s^2 + 37.9s + 52.8}{s^5 + 6.4s^4 + 25.6s^3 + 64s^2} .$$

This response plotted with MATLAB commands

```
n = [10.8 37.9 52.8]; d = [1 6.4 25.6 64 0 0];
w = logspace(-1,1);
bode(n,d,w)
```

is shown in Fig. 5.44. The step length looks about right.

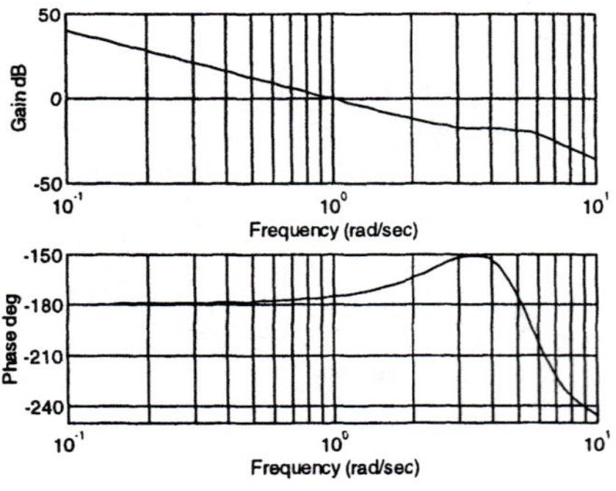

Fig. 5.44 Bode step adjustments

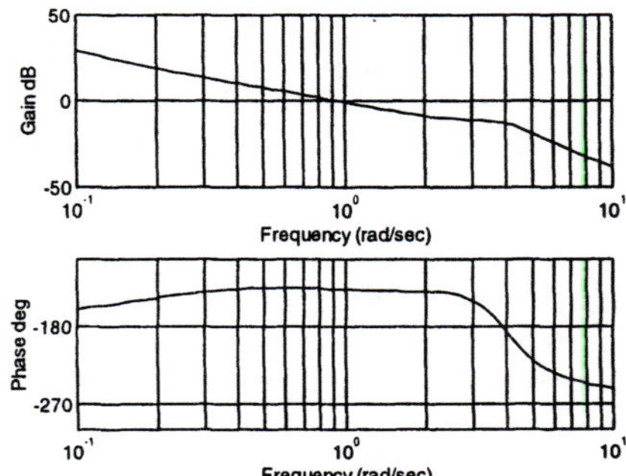

Fig. 5.45 Loop Bode diagrams

Now, a lead must be introduced into C_1 to reduce the slope at lower frequencies. With the lead,

$$T_2(s) = \left(\frac{3.3(s + 0.3)}{(s^2 + 4s)(s + 1)} + \frac{7.5}{s^2 + 2.4s + 16}\right)\frac{1}{s} .$$

With:

```
n1 = conv(3.3,[ 1 0.3]);
```

```
% during iterations, adjust the zero
% (try  also zero 0.75, pole 1.5)
d1 = conv([1 4 0], [1 1]);
% during iterations, adjust the pole
d2 = [1 2.4 16];
d1d2 = conv(d1,d2);
d = conv(d1d2,[1 0])
n = conv(n1,d2) + conv(7.5,d1)
% vectors have equal length since
% both polynomials are cubic
w = logspace(-1,1);
bode(n,d,w)
```

$T_2(s)$ is converted to the ratio of polynomials

$$T_2(s) = \frac{10.8s^3 + 46.41s^2 + 85.18s + 15.84}{s^6 + 7.4s^5 + 32s^4 + 86.9s^3 + 64s^2} \tag{5.23}$$

and plotted. The plot is shown in Fig. 5.45. Its shape is acceptable. The crossover frequency on the plot is 0.95 rad/sec. It remains now only to scale the response for the crossover frequency to be at 2200 by changing s to $s(0.95/2200) = s/2316$.

(2) We will use the already obtained solution to Problem 9. The compensator function

$$\frac{T_2(s)}{P(s)} = 601\frac{(s+0.88)(s^2 + 3.5s + 19.4)}{(s+0.22)(s+5.5)^2(s^2 + 10.7s + 157)} \tag{5.24}$$

or

$$\frac{T_2(s)}{P(s)} = \frac{601s^3 + 2632s^2 + 13510s + 10260}{s^5 + 21.9s^4 + 309.7s^3 + 2117.8s^2 + 5200.4s + 1044.8} \tag{5.25}$$

should be decomposed into a sum of transfer functions of lower order. Each such transfer function represents a link, and all the links are connected in parallel. There might be several options for such decomposition.

The function can be decomposed into the sum of partial fractions

$$\frac{T_2(s)}{P(s)} = \frac{r_1}{s+0.22} + \frac{r_{21}s + r_{22}}{(s+5.5)^2} + \frac{r_{31}s + r_{32}}{s^2 + 10.7s + 157} \tag{5.26}$$

The numerator of the fraction with a single pole can be found by assigning s the value of the pole and solving the resulting equation for the residue. During this exercise, all other fractions in the right-hand side can be neglected. Generally, the coefficients in the numerators of the fractions (including those with multiple poles) can be found by adding the fractions in (5.26) which results in a ratio of polynomials in s, and solving a system of linear equations that result from comparing the numerator coefficients of s at specific powers to those in (5.25).

With MATLAB, the function (5.26) can be decomposed into a sum of partial fractions as follows (the calculation method is ill-conditioned and requires high accuracy in the initial data):

```
num = [601 2632 13510 10260];
den = [1 21.9 309.7 2117.8 5200.4 1044.8];
[Res,Pol,K] = residue(num, den)
```

```
Res =                          Pol =                       K = []
    1.0e+002 *
    0.1598 - 0.2042i       -5.3460 +11.3430i
    0.1598 + 0.2042i       -5.3460 -11.3430i
   -0.1684 - 4.4410i       -5.4940 + 0.1396i
   -0.1684 + 4.4410i       -5.4940 - 0.1396i
    0.0172 - 0.0000i       -0.2200
```

Due to rounding errors, instead of the double real poles –5.5, a pair of complex poles appears, with small imaginary parts. The compensator transfer function is

$$\frac{T_2(s)}{P(s)} = \frac{1.72}{s+0.22} + \frac{-16.84 - j444.1}{s+5.494 - j0.1396} + \frac{-16.84 + j444.1}{s+5.494 + j0.1396} +$$
$$\frac{15.98 - j20.42}{s^2 + 5.346 - j11.343} + \frac{15.98 + j20.42}{s^2 + 5.346 + j11.343} .$$

The product of two fractions of the type

$$\frac{a+jb}{s-(c+jd)} \frac{a-jb}{s-(c-jd)}$$

is a ratio of two polynomials (first-order to second-order) with real coefficients. The polynomials can be found as follows:

```
num_prod = [2*a  (-2*(a*c + b*d))]
den_prod = [1 (-2*c) (c*c + d*d)]
```

When d is the result of calculation inaccuracy (for the double real poles) and can be neglected, then, approximately,

```
num_prod = [2*a  (-2*a*c)]
den_prod = [1 (-2*c) (c*c)]
```

After making this conversion, and neglecting the small imaginary parts of the double poles that we know must be real, we obtain the compensator transfer function

$$\frac{T_2(s)}{P(s)} = \frac{1.72}{s+0.22} + \frac{2.029s - 11.48}{(s+5.5)^2} + \frac{32s + 634}{s^2 + 10.7s + 157} .$$

(3) In the solution to Problem 9, the compensator function

$$\frac{T_2(s)}{P(s)} = 601 \frac{(s+0.88)(s^2+3.5s+19.4)}{(s+0.22)(s+5.5)^2(s^2+10.7s+157)} \tag{5.27}$$

can be presented as the product of two fractions,

$$\frac{T_2(s)}{P(s)} = \frac{601}{s+5.5} \frac{(s+0.88)(s^2+3.5s+19.4)}{(s+0.22)(s+5.5)(s^2+10.7s+157)} . \tag{5.28}$$

With:

```
num = conv([1 0.88],[ 1 3.5 19.4])
d1 = conv([1 0.22], [1 5.5]); den = conv(d1, [1 10.7 157])
```

the second fraction is converted to the ratio of two polynomials

$$\frac{s^3+4.38s^2+22.48s+17.072}{s^4+16.42s^3+219.414s^2+910.987s+189.97} \quad (5.29)$$

With MATLAB, the function (5.29) can be decomposed into a sum of partial fractions as follows (more digits are used here since the calculation method is ill-conditioned and sensitive to rounding errors):

```
num = [1 4.38 22.48 17.072];
den = [1 16.42  219.414  910.987 189.97];
[Res,Pol,K] = residue(num, den)

Res =                                   Pol =                           K = []
       0.3889 + 0.2973i                       -5.3500 +11.3304i
       0.3889 - 0.2973i                       -5.3500 -11.3304i
       0.2072                                 -5.5000
       0.0151                                 -0.2200
```

The sum of the two complex pole fractions of the type

$$\frac{a+jb}{s-(c+jd)}\frac{a-jb}{s-(c-jd)}$$

is a ratio of two polynomials with real coefficients which can be found as follows:

```
a = 0.3889; b = 0.2973; c = -5.35; d = 11.3304;
prod_num = [2*a   (-2*(a*c + b*d))]
prod_den = [1 (-2*c) (c*c + d*d)]
```

Finally, the compensator transfer function is

$$\frac{T_2(s)}{P(s)} = \frac{601}{s+5.5}\left(\frac{0.0151}{s+0.22}+\frac{0.2072}{s+5.5}+\frac{0.7778s-2.5758}{s^2+10.7s+157}\right)$$

which is the function of the parallel connection of three links preceded or followed by the link 601/(s + 5.5).

There are many options of the compensator's implementation. Some of them might be better suited for implementing multiwindow nonlinear controllers described in Chapter 13.

24 (a) Using (5.15) and (5.17), $f(z) = (16.67z + 10)(z - 1.667)$.
The C code is:

```
y = 10 * r;   r = x + 1.667 * r;   y += 16.67 * r;
```

Chapter 6

ANALOG CONTROLLER IMPLEMENTATION

This chapter explores a variety of issues concerning design and implementation of analog electrical compensators. Since the sensors' outputs and the actuators' inputs are most often analog electrical signals, it is convenient and economical to make the command summers and the compensators analog.

Operational amplifier circuits are considered: a summer, an integrator and a differentiator, leads and lags in inversion and noninversion configurations, a constant slope compensator, and compensators with complex poles, including computer controlled analog compensators.

Basic types of *RC* active filters that are employed in feedback system compensators are examined. *RC* circuit design in the element value domain is described, and the use of the *RC*-impedance chart is explained. Switched capacitor circuits are reviewed, with an example of a band-pass tunable compensator design.

Implementations of dead zone, saturation, and amplitude-window circuits are briefly discussed.

The most important issues of analog compensators breadboarding are surveyed.

Tunable compensators, *PID* and *TID*, are introduced, and also tunable compensators with one variable parameter.

Methods of loop gain and phase measurements are outlined.

This chapter is the last one in the introductory control course.

6.1 Active RC circuits

6.1.1 Operational amplifier

Industrial sensors of electrical, mechanical, hydraulic, thermal, and other variables produce, as a rule, electrical output signals. Most frequently, the input signals for the actuators are also electrical. Further, the commands are commonly generated as electrical signals. In many cases, these signals are analog, and the feedback summer subtracts the analog fed back signal from the analog command. In such cases, the compensator, prefilter, command feedforward, and feedback path links can be economically and easily implemented with active *RC* circuits. The main building block for these circuits is the operational amplifier.

The dc gain of an op-amp is typically 100 to 120 dB. Following the flat-response range, the amplifier gain drops linearly with a gain coefficient very close to f_T/f up until the ***unity gain bandwidth*** f_T, as shown in Fig. 6.1. Depending on the type of op-amp, the unity gain bandwidth can be from 100 kHz to 1 GHz. The amplifier phase lag is close to $\pi/2$ up to the frequency f_T. With a feedback circuit added, the system is stable if the phase lag of the feedback path is less than $\pi/2$ at all frequencies where the loop gain is more than unity.

Op-amps are used with large feedback to make the gain stable in time. The available gain therefore must be high, in anticipation of the feedback-induced gain reduction. For example, if the gain coefficient of an op-amp with feedback of 10 is required to be 50 at 10 kHz, then the op-amp gain coefficient with no feedback needs to be at least 500 at 10 kHz, which puts f_T at 5 MHz.

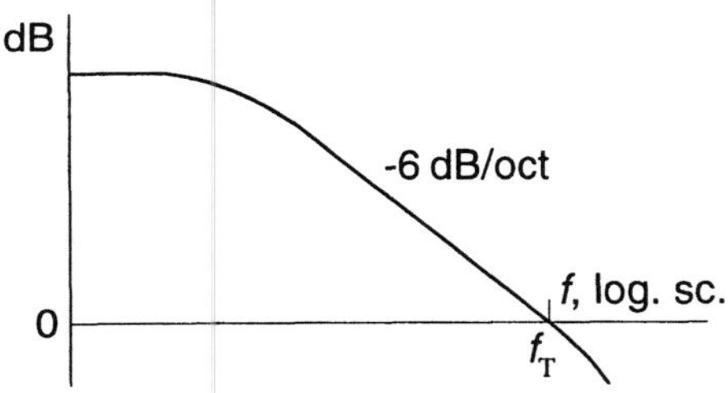

Fig. 6.1 Op-amp gain frequency response

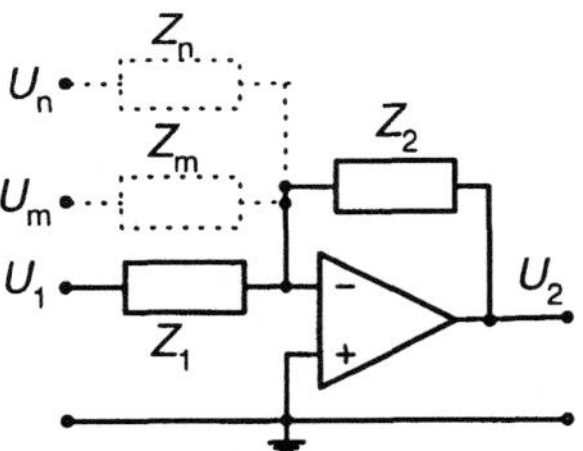

Fig. 6.2 Inverting amplifier and inverting summer (dotted lines)

The inverting configuration of an op-amp with feedback circuitry is shown in Fig. 6.2. Two-pole Z_1 connects the signal source to the amplifier input, and Z_2 is the feedback path impedance.

Over the bandwidth where the feedback is large, the error voltage across the op-amp input is very small compared with the input voltage U_1 and the output voltage U_2. Therefore, $U_1 \approx I_1 Z_1$ and $U_2 \approx I_2 U_2$, where I_1 and I_2 are the input and output currents. Next, since the input current of the op-amp itself is negligible, $I_2 \approx -I_1$. It follows that the transfer function of the inverting amplifier is

$$K = -\frac{Z_2}{Z_1}. \tag{6.1}$$

The input impedance of the inverting amplifier is Z_1 since at the op-amp input the voltage is the feedback loop error, i.e., it is very small, so that this node potential is very close to that of the ground. The amplifier can be used as a unity gain inverter (when $Z_1 = Z_2$) and as a summer amplifier combining signals from different sources, as is shown by the dotted lines, so that the output signal is $-(U_1/Z_1 + U_m/Z_m + U_n/Z_n)Z_2$.

Typically, $|Z_1|$ and $|Z_2|$ are chosen from 5 kΩ to 2 MΩ. The impedance Z_2 should not be too small or else the current in the impedance and the consumed power will be too big. The impedance Z_1, on the other hand, should not be too large since it reduces the signal at the amplifier input and increases the thermal noise whose mean square voltage at room temperature, according to the Johnson-Nyquist formula, is, in volts,

$$\overline{E} = 0.4 \cdot 10^{-8} \sqrt{\Delta f R}\ ;$$

here, Δf is frequency range in Hz, and R is the resistance in Ohms faced by the input port of the amplifier; R is the parallel connection of the input and feedback resistances.

6.1.2 Integrator and differentiator

An inverting integrator with transfer function $-1/(R_1 C_2 s)$ results when the impedances in the schematic diagram in Fig. 6.2 are chosen to be $Z_1 = R_1$, $Z_2 = 1/(sC_2)$. The Bode diagrams for the integrator are shown in Fig. 6.3. The slope of the closed-loop gain is −6 dB/oct over a wide frequency range.

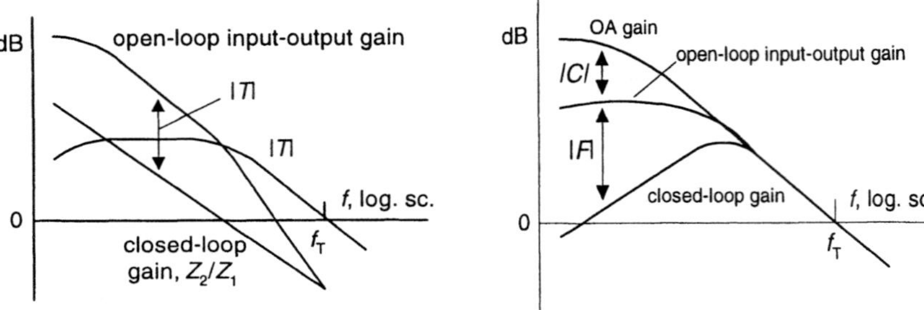

Fig. 6.3 Integrator gain responses **Fig. 6.4** Differentiator gain responses

When the open-loop gain is being calculated, the left end of the two-pole R_1 must be connected to the ground. Therefore, the open-loop gain is the product of the op-amp gain coefficient and the coefficient of the voltage divider $R_1/(R_1 + Z_2)$.

The loop gain is small at very low frequencies where the impedance of the feedback capacitor is very large. Therefore, the integrator does not perform well at these frequencies. That is, the integrator is not accurate when the integration time is very long.

At medium and higher frequencies, the gain coefficient about the feedback loop is

$$\frac{f_T}{f}\frac{R_1}{R_1 + Z_2}.$$

It is large and nearly constant over the frequency range where $R_1 < |Z_2| = 1/(\omega C_2)$. At higher frequencies, R_1 becomes dominant in the denominator and the loop gain decreases as a single integrator, with 90° phase stability margin.

An inverting differentiator can be implemented with $Z_1 = 1/(\omega C_1)$, $Z_2 = R_2$. The open- and closed-loop responses for the differentiator are shown in Fig. 6.4. The external feedback circuit for the differentiator is a low-pass filter which, together with the op-amp itself, produces a double integrator in the feedback loop. To provide some phase stability margin, a lead compensator C can be introduced in the feedback loop in front of the op-amp to reduce the gain at lower frequencies, as shown in Fig. 6.5.

It is seen in Fig. 6.4 that *the effective bandwidth of the differentiator is much smaller than that of the integrator.* This is one of the reasons why integrators but not differentiators are usually employed in analog computers.

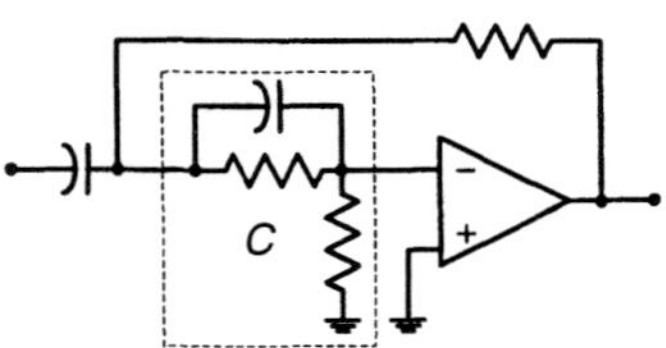

Fig. 6.5 Differentiator schematic diagram

6.1.3 Noninverting configuration

The unity gain amplifier shown in Fig. 6.6(a), which is the ***voltage follower*** described briefly in Section 1.3, Fig. 1.5, has the feedback path transmission coefficient $B = 1$. The schematic diagram of a more general noninverting amplifier configuration is shown in Fig. 6.6(b). The feedback path transfer function is $B = Z_1/(Z_1 + Z_2)$; the feedback amplifier transfer function is $1/B$, i.e.,

$$K = \frac{Z_2}{Z_1} + 1. \tag{6.2}$$

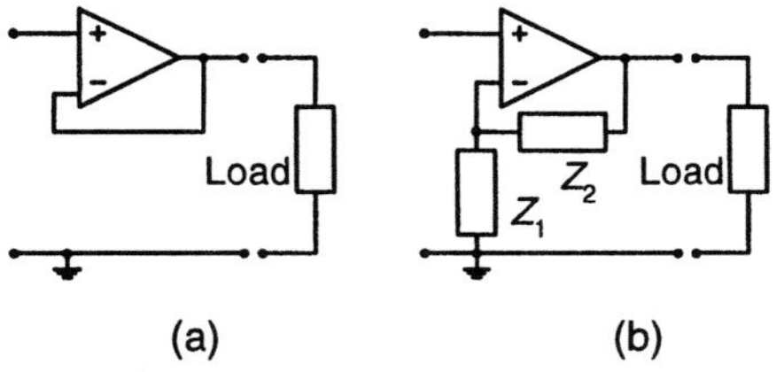

(a) (b)

Fig. 6.6 Follower (a) and non-inverting amplifier (b)

The fed back signal in both the inverting and non-inverting amplifier configurations is proportional to the output voltage. Therefore, this feedback loop stabilizes the output voltage, i.e., makes the output voltage nearly independent of various disturbances and of the load impedance variations. Hence, *the output impedance of the circuits in Fig. 6.6 is low over the range where the feedback is large*.

6.1.4 Op-amp dynamic range, noise, and packaging

The dynamic range of the op-amp is limited from above by the output voltage swing, and from below by the input noise and drift. The input noise is typically comparable with the thermal noise of a resistor of several hundred ohms to several kilo-ohms connected in series to the input. Therefore, external resistances in series with the op-amp input, like R_1 in Figs. 6.2 and 6.6(b), do increase the device noise. It is desirable to keep these resistances small, especially in the op-amp immediately following the feedback summer.

The input voltage dc offset (the internal parasitic dc bias) is typically in the 10nV to 1mV range, with the thermal drift in the 1nV/°C to 10μV/°C range.

While op-amp circuits are tested, neither of the input pins should be left open or else the voltage on the open pin will depend on the initial charge. Due to the extremely high impedance of the op-amp input, this charge will remain for an unpredictable time, thus producing confusion for the experimenter by altering the readings in an often irreproducable fashion.

The choice of f_T depends on the desired gain and the power consumption. As a rule, the higher f_T is, the larger must be the current in the transistors and consequently, the dc current consumed by the op-amp from the power supply.

Op-amps come packaged as single, double, or quad in one case. The standard pinouts for the double and the quad are shown in Fig. 6.7. The pinout is the same for the dual in line package (DIP) and surface mount (SM) package (there also exist much smaller SM packages).

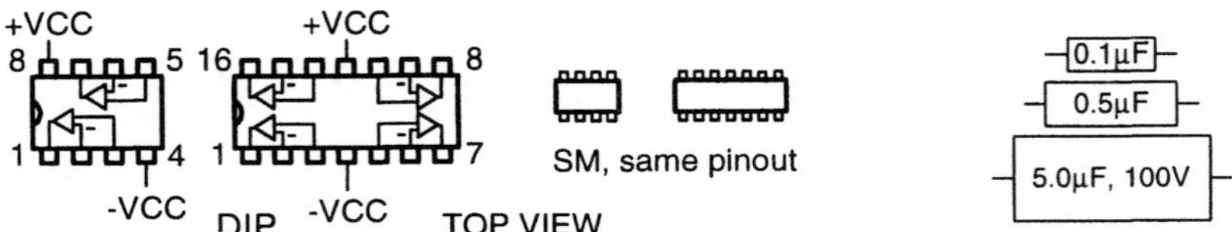

Fig. 6.7 Typical double and quad OA, DIP and SM packages

Fig. 6.8 Typical size of mylar capacitors

One quad op-amp is sufficient for all the needs of a typical analog compensator. Typically, it consumes an order of magnitude less power, occupies an order of magnitude smaller space, and costs an order of magnitude less than a digital microcontroller. Breadboarding and testing such an analog controller takes, typically, an order of magnitude less time than doing so with a digital controller.

The values of the *RC* constants correspond to the poles and the zeros of the transfer functions. For low-speed processes, these time constants can be large and the capacitors become bulky.

The higher the resistances are, the smaller the capacitors can be. The resistances are, however, limited. The resistance in series with the op-amp input is limited by the requirements of keeping the thermal noise below certain level, and the feedback resistance is limited by stray capacitances. Typically, resistor values should not exceed several megohms. Thus, for example, $R = 2\,\text{M}\Omega$ and $C = 0.5\,\mu\text{F}$ can produce a pole or a zero at a frequency of $f = 1/(2\pi RC) \approx 0.16\,\text{Hz}$.

The capacitance must be stable in time, like that of mylar capacitors. The typical size of mylar capacitors is shown in Fig. 6.8.

6.1.5 Transfer functions with multiple poles and zeros

With the feedback amplifiers shown in Fig. 6.6 where Z_1 and Z_2 are *RC* two-poles, transfer functions can be realized with multiple real poles and zeros.

Non-inverting lag and lead compensators are shown in Fig. 6.9(a) and (b), respectively.

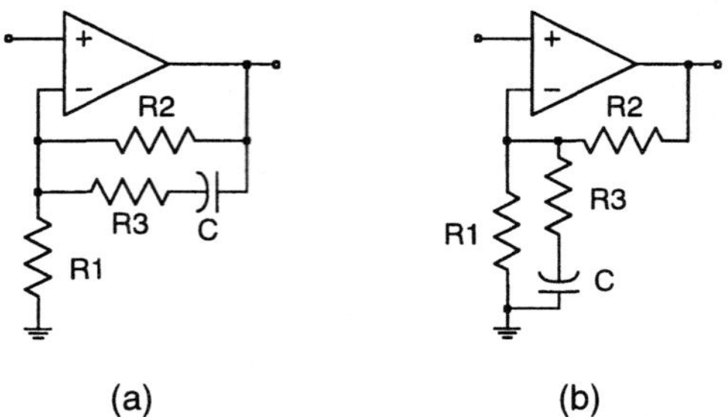

Fig. 6.9 Lag (a) and lead (b) implementation, non-inverting

Inverting lead compensators are shown in Fig. 6.10(a) and (b). An inverting lag compensator is shown in Fig. 6.10(c). (Design of these compensators will be discussed further in Sections 6.2.1, 6.2.2, and in solutions to Problems 7, 8, and 9 at the end of the chapter.)

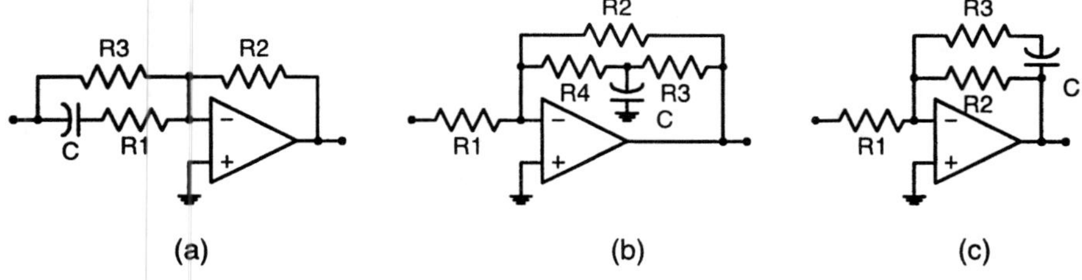

Fig. 6.10 Lead (a) and (b), and lag (c) implementation, inverting

Fig. 6.11 shows the implementation of transfer function $s^{-1/3}$ which uses approximation (5.2) shifted to cover the band from 1 to 100 Hz.

R8	555.2	C8	500
R7	563.9	C7	123.7
R6	423.3	C6	51.44
R5	308.13	C5	23.7
R4	185.0	C4	13.1
R3	39.43	C3	9.87
R2	599.0	*R* in kΩ	
R1	54.53	*C* in nF	

Fig. 6.11 Implementation of transfer function $s^{-1/3}$ over the band from 1 to 100 Hz

Fig. 6.12(a) shows the ***bridged T-circuit*** employed in front of an op-amp to implement complex zeros. The mutual compensation of the phase-advanced output signal of the upper path B_1 with the phase-delayed output signal of the path B_2 produces a broad notch on the Bode diagram.

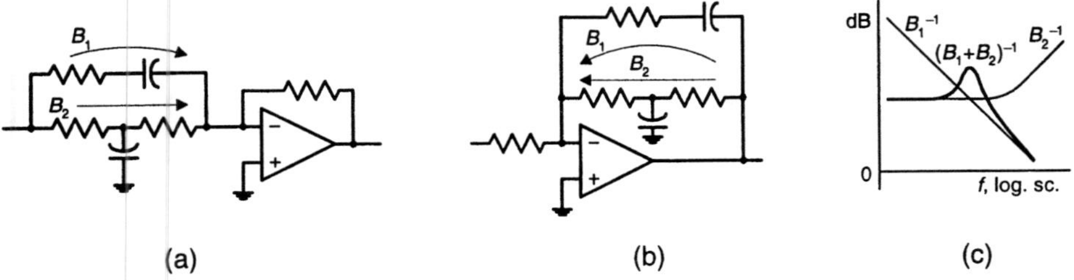

Fig. 6.12 Implementation of (a) complex zeros, and (b) and (c) complex poles

The bridged T-circuit in the feedback path shown in Fig. 6.12(b) allows implementation of a complex pole pair. The gain responses for the amplifier using the feedback path B_1, feedback path B_2, and both, are shown in Fig. 6.12(c).

Implementation of a complex-pole pair using parallel feedback paths is also shown in Fig. 6.13. (This circuit has been used as a part of the compensator for the 100 kV, 1.6 MW precision power supply for a klystron transmitter.) The feedback path B_2 dominates at lower frequencies. The path B_1 dominates at higher frequencies. At the crossing frequency, the output signals of the feedback paths have nearly opposite phase so that the total feedback path has a pair of complex conjugate zeros. As the result, the closed-loop transfer function possesses a complex pole pair. The quality factor Q depends on the difference in phase shift between the two paths. The difference can be adjusted by adding series or parallel resistors to the capacitors.

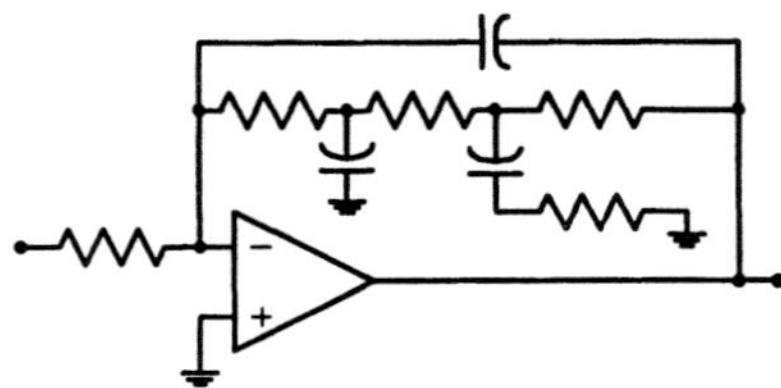

Fig. 6.13 Compensator with parallel feedback paths

6.1.6 Active RC filters

Fig. 6.14 shows the schematic for a unity-gain ***Sallen-Key*** second-order low-pass filter with the transfer function

$$K(s) = \frac{\omega_o^2}{s^2 + Q^{-1}\omega_o s + \omega_o^2} \tag{6.3}$$

where

$$\omega_o = \frac{1}{\sqrt{R_1 R_2 C_1 C_2}}, \text{ and } Q = \frac{1}{\omega_o (R_1 + R_2) C_2}. \tag{6.4}$$

This filter is well-suited to the implementation of low-Q complex poles like those required to make Bode steps in the examples in Sections 5.6 and 5.7. The pole frequency ω_o and the damping coefficient $\zeta = 1/(2Q)$ are prescribed. Two of the circuit elements can be chosen, and values for the remaining two can be found from the equations. For example, when the resistor values are initially chosen, $C_2 = 1/[\omega_o(R_1 + R_2)Q]$, and $C_1 = 1/(\omega_o^2 C_2 R_1 R_2)$.

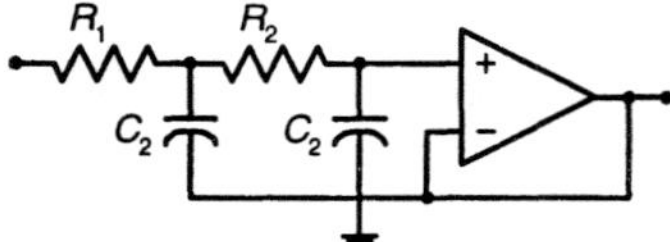

Fig. 6.14 Sallen-Key low-pass filter

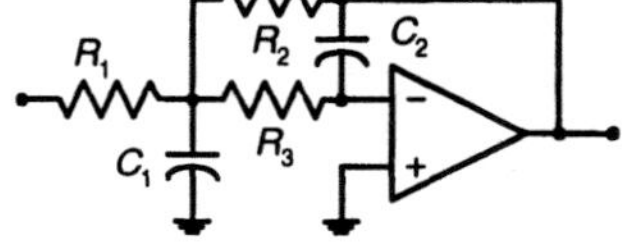

Fig. 6.15 Multiple feedback low-pass filter

A ***multiple feedback*** second-order low-pass filter, shown in Fig. 6.15, has reduced sensitivity to component parameter variations. It implements the transfer function

$$K(s) = \frac{-H_o\omega_o^2}{s^2 + Q^{-1}\omega_o s + \omega_o^2} = \frac{-(R_1 R_3 C_1 C_2)^{-1}}{s^2 + s(R_1^{-1} + R_2^{-1} + R_3^{-1})C_1^{-1} + (R_2 R_3 C_1 C_2)^{-1}}. \tag{6.5}$$

The low-frequency gain coefficient H_o is limited by the prescribed Q. Particularly, H_o must be chosen less than 100 when $Q \geq 1$, and less than 10 when Q approaches 10. (The element values of this filter become inconvenient when Q exceeds 10.) For this filter, two of the element values can be chosen and the remaining three calculated from the following equations:

$$|H_o| = \frac{R_2}{R_1} \quad (6.6)$$

$$\omega_o^2 = \frac{1}{R_2 R_3 C_1 C_2} \quad (6.7)$$

$$Q = \frac{\sqrt{\frac{C_1}{C_2}}}{\sqrt{\frac{R_3}{R_2}} + \sqrt{\frac{R_2}{R_3}} + \frac{\sqrt{R_2 R_3}}{R_1}} \quad (6.8)$$

Table 6.1 Chebyshev 1 kHz low-pass filter

p-p ripple, dB	1	2	3	4
R_1=R_2, kΩ	8.13	7.55	6.17	3.01
R_3, kΩ	12.8	12.8	12.3	12.6
C_1, nF	47	68	100	220
C_2, nF	4.7	4.7	4.7	4.7
Q	0.956	1.129	1.305	1.493
f_o	1.05	0.907	0.841	0.803

Table 6.1 gives the elements' values for the Chebyshev second-order low-pass filter with $H_o = -1$ and 1 kHz cut-off frequency. The filter response remains the same when all resistances are increased and all capacitances reduced by the same factor. The corner frequency can be changed by changing the product of the capacitances, and Q, by changing the ratio of the capacitances. The gain coefficient can be changed by changing R_1 and then adjusting the ratio of the capacitances to preserve the desired Q and changing both capacitances by some coefficient to preserve ω_o.

The circuit shown in Fig. 6.16(a) is often called a ***state-variable filter***. This is an analog computer consisting of a summer with gain adjustment resistors and two integrators. The block diagram is similar to that shown in Fig. 5.22. The circuit can mimic second-order linear differential equations describing low-pass, high-pass, band-pass, and band-rejection filters. This circuit enables reliable implementation of poles and zeros with Q up to 100. (An example of using band-pass compensators with state-variable filters will be given in Section 6.4.2.)

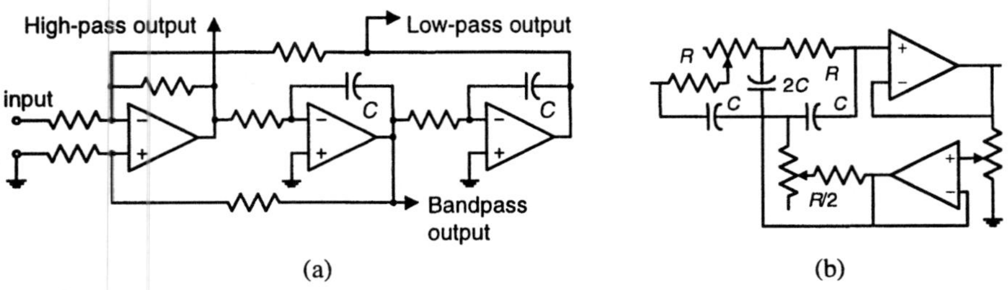

Fig. 6.16 (a) State-variable filter, (b) twin-T notch

The state-variable filters are available as ICs, with only four resistors to be added to set the cut-off frequency, Q, and the gain. There also exist IC filters in which all resistors are built in and are programmable or controlled from a computer port. These circuits combine the best of the analog and the digital worlds: they are easily reprogrammed, and they do not introduce the delay associated with sampling.

Programs for calculating element values and plotting frequency responses for Sallen-Key, multiple feedback, and state-space filters are available from many manufacturers (e.g., Burr-Brown, Harris, MAXIM, National Semiconductor).

A notch $(s^2 + \omega_o^2)/(s^2 + 2\zeta\omega_o + \omega_o^2)$ can be implemented with the twin-T bridge shown in Fig. 6.16(b). The resonance frequency $\omega_o = 1/(RC)$. The variable resistors allow the adjustment of ω_o and of the damping of the numerator (which must be 0). The potentiometer allows the adjustment of the denominator damping coefficient ζ over the range 0.01 to 0.16.

Example 1. Two cascaded notches with $\zeta = 0.1$ and slightly different resonance frequencies, 1% under and 1% over the resonance, can be used to reject a plant structural resonance by at least 40 dB over the range ±1.4% of the nominal resonance frequency, while introducing only 5° lag at the frequency two octaves below the resonance.

6.1.7 Nonlinear links

A saturation link can be implemented with an op-amp as shown in Fig. 6.17(a). This arrangement uses Zener diodes in the feedback path that have small differential resistance when the voltage across the diode exceeds the threshold. The 5 V saturation threshold shown in the figure results from the sum of the voltage drops across the open diodes: 4.3 V on the Zener with inverse polarity, and around 0.7 V on the open diode of the other Zener. The resulting saturation response, with the gain coefficient of –2, is shown in Fig. 6.17(c).

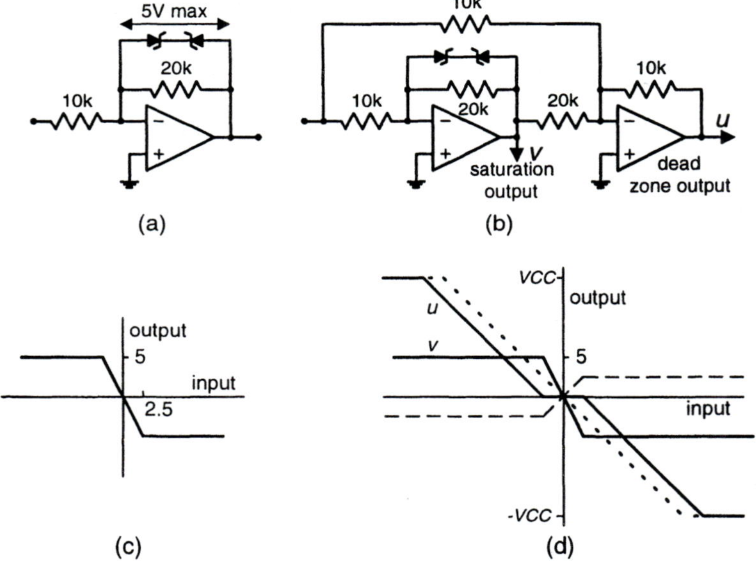

Fig. 6.17 Saturation link (a) and its characteristic (c), and a dead zone link (b) and its characteristic (d)

The dead zone link can be formed by summing the input of a saturation link with its inverted output. Fig. 6.17(b) shows an implementation of such a circuit, and the resulting response is shown in Fig. 6.17(d). The dotted line indicates signal transmission via the upper path, and the dashed line, via the lower path. The resulting characteristic is not a pure dead zone – it includes saturation at the VCC level. By using different resistors and Zeners, characteristics with different dead zones and saturation thresholds can be obtained.

Sometimes it is desirable to direct signals that have amplitudes within specified windows to separate outputs for further processing. To an extent the circuits shown in Fig. 6.17(b) do just that, for two windows. Fig. 6.18(a) exemplifies a circuit with three windows which directs the input signal into three outputs for further processing, depending on the signal amplitude. The circuit to combine signals via different amplitude windows, whose block diagram is shown in Fig. 6.18(b), can be designed in a similar way.

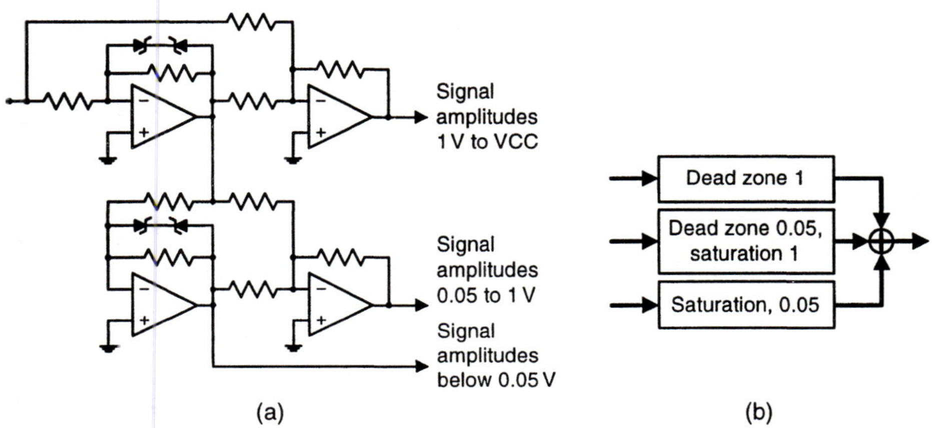

Fig. 6.18 Three-window (a) splitter and (b) combiner

Nonlinear dynamic links can be designed by combining nonlinear and linear links. For example, rate can be limited by placing a saturation link in front of an integrator and closing a tracking feedback loop with sufficient gain coefficient k as shown in Fig. 6.19. Rate limiters are often included in the command path to prevent the plant from being damaged by excessive velocity, and also to reduce the overshoot in the control system response to large commands. Further examples of nonlinear dynamic links will be given in Chapters 11 and 13.

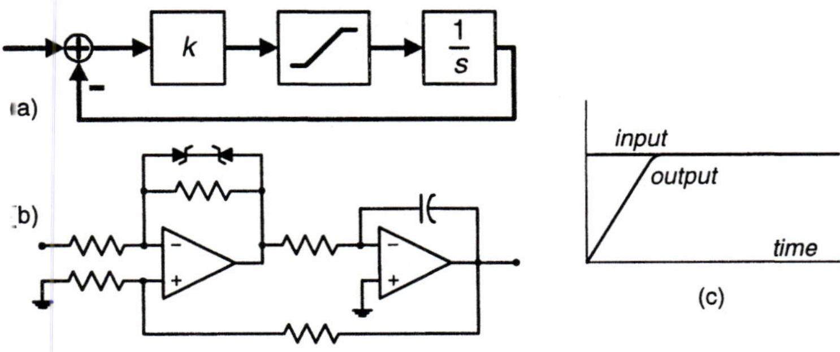

Fig. 6.19 Rate limiting follower, (a) block diagram, (b) schematic diagram, (c) ramp output following step input

6.2 Design and iterations in the element value domain

6.2.1 Cauer and Foster *RC* two-poles

RC two-poles are widely used as components of analog compensators. Not every function can be implemented as an impedance of an *RC* two-pole but only one whose poles and zeros are real, these poles and zeros alternating along the real axis of the *s*-plane, and the closest to the origin being a pole (this constitutes a part of R. Foster's theorem that generally considers *RC*, *RL*, and *LC* two-poles). Any *RC*-impedance function can be implemented in either of the Foster canonical forms shown in Fig. 6.20. The parallel branch form renders smaller total capacitance.

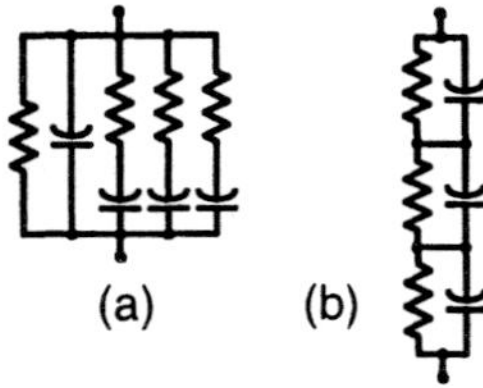

Fig. 6.20 Foster canonical forms of *RC* two-pole

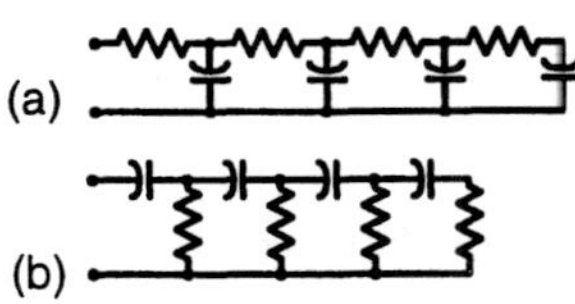

Fig. 6.21 Cauer (ladder) canonical forms of *RC* two-pole

(Notice that Foster's theorem also states that the impedance of a lossless (*LC*) system has alternating purely imaginary poles and zeros; this is the basis of collocated control discussed in Chapters 4 and 7.)

Also, any *RC*-impedance function can be implemented in any of the ladder (W. Cauer's) forms shown in Fig. 6.21. The element values are the coefficients of a chain fraction expansion of the impedance function.

The Foster and Cauer two-poles are employed in analog compensators in Figs. 6.9, 6.10, 6.11 and many others.

Example 1. Two parallel branches, the first a series connections of R_1 and C_1, and the second, a series connections of R_2 and C_2, are often placed in the feedback path of an op-amp to form a compensator making the loop Bode diagram steeper. The frequencies of the two zeros ω_{z1} and ω_{z2}, of the pole ω_{p1}, and the high-frequency asymptotic value of the impedance are expressed by the following equations:

$$\omega_{z1} = 1/(R_1C_1), \quad \omega_{z2} = 1/(R_2C_2),$$
$$\omega_{p1} = (R_1 + R_2)/(1/C_1 + 1/C_2),$$
$$Z(s)|_{s\to\infty} = R_1R_2/(R_1 + R_2)$$

from which the element values can be easily found.

In general, the capacitances and resistances for the Foster form in Fig. 6.20(a) can be calculated by expanding $1/[sZ(s)]$ into a sum of elementary first-order functions. The impedance frequency response of this two-pole with known elements in each branch R_iC_i can be easily found with SPICE. In MATLAB, the response can be found by first calculating the residues $r_i = 1/R_i$ and poles $p_i = -1/(R_iC_i)$, and then using

```
r = [r1 r2 .. ri ..];  p = [p1 p2 .. pi ..];
[num, den] = residue(r,p); bode(num,den)
```

Example 2. Impedance

$$Z(s) = \frac{1{,}000{,}000(s+100)(s+2000)}{s(s+500)}$$

can be implemented as an *RC*-circuit since the poles and the zeros are real, they alternate, and the lowest among them is a pole. We find the elements' values for the Foster form in Fig. 6.21(a) by expanding the function

$$\frac{Y(s)}{s} = \frac{1}{sZ(s)} = \frac{\text{num}(s)}{\text{den}(s)} = \frac{0.000001s + 0.0005}{s^2 + 2100s + 200{,}000}$$

into the sum of two components of the partial fraction expansion

$$\frac{\text{num}(s)}{\text{den}(s)} = \frac{r_1}{s - p_1} + \frac{r_2}{s - p_2} + \ldots$$

The MATLAB code

```
num = [0.000001 0.0005]; den = [1 2100 200000];
[r, p] = residue(num, den)
```

calculates the residues: 0.7895×10^{-6}, 0.2105×10^{-6}, and the poles: –2000, –100.

Next, by comparing a term in the partial fraction expansion of $Y(s)/s$ with the admittance of a series connection of a resistor and a capacitor

$$\frac{s/R}{s + 1/(RC)},$$

we identify $R = 1/r$ and $C = -r/p$. Therefore, the resistors and the capacitors are:

$R_1 = 1.266$ MΩ, $C_1 = 390$ pF, $R_2 = 4.75$ MΩ, $C_2 = 2.1$ nF.

While working in the laboratory, it is often more convenient to think in terms of the element values *R* and *C*, rather than in terms of poles and zeros. In this case, Cauer forms can be of use.

Example 3. Consider the circuit shown in Fig. 6.22(a).

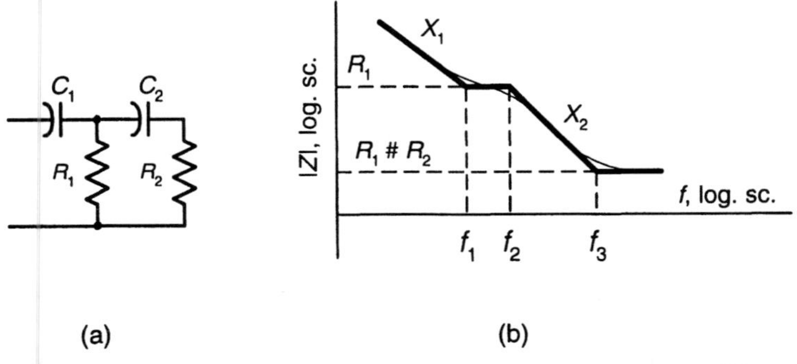

Fig. 6.22 Example of (a) an *RC* two-pole and (b) of its impedance modulus response

We use in this example $C_2 << C_1$. The capacitance C_1 is dominant at lower frequencies where its impedance $X_1 = 1/(2\pi fC)$ is much higher than R_1. Starting at frequency f_1 where X_1 reduces below R_1, the resistor R_1 becomes dominant. At even higher frequencies, starting with f_2, where $X_2 = 1/(2\pi f_2 C_2)$ equals $R_1 + R_2$, the impedance of the small capacitor C_2 becomes small enough to shunt R_1 and becomes dominant. Beyond the frequency f_3 where the impedance $X_2 = 1/(2\pi f_3 C_2)$ approximately equals R_2, the capacitances' impedances can be viewed as negligibly small, and the total impedance becomes $R_1 \# R_2 = R_1 R_2/(R_1 + R_2)$.

From this analysis it is seen that it is easy to identify the elements which need to be adjusted.

Example 4. If the total impedance of the circuit in Fig. 6.22(a) needs to be increased at specific frequencies, it is easy to see from the plot in Fig. 6.22(b) what needs to be done: at lower frequencies, C_1 needs to be reduced; over the interval $[f_1, f_2]$, R_1 needs to be increased; over the interval $[f_2, f_3]$, C_2 needs to be reduced; at higher frequencies, R_2 needs to be increased. In this way the loop response can be adjusted in the laboratory. The same method is convenient to use for loop adjustments with SPICE simulation.

When the final version of the design is being implemented, it might be desirable to reduce the values of the capacitors and to convert the Cauer form into the parallel Foster form. With some experience, it is also possible to start with and make the iterations using the Foster form.

6.2.2. *RC*-impedance chart

A chart for calculating the corner frequencies and for determining the values of resistors and capacitors is shown in Fig. 6.23. Using the chart is especially convenient in the laboratory environment where the accuracy of calculation need not be high, the number of iterations is substantial, and the computer might be on a far-away desk. The chart accuracy is approximately 10%, i.e., 1 dB, which is sufficient for compensator design (at least, for preliminary design). With this chart, many design questions can be instantly answered.

Example 1. If the source impedance is 100 kΩ and the load resistance is 400 kΩ, what is the series capacitance that will place the pole at 20 Hz? – Since the total contour resistance is 500 kΩ, the capacitance must have 500 kΩ reactive impedance at 20 Hz, i.e., it is 15 nF.

Example 2. If the source impedance is 20 kΩ, the load impedance is 20 kΩ, and the shunting capacitance is 1 μF, where is the pole frequency? – The reactive impedance of the capacitance at the pole frequency equals the parallel resistance of the source and the load which is 10 kΩ. Therefore, the pole frequency is at the crossing of the 1 μF line and the 10 kΩ line, i.e., it is 16 Hz.

Consider next the design of the inverting compensators depicted in Fig. 6.10.

Example 3. For the lead link in Fig. 6.10(a), what is the capacitance of the capacitor that is shunting the 100 kΩ resistor R_3, for the compensator zero to be at 200 Hz? – 7 nF. What resistance needs to be placed in series with this capacitor to make a pole at 1 kHz? – 20 kΩ.

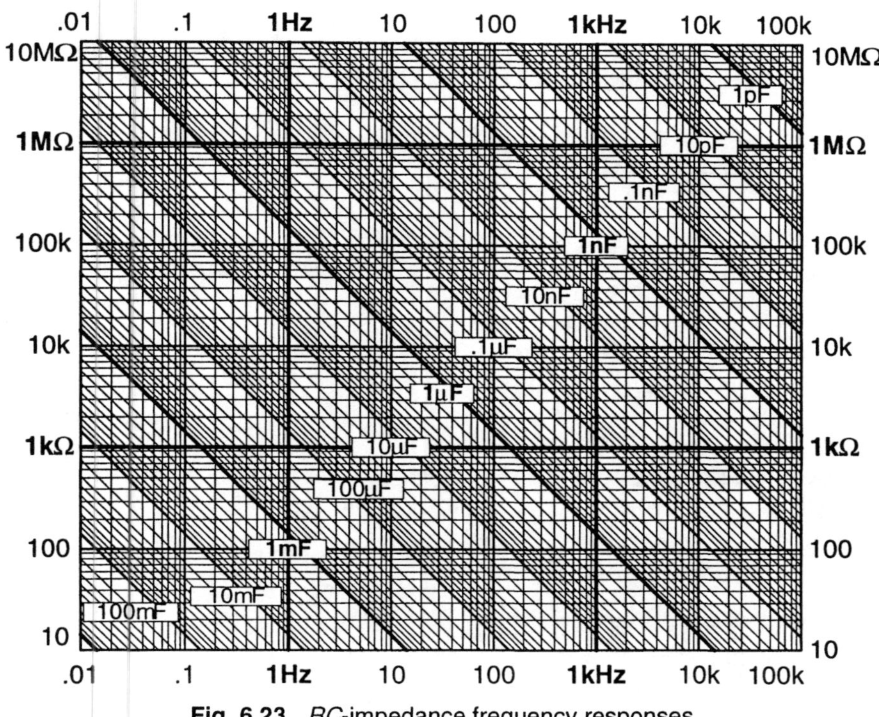

Fig. 6.23 *RC*-impedance frequency responses

Example 4. For the lag link in Fig. 6.10(c), what is the capacitance of the capacitor that is shunting the 100 kΩ feedback resistor R_2, for the compensator pole to be at 200 Hz? – 7 nF. What is the resistance R_3 that makes a zero at 1 kHz? – 20 kΩ.

Example 5. For the lead link in Fig. 6.10(b), the resistance at the connection point of the shunting capacitor is the parallel connection of R_3 and R_4. When $R_3 = R_4 = 200$ kΩ, it is 100 kΩ. What is the capacitance of the capacitor for the compensator zero to be at 200 Hz? – 7 nF. What is the resistance R_2 for the pole to be at 1 kHz? This resistor must provide the same transconductance of the feedback path at this frequency as the T-branch at the pole frequency, which is 5 times smaller than the transconductance at dc (since the ratio of the pole to the zero is 5). Therefore, $R_2 \approx 1$ MΩ.

In this way, design of the compensators is done in no time. Design of the compensators in Fig. 6.10 is left as an exercise.

6.3 Analog compensator, analog or digitally controlled

A multiplying D/A converter is an attenuator with a digitally-controlled attenuation coefficient. Using the converter, it is possible to change the parameters of an analog compensator via a computer command. This method combines the best of the digital and the analog worlds: the compensator is analog, without digital delay, yet its frequency response can be modified by software. Fig. 6.24 shows the application of a multiplying D/A converter (i.e., controllable attenuator) for changing the coefficients *P* and I in a compensator with transfer function $P + I/s$. There exist ICs for implementation of second-order transfer functions with built-in multiplying D/A converters that are controlled from an external parallel bus.

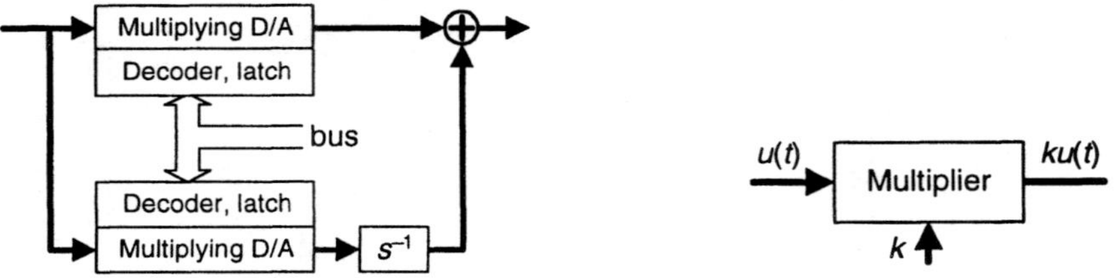

Fig. 6.24 Analog compensator, digitally controlled

Fig. 6.25 Gain regulation with a multiplier

Analog multipliers can be used to change the gain coefficient of a link under the control of an analog signal, as shown in Fig. 6.25 and Fig. 1.19.

6.4 Switched-capacitor filters

6.4.1 Switched-capacitor circuits

Since the size of capacitors in analog circuitry is limited by economy considerations, to achieve the required large time constants, the resistors must be big. However, when active RC circuits are implemented as silicon ICs, large resistors tend to occupy substantial real estate. As an alternative, the resistor can be replaced by a circuit that transfers the charge in small discrete steps. The resistor is imitated by charging a small capacitor C, f_S times per second, and then discharging it into the load using an electronic switch. This method of charge transfer (charge pump) is equivalent to a resistance $R = 1/(f_S C)$. Fig. 6.26 shows a switched-capacitor integrator with transfer function $f_S C_1/(sC_2)$.

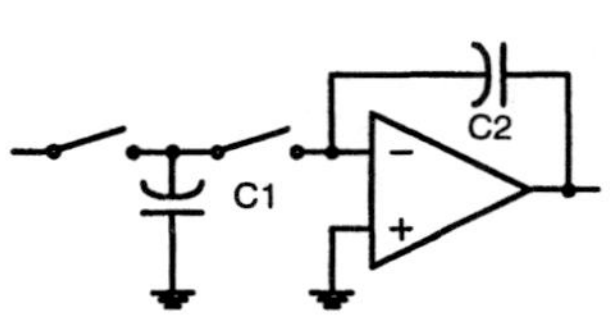

Fig. 6.26 Switched-capacitor integrator

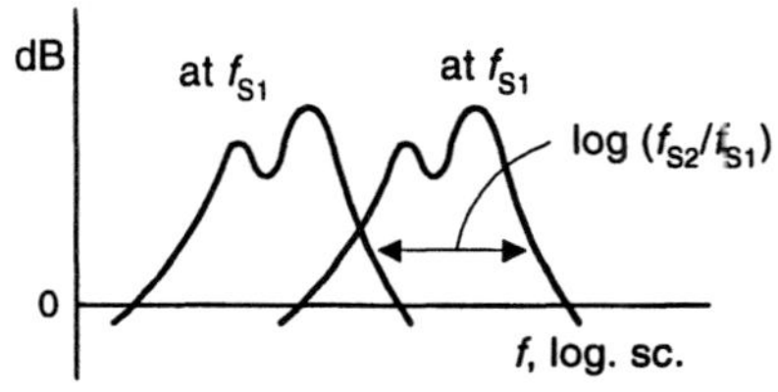

Fig. 6.27 Change in the frequency response with changing sampling frequency

Switched-capacitor active RC circuits use switched-capacitor integrators, summers, and amplifiers. When the switching frequency is varied, all poles and zeros of the transfer function will change by the same factor. Therefore, the frequency response of the switched-capacitor circuit will shift on the logarithmic frequency scale without changing its shape, as shown in Fig. 6.27. This makes the switched-capacitor filter easy to tune.

Switched-capacitor Chebyshev, Butterworth, and Bessel filters, and switched-capacitor biquads programmable from an external parallel bus or by connection to ground or to the VCC of some of the IC pins, are available from several manufacturers.

6.4.2 Example of compensator design

As an example of the application of switched-capacitor circuits in control compensators, we consider a 3-input, 3-output control system for rejecting the vibrations of a spacecraft camera's focal plane sensors. The sensors are cooled by a cold finger connected thermally and mechanically to a cryogenic cooler as shown in Fig. 6.28. The cooler's vibration is counteracted by three piezoelement actuators placed orthogonally between the cooler and its armature. This disturbance rejection feedback system is a kind of homing system, with zero command.

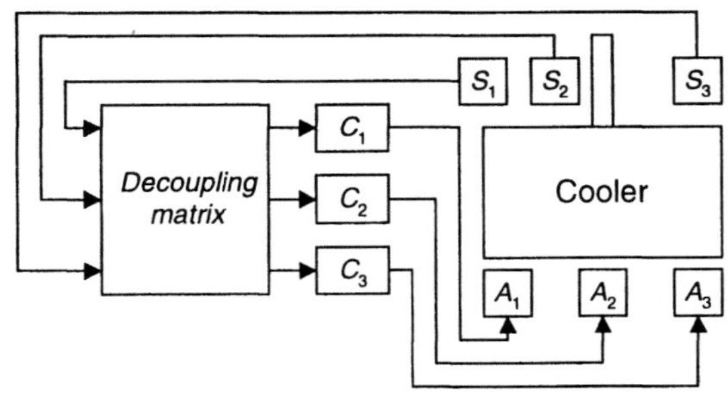

Fig. 6.28 Control loops for vibration suppression of a cryogenic cooler for a camera focal plane

The feedback bandwidth is limited by the flexible modes of the cooler armature. The modes' frequencies are kept over 300 Hz, although their exact values remain unknown. The modes' damping can be quite low so that the peaks on the Bode diagram can be up to 40 dB high. As can be calculated, if the feedback system is implemented as a low-pass system, the feedback of 100 cannot be implemented over a bandwidth wider than 5 Hz.

The frequency of the cooler operation is between 50 and 60 Hz. The frequency is controlled by a spacecraft computer in accordance with the temperature of the cooler. Therefore, in order to effectively reject the disturbances with frequencies in the 50 to 60 Hz range, the system should be the band-pass type.

Due to the system parameter variations, the reliable bandwidth of the feedback is only 3 Hz while the frequency of the cooler motor varies by 10 Hz. Therefore, the system needs to be designed as adaptive, with the feedback frequency response adjustable. Thus, the Bode diagram should be slightly shifted along the frequency axis without substantial variations of its shape – which can be easily implemented by changing the sampling frequency for the switched-capacitor compensator.

The three-axis sensor arrangement is shifted by 90° from the arrangement for the actuators. A decoupling matrix with constant coefficients was calculated and implemented as an analog circuit. With the matrix, experiments show that coupling from the ith piezoelement to the jth sensor is less by an order of magnitude than the coupling from the ith piezoelement to the ith sensor. Therefore, the three loops can be considered fairly independent.

The loop Bode diagrams for each of the three channels are identical, and shown in Fig. 6.29(a). The response is a band-pass version of a *PID* controller, with shallow slopes at the crossover frequencies.

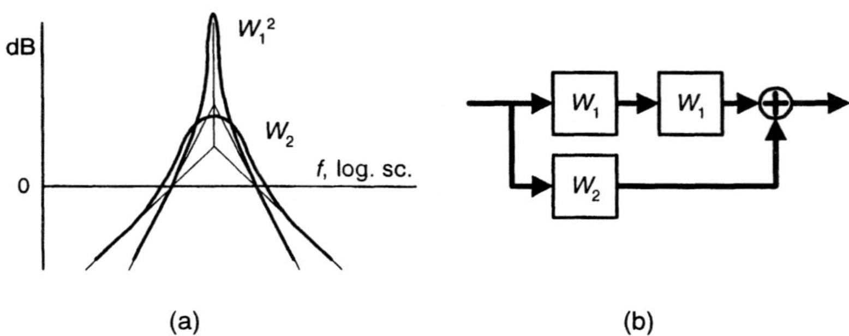

Fig. 6.29 Compensator frequency response (a) and block diagram (b)

The block diagram for the compensator is shown in Fig. 6.29(b). Here, W_1 is a bandpass filter with quality factor 30, and W_2 is a bandpass filter with quality factor 3. Two cascaded high-Q filters have steep roll-off with phase shifts of $\pm 180^\circ$. These filters dominate at frequencies in the functional band (approximately 2 Hz wide) as seen in Fig. 6.29(a). At the higher and the lower frequencies, the low-Q channel is dominant with shallower slopes of the Bode diagram and smaller phase shift which provides the desired stability margin.

Each of the second-order bandpass filters was implemented using a bandpass second-order switched capacitor filter IC. The sampling frequency, nominally of 10 kHz, was controlled from the flight computer to vary the central frequency, as required, within the range of 50 to 60 Hz.

6.5 Miscellaneous hardware issues

6.5.1 Ground

Practical implementation of analog compensators is simple and forgiving, and can be performed by people without a special background in electrical engineering. Still, some assorted hints and warnings are worth noting, as in the following.

Signal distortion, noise, and signal interference may result from improper grounding. The resistance of a ground wire or a ground plane (a grounded layer of a multilayer printed circuit board) is commonly very small, and for simplicity is commonly considered to be zero. This assumption is acceptable in most cases, but not in all cases. Consider, for example, a power supply for the circuitry with the schematic diagram shown in Fig. 6.30. The diagram includes a power down-transformer, a two-way diode rectifier, a Π-type *RC* low-pass filter to smooth the ripple of the output voltage, and a voltage stabilizer IC.

The ac current through the filter capacitors is relatively large, especially through the first capacitor. The currents flowing via the ground plane to the central tab of the transformer winding create voltage drops on the ground plane that are typically of several μV and can even reach several mV. To reduce the voltage variations on the analog ground, first, the transformer tap wire should be placed close to the capacitors, and second, the rectifier ground and the analog ground should be connected in a single point as shown in Fig. 6.30. Any extra connection between the grounds would create paths for the ac currents to flow via the analog ground and to produce voltage drops.

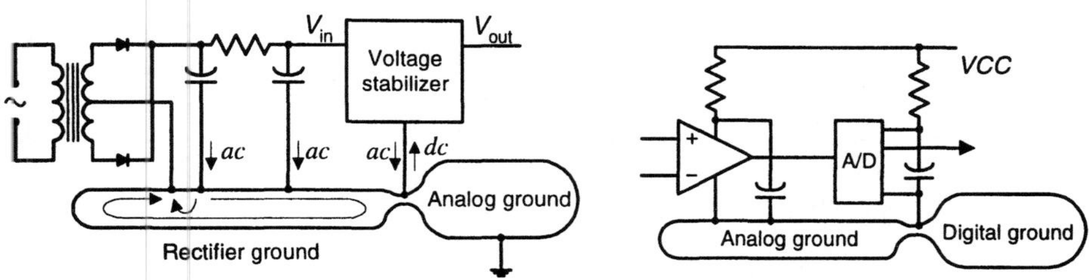

Fig. 6.30 Ground for voltage regulator **Fig. 6.31** Grounding for A/D converter

In systems employing digital controllers, commonly, the actuator and the plant of a control loop are analog, and the sensor in most cases is also analog. The digital part of the loop starts with an A/D converter following the sensor. The currents in the digital part of an A/D converter are typically several orders of magnitude larger than the lowest analog signal to be measured. Therefore, provisions should be made for the digital currents not to interfere with the analog input signal.

In particular, attention needs to be paid to maintaining proper ground configuration. The digital currents should be prevented from flowing over the analog ground and there producing voltage drops, otherwise different points of the analog ground would have different potentials. This is achieved by connecting the analog and digital grounds at one point only. Since commonly the analog and digital grounds are already connected within an A/D converter, this point must be the only one connecting the grounds as shown in Fig. 6.31.

The ac current also flows via the power supply VCC line, and care should be taken to avoid parasitic coupling between the grounds. The low impedance ac connection between the grounds via power supply lines must be eliminated. To do this, the blocking capacitors from the power lines should be connected to the proper ground points as shown in Fig. 6.31.

In the digital part of the circuit, it is commonly sufficient to connect the VCC line to the ground by capacitors placed close to the IC that consume large pulsed current from the power supply line. In the analog part of the circuit, low-pass filters are usually placed in the power supply line. The capacitors of the filters should be placed close to the IC or else at higher frequencies, the impedance of the wires might introduce parasitic feedback and coupling due to imperfect ground.

6.5.2 Signal transmission

In some control systems, the sensors are far away from the controller and need to be connected with long cables. The resulting communication link is subject to noise and electromagnetic interference. Fig. 6.32(a) shows signal transmission of the asymmetrical transmitter output voltage V_t to an asymmetrical receiver using common ground as the return wire. The ground potential at the input to the receiver differs by V_g from the transmitter ground potential because of various ac and dc currents from different sources flowing through the imperfect (non-zero impedance) ground wire. As a result, the received signal $V_r = V_t - V_g$ is corrupted with the ground noise V_g.

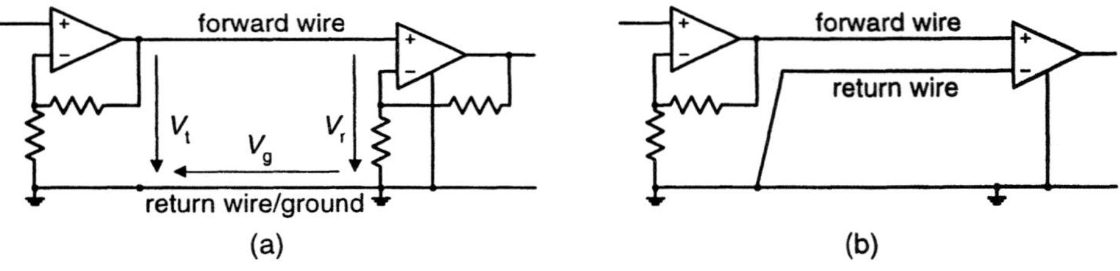

Fig. 6.32 Parasitic feedback due to common ground

To avoid errors caused by imperfect ground, differential amplifiers are used with a separate return wire which is only connected to the ground at the transmitter output as shown in Fig. 6.32(b). Precision amplifiers with differential inputs are called ***instrumentation amplifiers***. They are commonly composed of three op-amps or, better yet, purchased as complete ICs.

A simple differential amplifier with the gain coefficient of 2 can be built according to the schematic diagram in Fig. 6.33. The amplifier is an acceptable replacement for an instrumentation amplifier when the common signal component does not exceed half of the VCC (so it will not saturate the amplifiers) which is commonly the case.

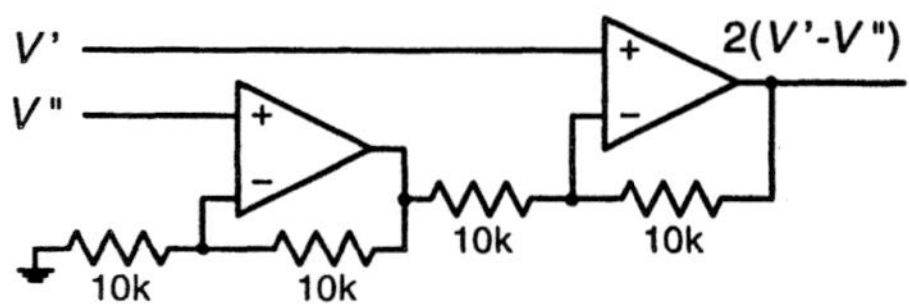

Fig. 6.33 Differential amplifier

Electromagnetic interference from various sources in an industrial environment into the signal wires could contaminate relatively small analog signals in the control loop, especially when the links are connected with rather long wires. Twisting the pair of wires reduces the difference in the voltages induced by external magnetic fields in each of the wires, and these voltages due to their opposite polarity cancel each other at the input to the differential receiver.

As shown in Fig. 6.34(a), bifilar series coils are employed for suppression of the interference at higher frequencies where it is rather easy to make the coil with high inductive impedance. For the signal, the coil impedance is small since the signal current flows through the two coil windings in opposite directions producing no magnetic field. For the interfering signal affecting only one wire, the coil impedance is high, thus suppressing the interference.

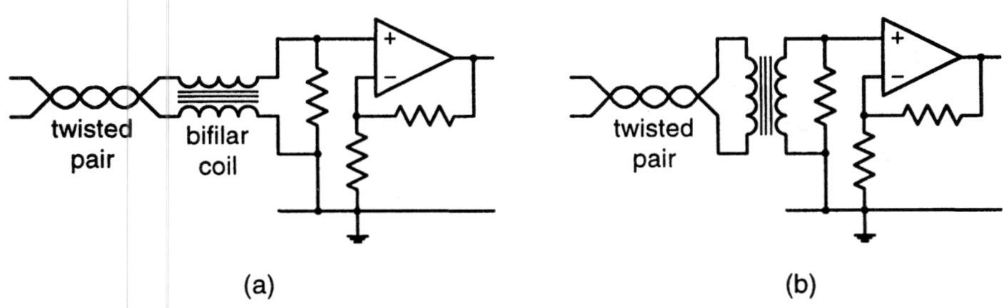

Fig. 6.34 Using (a) bifilar coil to reject high-frequency interference and (b) balun transformers to reject low-frequency and medium-frequency interference

To reduce interference at lower and medium frequencies while using asymmetrical receivers, balanced-to-unbalanced transformers (baluns) can be used as shown in Fig. 6.34(b). Since transformers do not convey dc, this method is only suitable for signals not containing dc components (like sensor signals in feedback systems for vibration suppression).

6.5.3 Stability and testing issues

The internal feedback in an op-amp is maximum when the op-amp is used as a follower. Some op-amps in this configuration are only marginally stable, especially at higher frequencies and when the wires are long thus introducing stray inductances and capacitances in the feedback loop. Placing a resistor in the feedback path as shown in Fig. 6.35 commonly provides sufficient stability margins.

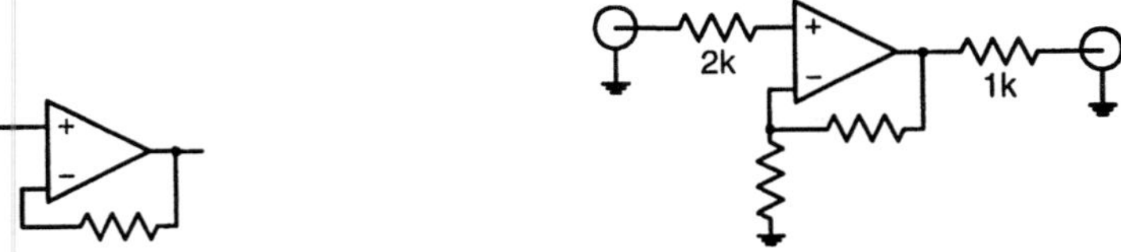

Fig. 6.35 Using a feedback resistor in the follower

Fig. 6.36 Reducing the effect of cables on an op-amp feedback loop

Amplifiers are often connected with the rest of the equipment with long cables. Since the cables' input impedances shunt the feedback for the op-amp, the cable capacitance and the resonances in the cable can make the op-amp unstable. The instability can be eliminated by the addition of series resistors at the input and output of the amplifier as shown in Fig. 6.36. The resistances cannot be too large since the input resistor introduces noise into the circuit, and the output resistor increases the amplifier output impedance. The resistances are commonly chosen from 1 to 2 kΩ.

An oscilloscope should be connected to the breadboard while the circuit is tested, tuned, or troubleshooted. Without an oscilloscope, many possible problems can be misunderstood. For example, some power amplifiers with f_T of only 2 MHz can oscillate at 50 MHz, if the wires connecting the IC are too long and their inductances form a resonance tank amplifying the signal in the parasitic feedback loop. Without an oscilloscope, when a signal analyzer is used, this oscillation might manifest itself by only a small change in the gain and some reduction of the output power. (But, at the same time, the IC gets hot, the power supply current increases, and the output signal decreases.)

On the breadboard, the input, the output, and some spare connectors should be firmly attached as shown in Fig. 6.37(a). The spare connectors will be used for troubleshooting, to test the signals at various nodes of the circuit.

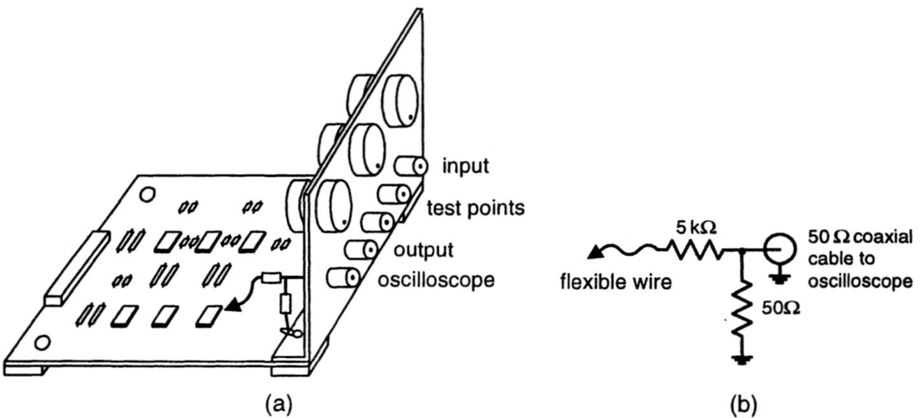

Fig. 6.37 (a) Breadboard with connectors and (b) oscilloscope matching termination

Signal reflections from mismatched ends of a coaxial cable going to an oscilloscope can corrupt the measurements. To eliminate the distortions, the cable should be matched at least at one side. The test fixture shown in Fig. 6.37(b) is convenient to have implemented on the board. It matches the 50 Ω cable and, at the price of attenuating the measured voltage 100 times, makes the input impedance of the probe 5 kΩ which is sufficiently high for most troubleshooting tasks.

While designing an experimental breadboard, it is a good practice to leave one of the op-amps as an extra available inverter for the purpose of changing the phase of the signal from the sensor – since it is quite common that the sensor output signal polarity is not known or will be changed during experimentation.

While closing control loops over some physical plant, especially when the plant is mechanical and can be damaged by high-amplitude oscillation, one should introduce potentiometers either in the paths from the compensators to the actuators, or in the paths from the sensors, to be able to introduce feedback gradually and to reduce it rapidly if an oscillation starts. Such potentiometers greatly improve the troubleshooting options when multiloop systems are tested.

6.6 *PID* tunable controller

6.6.1 *PID* compensator

Fig. 6.38 shows a block diagram of a feedback system with a ***PID compensator*** $I/s + P + Dqs/(s + q)$. The third term is a band-limited differentiation path approximating Ds for $\omega < q$. Here, the scalar parameters P, I, and D are to be determined (tuned) for a specific plant. A saturation link is commonly placed in front of the integral term (and, if the plant is a double integrator, also in front of the proportional term) to prevent windup (see Chapter 13).

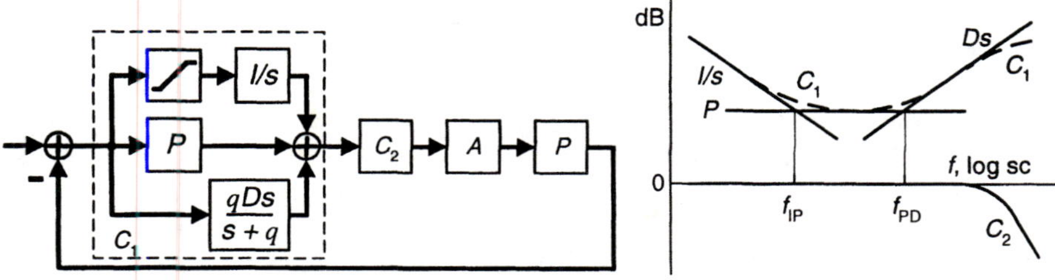

Fig. 6.38 *PID* controller block diagram

Fig. 6.39 Bode diagrams for the *PID* compensator

Fig. 6.39 shows Bode diagrams for each path of the compensator C_1 and the entire compensator. The component P dominates at midrange frequencies, the component I/s, at lower frequencies, and the component Ds, at higher frequencies. Typically, the distance between the corner frequencies f_{IP} and f_{PD} is 2 octaves or more, and the compensator transfer function has two real zeros corresponding to these frequencies. The pole q in the differentiator is typically chosen to be from $3\times2\pi f_{PD}$ to $10\times2\pi f_{PD}$.

The Bode diagram of a typical plant is convex, monotonic, and with increased slope at higher frequencies. If the plant response at higher frequencies is too shallow, a low-pass filter C_2 is commonly added to attenuate the sensor noise and to gain-stabilize the system at higher frequencies. The concave response of the *PID* compensator needs to be adjusted for the loop Bode diagram to have the desired slope.

In a system with the plant having a pole at the origin and two or more additional real poles at higher frequencies, typically, the P term dominates near f_b, the Ds term at frequencies over $4f_b$, and the I/s term at lower frequencies up to $f_b/4$ as shown in Fig. 6.40. Thus, P can be tuned for the feedback bandwidth, D for the phase stability margin (D-term phase advance partially compensates the lags of the high-frequency poles of P and C_2), and I, for disturbance rejection at lower frequencies. When a step test-command is used for closed-loop tuning, P is tuned for the rise-time/overshoot trade-off, D for the overshoot reduction, and the coefficient I is increased until it starts affecting the overshoot too much.

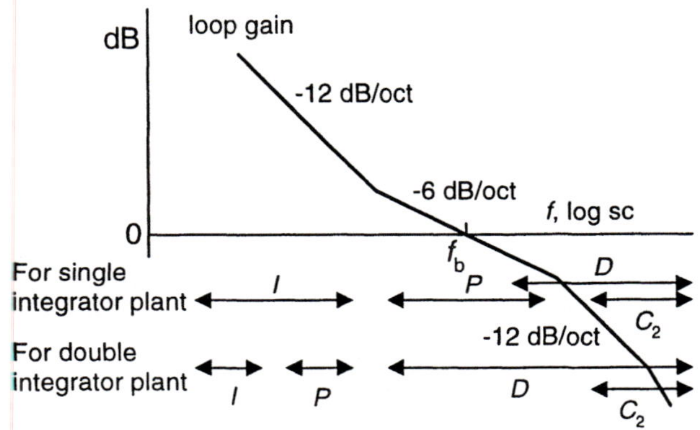

Fig. 6.40 Ranges of dominant terms in PID compensators

The *PID* controller for a double integrator plant uses the term *Ds* to reduce the slope of the loop Bode diagram at the crossover. When the plant has also several high-frequency poles or a large n.p. lag, the *PID* controller may be augmented by a fourth parallel branch with a band-limited double differentiator.

The *PID* controller is very popular. In fact, it is the industry standard for tunable controllers. The controller is easily tuned to provide robust and fairly good performance for a great variety of plants. It typically provides an acceptable transient response without a prefilter or command feedforward. Several automatic tuning procedures based on tuning the transient response are successfully and widely used. However, the performance of the controller is not optimal. With the same average loop gain at higher frequencies, the Bode-step responses provide larger feedback, and prefilters can reduce the overshoot.

6.6.2 *TID* compensator

In the *PID* controller, the vector diagram at the corner frequencies f_{PD} is as shown in Fig. 6.41(a). The output is larger by 3 dB than each of the components.

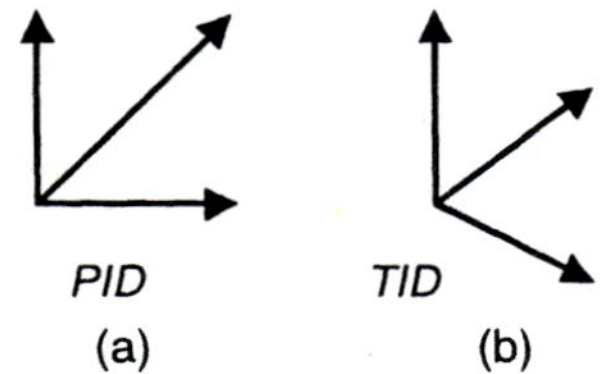

Fig. 6.41 Vector diagrams for corner frequencies for PID (a) and TID (b) compensators

As shown in Fig. 6.42, the zero-slope *P*-path can be replaced by the *tilted* response path $Ts^{-1/3}$ with constant gain slope −2 dB/oct, where *T* is a scalar tunable parameter. Implementation of the transfer function $s^{-1/3}$ was already described in Section 5.3 and exemplified in Fig. 6.11. Bode diagrams for the *T*, *I*, and *D* paths' gains are shown in Fig. 6.43. (For simplicity, the band-limiting pole in the differentiator is not shown in the digram).

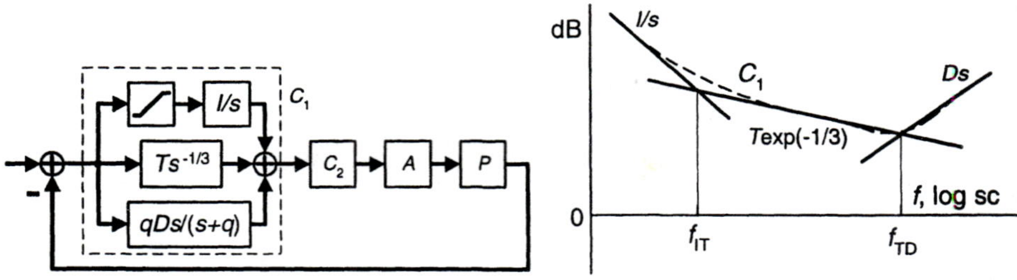

Fig. 6.42 *TID* controller

Fig. 6.43 Bode diagrams for *TID* compensator

The vector diagram in Fig. 6.41(b) shows the components forming the output signal of the ***TID compensator*** at the corner frequency f_{TD}. The output signal amplitude is the same as the amplitudes of the components. Because of this, the f_{TD} corner of the *TID* gain frequency response is sharper than the similar corner in the *PID* compensator. Due to this and due to the slope of the T-term gain response, the controller provides a better loop response.

Example 1. Consider *PID* and *TID* controllers for a first-order plant using a sensor whose transfer function possesses a triple real pole at 80 Hz. Limited by the sensor noise, the crossover frequency is $f_b = 20$ Hz. The Bode diagrams for the loop gain achieved with *PID* and *TID* compensators are shown in Fig. 6.44(a). The controllers have the same coefficient D (i.e., the same gain at higher frequencies and, therefore, nearly the same level of high-frequency noise).

At the typically critical frequencies of about half f_b, the feedback in the *TID* controller is 4 dB larger than that in the *PID* controller (although the *TID* feedback is still smaller than the feedback achievable with a Bode step).

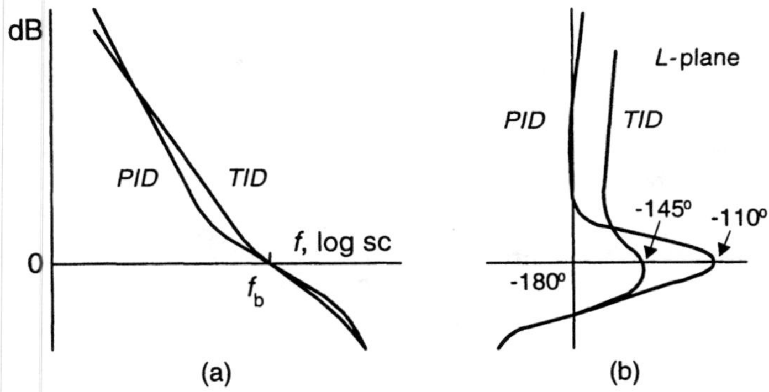

Fig. 6.44 Bode (a) and Nyquist (b) diagrams for *PID* and *TID* controllers

The L-plane open-loop diagrams for *PID* and *TID* control are shown in Fig. 6.44(b). It is seen that the *PID* phase stability margin near f_b is too large, 70°, and the phase margin is not excessive in the *TID*-compensator.

PID and *TID* controllers are both easy to tune when the plant parameters change, e.g. for a temperature control of an industrial furnace with variable payload. When the controller is not supposed to be tuned for each individual plant, preference should be given to a high-order compensator with a Bode step.

6.7 Tunable compensator with one variable parameter

6.7.1 Bilinear transfer function

Plant characteristics may depend on the varying environment. Examples are the dependence of an aircraft's dynamics on altitude, and the dependence of a telecommunication cable's attenuation on temperature. For the loop response to be optimal for each of the intermediate values of the environmental parameter, the compensator must be variable as shown in the block diagram for the resulting adaptive system in Fig. 6.45 (and in Fig. 9.2). We assume we know how the plant response varies with the changing environment (although not precisely, so that cancellation of the effects in an open-loop manner cannot be done).

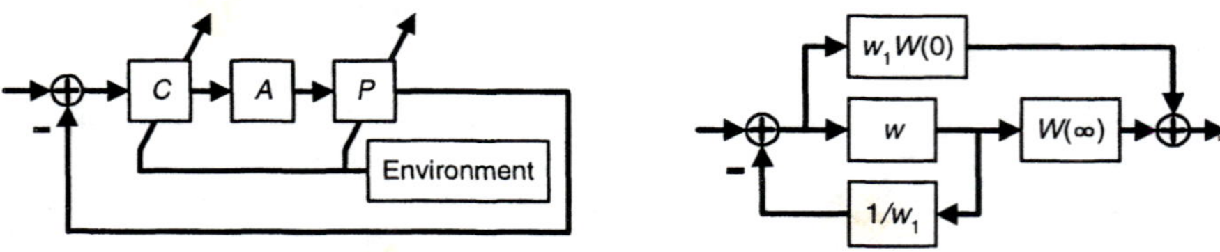

Fig. 6.45 Adaptive system block diagram

Fig. 6.46 Flowchart for transfer function dependent on a parameter

It would be convenient to modify the transfer coefficient (or immitance; this term defines both impedance and admittance) W of a linear system by only changing one parameter w. Here, w stands either for the variable transfer function of a unilateral link, or for immitance of a variable two-pole.

The function $W(w)$ is bilinear (a ratio of linear functions) [1] and as such can be expressed as

$$W(w) = \frac{w_1 W(0) + wW(\infty)}{w_1 + w} \tag{6.13}$$

If w represents the variable impedance of a two-pole, w_1 is the driving point impedance between the terminals to which the two-pole w is connected. If w designates the transfer coefficient of an amplifier, then $-1/w_1$ is the feedback path transmission coefficient for this amplifier.

The flowchart corresponding to (6.13) is shown in Fig. 6.46.

The responses of the variable compensator in Fig. 6.46 should be regulated smoothly. For this purpose, not all functions of the form (6.13) will do. For example, if it is desired to gradually change the slope of a Bode diagram from –6 dB to 6 dB, and we correspondingly choose $W(0)$ to be an integrator, and $W(\infty)$ a differentiator, as shown in Fig. 6.47, and only use a gain block w in series with the differentiator, then the response will be changed from that of an integrator to that of a differentiator with w changed from 0 to a large value, but, as is seen in Fig. 6.48, the frequency response of W in the intermediate position possesses a zero, i.e., the regulation is not smooth.

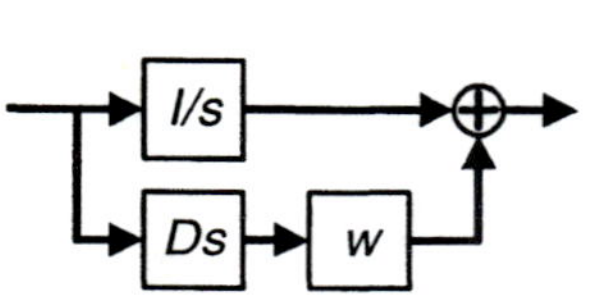

Fig. 6.47 Block diagram of a regulator

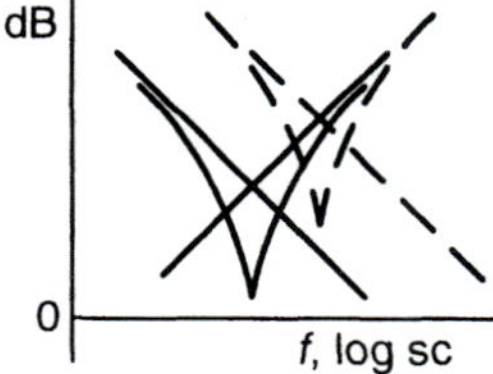

Fig. 6.48 Frequency responses of a regulator

6.7.2. Symmetrical regulator

Smooth regulation can be obtained with Bode symmetrical regulators. Regulation is called symmetrical with respect to the nominal value w_o of the variable parameter w when the maximum relative deflections of w from w_o, up and down, cause symmetrical (in dB) variations in W as shown in Fig. 6.49, i.e., when the regulation has the following property:

$$Q = \frac{W(\infty)}{W(w_o)} = \frac{W(w_o)}{W(0)} \tag{6.14}$$

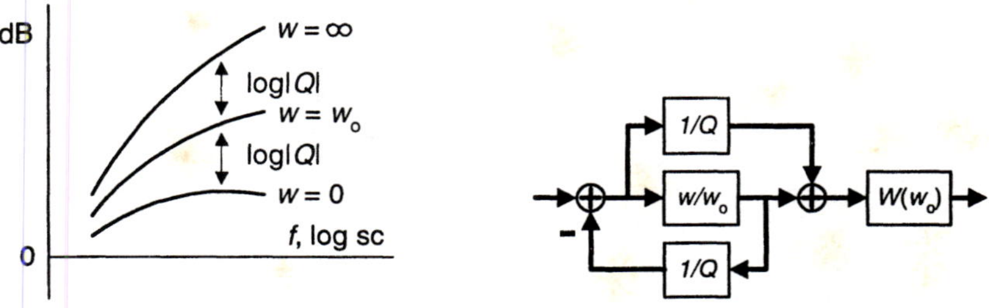

Fig. 6.49 Regulation frequency responses of a symmetrical regulator

Fig. 6.50 Flowchart for a symmetrical regulator

By substituting this expression into (6.13) we have

$$w_o = w_1/Q, \tag{6.15}$$

and

$$W = W(w_o)\frac{1+(w/w_o)Q}{(w/w_o)+Q}. \tag{6.16}$$

The flowchart for the symmetrical regulator is shown in Fig. 6.50.

The gain of the regulator

$$20\log|W| = 20\log|W(w_o)| + 20\log\left|\frac{1+(w/w_o)Q}{(w/w_o)+Q}\right| \tag{6.17}$$

changes gradually with variations in w. When $w = w_o$, the second component is 0. The second component retains the value but changes the sign when w_o is switched from 0 to ∞ as illustrated in Fig. 6.49. The regulation is exactly symmetrical for these maximum variations of the gain response. It can be shown that in the Taylor expansion of the second component, the first-order term dominates to the extent that the gain $20\log|W|$ depends on w/w_o nearly linearly over the regulation range more than 20 dB.

The regulator can be used in the compensator for changing the loop Bode diagram in accordance with changing system requirements, for solving the trade-offs between the available disturbance rejection and the output noise. It can be used to compensate the effects of known plant parameter variations in adaptive systems (adaptive systems are studied in Chapter 9). Also, a nonlinear element can be used in place of the variable element w. In this case, a nonlinear dynamic link with desirable properties can be built. Such nonlinear links can be employed in compensators for enhancing the system performance as will be discussed in Chapters 10, 11, and 13.

6.7.3 Hardware implementation

A hardware implementation of such a regulator is shown in Fig. 6.51(a). The possible implementations of the block $1/Q$ as low-pass and high-pass RC filters are shown in Fig. 6.51(b) and (c).

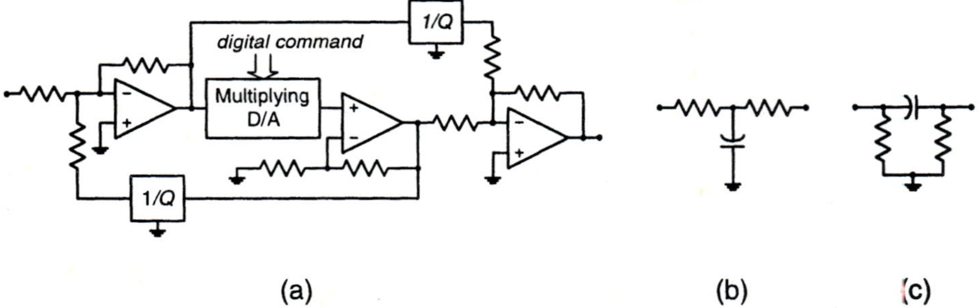

Fig. 6.51 (a) Symmetrical regulator implementation with (b) low-pass and (c) high-pass *RC* filters as $1/Q$

The regulator responses with the high-pass filter are shown in Fig. 6.52. The regulator can be used to vary the compensator response when the plant response varies mostly at higher frequencies. Additional examples are given in Sections 11.7, 13.5.

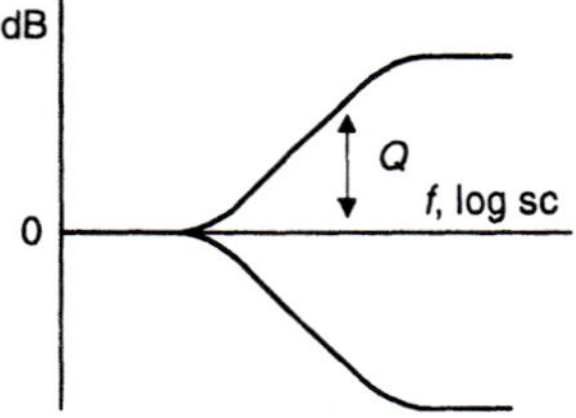

Fig. 6.52 Frequency responses for the symmetrical regulator

Regulation of frequency responses can be performed also with digital FIR (finite impulse response) filters which can be made to approximate any desired frequency response. However, using the filter for gradual response changes requires changing several coefficients which is less convenient.

6.8 Loop response measurements

To design and to verify the loop frequency response, one has to calculate or measure the plant response. The choice between the plant simulations and the actual plant measurements might depend on the plant size and accessibility. For example, if the plant is big or expensive (like a spacecraft or heavy machinery) or, if it is very small like a microwave feedback amplifier or a micromachined mechanical device which is difficult to access, then simulation of the plant should play a major role in the feedback loop design. In this case the plant model needs to be accurate, and substantial efforts are justified to develop this model. At the beginning of the project, when this model is not yet completed, simpler models must be developed for preliminary control design, to be able to answer the questions crucial for the plant design. On the last stages of the project, when the plant is built, it can be measured and the feedback loop tested. In all cases, the compensator must be tested first, before testing the entire loop.

On the other hand, when the plant is inexpensive and easily measured, the compensator (analog or digital) can be designed well directly in the laboratory while working with the real plant, by measuring the loop response and adjusting the elements of analog compensators or coefficients in digital compensators.

The loop response can be measured when the feedback loop is open as in Fig. 6.53, or while the system is in the closed-loop configuration. While the loop response is being tested, the loop can be broken, theoretically, at any cross section. In practice, the cross section must be carefully chosen such that the signal at this cross section is large enough for convenient measurements but not excessively large to make it difficult to generate such a signal. Typically, the loop is broken after the first stage C_1 of the compensator, as shown in Fig. 6.53.

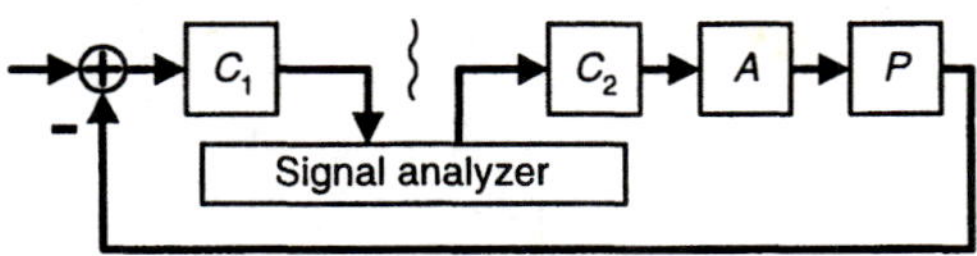

Fig. 6.53 Breaking the feedback loop

The choice of the cross section for breaking the loop also depends on how easy is it to simulate the load for the loop output. For conventional control loops with separate compensators this is not an issue, but it can be a difficult problem in some systems, especially at microwave frequencies or in magnetics.

Measurements of the open-loop response in the open-loop configuration are not always convenient or even possible. The feedback loop transfer function measurements can be complicated by parasitic feedback loops from the loop output to the loop input (especially at higher frequencies). Also, some plants cannot be used without feedback – they could be unstable, self-destructive, or dangerous to deal with. The control loops about such plants need to be measured in the closed-loop configuration – however, perhaps with reduced feedback.

Loop response measurements in the closed-loop configuration require that the closed-loop system be stable. If it is not, it still might be stable when the feedback is reduced by an extra attenuator, i.e., the system is closed with smaller feedback. Injection of the signal into the feedback loop for the purpose of the open-loop response measurements with the loop kept closed is shown in Fig. 6.54. The loop transfer function is the ratio of the signal at input 2 to the signal at input 1 (the latter is applied to the "reference" input of the analyzer).

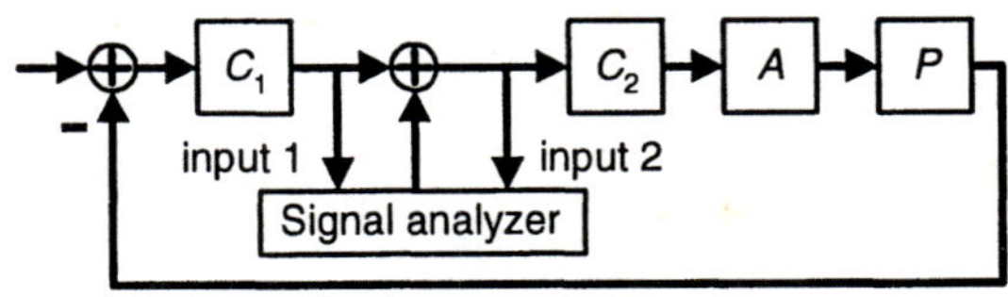

Fig. 6.54 Loop response measurements with the loop closed

A practical circuit for the signal injection is shown in Fig. 6.55. The variable resistor allows reducing the loop gain, and the *RC* lead is used to increase the test signal level at higher frequencies – as will be discussed below. The capacitance value $C = 2{\cdot}10^{-5}/f_b$ places the zero of this lead at approximately $0.08f_b$, and the pole at approximately $4f_b$.

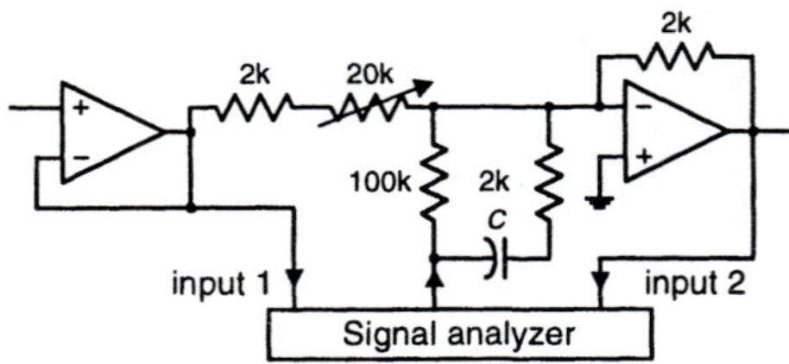

Fig. 6.55 Summer for injecting test-signals into the closed loop

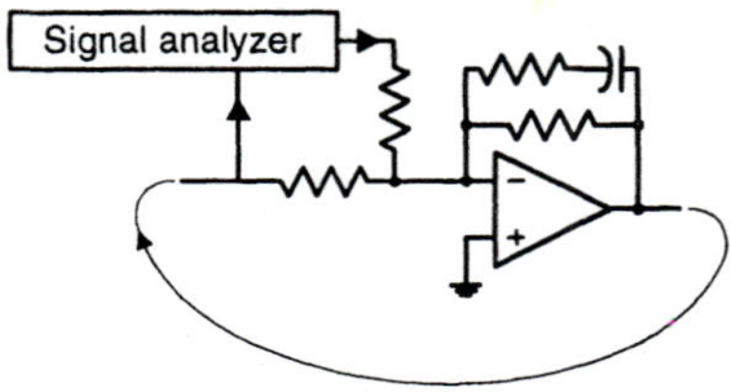

Fig. 6.56 Signal injection using an already existing inverting amplifier

If an analog compensator already includes an inverting op-amp, the signal injection can be implemented with a few passive elements as shown in Fig. 6.56, using the op-amp as a summing amplifier.

Signal analyzers typically use two types of test-signals: swept-frequency sinusoidal and pseudorandom with subsequent fast Fourier transform (FFT) and averaging. Both methods are quite suitable for the loop response measurements. At higher frequencies, sinusoidal excitation is easy to implement and is appropriate. At low frequencies, FFT might provide better accuracy or shorter time of measurements. Still, low-frequency response measurements often take longer than desired, especially when the test is performed in the field, or some equipment normally in operation needs to be shut down to perform the test. Part of the problem is that when the test-signal is employed with constant amplitude of sinusoidal signals at all frequencies, or with constant spectral density of pseudorandom signals, the dynamic range of the signals at the output of the loop becomes very large, since the loop gain is many tenths of dB in the functional feedback band and –20 dB at higher frequencies still of interest for the stability margin test. Then, the signal amplitude chosen so as not to overload the actuator at lower frequencies becomes too small at higher frequencies, and data recovery from the noise requires multiple runs and averaging.

An extra linear link with the gain increasing with the frequency, such as that shown in Fig. 6.57, can be introduced at the signal analyzer generator output to reduce the dynamic range of the loop output signal and correspondingly to reduce the required time for accurate measurements.

In the absence of a signal analyzer, the loop gain response can be measured point-by-point, at discrete frequencies, with a signal generator and a two-input oscilloscope. The input signal of the loop is applied to one of the oscilloscope *Y*-inputs, and the loop output is applied to the other *Y*-input. The phase difference between the signals can be seen on the scope screen with sufficient accuracy.

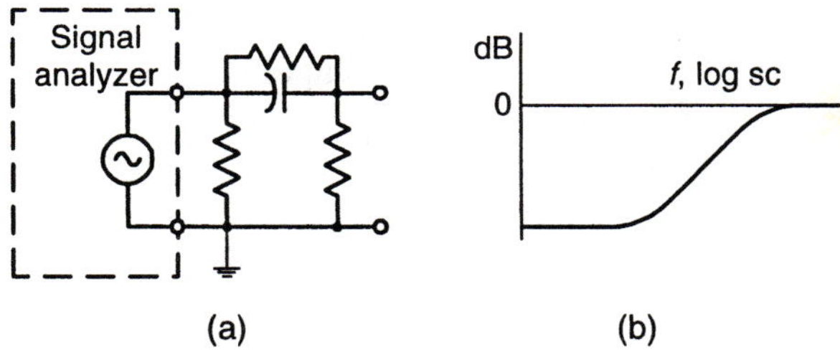

Fig. 6.57 (a) Gain response corrector to be placed in cascade with the signal source to improve the measured signal-to-noise ratio, and (b) the corrector gain-frequency response

With a sweep generator and a scope, an addition of the circuitry shown in Fig. 6.58 can mechanize the loop response measurements. Here, two amplifiers (optional) increase the levels of the signals at the input and output of the feedback loop. The amplifier gain can be made to increase with frequency in order to reduce the dynamic range of the signals.

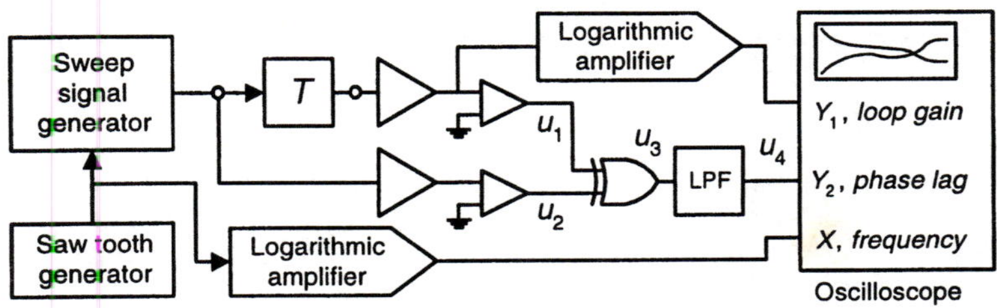

Fig. 6.58 Set for loop response measurements

The two single-supply comparators convert the sinusoidal signals to rectangular single-polarity signals u_2 and u_1 as illustrated in Fig. 6.59. The Exclusive OR serves as phase detector. Its output signal u_3 is nonzero only when only one of the input signals is nonzero as shown in Fig. 6.59. The average value u_4 of the signal u_3 is proportional to the phase lag between u_2 and u_1. The low-pass filter producing u_4 can be a Butterworth active *RC* filter.

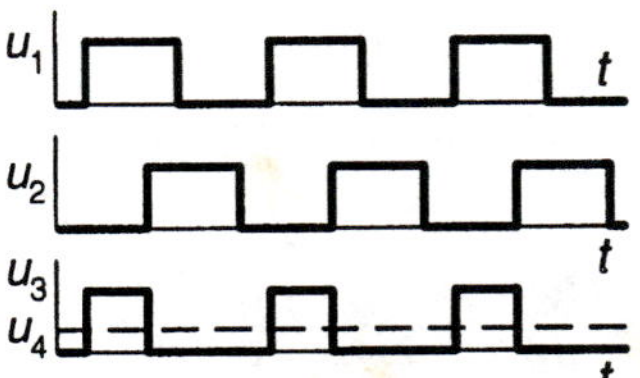

Fig. 6.59 Time diagrams for the phase detector signals

When the plant transfer function is relatively simple, the plant can be identified by its response to the step function applied to its input. This method is often employed for tuning *PID* controllers for chemical processes as is discussed below.

While the process is going on open-loop, the output is made steady by adjusting the command manualy. Then, a small step is applied directly to the plant input as shown in Fig. 6.60(a). This step produces an increment in the plant's ouput (controlled variable) as shown in Fig. 6.60(b). From the response, the delay time t_d and the rate r are determined. The parameters in the *PID* controller written in the form

$a[1 + 1/(bs)](1 + cs)/(1 + cs/d)$ are set to be: $a = 1/(rt_d)$, $b = 5t_d$, $c = t_d/2$, $d = 4$, and then further adjusted experimentally for good closed-loop performance. This open-loop Ziegler-Nichols tuning procedure is not universal and does not necessarily lead to good tuning of PID controllers for all possible plants.

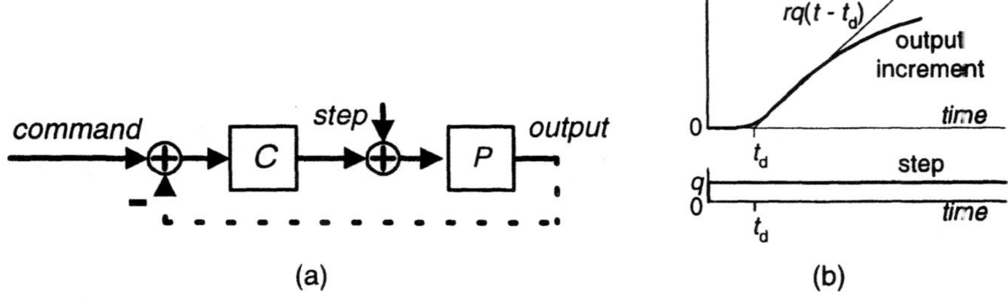

Fig. 6.60 Plant transient response measurements, (a) block diagram and (b) transient response

A more general approach is to create a plant model with a transfer function having a couple of real poles and zeros, and adjust the values of the model poles and zeros with Matlab to obtain the same transient response as that of the real plant. Then, using this plant model, a compensator can be designed that provides the desired loop response.

6.9 Problems

1 (a) Design an amplifier with gain coefficient –25 using an op-amp in the inverting configuration. The amplifier will be employed in the compensator for a control system with 1 kHz feedback bandwidth. Choose f_T, and choose resistors such that the power consumption by the feedback resistor from the 12 V power supply does not exceed 3 mW.
(b) Do the same for the gain coefficient –100, f_b = 100 kHz, power not exceeding 0.2 mW.

2 (a) Discuss the accuracy of implementation of a transfer function as an op-amp with *RC* impedances.
(b) Define the required tolerances of resistors and capacitors.

3 Draw a schematic diagram for an op-amp circuit implementing the function:
(a) $-ax - by - c$;
(b) $+ax + by + c$;
(c) $-ax + by - c$;
(d) $-ax + b\int y dt - c$;
(e) $-ax + bdy/dt - c$;
(f) $\int(-ax - by - c)dt$.

4 Draw a schematic diagram and specify the values of the circuit elements (use a 100kΩ feedback resistor) for a practical op-amp circuit implementing the function:
(a) $-0.5x - 0.2y - 3$;
(b) $+2x + 8y + 0.2$;
(c) $-3x + 6y + 1$;
(d) $-4x + 4\int y dt + 0.3$;
(e) $-2.72x + 0.5dy/dt + 0.1$;
(f) $\int(-6x + 0.4y - 2)dt$.

5 Using 20 kΩ feedback resistors in op-amp summers, draw a schematic diagram for an analog implementation of decoupling matrix:

(a) $x = 0.5038x' + 0.0443y' + 0.0772z'$
$y = -0.0235x' + 0.4796y' - 0.0215z'$
$z = -0.0094x' - 0.0243y' + 0.5291z'$

(b) $x = -0.2x' + 0.05y' - 0.1z'$
$y = 0.1x' + 0.6y' + 0.01z'$
$z = 0.001x' - 0.002y' - 0.7z'$

(c) $x' = 3x + 0.4y + 0.3z$
$y' = 0.3x - 2.1y + 0.2z$
$z' = 0.04x + 0.1y + 1.9z$

(d) $x' = 2x + 0.1y + 0.1z$
$y' = 0.1x - 3.1y + 0.1z$
$z' = 0.04x + 0.4y + 1.9z$

(e) $x' = x + y - z$
$y' = -x + y + z$
$z' = x - y + z$

(f) $x' = -x + y + 0.5\int y dt + z$
$y' = x + y + z$
$z' = x - 0.1\int x dt + y - z$

6 Choose the type of op-amp for applications requiring low noise in the bandwidth: (a) 10 Hz; (b) 100 Hz; (c) 1 MHz.

7 Choose the type of op-amp to use when the output power needs to be up to 3 W and the frequency band of interest is: (a) 0 to 100 kHz; (b) 0 to 1 MHz; (c) 100 Hz to 1 kHz; (d) 1 kHz to 10 kHz.

8 In the compensators using the inverting op-amp configurations in Fig. 6.10(a),(c), the maximum gain (at lower frequencies for the lag and at higher frequencies for the lead) must be 40 dB. Choose $R_2 = 1$ MΩ and find the remaining element values for:
(a) pole $f_p = 5$ Hz, zero $f_z = 10$ Hz;
(b) pole $f_p = 12$ Hz, zero $f_z = 20$ Hz;
(c) pole $f_p = 300$ Hz, zero $f_z = 1000$ Hz;
(d) pole $f_p = 15$ Hz, zero $f_z = 40$ Hz;
(e) pole $f_p = 27.2$ Hz, zero $f_z = 120$ Hz;
(f) pole $f_p = 800$ Hz, zero $f_z = 3000$ Hz.

Find the elements analytically (start with resistors), or with MATLAB, or with the chart in Fig. 6.23.

9 Use a non-inverting op-amp with *RC* feedback according to Fig. 6.9(b) to implement a lead link, assuming $R_2 = 500$ kΩ and dc gain 20 dB with
(a) $f_z = 100$ Hz, $f_p = 300$ Hz;
(b) $f_z = 50$ Hz, $f_p = 150$ Hz;

(c) $f_z = 10$ Hz, $f_p = 100$ Hz;
(d) $f_z = 150$ Hz, $f_p = 400$ Hz;
(e) $f_z = 27.2$ Hz, $f_p = 100$ Hz;
(f) $f_z = 60$ Hz, $f_p = 600$ Hz.

10 Use a non-inverting op-amp with *RC* feedback according to Fig. 6.9(a) to implement a lag link, assuming $R_1 = 10\ \text{k}\Omega$ and dc gain 30 dB with
(a) $f_p = 300$ Hz, $f_z = 1000$ Hz;
(b) $f_p = 150$ Hz, $f_z = 500$ Hz;
(c) $f_p = 3$ Hz, $f_z = 10$ Hz.

11 Find a transfer function with three poles and two zeros approximating the gain response slope –4 dB/oct over the range 3 to 30 Hz. The gain at dc must be 26 dB, and the high-frequency asymptote must be –6 dB/oct. Design a compensator with this transfer function using an op-amp in the inverting configuration. Use an initially asymptotic Bode diagram, then (a) simulate the response in MATLAB or SPICE and (b) adjust the positions of poles and zeros (if using MATLAB) and then calculate the element values, or directly adjust the values of the circuit elements (if using SPICE).

12 Use the op-amp with *RC* feedback shown in Fig. 6.12(a) to produce a broad notch, using a real pole and a real zero, and a pair of complex zeros. Make simulations with SPICE.

13 Use the FILTER1 program (available from Burr-Brown) to design the following Sallen-Key filters:
(a) a low-pass second-order link with cut-off frequency of 200 Hz and $\zeta = 0.3$;
(b) Butterworth 3rd-order with cut-off at 1 kHz;
(c) Chebyshev 5th-order with 0.4 dB peak-to-peak ripples in the passband and cut-off at 10 kHz;
(d) Butterworth 5th-order with cut-off at 30 Hz.
Find filter elements, simulate the filter performance, and plot the gain response.

14 Using the FILTER2 program (available from Burr-Brown) to design the following multiple feedback filters:
(a) a single second-order link with 200 Hz cut-off frequency and $\zeta = 0.2$;
(b) Butterworth 3rd-order;
(c) Chebyshev 5th-order with 0.4 dB peak-to-peak ripples in the passband;
(d) Butterworth 5th-order.
Find filter elements, simulate the filter performance, and plot the gain response. Find sensitivities to the filter elements.

15 Some anti-aliasing filter ICs pass the dc input directly to the output, without contaminating the dc signal by the dc drift of the employed op-amp. For what applications are such filters especially suitable?

16 (a) Implement a Bode step with the Sallen-Key lowpass filter shown in Fig. 6.14 with the data in Example 1 in Section 6.7, using the method illustrated in Fig. 6.19(a).
(b) Do it for the loop with $f_b = 500$ Hz, main Bode diagram slope –10 dB/oct, gain stability margin 10 dB, high-frequency asymptotic slope –18 dB/oct.
(c) Design a Nyquist-stable loop Bode diagram, with the lower and the upper Bode steps at –10 dB and 15 dB levels respectively, using Sallen-Key filters.
Use the FILTER1 program from Burr-Brown, or use MATLAB simulation for second-

order links with appropriate damping.

17 Use the FILTER1 program to study sensitivity of Sallen-Key filters. Show that plenty of margin exists when using 1% resistor tolerances, when the filter is used to implement a Bode step.

18 Prove that the equivalent resistance of a charge pump is $R = 1/(Cf_S)$.

19 Use SPICE to simulate the circuits in Figs. 6.18 and 6.19. Discuss how to make these simulations with MATLAB and SIMULINK.

20 (a) Using the plot in Fig. 6.23, draw the frequency response of the magnitude of the impedance of the *RC* circuit in Cauer form in Fig. 6.20(b), with capacitances (starting from input) 0.1 mF, 10 nF and resistances 100 kΩ, 5 kΩ (you can draw the response directly on a copy of the chart);
(b) make the plot with SPICE;
(c) make the plot with MATLAB after deriving the equations describing the circuit.

21 Using the plot in Fig. 6.23, find:
(a) where is the pole of impedance of the parallel connection of 1 mF and 1 kΩ;
(b) where is the zero of impedance of the series connection of 10 nF and 10 kΩ;
(c) what is the impedance of the series connection of 20 kΩ and 4 μF at 300 Hz;
(d) what capacitor needs to be connected in series with a 50 kΩ resistor to make a zero at 800 Hz;
(e) what resistor needs to be connected in parallel with a 2 μF capacitor to move the pole from 0 Hz to 30 Hz;
(f) *R* and *C* for the series *RC* two-pole connected in parallel to 100 kΩ feedback resistor of an op-amp in inverting configuration; the transfer function pole is at 200 Hz and the zero, at 800 Hz.

22 Choose ICs for the circuit in Fig. 6.58, then build and test the device.

23 Simulate a control loop for pointing a spacecraft with flexible solar panels using thrusters (in pulse-width modulation mode, i.e., with saturation type of characteristic) using the compensators in Fig. 6.12.

24 Make SPICE simulation for the circuit in
(a) Fig. 6.12(a);
(b) Fig. 6.12(b);
(c) Fig. 6.13.

25 Use MATLAB or SIMULINK to solve Problem 24.

Answers to selected problems

1 (a) At 1 kHz the closed-loop gain coefficient is –25, and the feedback must be 30 dB, i.e., the open-loop gain coefficient must be 750. Therefore, $f_T > 750 \times 1$ kHz = 750 kHz; $R_2 > 50$ kΩ, and $R_1 = 50/25 = 2$ kΩ.

2 (a) When impedance of a series feedback branch changes by +5%, and of the parallel branch, by –5%, the gain coefficient increases by, approximately, 10%. The angle of an RC impedance is within the [0, –90°] interval. Because of this, the sensitivity of the modulus of the impedance to a resistance or a capacitance within the circuit is less than 1.

3 (d) The summer is shown in Fig. 6.61.

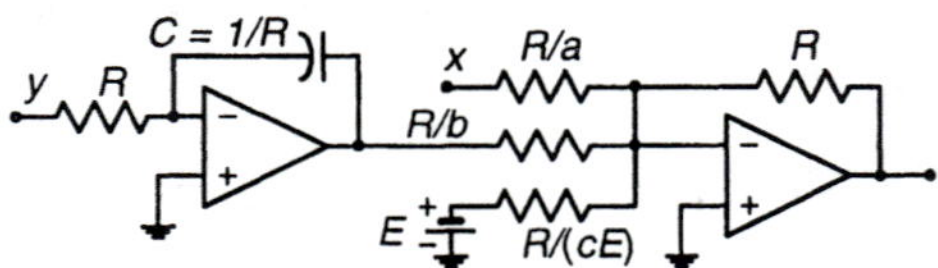

Fig. 6.61 Implementation of function $-ax + b \times$[integral of (y)] $- c$

5 (a) The schematic diagram is shown in Fig. 6.62. The resistances are 20 kΩ/(*coefficient in the relevant equation*).

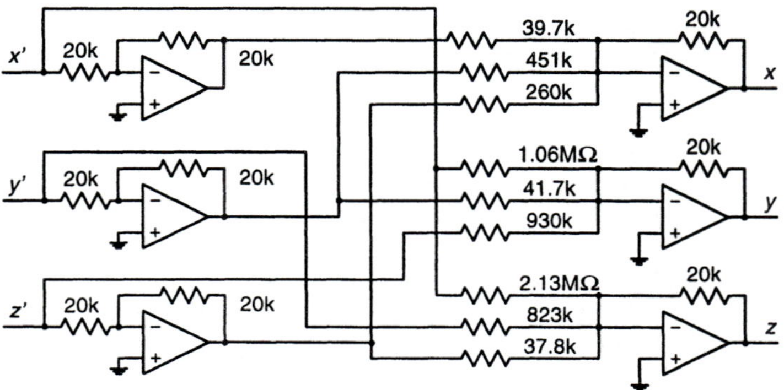

Fig. 6.62 Schematic diagram for a decoupling matrix implementation

8 (a) Since $f_z > f_p$, this is a lag link and the schematic in Fig. 6.10(c) can be used, and the gain coefficient at zero frequency must be 100. Then, $R_1 = 10$ kΩ. At infinite frequency, the gain coefficient is $100f_p/f_z = 50$. From here, R_3 # $R_2 = 500$ kΩ, i.e., $R_3 = 1$ MΩ. The capacitance $C = 16$ nF is found from the condition $1/(2\pi f_z C) = R_3$.

9 (a) From the required gain coefficient 10 at zero frequency, $R_1 = R_2/11 = 45.5$ kΩ. At infinite frequency, the gain coefficient is 3 times larger (because the ratio of the pole to the zero frequencies is 3), i.e., it is 30. From this, the parallel connection of resistances R_1 and R_3 is R_1 # $R_3 = R_2/31 = 16.1$ kΩ and $R_3 = 16.1R_1/(16.1 + R_1) =$ 25 kΩ (we are designing a forward path compensator so there is no need for extreme accuracy of calculations). The capacitance can be found from the equality $1/(\omega C) = R_3$ at the frequency of the pole of the compensator transfer function which is the frequency of the zero of the shunting impedance in the feedback path; $C = 21\times10^{-9}$ F, i.e., $C \approx 20$ nF.

10 (a) At zero frequency, the gain coefficient is $31.6 = (R_2 + R_1)/R_1$ wherefrom $R_2 = 306$ kΩ. At infinite frequency, the gain coefficient is $31.6f_p/f_z = 9.48$. Thus, $9.48 = (R_3$ # $R_2 + R_1)/R_1$. From here, R_3 # $R_2 = 848$ kΩ, and $R_3 = 477$ kΩ. The capacitance is found from the condition that at the pole frequency $R_3 + R_2 = 1/(2\pi f_p C)$, wherefrom $C = 400$ pF.

Chapter 7

LINEAR LINKS AND SYSTEM SIMULATION

Two approaches are presented for modeling systems composed of electrical, mechanical, thermal and hydraulic elements: describing the system elements and the topology of their connections, and deriving mathematical equations for the system.

It is emphasized that to reduce the plant uncertainty (and therefore, to increase the available feedback), the actuator output impedance needs to have a specific value. The use of local feedback loops to modify this impedance and, generally, the effect of feedback on impedance are considered, including parallel, series, and compound feedback.

Equivalent block diagrams are developed for chain connection of two-ports, like drivers, motors, and gears, to simplify system modeling and to make it structural.

Several issues are considered important for feedback control: flexible structures, collocated and noncollocated control, sensor noise, and the effect of feedback on the output noise.

The feedback system equations are analogous to the equations describing parallel connection of two links and parallel connection of two two-poles. These analogies allow the theory developed for two-pole connection to be applied to feedback systems, as will be explored in Chapters 10 and12. The chapter ends with a brief demonstration of the specifics of linear time-variable systems.

For a shortened control course, Sections 7.2.2, 7.2.3, 7.4.2–7.4.5, 7.5, 7.8.2, 7.10, and 7.11 can be omitted.

7.1 Mathematical analogies

7.1.1 Electro-mechanical analogies

The first step in simulating a system behavior on a computer is to generate the equations describing the system. This may be done either by the user or, preferably, by a computer. In the latter case, parameters describing the system's elements and the topology of their connections constitute the simulation program input file.

Control engineers deal with systems that are part electrical and part mechanical, thermal, etc. Converting such a system to an equivalent system containing only one kind of variable and described by only one kind of physical law can facilitate both the preliminary back-of-the-envelope analysis and the write-up of the computer input file. Also, the conversion allows the use of programs which were initially supposed to be used for the analysis of electrical circuits, like SPICE, as universal control analysis and simulation tools.

There exist several mathematical analogies between mechanical and electrical systems. Among them, the most useful are those which preserve power, i.e., they convert the product (voltage × current) into the product (velocity × force). We will most often use the ***voltage-to-velocity*** analogy relating voltage U to velocity V, and current I to force F.

Example 1. Figs. 7.1 and 7.2 show an example of the application of this analogy. It is seen that the topology of the diagram is conveniently preserved, the inductors are replaced by springs with spring coefficients k_1, k_3, and the capacitors, by rigid bodies specified by their masses M_2, M_4.

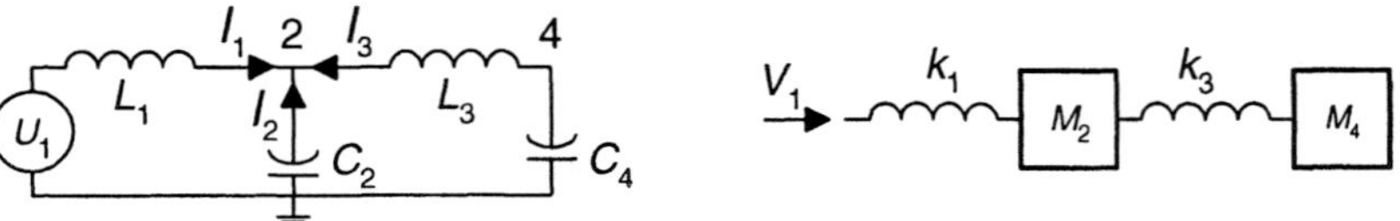

Fig. 7.1 Electrical circuit schematic diagram

Fig. 7.2 Mechanical circuit schematic diagram*

For nodes 2 and 4, Kirchhoff's equations correspondingly are

$$I_1 + I_2 + I_3 = 0,$$

$$U_4 = I_3/(sC_4)$$

where

$$I_2 = U_2 sC_2.$$

For bodies 2 and 4, Newton's equations are

$$F_1 + F_2 + F_3 = 0,$$

$$V_4 = F_3/(sM_4)$$

where

$$F_2 = V_2 sM_2.$$

The equality of the sum of currents to zero at a node corresponds to the equality of the sum of forces to zero at a rigid body (taking into account D'Alembert's force). Similarly, the equality of the sum of the voltages (relative potentials) about a contour to zero reflects the equality to zero of the sum of relative velocities about the contour. Zero voltage of the electrical "ground" most commonly translates to zero velocity of the inertial reference. Resistance converts into the inverse of the viscous damping coefficient, capacitance into mass, inductance into the inverse of the stiffness coefficient. Generally, electrical impedance converts into mechanical mobility. Since mobility is the ratio of relative velocity to force, the mobility of a difficult to move and/or heavy load by a given force is small**.

Table 7.1 summarizes the analogy for translational and rotational motions.

Example 2. Consider the active suspension strut diagrammed in Fig. 7.3(a). The strut consists of a spring, a dashpot (a device providing viscous friction), and a linear motor (voice coil) connected in parallel and driven by an electrical amplifier in accordance with the information obtained from the force sensor (load cell). The

---

* To accentuate the analogy, the springs are shown similar to inductors, and not similar to resistors as is conventional in drawing mechanical diagrams.

** The term "mechanical impedance" is sometimes used to mean the mobility, and sometimes the inverse of the mobility.

Table 7.1 *Voltage-to-velocity* analogy

electrical	*translational*	*rotational*
voltage U (or V)	relative velocity V	relative angular velocity Ω
current I	force F	torque τ
power $P = UI$	power $P = VF$	power $P = \Omega\tau$
impedance $Z = U/I$	mobility $Z = V/F$	mobility $Z = \Omega/\tau$
at a node, $\Sigma I = 0$	at a rigid body, $\Sigma F = 0$	at a rigid body, $\Sigma\tau = 0$
about a contour, $\Sigma U = 0$	along a contour through a sequence of connected bodies, $\Sigma\Delta V = 0$	along a contour through a sequence connected bodies, $\Sigma\Delta\Omega = 0$
capacitance C	mass M	moment of inertia J
inductance L	1/(stiffness coefficient k)	1/(angular stiffness)
resistance R	1/(viscous damping coefficient B)	1/(angular viscous damping coefficient)

vibrations of the base should be prevented from shaking the payload. For the suspension strut, the mobility is defined as the ratio of the difference in velocities at its ends to the force. The mobility must be large at frequencies where the vibration should not pass through. At the same time, the mobility should be low enough at lower frequencies in order for the motion of the base to be conveyed to the body of interest. Thus, the notion of mobility is particularly applicable to the design of suspension and vibration isolation systems. Fig. 7.3(b) exemplifies the modulus of the mobility with and without introducing feedback to make the isolation better.

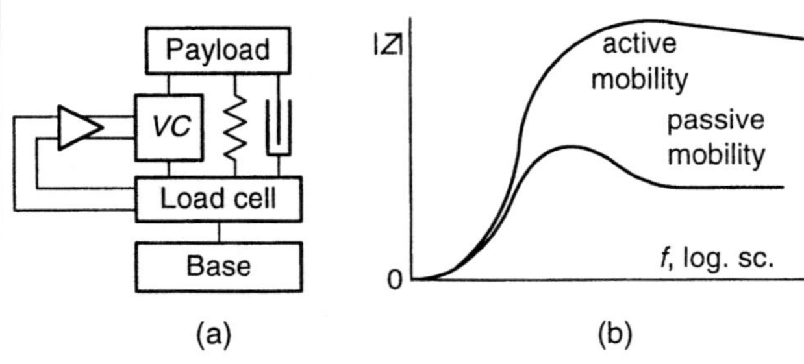

Fig. 7.3 (a) Active suspension strut and (b) its mobility frequency response

For the analysis of mechanical systems with many degrees of freedom at the joints such as ball joints, using the electro-mechanical analogy gets more difficult (although it is still possible). For complicated mechanical systems, specialized simulation programs should be used (like ADAMS ® or SD FAST ®). However, in the great majority of practical cases, control engineers only deal with slides or pin joints, i.e., with the joints characterized each by only two variables, the position and the force or the angle and the torque. For these systems, using the analogy of a mechanical joint to a joint of cascaded electrical two-ports is of great help.

The *voltage-to-velocity* (i.e., *current-to-force*) analogy described above is especially convenient when the system includes electromagnetic transducers like electrical motors and solenoids, where *current* creates a magnetic field which in turn produces *force*. In piezoelectric and electrostatic transducers (the latter are used in micromachined devices), the applied *voltage* produces *force*, and the analogy shown in Table 7.2 may appear more convenient.

Table 7.2 *Voltage-to-force* analogy

electrical	*translational*	*rotational*
voltage U (or V)	force F	torque τ
current I	velocity V	angular velocity Ω
charge q	displacement x	angle θ
power $P = UI$	power $P = VF$	power $P = \Omega\tau$
impedance $Z = U/I$	inverse of mobility $Y = F/I$	inverse of mobility $Y = \tau/\Omega$
at a node, $\Sigma I = 0$	along a contour through a sequence of connected bodies, $\Sigma\Delta V = 0$	along a contour through a sequence of connected bodies, $\Sigma\Delta\Omega = 0$
about a contour, $\Sigma U = 0$	at a rigid body, $\Sigma F = 0$	at a rigid body, $\Sigma\tau = 0$
capacitance C	1/(stiffness coefficient k)	1/(angular stiffness)
inductance L	mass M	moment of inertia J
resistance R	viscous damping coefficient B	angular viscous damping coefficient

7.1.2 Electrical analogy to heat transfer

The electrical analogy to conductive heat transfer is shown in Table 7.3. The heat flow is measured in Watts or Btu/sec, 1 kW corresponding to 1.055 Btu. The Fourier law for heat flow is analogous to Ohm's law.

Table 7.3 *Voltage-to-temperature-difference* analogy

electrical	*thermal*
voltage (potential difference) U (or V)	temperature difference ΔT
current I	heat flow g
power $P = UI$	ΔTg
impedance $Z = U/I$	thermal impedance $Z_T = \Delta T/g$
at a node, $\Sigma I = 0$	at a body, $\Sigma g = 0$ (not counting heat accumulation in C_T)
capacitance C	thermal capacitance C_T which is the product of the specific heat c and the mass M
resistance $R = dU/dI$	thermal resistance $R_T = dT/dg$

Heat also can be transferred by radiation and convection. Since radiation increases as the fourth power of the temperature difference between the bodies (or between the body and the environment), it can be represented by a nonlinear current source. For heat transferred via convection, the equivalent thermal resistance depends on many parameters including the geometry, the area, and the air flow.

The analogy is exemplified in Fig. 7.4 by an equivalent electrical circuit employed to simulate the temperature control of the Cassini spacecraft narrow view photocamera. Here, H_2 and H_1 represent heaters attached to the secondary and primary mirrors, respectively. The temperatures T_2, T_c, and T_1 represent absolute temperatures of the secondary mirror, the case, and the primary mirror. The capacitances are the products of the bodies' specific heats and masses, and the resistors characterize the thermoconductivities between the bodies. The nonlinear sources represent the heat radiation into free space from the mirrors and the case:

$$G_2 = 0.507 \times 10^{-9} \times T_2^4,\ G_c = 0.5 \times 10^{-9} \times T_c^4,\ G_1 = 0.1 \times 10^{-9} \times T_1^4.$$

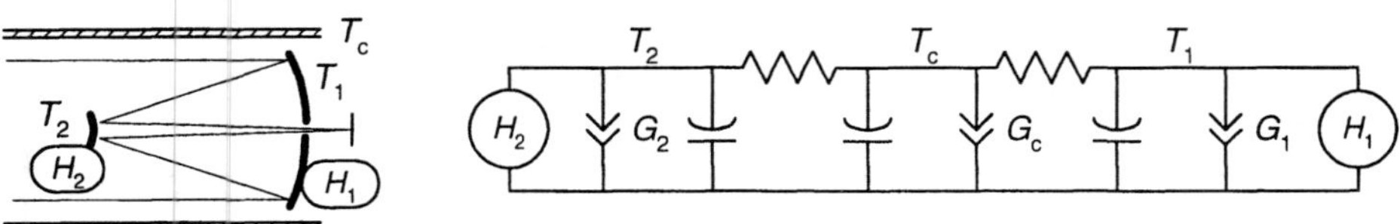

Fig. 7.4 Heat transfer analysis for a spacecraft photocamera

This circuit was simulated using SPICE – the model also incorporated the driver and compensator. With this approach, the frequency and time responses can be plotted without derivation of the plant equations.

7.1.3 Hydraulic systems

Example 1. A hydraulic system is exemplified in Fig. 7.5. Here, the reservoir of volume v_1 is filled with a liquid kept at pressure h_1. The sum of the liquid flow at each node is 0. The pipes connecting the reservoir to the cylinder with current volume v_2 and pressure h_2 present resistances R_1, R_2 to the liquid flow (we consider the volume of the liquid in the pipe insignificant, or else the system must be considered as a system with distributed parameters). The valve controls the liquid flow by introducing extra resistance in the pipe, or switching it on and off. The valve is operated by an electromagnet fed from the driver amplifier of the controller. The plunger stem moves a mechanical load with mobility Z_L.

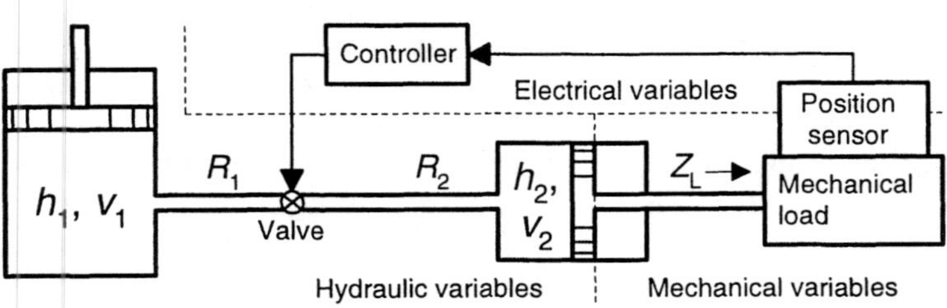

Fig. 7.5 Hydraulic system example

mechanical variables. The power is the same to the left and to the right sides of the vertical dashed line. The hydraulic and mechanical diagrams can be converted to their electrical equivalents and connected by an ideal transformer that preserves the power and provides appropriate correspondence between the hydraulic and the mechanical variables. Then, this system can be simulated with a program for electrical circuits analysis.

An electrical analog to a hydraulic system is described in Table 7.4.

Table 7.4 *Voltage-to-difference-in-pressure* analogy

electrical	*hydraulic*
voltage (potential difference) U	pressure difference Δh
current I	flow of liquid, volume per sec, q
power $P = UI$	power $P = \Delta hq$
impedance $Z = U/I$	impedance $Z = \Delta p/\Delta h$
at a node, $\Sigma I = 0$	at a node, $\Sigma q = 0$
capacitance C	volume v
resistance $R = dU/dI$	resistance $R = dh/dq$

When the flow of liquid in a pipe is laminar, q is proportional to h, and the resistance

$$R = dh/dq$$

does not depend on h. However, in most applications, the flow of liquid is turbulent. In this case the dependence of q on h is nonlinear, and the resistance is

$$R \approx \frac{2(\Delta h)^2}{q}. \tag{7.1}$$

The design problems and constraints are often similar for various systems and devices.

Example 2. Fig. 7.6 demonstrates the analogy of (a) an electrical amplifier to (b) a hydraulic amplifier. The electrical amplifier contains two parallel branches, each passing a single polarity signal (which is the ***class B*** mode of operation). A small dead zone in input voltage prevents the waste of power supply current in the quiescent mode, i.e., when the input signal is 0. The summer combines the output currents of the branches. The input-output characteristic for the amplifier shown in Fig. 7.6(c) is not quite linear, and, particularly, contains a dead zone. By application of large feedback (not shown in Fig. 7.6) the input-output characteristic is made much closer to linear, and dead zone is reduced many times (recall Section 1.7).

The hydraulic amplifier input is the position of the ***spool valve***. The valve directs the liquid under pressure to the appropriate side of the plunger, thus providing output force at the plunger stem. The dead zone should be sufficiently large to reliably stop the power waste when the input signal is zero. Still, the zone must be small enough for the input-output characteristic to be close to linear.

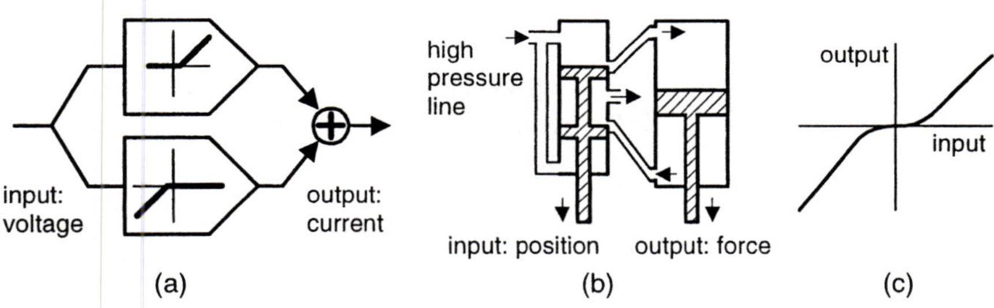

Fig. 7.6 (a) Electrical class B amplifier, (b) spool valve hydraulic mechanical amplifier, and (c) the electrical amplifier's input-output characteristic

This provision is especially difficult to implement in hydraulic amplifiers. The input-output characteristic of the spool valve amplifier is different for different speeds of the action. When the output motion is slow, even a small opening suffices to provide nearly full available pressure on the plunger, but if the output motion is fast, a small opening is insufficient to supply the liquid to the output cylinder. Thus, the input-output characteristic is steep for slow motion, but the gain decreases for faster motions. These variations in the actuator nonlinearity and dynamics limit the available accuracy of a control system using such an actuator. The dynamics and linearity can be improved by using local feedback about the actuator. This feedback can be implemented using a pressure sensor (or a position sensor, depending on the application) with electrical output, a compensator, a power electrical amplifier (motor driver), and an electrical motor moving the spool valve shaft. Such integrated actuators are commercially available, and their input is an analog or digital electrical signal.

7.2 Junctions of unilateral links

7.2.1 Structural design

With structural design, the system is composed of subsystems which are relatively large functional parts with rather simple interconnections between them. Structural design simplifies the analysis and design, and facilitates having several people work together on the same project. Since the specifications for each of the subsystems are tailored such as to minimize the requirements, it prevents the subsystems' overdesign (for example, the accuracy requirements for the links in the forward path can be much relaxed from the required accuracy for the summer and feedback path links). It simplifies verifying system performance, making design trade-offs, and troubleshooting. It facilitates redesigning the system, since, as long as the requirements for the subsystem are defined, each subsystem can be refined and redesigned independently of the others.

The subsystems are composed of smaller subsystems, etc. For example, the compensators themselves are built from simpler links cascaded or connected in parallel.

The interface between the subsystems should be as simple as possible. Between cascaded blocks in the block diagrams, the interface is the simplest: a single variable serves as the output of the preceding link and the input of the following links. However, in most physical links, we have to include into consideration the effects of ***loading***, i.e., the effects of the following link on the output variable of the previous link. The effects of loading are conveniently described using the notion of impedance.

7.2.2 Junction variables

Any electrical or mechanical source can be equivalently represented in Thevenin's and Norton's forms shown in Fig. 7.7. Here, E is the open-circuit voltage, or emf, I_S is the short-circuit current, V_f is the free (unloaded) velocity, and F_B is the brake force. When the internal source impedance Z_S is small, the output voltage (or velocity) does not depend on the value of the load impedance and therefore is called source of voltage U (or velocity V); when the source impedance is very large, the source is called source of current I or force F.

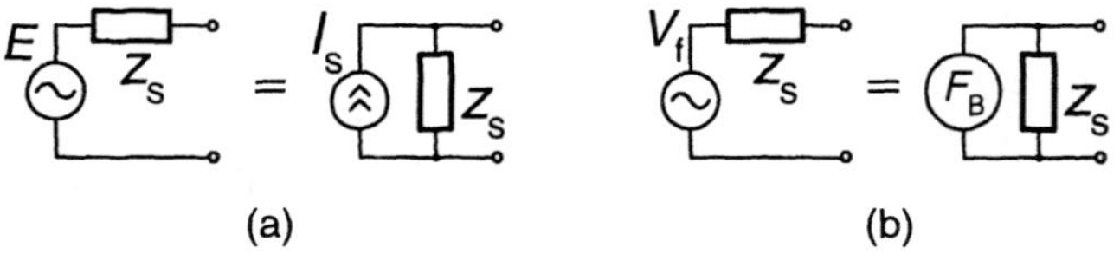

Fig. 7.7 Thevenin and Norton representations of (a) electrical and (b) mechanical signal sources (for the voltage-to-velocity analogy)

While the notion of driving point impedance is universally employed in the design of electrical circuits, in mechanical systems design, mechanical driving point impedance (or mobility) is used less frequently. This is probably so because at a mechanical junction, there could be more than two variables to deal with. For example, 3 angles and 3 torques need to be considered at a ball joint. Nevertheless, as mentioned before, the most common mechanical junctions are the simple ones, like a slide or a pin with one position (or rotation) variable and one force (or torque) variable. These junctions are mathematically similar to connections of electrical two-node ports. The notion of mobility facilitates the analysis of such systems.

Fig. 7.8 shows a preceding link (two-port) loaded at the input impedance of the following link. The voltage at the junction is

$$U = \frac{EZ_L}{Z_L + Z_S} \tag{7.2}$$

and the current is

$$I = \frac{I_S Z_S}{Z_L + Z_S}. \tag{7.3}$$

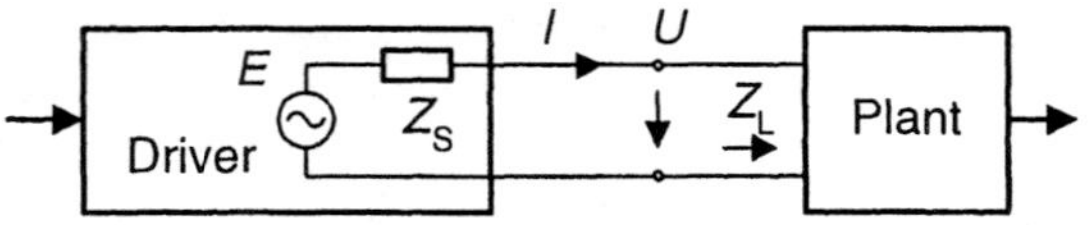

Fig. 7.8 Variables at the links' joint

For mechanical systems,

$$V = \frac{V_f Z_L}{Z_L + Z_S}, \tag{7.4}$$

$$F = \frac{F_B Z_S}{Z_L + Z_S}. \tag{7.5}$$

When the driver's output impedance (mobility) Z_S is much lower than the input impedance (mobility) Z_L of the following link, the effect of loading can be neglected and $U \approx E$. For instance, the output impedance of an operational amplifier is typically much lower than the load impedance due to the application of voltage feedback.

7.2.3 Loading diagram

The voltage across the terminals of the signal source with resistive load shown in Fig. 7.9(a) is

$$U = E - R_S I. \tag{7.6}$$

For a rotary mechanical actuator (motor), similarly, the angular velocity is

$$\Omega = \Omega_f - R_S \tau, \tag{7.7}$$

where Ω_f is the ***free running*** (no load) relative angular velocity. When τ equals the ***brake torque*** (or ***stall torque***) $\tau_b = \Omega_f / R_S$, the angular velocity becomes zero.

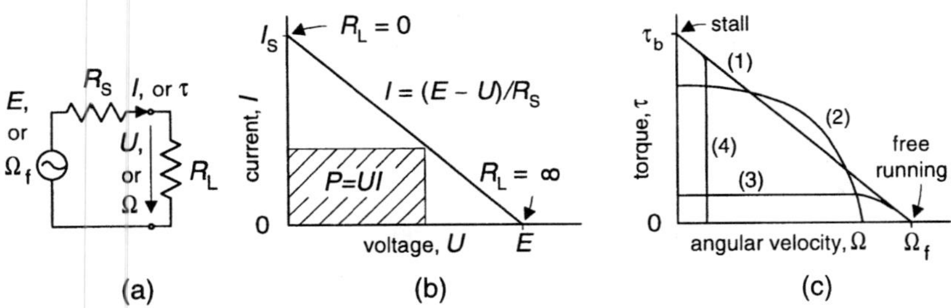

Fig. 7.9 Schematic diagram (a) and loading diagrams for (b) an electrical linear signal source and (c) for motors: (1) permanent magnet motor, (2) inductive motor, (3) motor with torque feedback, (4) motor with velocity feedback

The ***loading diagram*** shown in Fig. 7.9(b) reflects equation (7.6). Resistance R_S determines the slope of the ***loading line*** $I = (E - U)/R_S$. The smaller the resistance, the steeper the line. The output power UI varies with the load resistance. It becomes zero when R_L is very large and, therefore, the current is small, and also when R_L is very small so that the output voltage is small.

Fig. 7.9(c) shows the loading diagram (1) for a permanent magnet motor described by linear equations (we do not call the motor "linear" since the term ***linear motor*** is reserved for translational motion motors). The curve (2) exemplifies a nonlinear loading diagram for a motor with a flux winding. The slope of this curve defines the differential output resistance of the motor. The power, i.e., the product of the angular velocity and the torque, is defined similarly to the power in Fig. 7.9(b).

The loading curve can be changed by application of feedback about the motor – although, certainly, the output power cannot be increased. Large torque feedback maintains the torque constant as shown by the curve (3). The torque does not depend on the angular velocity and, therefore, does not depend on the load, within the range of the motor torque and velocity capabilities. The output mobility of the motor with torque feedback is large as can be seen from the slope of the loading line. The loading line for a motor with velocity feedback is shown by the curve (4); the output mobility of the motor is very low.

Output impedance (mobility) is an important parameter of drivers and motors, and its implications will be discussed in the next section.

7.3 Effect of the plant and actuator impedances on the plant transfer function uncertainty

Plant parameter uncertainty affects the accuracy of the closed-loop transfer function in two ways. First, directly: the variations in the closed-loop transfer function are feedback times smaller than the plant transfer function variation. Second, indirectly: the smaller the plant parameter variations, the smaller can be the stability margins and the larger is the feedback that can be implemented – as has been shown in Chapters 4 and 5 and will be further discussed in Chapter 9. In the hypothetical case that the plant is perfectly known (the so-called full state feedback described in Chapter 8), infinite feedback is available. When, on the other hand, the plant is completely unpredictable (as in the extreme case described in *Alice in Wonderland*: "–... you can balance an eel on the end of your nose"), no feedback control can be implemented. Reducing the plant transfer function uncertainty is an issue of high priority in feedback system design.

As will be shown in the examples below, the plant transfer function uncertainty depends to a large extent on the value of the actuator output mobility. Therefore, an actuator with an appropriate output mobility must be used.

Example 1. Consider an actuator driving a plant whose mass M is uncertain. The control system output variable (controlled variable) is the plant velocity. Let's consider two extreme cases of the output mobility of the actuator shown in Fig. 7.10 (a) and (b), respectively. When this mobility is much larger than the load mobility, the actuator can be viewed as a force source, and when the actuator's mobility is much smaller than the load mobility, the actuator can be viewed as a velocity source.

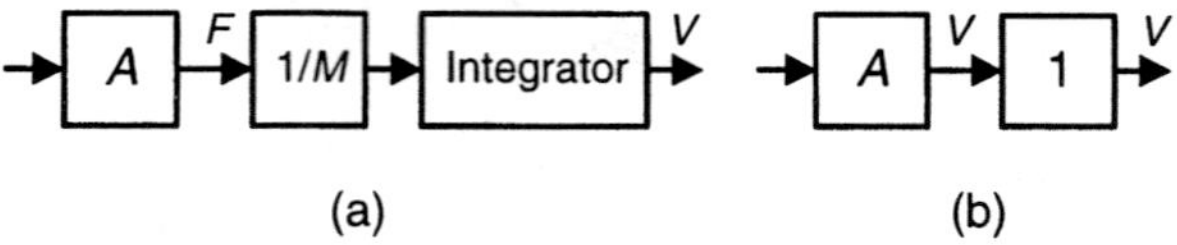

Fig. 7.10 Actuators with (a) constant force and (b) constant velocity driving a rigid body plant

When the actuator is a force source, the plant acceleration F/M and the plant velocity are uncertain due to the uncertainty of M. In contrast, when the actuator is a velocity source, the plant velocity equals the actuator velocity, and the plant transfer function is simply 1.

The actuator output mobility can be modified by application of local feedback about the actuator.

Example 2. Consider the plant depicted in Fig. 7.11. (The rotational version of this system is the problem of pointing a camera mounted on the end of a flexible boom attached to a spacecraft.) The actuator is placed between the second body and the spring, and the position of the second body is controlled. If the actuator is a force source, i.e., its output impedance is very large, the suspension resonance does not appear on the plant transfer function response: the second body only sees the force F, and this force does not depend on the spring parameters. On the other hand, a velocity source actuator results in a profound resonance on the control loop response.

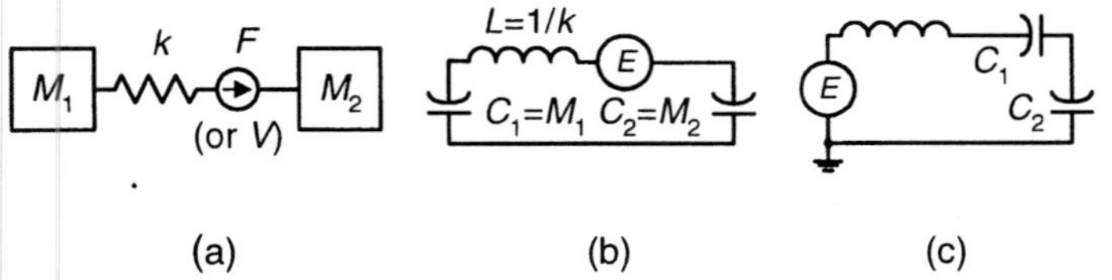

Fig. 7.11 (a) Actuator and mechanical plant, (b) its equivalent electrical circuit diagram, and (c) its modification

Notice that in the electrical circuit diagram, the order of the elements connected in series can be changed. Diagram (b) can be transformed into the form (c) where one of the terminals of the source E is grounded and the capacitor C_1 is ***floating*** (not connected to the ground). Similar modifications of mechanical diagrams are not as evident. Conversion from a mechanical to the equivalent electrical circuit might therefore simplify the analysis.

While choosing the actuator output impedance, attention also must be paid to reduction of disturbance transmission from the base to the object of control, if this is a problem.

7.4 Effect of feedback on the impedance (mobility)

7.4.1 Large feedback with velocity and force sensors

Examples already considered showed that feedback has a profound effect on impedance. *The output impedance (mobility) is defined by the type of the sensor used. What is measured by the sensor, is stabilized and made equal to the command or reference.* Hence, if the sensor measures the output *voltage*, the voltage is stabilized, and the feedback system output is a voltage source which means its internal (output) impedance is 0 (certainly, not exactly 0, but very small). If a *current* sensor is employed, the feedback system output represents a current source, i.e., the output impedance is high. If force feedback is employed, a force source is obtained. If velocity feedback is employed, a velocity source is obtained.

If both force and velocity sensors are used, and a linear combination of their readings is fed back, the feedback is called ***compound***. Compound feeedback is employed in most biological and many robotic motion control systems. With compound feedback, the output mobility of the control system at the joint of the actuator and the load can be made as desired, and damping can be introduced which makes smoother the interaction between various feedback loops.

A feedback loop about an actuator is depicted in Fig. 7.12. Two sensors are employed, one measuring the force and the other measuring the velocity. The outputs of the sensors are combined and fed back. A link Z is included in the force sensor path.

The output of the link Z must have the dimension of velocity so that it can be summed with V. Therefore, the dimension of Z is mobility.

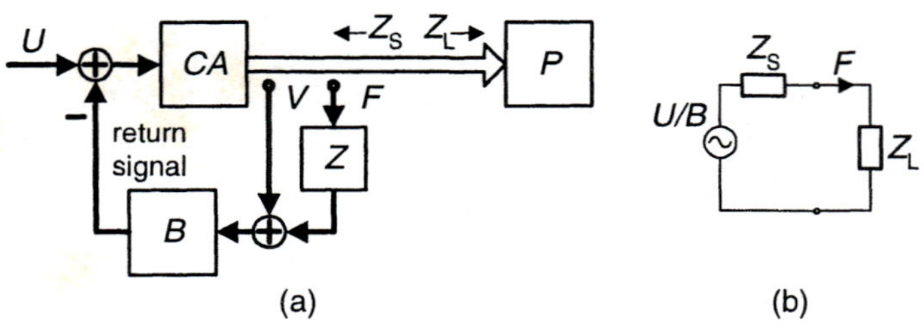

Fig. 7.12 (a) Feeding back a linear combination of sensor readings and (b) an equivalent circuit diagram for the junction

Since the velocity $V = FZ_L$, the return signal is $BF(Z_L + Z)$. If the feedback is large and therefore the error is small, the return signal equals the command U. From here,

$$F = U/[B(Z_L + Z)]\ . \tag{7.8}$$

The same expression follows from the analysis of the circuit diagram in Fig. 7.14(b) where the source with electromotive force U/B and internal impedance Z is loaded at impedance Z_L. From this comparison we conclude that *large compound feedback makes the actuator output impedance (mobility) equal to the transfer function of the link Z*.

In electrical systems, a linear combination of the current and voltage at the junction can be obtained using Wheatstone bridge circuitry as will be shown in Section 7.4.5.

7.4.2 Blackman's formula

A two-wire connection where the current in one (direct) wire equals the negative of the current in the second (return) wire, is called a ***port***. The system shown in Fig. 7.13 comprises a single-directional two-port network and a three-port linear system. Z_L is the impedance of an external load that can be connected to the port with the output impedance Z.

The feedback return ratio in the system certainly depends somehow on the impedance Z_L (examples will be provided later) so that $T \equiv T(Z_L)$. By this definition, when $Z_L = 0$, the return ratio is $T(0)$, and when $Z_L = \infty$, the return ratio is $T(\infty)$.

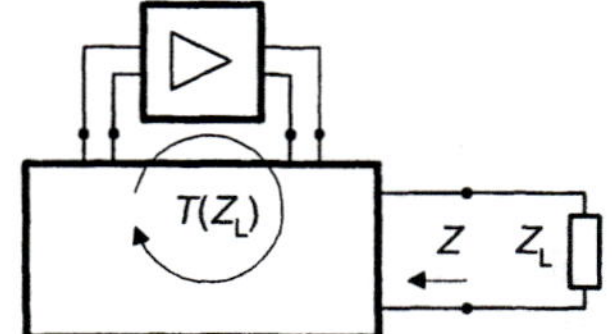

Fig. 7.13 Driving point impedance of a feedback system

The input impedance of a feedback system can be calculated using ***Blackman's formula***,

$$Z = Z_o \frac{T(0) + 1}{T(\infty) + 1} \tag{7.9}$$

where Z_o is the impedance without the feedback, i.e., with the feedback loop disconnected. (The proofs are given in Appendices 6 and 7.) The formula expresses the driving point impedance via three quantities that are much simpler to estimate and calculate: the impedance without the feedback, and the feedback in the cases when the system is simplified by setting load impedance to 0 or ∞.

7.4.3 Parallel feedback

The ***parallel feedback*** block diagram is shown in Fig. 7.14(a). Here, the output of the forward path, the input of the feedback path, and the load are all connected in parallel. The fed back signal is therefore proportional to the voltage across the load. This is why parallel feedback is also called voltage feedback.

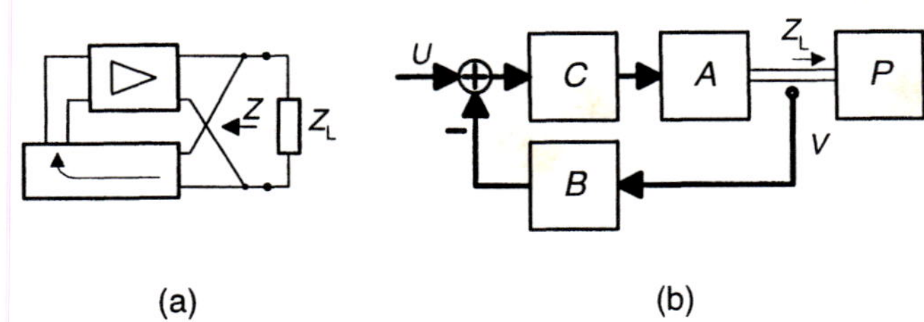

Fig. 7.14 Parallel feedback, (a) electrical and (b) mechanical

In a parallel feedback circuit, reducing the load impedance (mobility) to 0 disconnects the feedback loop and makes $T(0) = 0$, and Blackman's formula reduces to

$$Z = \frac{Z_o}{T(\infty)+1} . \tag{7.10}$$

Hence, *negative parallel feedback reduces the impedance*. Large parallel feedback can be used to make nearly perfect voltage sources, such as the one described in Chapter 1 and depicted in Fig. 1.5.

The mechanical analog to parallel feedback is rate feedback. Such feedback is shown in Fig. 7.14(b). It uses as a sensor a tachometer motor winding, an optical encoder, or some other velocity sensor. The feedback makes the actuator a velocity source, the velocity proportional to the input signal and independent of the loading conditions. Velocity feedback can be employed, for example, in the motor which propagates the tape in a tape recorder. The feedback maintains the tape velocity constant and independent of the tape tension.

Example 1. A driver amplifier with high input and output impedances has transconductance Y (i.e., when voltage V is applied to the input, the output current is YV). The output terminal is connected to the input terminal via a feedback two-pole with impedance Z_B. The signal source impedance is Z_S. The feedback path transfer function, from the output current to the input voltage, is YZ_S (when no load is connected to the output). Then, the output impedance of the amplifier is

$$Z = Z_o / [T(\infty) + 1] = (Z_B + Z_S)/ (YZ_S + 1) .$$

7.4.4 Series feedback

A ***series feedback*** diagram is shown in Fig. 7.15(a). Here, three ports are connected in series: the load, the output of the forward path, and the input of the feedback path, so that the same current flows through these three ports. The feedback signal is proportional to this current, i.e., to the load current. This is why series feedback is also called current feedback. In this circuit, increasing the load impedance to ∞ disconnects the feedback loop and makes $T(\infty) = 0$. Then, Blackman's formula reduces to

$$Z = Z_o\,[T(0) + 1]\,. \tag{7.11}$$

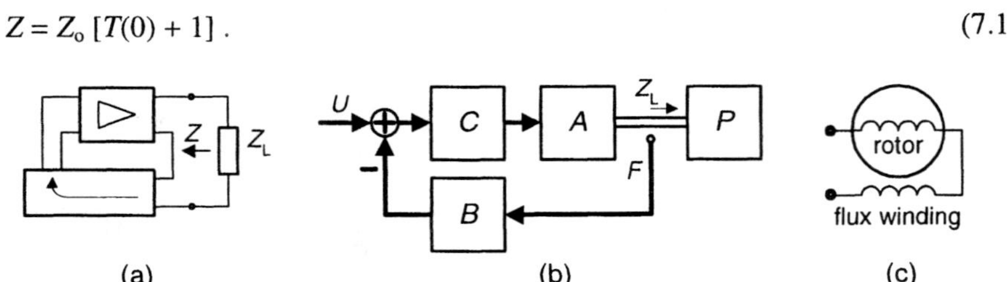

Fig. 7.15 Series feedback: (a) electrical, (b) mechanical; (c) series motor

When series feedback is negative, it increases the impedance. Nearly perfect current sources can be made using large series feedback. Such a current source was depicted in Fig. 1.29. In mechanical systems, a force or a torque source can be made using series feedback from a load cell or another kind of force or torque sensor. The torque sources are commonly used in pulleys, providing proper constant tension in the cable.

In series electrical motors, the flux winding is connected in series to the rotor as shown in Fig. 7.15(c). The effect of the flux winding can be described by the following internal feedback mechanism. Increasing the load, i.e., increasing the torque, increases the rotor current and, therefore, the flux winding current. This increases the voltage drop on the flux winding and, therefore, reduces the voltage on the rotor winding, thus maintaining the torque nearly constant and independent of the load.

7.4.5 Compound feedback

Compound feedback is the type of feedback when both $T(0)$ and $T(\infty)$ are nonzero. When both $|T(0)|$ and $|T(\infty)|$ are much larger than 1, then the impedance

$$Z = Z_o \frac{T(0)+1}{T(\infty)+1} \approx Z_o \frac{T(0)}{T(\infty)} = Z_o \frac{CAPB(0)}{CAPB(\infty)} = Z_o \frac{B(0)}{B(\infty)} \tag{7.12}$$

does not depend on the forward path gain. Compound feedback differs in this respect from series feedback and from parallel feedback.

Compound feedback stabilizes not the voltage or current at the system's output but the system's output impedance and the emf (and therefore, the short circuit current). In mechanical rotational systems, compound feedback stabilizes the output mobility, the brake torque, and the free run velocity.

Example 1. A circuit diagram for an amplifier with compound feedback at the output and series feedback at the input is shown in Fig. 7.16. The output voltage is connected to the input of the feedback path via the voltage divider R_1, R_2. The current is sensed by the small series resistor R_3. The total signal at the input to the B-circuit is approximately $U_L R_2/(R_1 + R_2) + I_L R_3$. Therefore, when the feedback is large, according to the rule developed when we were considering the system in Fig. 7.12, or directly from (7.11), the output impedance is $R_3(R_1 + R_2)/R_2$.

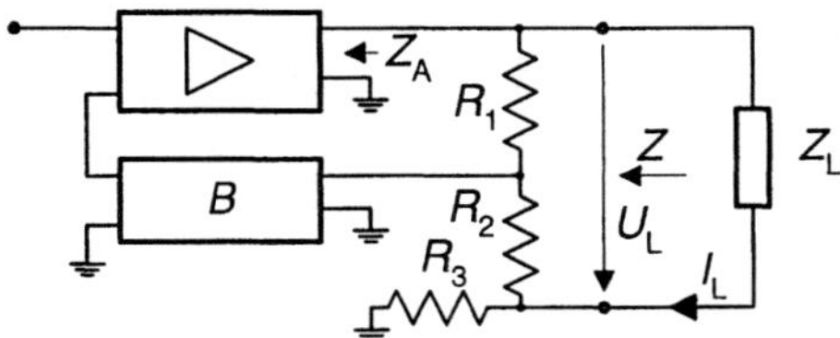

Fig. 7.16 Resistive compound feedback at the output of an amplifier

The circuit in Fig. 7.16 is often referred to as an implementation of ***bridge feedback*** since resistors R_1, R_2, R_3 and the output impedance of the amplifier Z_A constitute a Wheatstone bridge. The load impedance is connected to one diagonal of the bridge, and the B-circuit input, to another diagonal. In high-frequency amplifiers, bridge feedback with Wheatstone or transformer bridges is often employed to make the output impedance equal to 75 Ω or 50 Ω to match a cable, a filter, or an antenna.

Example 2. When compound feedback employs angular velocity and torque sensors, the output of a servo motor imitates a damper of desired value. This way a flexible plant this motor is driving can be damped and the control accuracy improved. The drawback of this method is the need to use velocity and torque sensors which may be relatively expensive. Using only a rate sensor with resulting rate feedback and a driver amplifier with an appropriate output impedance is commonly sufficient for most practical applications.

Example 3. When power losses in a motor are small, the motor output mobility is proportional to the output electrical impedance of the driver amplifier – as will be shown in Section 7.6.2. Then, to obtain the required output mobility of the motor, it is sufficient to implement compound feedback in the driver amplifier, which only requires several resistors. The smaller the motor winding resistance, the better this method works.

Blackman's formula allows calculation of the driving point impedance at any specified port. Until now, we only considered the output port of the actuator – which is the most important for the control system designer. However, it can be any other port, for example, the input port of a feedback amplifier. While the impedance at a specified port is being calculated, Z_o is the impedance without feedback at this port, and $T(0)$ and $T(\infty)$ are the return ratios with, respectively, this port open or shorted.

7.5 Effect of load impedance on feedback

We have already seen that the load impedance affects the feedback in the cases of series and parallel feedback. Generally, the return ratio can be expressed as a bilinear function of Z_L [9] (see also Appendix 8):

$$T(Z_L) = \frac{Z_o T(0) + Z_L T(\infty)}{Z_o + Z_L}. \tag{7.13}$$

It is easy to check that when $Z_L = 0$, then from (7.13), $T(Z_L) = T(0)$, and when $Z_L = \infty$, then $T(Z_L) = T(\infty)$.

The feedback dependence on the load makes it more difficult to design systems where the load impedance is uncertain or could vary, especially when the plant is flexible and the resonance frequencies are not well known. It is preferable in this case to have the feedback independent of the load. The general condition for this is

$$Z = Z_o, \tag{7.14}$$

since in this case, according to Blackman's formula, $T(0) = T(\infty)$, and therefore from (7.13), $T(Z_L) = T(0) = T(\infty)$. In order to implement this condition, a local loop about the actuator can be made. This loop provides the desired value of the output impedance of the actuator, this impedance (mobility) being Z_o for the main loop over the plant as shown in Fig. 7.17. Such a combination of local and main feedback is called ***balanced bridge feedback***.

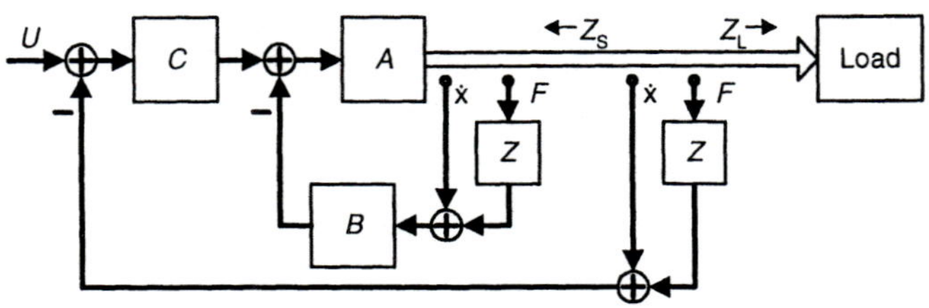

Fig. 7.17 Balanced bridge feedback

7.6 Flowchart for chain connection of bidirectional two-ports

7.6.1 Chain connection of two-ports

When two-ports are bidirectional, like electrical motors that can also be used as dynamos, or like transformers and gears, the load for a link depends on the load for the following link. For example, the input impedance of an electrical motor is affected by the input mobility of the gear connected to the motor, and the gear input mobility is affected by its load.

A general linear two-port is described by a pair of linear equations relating its two input variables and two output variables, like the following equations:

$$I_2 = aI_1 + dU_2 \tag{7.15}$$

$$U_1 = cI_1 + bU_2$$

Equations (7.15) can be represented by the block diagram in Fig. 7.18. The inside of the diagram for each bidirectional two-port is represented by four unidirectional links.

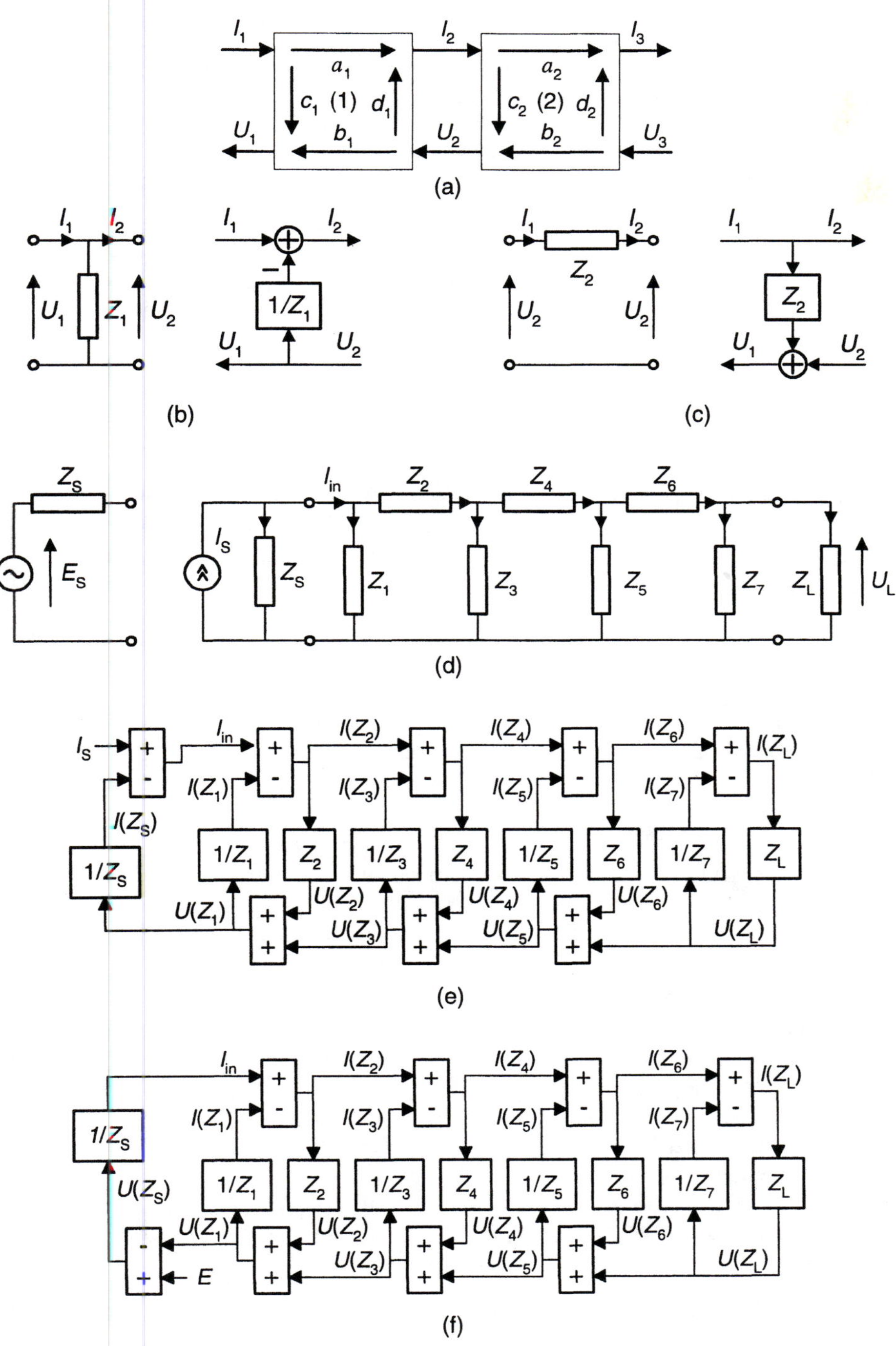

Fig. 7.18 Flowcharts of (a) a cascade connection of two-ports, (b) a parallel impedance two-port, (c) a series impedance two-port, (d) a ladder network; (e),(f) block diagrams for the ladder network with current and voltage signal sources

The meanings of the coefficients a, b, c, d can be understood from the boundary conditions. The coefficient a is the current gain coefficient under the condition that $U_2 = 0$, i.e., under the condition that the output port is shorted. The coefficient b is the reverse direction voltage gain coefficient under the condition that $I_1 = 0$, i.e., that the left port is open. The coefficient d is the output conductance under the condition that the left port is open. The coefficient c is the input impedance under the condition that the output port is shorted.

Example 1. An electrical two-port consisting of shunting impedance Z_1 and the related flowchart are shown in Fig. 7.18(b). The equations of the two-port are

$$I_2 = I_1 - (1/Z_1)U_2 \quad \text{and} \quad U_1 = U_2.$$

Example 2. An electrical two-port consisting of series impedance Z_2 and the related flowchart are shown in Fig. 7.18(c). The equations of the two-port are

$$I_2 = I_1 \quad \text{and} \quad U_1 = Z_2 I_1 + U_2.$$

Example 3. A ladder electrical two-port loaded at Z_L and excited by current source I_S with internal impedance Z_S is shown in Fig. 7.18(d).

The transimpedances (the ratios of the output voltages to the input currents) for the networks consisting of 3 and of 5 branches, are, correspondingly,

$$\frac{Z_1 Z_3}{Z_1 + Z_2 + Z_3} \quad \text{and} \quad \frac{Z_1 Z_3 Z_5}{(Z_1 + Z_2)(Z_3 + Z_4 + Z_5) + Z_3 Z_5}.$$

For the most frequently encountered network consisting of three branches, the numerator *numtr* and denominator *dentr* of the transimpedance are expressed in terms of the numerators and denominators of the branch impedances n_i, d_i as shown below:

$$numtr = n_1 n_3 d_1 d_2 d_3\,, \quad dentr = n_2 n_3 + n_1 n_3 + n_1 n_2$$

(when using MATLAB, an appropriate number of zeros must be added in front of the vectors to be added).

The flowchart of the network can be obtained by cascading flowcharts for parallel and series branches. The resulting block diagram is shown in Fig. 7.18(e); here, $I(Z_n)$ is the current flowing through impedance Z_n, and $U(Z_n)$, the voltage across impedance Z_n. As is seen, the summers in the upper row represent Kirchhoff's law for the currents in the nodes, and the summers in the lower row represent Kirchhoff's law for the voltages about the elementary contours. Bode diagrams of the flowcharts transfer functions can be obtained with SIMULINK using commands described in Section 7.7.

When the numerator order n of a link transfer function is higher than the order of the denominator, SIMULINK might not be able to find the solution. In this case as^n can be added to the denominator with a sufficiently small coefficient a.

With current and voltage summers, any circuit (not only the ladder) can be modeled with a flowchart and modified, if required, by changing only a few links.

Example 4. In the previous example, the signal source is equivalently replaced by a source consisting of emf $E = I_S Z_S$ in series with impedance Z_S. The resulting block diagram is shown in Fig. 7.18(f). A further example is given in Appendix 13.12.

7.6.2 DC motors

An electrical dc motor is an electromechanical transducer. The current I and the voltage U characterize the electrical side of the motor model. The generated torque τ is applied to the mechanical load which is rotated with angular velocity Ω.

In application to the motor, equations (7.15) can be rewritten as follows:

$$\tau = kI - \Omega/Z_m$$
$$U = rI + k\Omega \tag{7.16}$$

and reflected in the flowchart in Fig. 7.19. Here, k is the electromechanical conversion coefficient, i.e., the ratio of the break torque (i.e., the torque when the velocity $\Omega = 0$) to the applied current. As follows from the first equation, Z_m is the motor output mobility measured while the electrical winding is open, i.e., when the current is 0. This mobility reflects the viscous friction in the bearings and the moment of inertia of the rotor, $Z_m = 1/B + 1/(Js)$.

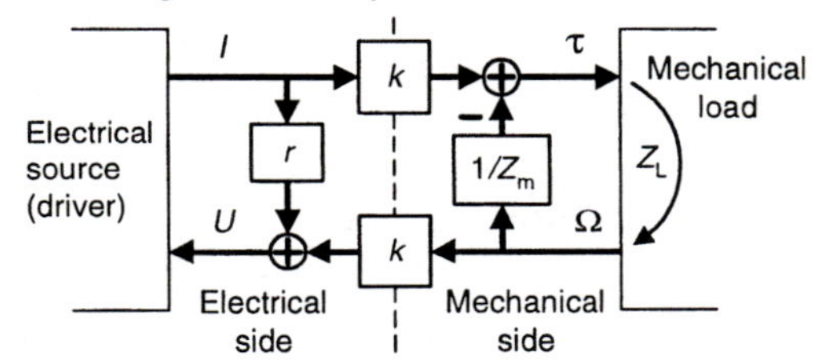

Fig. 7.19 Electrical motor flowchart

Z_L is the load mobility. The transfer function from the output of the link k in the forward path to the input of the second link k can be found using (1.3) as

$$\frac{Z_L}{1+Z_L Z_m} = \frac{Z_L Z_m}{Z_L + Z_m}.$$

This is the parallel connection of the load and motor mobilities.

The input voltage U equals the sum of the voltage drop rI on the winding resistance r and the ***back electromotive force*** $k\Omega$. (The reactive component of the winding impedance is most frequently neglected since it is typically much smaller than the resistance, but in some cases this reactance must also be accounted for.)

If the mechanical losses, the rotor inertia, and the winding resistance are neglected, then $\tau = kI$ and $U = k\Omega$ so that $\tau\Omega = IU$, i.e., the power is converted from electrical to mechanical forms without losses.

In most low-power feedback control applications, brushless motors are used with a permanent magnet rotor. The stator windings (phases) are switched in accordance with the information about the angle position of the rotor obtained from position sensors based on the Hall effect or from optical encoders. The encoder consists of a disk on the motor shaft with specific patterns printed on it and several photosensors placed on the stator and separated by specific angles to read the printed information. The encoders can also serve as angle sensors for the feedback path. The sensors are described in more detail in Section 7.9.1.

A permanent magnet motor can be driven by sinusoidal ac current generated by dc to ac inverter. The phase and the frequency of the ac current are controlled by a separate feedback loop, using either a rotor angle sensor or the information about the rotor angle extracted from the currents and voltages in the motor windings.

The periodic dependence of the coefficient k on the angle of rotation causes periodic variations in the motor torque. To model this effect in the system's block diagram, a parallel branch can be added to the k path. This branch includes a multiplier to whose second input the shaft angle is applied via the function $ak\sin n\varphi$, where a is the relative amplitude of the torque variations, φ is the shaft angle, and n the number of torque ripples per rotation. A similar branch added in parallel with the path $1/Z_m$ in Fig. 7.19 can be used to model the holding torque in step motors.

Higher power dc motors have both rotor and stator windings. The stator winding is often referred to as a flux winding. The motor can be controlled by varying the current in either winding or in both windings.

7.6.3 Motor output mobility

Using the flowchart in Fig. 7.19, the motor output mobility $Z_{out\ mot}$ can be calculated as the inverse of the transmission from Ω to τ. The link $1/Z_d$, which is the inverse of the output impedance of the driver amplifier, should be connected from U to I. The mobility is

$$Z_{out\ mot} = (Z_d + r)/k^2 \tag{7.17}$$

(Derivation of this formula is requested in Problem 21.) When r is relatively small and a voltage driver is used, the actuator output mobility is low and the actuator approximates a velocity source. When the plant is driven by a velocity source, the angular velocity is constant and independent on the load and friction.

The effect of r can be compensated using a small sensing resistor in series with the motor. The voltage drop on this resistor is proportional to the voltage drop on r. By amplifying this voltage and applying it with proper phase to the input of the driver amplifier (i.e., by making a compensating feedback loop), an extra voltage at the output of the driver amplifier is created which has the same amplitude and opposite phase compared to the voltage drop on r, thus compensating the effect of r. The sensing resistor should be temperature dependent, or some additional circuitry should be added to compensate for the temperature dependence of the winding resistance.

The same circuit can be analyzed and designed with Blackman's formula by creating a driver amplifier with the output impedance equal to $-r$, thus making $Z_d + r = 0$ and $Z_{out\ mot} = 0$.

The feedback bandwidth in the compensating loop is limited by the winding inductance. The system stability can be analyzed with the multiloop Bode-Nyquist criterion, and/or with the Bode-Nyquist criterion for connections of two-poles [2].

7.6.4 Piezoelements

Piezoelement actuators can be analyzed using the following equations:

$$\begin{aligned} F &= aU + dV \\ I &= cU + bV \end{aligned} \tag{7.18}$$

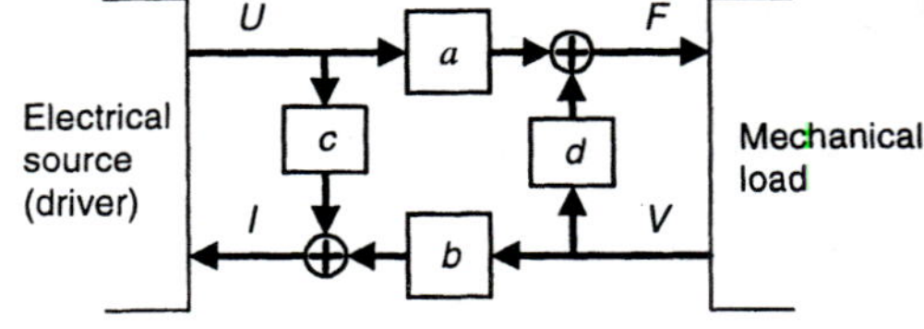

Fig. 7.20 Piezoactuator flowchart

reflected in the flowchart shown in Fig. 7.20. The coefficient a is measured under the condition of zero output velocity, i.e., when the output is clamped. It is the ratio of the clamped force to the incident voltage. The coefficient b is the reverse direction transmission coefficient from V to I while the input port is shorted.

The coefficient d is the inverse of the output mobility under the condition that the input port is shorted. The coefficient c is the input port electrical conductance under the condition that the output port is clamped.

Both the electromagnetic actuators and the piezoactuators possess some hysteresis due to the ferromagnetic and the piezoelectric material properties. Most often, the hysteresis is small and can be neglected. If not, it can be modeled by introducing hysteresis links (described in Chapter 10) into the elementary links in the diagrams in Figs. 7.19 and 7.20.

7.6.5 Drivers, transformers, and gears

The input impedances of driver amplifiers are typically high, i.e., the amplifiers are ***voltage-controlled***. The amplifiers are characterized by their transconductance $Y_T = I_S/U_1$ (measured with zero impedance load) or their voltage gain coefficient $K = E/U_1$ (measured with no load); here, U_1 is the voltage at the driver's input, and U_2 is the voltage at the driver's output. Evidently, $K = Y_T Z_d$ where Z_d is the output impedance of the driver. The driver amplifier flowcharts are shown in Fig. 7.21.

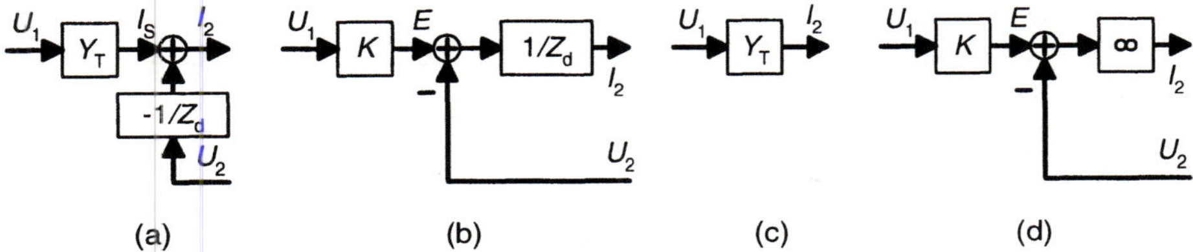

Fig. 7.21 Flowcharts for (a), (b) a driver amplifier with output impedance Z_d, (c) a current driver, and (d) a voltage driver

Example 1. In Fig. 7.21(b), the emf at the output of the driver $E = U_1K$. The motor angular velocity can be calculated from the equations obtained by cascading the flowcharts for the driver and the motor in Fig. 7.19. With the motor mobility neglected or included into the load, the angular velocity Ω and its sensitivity to variations in the load mobility Z_L are

$$\Omega = \frac{\frac{E}{k}}{1+\frac{Z_d + r}{Z_L k^2}}$$

and

$$S_\Omega = \frac{\frac{d\Omega}{\Omega}}{\frac{dZ_L}{Z_L}} = \frac{1}{1+\frac{Z_L k^2}{Z_d + r}}.$$

It is commonly desired that velocity in the velocity controlled systems be less dependent on the load mobility Z_L (which includes the uncertain friction and load dynamics), i.e., $|S_\Omega|$ needs to be small. The sensitivity depends on Z_d. Therefore, it is important to choose and implement Z_d properly. The sensitivity becomes small when

$|Z_d + r|$ is small (recall Example 1 from Section 7.3, the description of the effect of r in Section 7.6.2, and the effect of the actuator impedance described in Section 7.5).

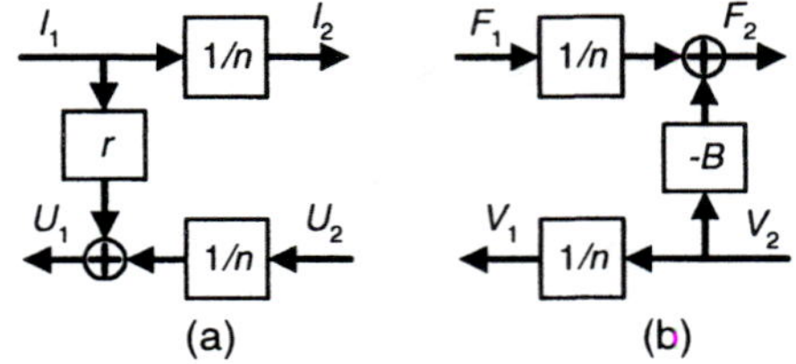

Fig. 7.22 Flowcharts for (a) an electrical 1:*n* transformer and (b) a mechanical gear

A flowchart for an electrical transformer with the turn ratio n is shown in Fig. 7.22(a), and a flowchart for a mechanical gear with the velocity ratio n, in Fig. 7.22(b). The resistance of the primary winding being r_1 and of the secondary, r_2, the total equivalent resistance of the primary is $r = r_1 + r_2/n^2$. For a mechanical gear box, the equivalent viscous friction can also be lumped into one coefficient B.

The composite flowchart of a driver and a motor with an attached gear box is a cascade connection of the three corresponding flowcharts.

Example 2. The driver has voltage gain coefficient 10 and output impedance 0.25 Ω. The motor winding impedance is $20 + 0.0003s\,\Omega$ and the motor constant is 0.1 Nm/A. The motor rotor's moment of inertia is 0.02 Nm². The gear amplifies the motor torque twice, i.e., the gear ratio, the load angle to the motor angle (or, the load angular velocity to the motor angular velocity), is 1:2. The losses in the gear and bearings make $B = 0.01$ Nm/(rad/sec). The load's moment of inertia is 0.2 Nm², i.e., the load mobility is $5/s$. The model for the driver, motor, gear, and load assembly shown in Fig. 7.23 is a cascade connection of the models in Figs. 7.21, 7.19, 7.22, and the load.

Since the output impedance of the driver and the impedance of the motor winding are connected in series, the model can be simplified by changing the value of the link $1/0.25 = 4$ to $1/(20.25 + 0.003s)$ and eliminating the link $20 + 0.003s$.

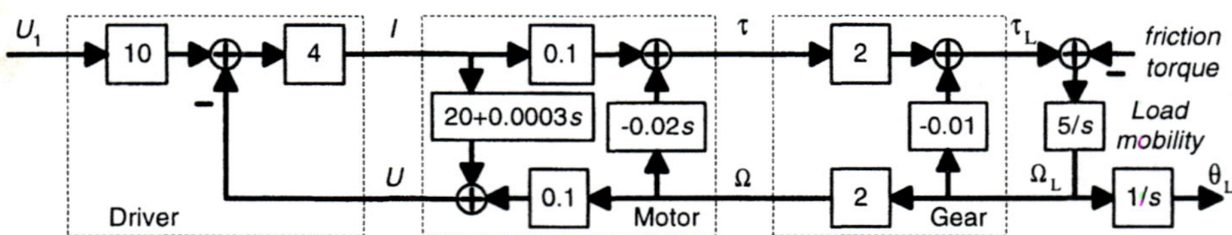

Fig. 7.23 Block diagram of cascade connection of driver, motor, gear, and load

Example 3. The actuator of the previous examples applied to the plant which is not rigid: the torque τ_L minus *friction torque* is applied to the body with moment of inertia J_1 (which reflects the inertia of the gear and the motor), and an antenna with moment of inertia J_3 is connected to the gear via a shaft with torsional stiffness coefficient k_2.

Therefore, the load mobility $5/s$ in Fig. 7.23 must be replaced by the mobility Z which is the input impedance of the equivalent electrical ladder Π-type network:

$$Z = \frac{1}{J_1 s + \dfrac{1}{\dfrac{s}{k_2} + \dfrac{1}{J_3 s}}} = \frac{\dfrac{1}{J_1}\left(s^2 + \dfrac{k_2}{J_3}\right)}{\left[s^2 + k_2\left(\dfrac{1}{J_1} + \dfrac{1}{J_3}\right)\right]s}.$$

This mobility has two zeros, $\pm j\sqrt{k_2 / J_3}$, and three poles: $\pm j\sqrt{k_2(J_1 + J_3) / J_1 J_3}$ and 0. The pair of purely imaginary poles will bring about a pair of complex poles in the transfer function Θ_L/U_1 where Θ_L is the angular velocity of the output of the gear. These poles will be substantially damped by the output mobility of the actuator.

7.6.6 Coulomb friction

Coulomb friction is modeled by the block shown in Fig. 7.24(a). The dependence of the friction force F_{Coulomb} on the velocity V is shown in Fig. 7.24(b).

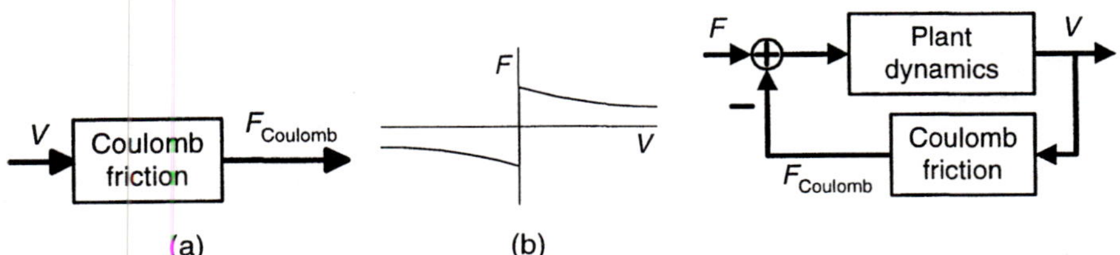

Fig. 7.24 Coulomb friction model and characteristic

Fig. 7.25 Model of a dynamic system with friction

The friction model is commonly incorporated in the plant model as a feedback path shown in Fig. 7.25. In this composite plant model, F is the force applied to the plant, and the difference between this force and the Coulomb friction force is applied to the plant dynamics (the summer is also shown in Fig. 7.23). When the actuator is not a pure force source, the inverse of the actuator output mobility can be included in the model as a feedback path in parallel with the Coulomb friction link.

In Fig. 7.22, the model of a gear was shown with viscous friction B. Instead of this block (or, parallel to this block), a Coulomb friction link can be placed. The Coulomb friction link can be also placed parallel to the link $1/Z_m$ in Fig. 7.19. It is seen that when Z_m is small, the link $1/Z_m$ will dominate and the effect of Coulomb friction will be negligible. It is seen therefore that the effect of Coulomb friction greatly depends on the actuator's output mobility.

Example 1. When a rigid body is dragged over another rigid body with a rough surface by a soft spring, Coulomb friction causes oscillation (with some re-arrangement of coordinates, this is the case of playing a violin). In this case the actuator's output impedance is large, and the actuator is a nearly pure force source. However, when the source is that of velocity, evidently, no oscillation occurs.

7.7 Examples of system modeling

A system model can be described either by the system elements and the topology of their connections (as in SPICE), or by mathematical equations (as in MATLAB and many other computer languages). The use of flowcharts simplifies the organization of the mathematical description of the system. The equations expressed in the flowcharts can be entered in the input file of a computer simulation program via a graphical interface (as in SIMULINK ® and some other control software packages).

Example 1. A SIMULINK-like model of a control system using a brushless dc permanent magnet motor is shown in Fig. 7.26.

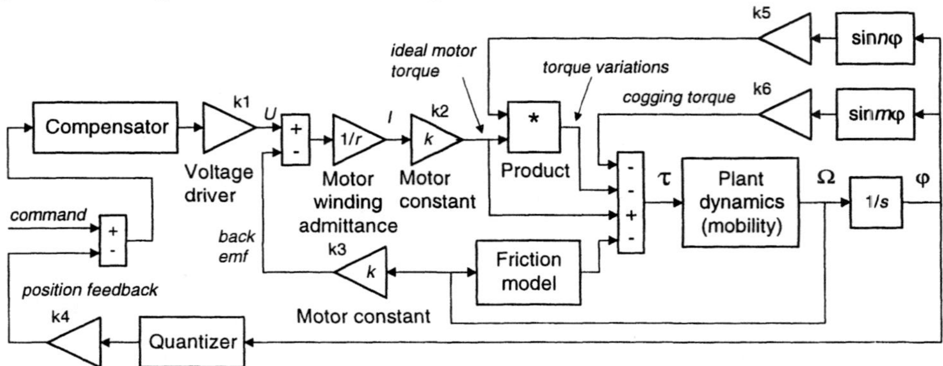

Fig. 7.26 Block diagram of a control system

The system includes an input summer, a compensator with voltage output, a voltage driver as a voltage controlled source of voltage U, a brushless permanent magnet dc motor with motor constant k, a plant with specified mobility, a friction model, a position (angle) feedback including a quantizer (since an optical encoder used as the angle sensor), and the models of the motor torque variations and cogging as periodic dependencies on the output angle φ. The torque τ is applied to the plant, the output angular velocity is Ω. The back electromotive force is subtracted from the driver's output voltage U, and, divided by the motor winding resistance r, produces the winding current I. The torque τ is the ideal motor torque (current multiplied by the motor constant) from which the torque variations, cogging torque, and friction torque are subtracted. SIMULINK can provide the output time-response to the input signal. The SIMULINK analysis tools (signal source; oscilloscopes; plotters; multiplexers to provide data for the workspace) are not shown in Fig. 7.26. Their use is described in the manuals.

A loop Bode diagram can be found as follows: disconnect the feedback path, attach inport 1 (from the Simulink "Connections" library) to the command input, outport 1 to the loop output, and type in the MATLAB command window:

```
[A,B,C,D]=linmod('file_name'); bode(A,B,C,D,1)
```

The closed-loop frequency response can be obtained by applying the same program to the system with reconnected feedback path. The meaning of the matrices A,B,C,D is explained in Chapter 8.

Example 2. The two methods of system description are illustrated below with an example of a vibration isolation system shown in Fig. 7.27. A voice coil actuator with a load cell sensor is placed between two flexible bodies with mobilities Z_1 and Z_2. The vibration source is on the second body, and the actions of the voice coil should reduce the vibrations of the first body. The load cell together with its amplifier has sensitivity 1 V/N. The voice coil is characterized by the coil resistance $r = 4\,\Omega$ and the electromechanical coefficient $k = 3$ N/A. The coefficient's equivalent is a lossless down-transformer with the turn ratio of 3 (the ratio of the primary to the secondary windings).

Fig. 7.28 shows the equivalent electrical schematic diagram. All currents and voltages to the left of the vertical dashed line represent physical electrical variables in amperes and volts; to the right of the line, they represent forces in newtons and velocities in m/sec, correspondingly. The load cell is therefore represented by an ampermeter, in this case, a current-controlled voltage source. The schematic diagram can now be coded into the input file of SPICE.

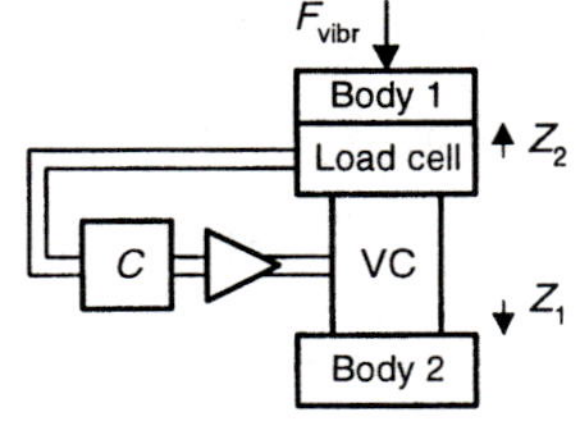

Fig. 7.27 Vibration isolation system

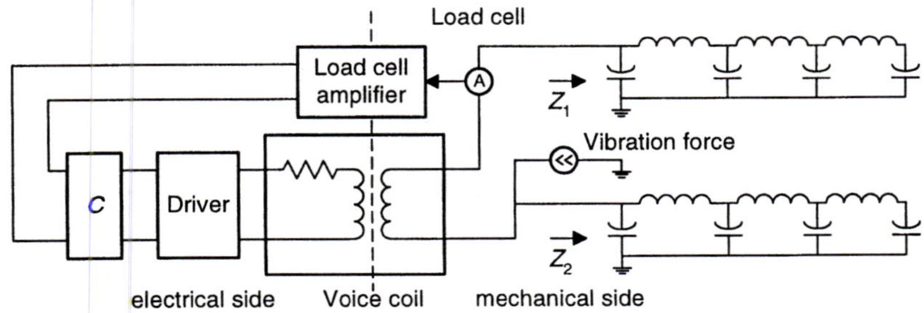

Fig. 7.28 Equivalent schematic diagram for a vibration isolation system

Fig. 7.29 shows a flowchart description for the same system. It can be used, for example, with SIMULINK. Here, Z_{out} is the output impedance of the driver amplifier.

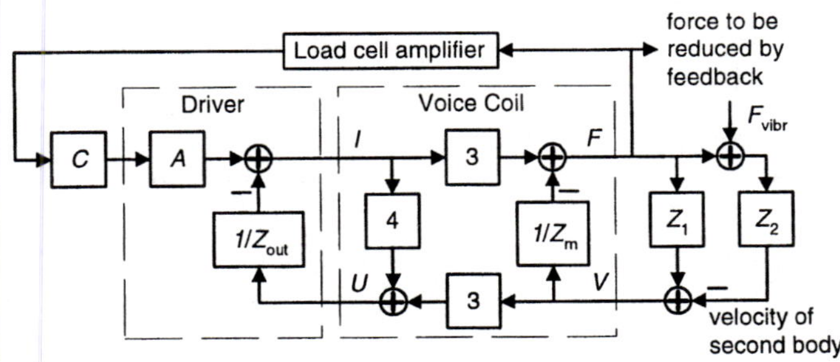

Fig. 7.29 Flowchart description of a disturbance isolation feedback system

The choice of the modeling method depends on the availability of simulation software, on the problem specifics, on the system links, and, to a large extent, on the designer's personal preferences. If, for example, the driver amplifier is already designed and should not be changed, the compensator *C* is implemented in software, and the designer is a mechanical engineer, then probably flowchart simulation with SIMULINK will take less time than simulation with SPICE. If, however, the driver amplifier needs to be designed and optimized simultaneously with the compensator, the compensator is analog, and the designer is an electrical engineer, then probably the design and simulation is easier to perform using an equivalent electrical schematic diagram and SPICE.

7.8 Flexible structures

7.8.1 Impedance (mobility) of a lossless system

The ***Foster theorem*** states that the zeros and poles of a driving point impedance (mobility) of a lossless system are purely imaginary and alternate on the $j\omega$-axis, as shown in Fig. 7.30. At zero and at infinite frequencies, there could be either a pole or a zero.

Mobility frequency responses of non-dissipative flexible plants are exemplified in Figs. 4.36, 7.31. The angle of the mobility is -90° where $|Z|$ is falling, and 90° where $|Z|$ is rising. The low-frequency asymptote and the high-frequency asymptote reflect $|Z|$ being either proportional or inversely proportional to frequency. It is seen that in Fig. 7.31(a), the high-frequency asymptote is shifted from the low-frequency one. The low-frequency asymptote is determined by the rigid body mode, i.e., by the masses of all the bodies connected together by springs which are considered stiff at very low frequencies. The high-frequency asymptote is determined solely by the body to which the incident force is applied. All other bodies are disconnected from this body since the springs' mobilities become very large at high frequencies.

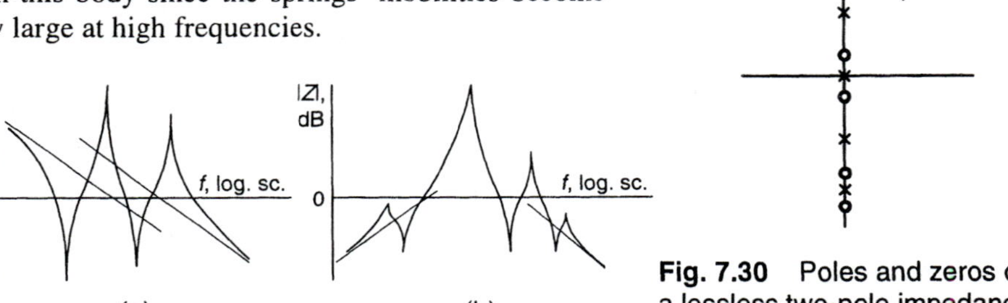

Fig. 7.30 Poles and zeros of a lossless two-pole impedance

Fig. 7.31 Driving point impedance (mobility) of a lossless system, logarithmic scale:
(a) having a pole at zero frequency and a zero at infinite frequency,
(b) with zeros at zero frequency and infinite frequency, and a prominent suspension mode

It can be calculated that each additional zero-pole pair lifts the high-frequency asymptote by the square of the ratio of the pole to the zero frequencies. This ratio commonly depends on the mass participation in the flexible mode. If, for example, an actuator drives a massive main body from which a small additional mass is suspended on a spring, the pole-to-zero ratio of the flexible mode equals the square root of the ratio of the sum of the masses to the mass of the main body. The smaller the mass of the smaller body, the closer is the pole to the zero and the smaller is the shift in the asymptote.

Structural damping displaces the poles and zeros to the left of the $j\omega$-axis. As a result, the peaks and valleys on the frequency response of $|Z|$ become smoothed. Still, the losses in mechanical systems without special dampers can be so small that the peaks reach 40 and even 60 dB over the smoothed response of the mobility.

7.8.2 Lossless distributed structures

Distributed structures can be approximated by lumped element structures with a large number of elements as shown in Fig. 7.32(a) and (b).

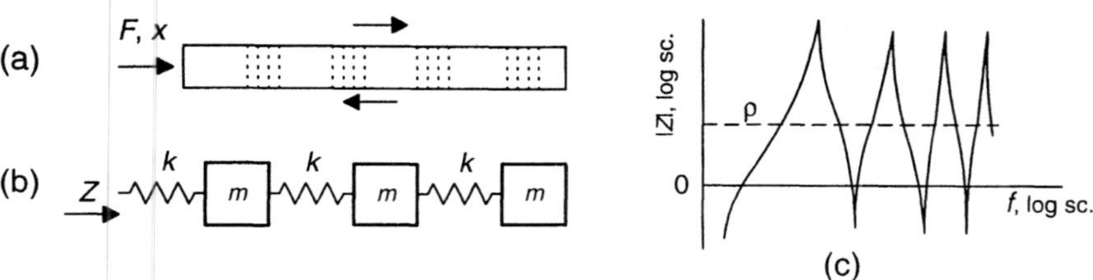

Fig. 7.32 (a) Distributed structure, (b) its equivalent representation with lumped parameters, (c) its driving point mobility

Correspondingly, the impedance (mobility) of a distributed structure $Z(s)$ possesses an infinite number of poles and zeros, as illustrated in Fig. 7.32(c).

These poles and zeros can be viewed as produced by interference of the incident signal and the signal reflected back from the far end of the structure. The frequencies of high-frequency resonances are very sensitive to small variations of the structure parameters, and therefore, in physical systems, are to a large extent uncertain. This uncertainty becomes a critical factor in limiting the bandwidth of the available feedback, as has been discussed in Chapter 4.

The ***wave impedance*** (***mobility***) is the input impedance (mobility) of the structure extended to infinity so that no signal reflects at the far end and returns back to the input. The wave impedance of a uniform lossless distributed structure is the resistance $\rho = 1/\sqrt{kM}$, where k and M are the stiffness and the mass per unit length of the structure. The plot of |Z| in logarithmic units is commonly nearly symmetrical about the wave impedance as shown in Fig. 7.32(c).

Matching at the far end means loading the structure at some resistor (or, if mechanical, at a dashpot) with the impedance (mobility) equal to the wave impedance (mobility) of the structure. In this case, we can think of the structure as extending to infinity, and the input impedance (mobility) of the structure equaling the wave impedance (mobility).

Matching between the signal source and the structure is provided when the signal source output mobility equals the wave mobility of the structure. Matching at either end of the structure fully damps it (since the resonances can be considered a result of interference of the waves reflected at the ends of the structure, and if at either end no reflection occurs due to matching, no resonances take place).

Example 1. A motor with motor constant k is employed to rotate a spacecraft solar panel having moment of inertia J to keep the panel perpendicular to the direction to the sun. The solar panel torsional quarter wave resonance frequency is f_r. The spacecraft attitude control can be improved by damping this resonance. This can be achieved by making the motor output mobility approximately equal to the solar panel wave mobility ρ. The desired motor output mobility can be created by compound feedback in the driver.

We calculate the mobility using the voltage-to-velocity electro-mechanical analogy. In electrical transmission lines, the phase is $2\pi fx(lc)^{1/2}$ where x is the length and l and c are the inductance and capacitance per unit length. At the quarter wave resonance, $2\pi f_r x(lc)^{1/2} = \pi/2$ wherefrom $(lc)^{1/2} = 4xf_r$. Then, the wave impedance is $\rho = (l/c)^{1/2} =$

$1/(4cxf_r)$. The cx is the full capacitance of the line, which is analogous to the moment of inertia J. Hence, $\rho = 1/(4Jf_r)$ and the driver output mobility is $1/(4Jk^2f_r)$.

7.8.3 Collocated control

Frequently, the actuator is a pure source of force (or torque, or current), and the feedback sensor is connected to the same port of the plant but reads the variable which is related to the actuator variable by the plant driving point impedance (mobility). The plant transfer function is in this case the plant driving point impedance or admittance (mobility or the inverse of mobility). Such control is called ***collocated***, as was already mentioned in Section 4.3.6.

The driving point impedance (mobility) of a passive plant is ***positive real*** (***p.r.***), and the phase angle of the function is constrained within the [−90°, 90°] interval. (Properties of p.r. functions are reviewed in Appendix 2.) This feature greatly simplifies the controller design and the stability provision.

Example 1. In the mechanical arrangement shown in Fig. 7.33(a), the actuator is a torque source and the sensor measures angular velocity $d\varphi_1/dt$ (or a linear operator of the velocity, such as position or acceleration).

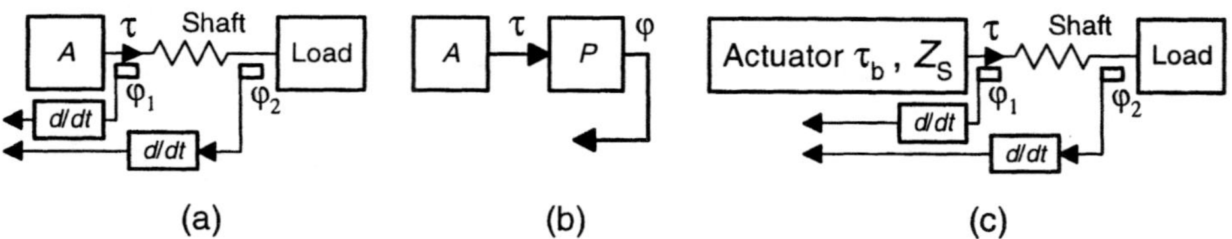

Fig. 7.33 (a) Actuator is a torque source, control is collocated with any of the sensors, φ_1 or φ_2, providing the fed back signal; (b) block diagram of collocated control; (c) the case is considered in Section 7.8.4; the output mobility of the actuator is finite; control is collocated when sensor φ_1 is used and is non-collocated when sensor φ_2 is used

In the related block diagram shown in Fig. 7.33(b), the plant transfer function $P = \varphi/\tau$ is the mobility of the mechanical system.

Since the torque is the same before and after the shaft flexibility, the control remains collocated even when the sensor φ_2 is used. (The control will not be collocated if there will be an additional rigid body with substantial moment of inertia at the point of the torque application.)

7.8.4 Non-collocated control

Contrary to the driving point lossless mobility, transfer functions between different ports of a passive lossless structure may have several consecutive poles or zeros, and the phase shift of the transfer functions is not constrained. As has been shown in Chapter 4, this phase shift reduces the available feedback.

The sensor and the actuator cannot always be placed exactly at the same location. For example, for better control accuracy, the sensor might be placed closer to the load. If the plant were an ideal rigid body, or the actuator an ideal force (or torque) source, the control would still be collocated. Otherwise, as shown in Fig. 7.33(c), the control is collocated when the first sensor is used and is not collocated, i.e., the actuator and plant transfer function is not, generally, p.r., when the second sensor is used. In this case, the control can be called collocated only as an approximation of the reality, the approximation being valid only over a limited frequency bandwidth.

With respect to the flexibility of the shaft connecting the motor and the load, misunderstanding frequently arises between the designer of the mechanical structure and the control loop designer about the meaning of the statement: "Higher frequency structural modes are *much* higher than the frequency range of operation." The structural designer is mostly concerned with the structural soundness over the working frequency range and might underestimate the following factors: (a) the structural mode frequency is proportional to only the square root of the stiffness, and (b) the control feedback loop bandwidth must be *many times* wider than the functional band of the system (as has been shown in Chapters 4 and 5).

Trying to make the structure as inexpensive and as lightweight as possible, structural designers tend to design structures with structural modes falling within the feedback bandwidth which prevents the control system from achieving high accuracy. This is why *the control system designer/dynamicist needs to be involved in the design process before the mechanical structure design is completed, and he must be able to give his recommendations for the required changes in the structure and to evaluate the proposed solutions in real time during the meetings with the structural designer*.

As has been shown in Section 4.3.6, the plant flexible modes restrict the available feedback. Introducing structural damping can greatly improve the control system performance.

The dampers can be built using Coulomb friction, hydraulic energy dissipation, or eddy currents. Using the dampers allows a substantial reduction of the effects of structural modes. The dampers are, however, relatively expensive, and to constrain the system cost, weight, and dimensions, using the dampers must be well justified by the available performance improvement.

7.9 Sensor noise

7.9.1 Motion sensors

7.9.1.1 Position and angle sensors

For position (or angle) variables' measurements, three categories of sensors are commonly employed: (a) position sensors, (b) rate (i.e., velocity) sensors, and (c) accelerometers.

For position and angle sensors, the potentiometers, the linear variable differential transformers (LVDT), the resolvers, the optical encoders, the laser interferometers, and the star trackers are representative.

An electrical potentiometer with the tap moved mechanically by the plant can provide high resolution and the position reading accuracy and linearity of 0.1%. Less accurate potentiometers are universally used in small servos for radio-controlled toys, with resolution still better than 0.5°.

The LVDT shown in Fig. 7.34(a) measures the position of the plant with respect to the base. It has three windings: the two symmetrical windings to which a signal from a generator is applied in opposite phases and the winding providing the output voltage. The voltage is amplified and applied to a synchronous detector the output of which is a product of this voltage and the generator voltage. The dc component of the detector's output is proportional to the third winding displacement from the central position.

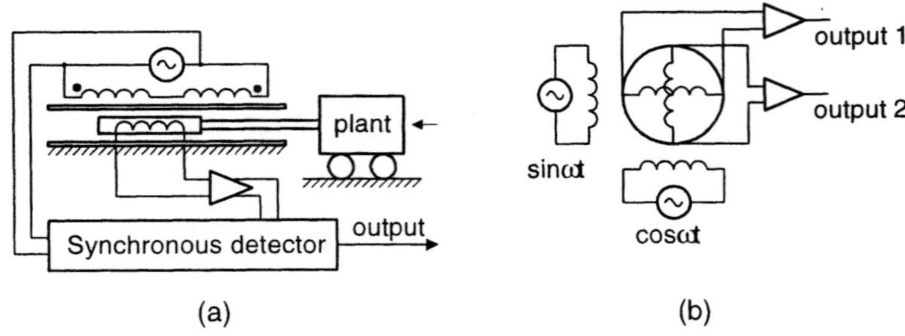

Fig. 7.34 Position sensors: (a) LVDT, (b) resolver

The maximum stroke of common LVDTs is from 0.1" to 1" with pretty good linearity, and the resolution of precision LVDTs can be 1 microinch. There also exist rotary versions of the device. Specialized ICs are available which incorporate all necessary electronics.

The ***resolver*** shown in Fig. 7.34(b) is a rotating transformer. It has two stator windings and two rotor windings. To a pair of these windings an ac signal is applied in quadrature. From the signal induced on the two other windings, the angle of rotation can be determined with high accuracy.

The optical encoder was already briefly described in Section 7.6.2. The encoders can be ***absolute***, with complicated patterns on the optical disk which at any time give full information about the shaft angle, or ***incremental***, with simple patterns of alternating transparent and black lines. The incremental encoders must be accompanied by some electronics keeping track of the counts. ***Quadrature*** incremental encoders have two 90°-shifted readers. This improves the accuracy by a factor of 2 and also gives information about the direction of rotation. Interpolation of the data from the readers additionally improves twice the angle reading accuracy.

The ***laser interferometer*** compares the phases of the incident laser light with the light reflected from a mirror placed on the target, and counts the fringes of the interference when the distance changes gradually. The interferometer can measure large distances with nanometer accuracy. Less accurate and less expensive interferometers use modulated light beams.

The ***star tracker*** is a small precision telescope equipped with an image recognition system.

7.9.1.2 Rate sensors

The most often used rate (velocity) sensors are the ***tachometer*** and the gyroscope. The tachometer is a dynamo mounted on the same shaft as the motor (or, there can be only a separate tachometer winding on the motor's rotor or stator; the motor windings in brushless motors which are disconnected from the driver during certain rotation angle intervals, can be used as tachometer windings). The emf on the tachometer winding is proportional to the motor angular velocity. In contrast to the resolver, the tachometer is not able to detect the angle of rotation when the rotation rate is very low since in this case the signal on the tachometer winding is below the noise level.

The ***rate-gyro*** is a gyroscope with a position servo loop. An electrical winding generates a torque preserving the relative position of the gyro to the base. The current in the winding is the output signal of the gyro; this current is proportional to the base angular velocity (i.e., the rate of the base rotation). The rate signal is analog. A/D conversion simplifies the data interface with the rest of the system, but it loses some higher-frequency information. For these reasons, it is common to use both the analog and digital outputs from the gyro.

7.9.1.3 Accelerometers

The accelerometers use electromagnetic, piezoelectric, piezoresistive, and tunnel-effect devices.

In an electromagnetic accelerometer, the magnetic ***proof mass*** is suspended on a spring. Motion of the base relative to the proof mass produces electromotive force in a coil surrounding the proof mass which is mechanically connected to the base. The proof mass position relative to the base is kept constant by a servo feedback loop which applies electromagnetic force to the proof mass. This force creates acceleration nearly equal to the base acceleration so the proof mass remains still relative to the base. The value of the compensating force in accordance to Newton's second law is the measure of the acceleration.

There exist many different types of accelerometers using a suspended proof mass the position of which is measured by some sensor and, via a servo loop, kept constant by some force. The signal producing this force (e.g., current in the coil or a voltage producing electrostatic force) serves as the measure of the acceleration. An example of an accelerometer control loop is given in Section 11.9.

A set of three orthogonal accelerometers can be used to determine the vector of gravity force and, therefore, the inclination of a vehicle.

7.9.1.4 Noise responses

Ideally, when the output data is post-processed, motion sensors are interchangeable: suitable integrations and differentiations allow us to translate between positions, rates, and accelerations. The principal difference between them is their dynamic accuracy characteristics, or, with the frequency domain characterization, their noise spectral density responses. The noise power and the mean square error can be calculated by integration of the spectral density with linear frequency scale over the bandwidth of interest.

Position sensors typically give an accurate steady state value of the position, but their accuracy decreases when the position changes rapidly. For example, it takes considerable time to accumulate enough light to identify dim stars in a star tracker, so that this sensor cannot react fast to the position changes of the spacecraft on which the star tracker is mounted. On the other hand, gyros have ***drift*** (slow and gradual change in the reading caused by device imperfections) and therefore do not determine the position accurately after some time passes from the initial setup. Gyros are worse in position determination than the star trackers at frequencies from 0 to 0.01 Hz, but get better at higher frequencies. At even higher frequencies (say, from 15 Hz and up) the gyro noise increases and the accelerometers become superior.

The frequency responses of typical sensor noise spectral density converted to position data are shown in Fig. 7.35.

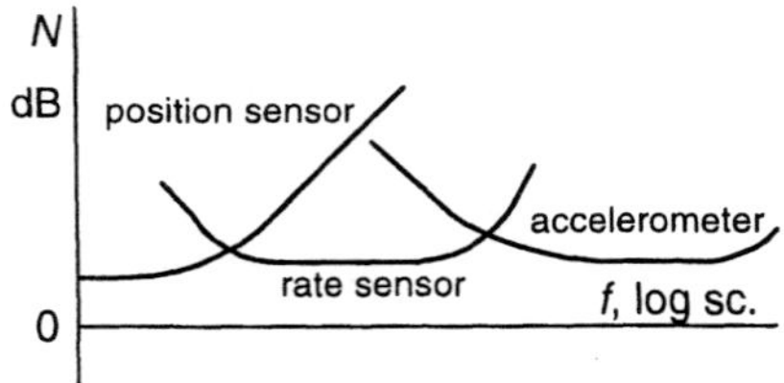

Fig. 7.35 Frequency response of sensor noise normalized to the sensor input

7.9.2 Effect of feedback on the signal-to-noise ratio

The most common noise sources in control systems are those of the error amplifier (which is the first amplifier after the feedback summer), of resistors, and of sensors. The noise is commonly characterized by its spectral density.

A feedback loop which reduces both the signal source and the noise source effects does not change the ratio of the signal to the noise at the system output. Thus, it is possible and often convenient to examine the signal-to-noise ratio as if there were no feedback.

There is a caveat here: when the system performance is compared with and without feedback, the system should not be changed in any other aspect. Particularly, when the output effect of the sensor noise is calculated, the transmission coefficient from the sensor to the system output should remain unchanged. Therefore, the feedback loop should be disconnected between the system output and the sensor input, not between the sensor output and the feedback path. The difference is shown in Fig. 7.36.

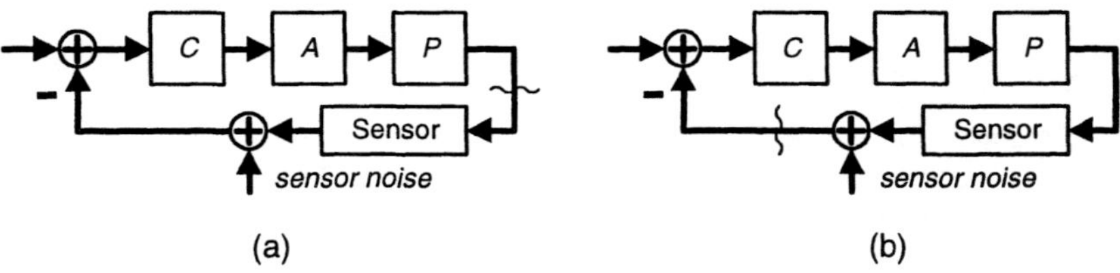

Fig. 7.36 Disconnecting the feedback loop while (a) preserving the signal-to-noise ratio at the system output, (b) changing the ratio

As was mentioned in Section 1.1.2, the loading for the disconnected ports must be preserved while disconnecting the feedback path for appropriate comparison of the system with and without feedback.

Example 1. Fig. 7.37 shows an amplifier with the feedback path from the emitter of the outer stage to the emitter of the input stage. The feedback loop can be disconnected by setting $R_2 = \infty$. This reduces the signal-to-noise ratio – not because of change in the feedback, but because the resistance in the input contour increases which reduces the signal and increases the noise. When the feedback is eliminated by setting $R_1 = 0$, the signal-to-noise ratio improves since in this case the resistance in the input contour decreases.

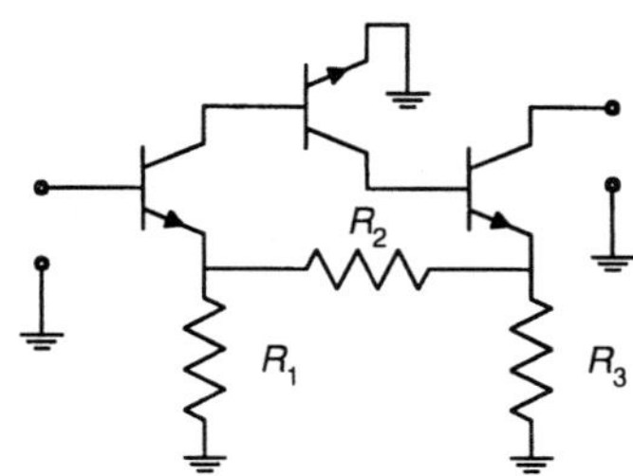

Fig. 7.37 Amplifier with an emitter feedback path

If, however, both disconnected ports of the feedback path are loaded onto loads equal to the loads that each port sees when the feedback path is closed, the signal-to-noise ratio remains the same with or without the feedback.

7.10 Mathematical analogies to the feedback system

7.10.1 Feedback-to-parallel-channel analogy

The summer in Fig. 7.38(a) implements the equation $E = U - TE$. From here it follows that $U = E + TE = (T + 1)E = FE$ as diagrammed in Fig. 7.38(b).

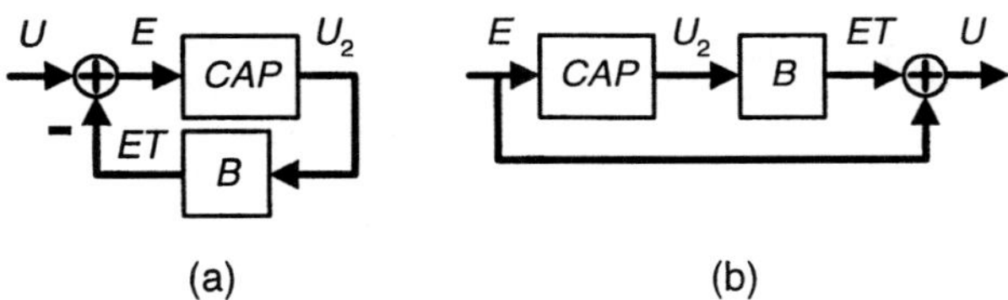

Fig. 7.38 Analogy between (a) a feedback system and (b) a system with two parallel forward paths

The formulas (1.1), (1.2), and (1.3) remain valid for this system which has no feedback and represents the parallel connection of two paths. When $|T + 1| > 1$, introducing the channel T increases $|U|$ and reduces the ratio E/U. When $|T| >> 1$, then $U \approx ET$ and $U_2/U \approx -1/B$. In this way, all the features of the feedback equations (1.1), (1.2), and (1.3) are apparent.

This analogy can be employed to analyze or simulate responses of system (a) when this system is unstable and system (b) is stable. We will use this analogy in Chapter 12.

7.10.2 Feedback-to-two-pole-connection analogy

Equations (1.1), (1.2), and (1.3) also describe the signals in the two-pole connections depicted in Fig. 7.39(a). The transfer functions of the links in block diagram (b) recite Ohm's law: voltage U_1 applied to the first two-pole causes current I, and this current applied to the second two-pole produces voltage drop U_2. The contour equation $U_1 = U - U_2$ reflects the summer in the feedback loop.

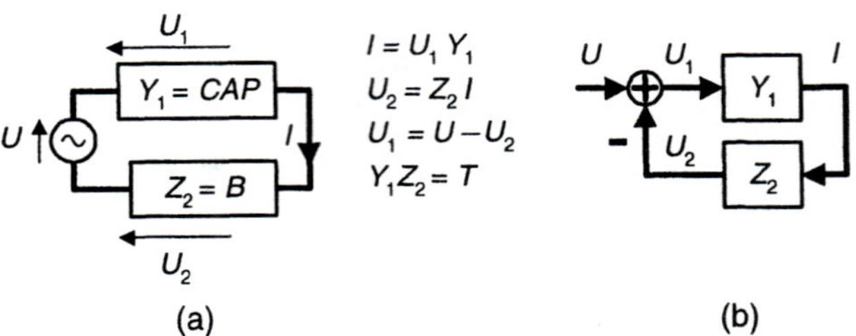

Fig. 7.39 Analogy between (a) a two-poles' connection and (b) a feedback system

This large feedback condition here, $|T| >> 1$, is that $|Z_2| >> |1/Y_1|$, i.e., the impedance of the second two-pole is much larger in magnitude than the impedance of the first two-pole. As a result, the second two-pole can be neglected when considering the current calculations: $I = U/(1/Y_1 + Z_2) \approx U/Z_2$. This is analogous to the closed-loop transmission of a system with large feedback where the output is the input divided by the feedback path transfer function.

The analogy can be employed to use the passivity condition of a network of electrical two-poles for stability analysis of feedback systems. We will use this analogy in Chapter 10.

7.11 Linear time-variable systems

Linear time-variable links (LTV) are described by linear equations whose coefficients explicitly depend on time. When the signal applied to the input of an LTV link is sinusoidal, the output, contrary to the case of a linear time-invariable link (LTI), is not necessarily sinusoidal, and might contain higher harmonics. When several sinusoidal components are applied to the link input, the output contains intermodulation products.

LTV links of digital compensators have already been analyzed in Section 5.10.7. In this section we will consider Mathieu's equation

$$d^2y/dt^2 + (a + 2\varepsilon\cos(t))y = 0 \tag{7.19}$$

which is representative of some LTV systems that might be encountered in practice. If $\varepsilon = 0$, this equation describes an LTI lossless resonator, the solution being a sinusoid with the angular frequency $\sqrt{a}$. The solution is on the boundary between self-oscillation and exponential decay. The time-variable coefficient $2\varepsilon\cos t$ changes the system behavior: some combinations of ε and a lead to solutions which are exponentially rising with time, and other combinations introduce damping into the system. The Ince-Strutt stability diagram shown in Fig. 7.40 depicts the areas of stability and instability in the plane of the equation parameters.

A feedback system described by Mathieu's equation is shown in Fig. 7.41(a). Intermodulation of the signal harmonics in the LTV link $2\varepsilon\cos t$ produces certain components at its output. Addition of these components to the signal passing through the LTI link a alters the phase of the signal at the summer's output. When the coefficient ε is large, the system is unstable with nearly all possible a, as seen from the stability diagram in Fig. 7.40.

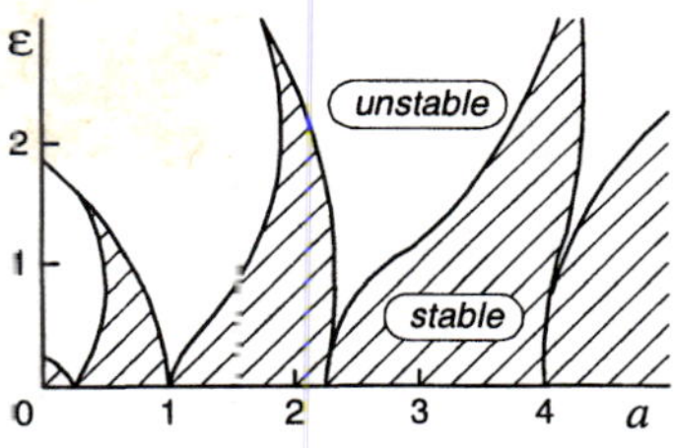

Fig. 7.40 Ince-Strutt diagram

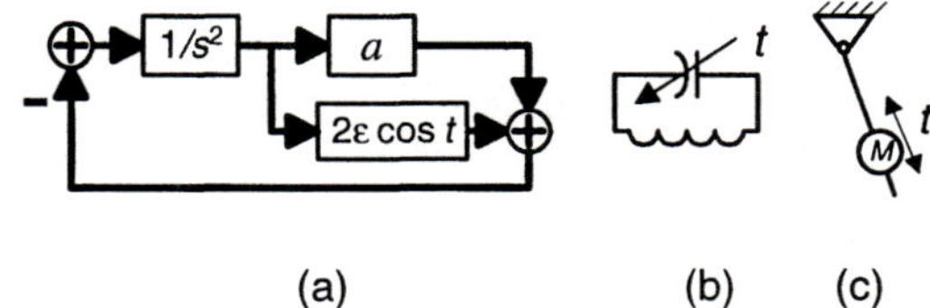

Fig. 7.41 (a) Feedback system, and (b) and (c) resonators described by Mathieu's equation

An electrical circuit equivalent to the feedback system is shown in the diagram in Fig. 7.41(b), and an equivalent mechanical system, in Fig. 7.41(c). The time-variable term periodically changes the resonance frequency of the resonator. This pumps the energy, on average over the length of the cycle, in or out of the resonator. For example, when the resonator capacitance is reduced while the charge is preserved, the voltage on the capacitor increases and the stored energy increases (the energy is proportional to the square of the voltage). The resonance frequency can be changed and the energy pumped into the pendulum resonator (c) by moving the center of mass of the swinging body up and down.

The phase shift for the passing signal in periodical LTV links depends on the phase of the incident signal. Therefore, while analyzing stability conditions, the worst possible case needs to be considered among all possible phases of the incident signal. This results in some uncertainty range in the link phase shift. However, the uncertainty of the loop transfer function reduces the potentially available feedback. Therefore, when controlling LTI plants, it is generally appropriate to use LTI compensators. When the compensators are LTV like in digital systems described in Chapter 5 or in the systems where the compensator parameters are varied in search for a maximum of a certain performance index, the available feedback is reduced.

To increase the feedback about an LTV plant, a controller can be chosen to be LTV in such a manner that the loop transfer function is less dependent on time (i.e., when the plant gain increases, the compensator gain decreases accordingly).

In adaptive systems (see Chapter 9), the compensators are LTV, but typically, the rate of the compensator variations in time is chosen much lower than the rate of changes in the critical part of the signals and the system dynamics (i.e., the dynamics that limit the value of the feedback and affect substantially the stability margins). In this case, the LTV link for the purpose of stability analysis can be considered LTI. Such links are called quasi-static.

For small signal deviations from the current value, nonlinear links can be seen as LTV links. We will use this approximation in Chapter 12 for the analysis of process stability.

7.13. Problems

1 Why is it convenient to preserve power while choosing the type of electromechanical analogy?

2 How many independent variables are at
(a) a ball joint,
(b) a pin,
(c) an *x-y* positioning system,
(d) sliding planes (e.g., one plate sliding arbitrarily on the surface of another plate),
(e) a two-wire junction of two electrical circuits,
(f) a three-wire junction of two electrical circuits,
(g) a four-wire junction of two electrical circuits.

3 What are the equivalents of Ohm's law for translational and rotational mechanical systems? For thermal systems?

4 What are the equivalents of Kirchhoff 's laws for mechanical systems (consider two analogies) and for heat transfer systems?

5 Draw the electrical analog circuit for the translational mechanical system in Fig. 7.42(a), the rotational system in Fig. 7.42(b) with torsional stiffnesses of the shafts k_1 and k_2, and for the thermal system in Fig. 7.42(c).

6 Plot frequency responses of transfer functions:
(a) V_2/F for the system in Fig. 7.42(a) for the case $M_1 = 100$, $k_1 = 2$, $M_2 = 5$;
(b) V_2/F for the system in Fig. 7.42(a) for the case $M_1 = 50$, $k_1 = 0.2$, $M_2 = 50$;
(c) Ω_3/t for the system in Fig. 7.42(b) for the case $J_1 = 20$, $J_2 = 3$, $J_3 = 12$, $k_1 = 0.1$, $k_2 = 0.03$. Derive the function using Lagrange equations or equations corresponding to an equivalent electrical circuit, and use MATLAB, or use SPICE.
(d) Same as (c) for the case $J_1 = 10$, $J_2 = 24$, $J_3 = 2$, $k_1 = 0.01$, $k_2 = 0.02$.
(e) T_c/P for the system in Fig. 7.42(c) for the case $R_T = 2.72$, $C = 100$.

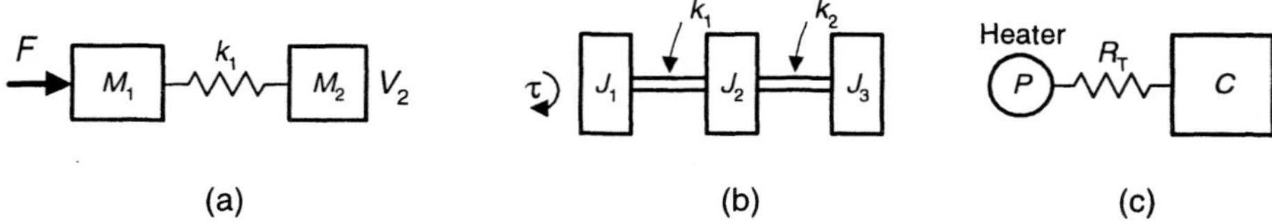

Fig. 7.42 Examples of dynamic systems

7 Draw an equivalent electrical circuit for cooling a power IC with a heat sink.

8 Using equations (7.2)–(7.5), show that if $Z_L = 0$, then $I = I_S$, and when $Z_L = \infty$, $U = E$.

9 In Fig. 7.9, the loading curve is expressed as $I = (E - U)/R_S$. Express I as a function of R_L.

10 The plant is a rigid body, $M = 50$ kg. The viscous friction coefficient is 0.01. The actuator output mobility is (a) 1(m/sec)/N; (b) 5 (m/sec)/N; (c) 10 (m/sec)/N. Use MATLAB to plot the frequency response of the actuator together with the plant. What is the plant transfer function uncertainty at 10 Hz?

11 The plant transfer function is the ratio of the output velocity to the force applied to a rigid body, $1/(sM)$, 20 kg < M < 50 kg. The actuator output mobility is (a) 1(m/sec)/N; (b) 5(m/sec)/N; (c) 10(m/sec)/N. Use MATLAB to plot the frequency response of the actuator with the plant. What is the response uncertainty?

12 The actuator is driving a rigid body, 20 kg < M < 50 kg, via a spring with stiffness coefficient $k = 1$. The actuator output mobility is (a) 1(m/sec)/N; (b) 5 (m/sec)/N; (c) 10 (m/sec)/N. Use MATLAB to plot the frequency response of the plant velocity to the input of the actuator, for cases of the maximum and minimum mass of the plant. Make a conclusion about the effect of the actuator impedance on the plant uncertainty.

13 Apply Blackman's formula to the calculation of the actuator output mobility in Fig. 7.12.

14 Calculate the input and output impedances of the circuits diagrammed in Fig. 7.43; the amplifier's voltage gain coefficient is 10,000, its input impedance is ∞, and its output impedance is very low. For circuit (d), while calculating the input resistance, initially, disregard R_3, calculate the impedance of the simplified circuit, and then simply add this resistance to the obtained result.

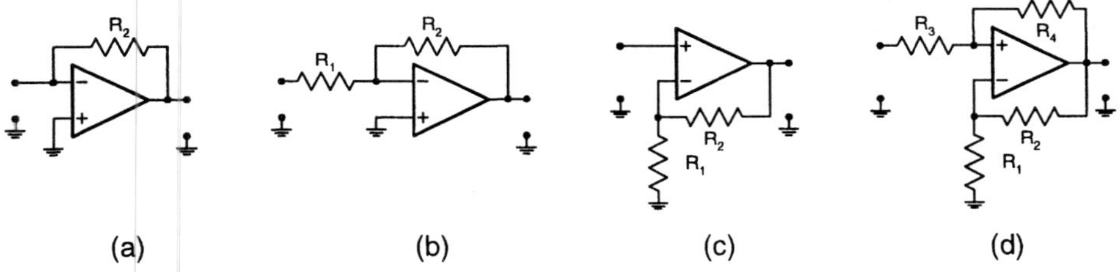

Fig. 7.43 Examples of feedback amplifiers

15 Determine the output mobility of the motor (without the main feedback loop about the plant) of the feedback systems shown in Fig. 7.44. The driver's output impedance is small for voltage drivers and large for current drivers.
Find the output mobility (at the load) for cases (c) and (d).

16 The output impedance of the feedback amplifier shown in Fig. 7.16 must be matched to the 50 Ω load. Find the resistors R_1, R_2, and R_3 such that signal losses in these resistors are not to be excessive and at the same time the attenuation of the circuitry in the direction of the feedback path is not too large (make an engineering judgment).

17 The velocity sensor gain coefficient is 1 V/(m/sec), the force sensor gain coefficient is 3 V/N (the outputs of both sensors are in volts). The sensor outputs are combined to provide large compound feedback about the actuator. What is the output mobility of the actuator? What are the outputs of the sensors when the actuator output is (a) clamped, (b) unloaded?

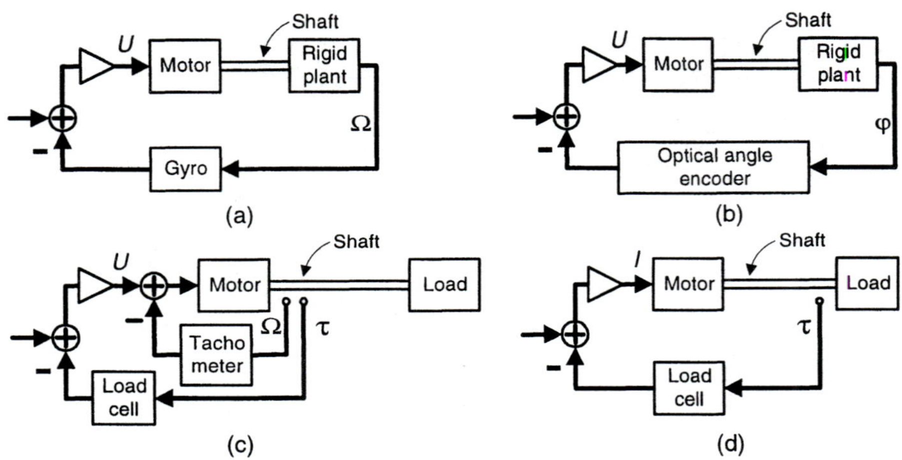

Fig. 7.44 Examples of mechanical feedback systems

18 The output mobility of a translational actuator is 40 (m/sec)/N. The return ratio is $100/(s+30)$ when only a force sensor is employed (i.e., when the output is clamped). The return ratio is (a) $300/(s+200)$, (b) $30/(s+10)$, (c) $8/(s+0.3)$, when only a velocity sensor is employed (i.e., when no load is connected to the actuator). Find the loop transfer function when the load is a 20 kg mass. Plot the Bode diagram.

19 In the specifications of a brushless motor, it is written: "$k_t = 72$ oz × in/A, $k_e = 0.5$ V/sec^{-1}." Make a good guess about what these parameters are, and how they correlate with the flowchart in Fig. 7.19 where there is only one k.

20 (a) A permanent magnet motor has $k = 0.2$ N × m/A. The driver output impedance is $R_S = 15\,\Omega$, the load mobility is $R_L = 0.8$ (rad/sec)/(N × m), the free run angular velocity is $\Omega_f = 100$ rad/sec. What is the voltage of the source? The break torque? The torque?

(b) Same for $k = 0.1$ N × m/A, $R_S = 25\,\Omega$, $R_L = 1.8$ (rad/sec)/(N×m), $\Omega_f = 200$ rad/sec.

(c) Same for $k = 0.24$ N × m/A, $R_S = 6\,\Omega$, $R_L = 0.6$ (rad/sec)/(N×m), $\Omega_f = 400$ rad/sec.

(d) Same for $k = 0.5$ N × m/A, $R_S = 2.5\,\Omega$, $R_L = 0.4$ (rad/sec)/(N×m), $\Omega_f = 850$ rad/sec.

21 Derive the expression (7.17) for the output mobility of a motor.

22 Draw a flowchart of a permanent magnet electrical motor that produces torque of 10 Nm per 1 A of the current in the winding, the winding resistance being $10\,\Omega$. Calculate the back emf and the output mobility of the motor when the source impedance (i.e., the output impedance of the driver) is $3\,\Omega$ and the current is 2 A, the load mobility is 0.12 (rad/sec)/(Nm), and the mechanical losses in the motor are negligible.

23 Draw a flowchart of cascade connection of a driver, a motor, a gear, and an inertial load for the following data:

(a) The driver has the voltage gain coefficient 5 and the output impedance $0.5\,\Omega$.

The motor winding resistance is 2 Ω and the motor constant is 0.6 Nm. The motor rotor's moment of inertia is 0.5 Nm^2. The gear amplifies the motor torque four times, i.e., the gear ratio, the load angle to the motor angle (or, the load angular velocity to the motor angular velocity), is 1:4. The losses in the gear and bearings make $B =$ 0.04 Nm/(rad/sec). The load's moment of inertia is 0.4 Nm^2, i.e., the load mobility is $2.5/s$ (rad/sec)/(Nm).

(b) The driver has the voltage gain coefficient 30 and the output impedance 5 Ω. The motor winding resistance is 30 Ω and the motor constant is 0.15 Nm. The motor rotor's moment of inertia is 0.15 Nm^2. The gear amplifies the motor torque ten times, i.e., the gear ratio, the load angle to the motor angle (or, the load angular velocity to the motor angular velocity), is 1:10. The losses in the gear and bearings make $B =$ 0.08 Nm/(rad/sec). The load's moment of inertia is 0.05 Nm^2, i.e., the load mobility is $20/s$ (rad/sec)/(Nm).

24 (a) Design a control system for a motor similar to that shown in Fig. 7.26 where $k = 0.3$, the motor winding resistance is 4, the plant transfer function is $55/s$, the sensor is 12 bit encoder, the torque variation model path is $0.1 \sin 4\theta$, the cogging model path is $0.05 \sin 4\theta$, and the crossover frequency is 12 Hz. (The values are in m, kg, sec, rad, N, Ohm.) Use a current driver. Use SIMULINK. Design the compensator and plot Bode and Nyquist diagrams with the function `linmod`. (Make the design using a PID controller and the prototype with Bode step given in Chapter 4.) Plot the output time history in response to step- and ramp commands in position using different scales, for initial part of the responses and for estimation of the accuracy of the velocity during the ramp command. Study the effects of Coulomb and viscous friction, the cogging, and the motor torque variations on the accuracy of the output position and velocity.
(b) Do the same when $k = 0.2$, the sensor is a 16 bit encoder, the torque variation path is $0.06 \sin 8\theta$, the cogging path is $0.02 \sin 8\theta$, the crossover is 20 Hz, and the driver is a current source.
(c) Do the same when $k = 0.9$, the sensor is a 10 bit encoder, the torque variation path is $0.04 \sin 16\theta$, the cogging path is $0.05 \sin 16\theta$, the crossover is 6 Hz, and the driver is a voltage source.

25 What is in common between the driving point impedance and the transfer functions of a linear system: the poles, or the zeros, or both?

26 What are the properties of the driving point impedance of a passive lossless system? Can the derivative of the modulus of the impedance on frequency be negative? What happens when the losses are small but not zero?

27 Indicate which of the following functions can be an impedance of a lossless two-pole, and plot the frequency response with MATLAB:
(a) $(s^2 + 2)(s^2 + 4)/[(s^2 + 3)(s^2 + 5)s]$;
(b) $(s^2 + 2)(s^2 + 3)/[(s^2 + 4)(s^2 + 5)s]$;
(c) $(s^2 + 2)(s^2 + 40)/[(s^2 + 3)(s^2 + 5)s]$;
(d) $(s^2 + 4)(s^2 + 40)s/[(s^2 + 2)(s^2 + 20)]$;
(e) $(s^2 + 4)(s + 40)s/[(s + 2)(s^2 + 20)]$.

28 Will the signal-to-sensor-noise ratio at the system output change when the feedback path is disconnected? How much?

29 (a) A capacitor and an inductor are connected in parallel. What is the equivalent of the return ratio?

(b) Same question about the parallel connection of a capacitor and a resistor;
(c) Same question about the parallel connection of a resistor and an inductor.

30 In the system composed of two parallel paths in Fig. 7.39(a), what is the analogy to the closed-loop transfer function for the system Fig. 7.39(b)?

31 What is the condition equivalent to large feedback, in the connection of two two-poles in parallel? How to explain the effect without using feedback theory, using Ohm's law?

32 Redo the equations and block diagrams in Figs. 7.38 and 7.39 for the case of $B = 1$.

Answers to selected problems

1 Since power losses in good motors are small, we can directly relate mechanical variables at one port to electrical variables at the opposite port, and the two-port output impedance is nearly proportional to the output impedance of the preceding link, and its input impedance is nearly proportional to the input impedance of the following link.

2 (a) Three angles (three degrees of freedom) and three torques

7 The diagram is shown in Fig. 7.45. The collector temperature must be below a certain specified temperature. When one is calculating the average collector temperature, thermal capacitances of the case and heat sink can be neglected but they need to be taken into account when calculating the collector temperature during power bursts. The nonlinear resistance of radiation and convection cooling of the heat sink is depicted by a nonlinear power source that can be specified in SPICE by the required mathematical expression.

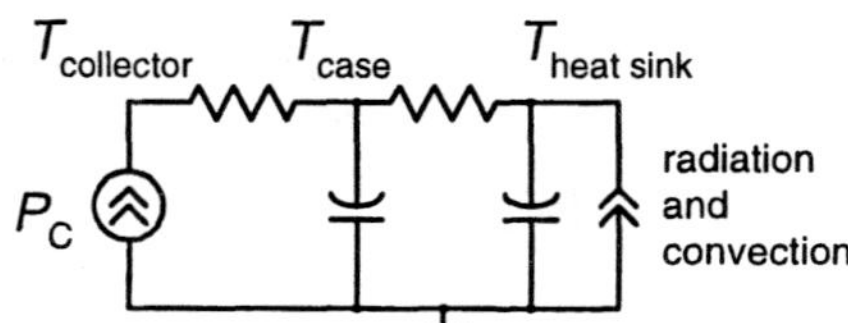

Fig. 7.45 Electrical equivalent to heat sink

14 (a) The input impedance is: $Z_o = R_2$, $T(0) = 0$, $T(\infty) = 10{,}000$, $Z \approx R_2/10{,}000$.

27 (a) The zeros and poles alternate, therefore this function can be a driving point impedance.

Chapter 8

INTRODUCTION TO ALTERNATIVE METHODS OF CONTROLLER DESIGN

This chapter surveys several important design methods and compares them with the classical Bode approach presented in the previous chapters. The methods discussed in this chapter use linear time-invariable compensators and produce linear control law which is optimal according to some performance index. These alternative methods may be encountered in industry, and software packages for many of them are readily available. The treatment here is cursory, with brief developments of the basic ideas.

8.1 QFT

The term ***Quantitative feedback theory*** (***QFT***) has been coined by Isaac Horowitz, the major contributor to the theory. (Some of his contributions to control theory have already been reflected in this book.) QFT is a frequency-domain design methodology which considers Bode methods to be a part of, and the basis for, QFT. QFT relies on simplified relationships between the frequency and time domains, uses prefilters and loop compensation to provide the desired closed-loop responses, considers sensor noise issues and actuator nonlinearities, and provides sufficient stability margins. Most of these issues have already been addressed in the previous chapters of this book. The QFT theorists aim to extend the Bode methods to handle performance issues more precisely, and they augment them with additional formalizations, somewhat different problem statements, and extensions to cover MIMO cases, linear time-variable plants, and nonlinear problems.

For simplicity we consider the QFT design of a single-loop tracking system. The design begins by determining an acceptable set of input-output transfer functions which satisfy the tracking performance requirements. This set is defined by upper- and lower-bounding frequency responses. The idea is to design loop compensation and a prefilter so that the input-output transfer function remains between these bounding responses for all possible plant parameter variations. (Disturbance rejection requirements can be handled similarly.) Since the loop compensation and a prefilter can be implemented with negligible uncertainty, the design focuses on the *variations* of the closed-loop gain due to plant parameter variations. The QFT specification for the design of the loop compensation takes the following form: at each frequency ω_i of a certain set, the variation in the closed-loop gain should not exceed a_i dB for all possible plants defined by the uncertainty ranges of the plant parameters. The tolerances a_i are the gains spanned by the upper- and lower-bounding responses at the frequencies ω_i.

To satisfy the specification, it is first necessary to calculate the plant transfer function for all possible parameter values at each of the frequencies ω_i. With the allowable parameter variations, the plant transfer function maps to an area on the L-plane, which happens to be P-shaped and the same at all frequencies in the example shown in Fig. 8.1. The shape is characteristic of the effects of parameter variations on the plant transfer function and is referred to as the ***plant template***. (The actuator is included in the plant.) The compensator transfer function at each frequency is defined by shifting the plant template to a proper location. With the template in a particular location on the Nichols chart, the gain curves indicate whether the variations in $|M|$

satisfy the QFT design requirement. If not, the template is shifted until the difference between the minimum and maximum gain is exactly a_i. In the example shown in Fig. 8.1, the original gain variation is 6 dB. Suppose that the tolerance is $a_i = 1$dB. From the lines on the Nichols chart it is evident how the template must be shifted. There is a continuum of such shifted templates which satisfy the design requirement, and the edges or corners of the shifted templates with minimum closed-loop gain form the ***minimum performance boundary*** $B(\omega_i)$ as shown in Fig. 8.1.

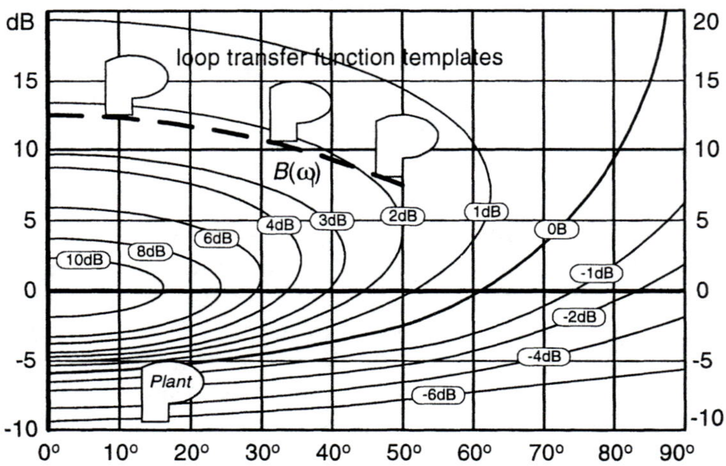

Fig. 8.1 Plant templates on the Nichols chart forming the minimum performance boundary $B(\omega_i)$

For each of the frequencies ω_i at which the system requirements are specified, the boundaries $B(\omega_i)$ must be plotted on the L-plane, as shown by the dashed lines in Fig. 8.2.

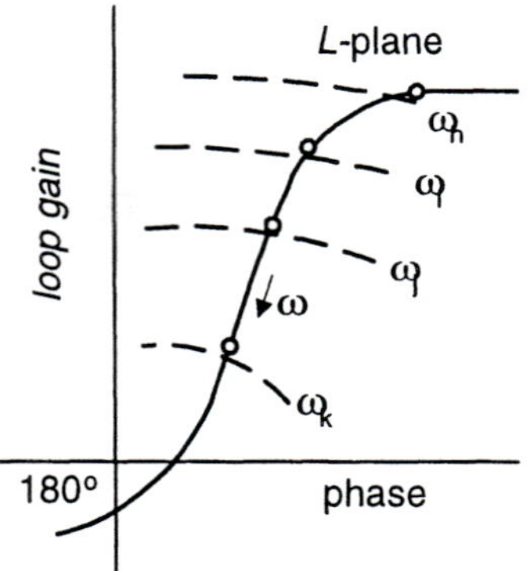

Fig. 8.2 Boundaries on the L-plane

An additional high-frequency L-plane bound is included to guarantee system stability and robustness. With the boundaries in place, the next step is to search for a rational compensator transfer function such that the loop gain at each of the frequencies ω_i will be just over the minimum performance boundary. At frequencies near the zero-dB crossing and higher, the compensator gain is shaped to follow the stability boundary. The design is performed by trial and error or by using specialized software. Finally, a prefilter is synthesized which corrects the input-output response to achieve the desired response. (Remember, the prefilter's uncertainty contribution is negligible.)

It can be shown that a solution to the QFT problem always exists, although the resulting feedback bandwidth may be unacceptably large. Generally, the best design is taken as that which has the smallest feedback bandwidth while satisfying the minimum performance and stability boundaries.

The QFT design philosophy departs from the Bode approach in the following areas:

- The Bode approach is to maximize the performance (feedback) while satisfying the constraints on the high-frequency loop asymptote due to plant features and parameter variations, and high-frequency noise. QFT pursues the inverse problem of providing the minimum acceptable performance while minimizing the feedback bandwidth. The QFT-designed system is not the best possible, but rather is just good enough to satisfy the closed-loop response and disturbance rejection specifications. We prefer the Bode approach for the following reason: the cost differential between a substandard controller and the very best available is generally insignificant compared with the cost of the system. The controllers differ only by several resistors and capacitors or a few lines of code, and perhaps a few days of work by the control engineer (if he uses the Bode approach). Improving the control law might also relieve some of the requirements on other components of the device, making the entire system better and cheaper to manufacture. This may in turn affect decisions made about the development of the next generation of the system. It makes little sense to lower the system performance just to reduce the feedback bandwidth. The Bode approach identifies the constraints on the bandwidth upfront, rather than minimizing the bandwidth and determining later whether it is still too high. (Whatever the philosophical difference, there is little doubt that engineers well trained in QFT can resolve these trade-offs and design high-performance controllers.)
- The QFT design is focused on satisfying the performance specifications for the worst-case plant parameters, and neglects optimizing the nominal performance. This may or may not be an advantage.
- With the multiple templates to be calculated and plotted, QFT design is far more complex than Bode design.

QFT methods have been developed to handle the design of MIMO systems with stable and unstable, time-invariable and time-variable, linear and nonlinear plants.

8.2 Root locus and pole placement methods

Another category of controller design methods focuses on the location in the complex plane of the roots of the closed-loop transfer function.

The ***root-locus method*** uses plots of the root loci to choose the constant multiplier of the loop gain coefficient and to design additional loop compensation. An elaborate set of rules exists for constructing the root loci from the open-loop transfer function. Today, the root locus *analysis* is usually performed by computer.

Example 1. Consider a control system with plant $P(s) = 100/s^2$, actuator $100/(s + 100)$ and lead compensator $C(s) = 10(s + 3)/(s + 30)$.

With the loop open, the poles of the system are just those of $T = CAP$, i.e., a double pole at the origin and real poles at –30 and –100 rad/sec. For the purpose of this analysis, suppose that there is a variable gain coefficient k in the loop which is gradually increased from 0 to 1. As the gain coefficient is increased, the poles move from their open-loop positions to their closed-loop positions. Fig. 8.3 shows the root loci for our example.

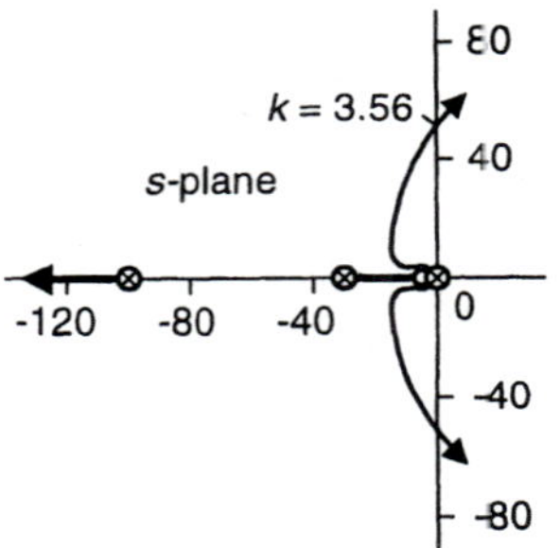

Fig. 8.3 Root loci for a feedback system

The loci can be continued by increasing the gain coefficient past the nominal closed-loop value of 1. In the example, when k reaches 3.56, some poles cross over into the right half-plane. The gain stability margin is therefore $20\log(3.56) = 11$dB. The robustness of the system is still difficult to determine from the root loci. The guard-point gain stability margin is apparent, but the phase stability margin and even the guard-point phase margin are not. It might be guessed that the distance of the roots from the $j\omega$-axis would be a good indicator of system robustness, but this is not always the case. A practical counter-example is an active *RC* notch-filter, where the root locus passes very close to the $j\omega$-axis but the system is quite robust.

It is also not evident from the root locus whether the system is well designed. In fact, it is not. It would be instantly seen from the Bode diagram that the pole and the zero in the compensator are in wrong places, and the phase and gain guard-point stability margins are not balanced.

The compensator design proceeds by trial and error, searching for compensation and a suitable loop gain which brings the closed-loop poles into desirable locations on the s-plane. What are desirable pole locations? Usually the system is examined for a pole or pole pair which is "dominant," meaning that the step-response of the closed-loop system resembles the step response of a system with just this pole or these poles. The design goal is to move the dominant poles into areas on the s-plane with sufficient distance from the origin for the transient response to be fast, and with sufficient damping (i.e., not too close to the $j\omega$-axis) to prevent excessive overshooting. (The possibility of a prefilter is not factored into the design.) Meanwhile, other system poles must be monitored for stability.

When the designer makes an *a priori* decision about the precise location of the system poles, the method is sometimes referred to as ***pole-placement***, although this label is often reserved for state-variable feedback control of MIMO systems as will be discussed below. A common choice is to place the closed-loop poles in a Butterworth filter configuration.

A major inadequacy of the root-locus design method is that it does not allow the designer to judge how close the system performance is to the best available. In addition, no convenient rules exist for designing good high-order compensators. Another problem is the complete lack of visibility into low-frequency disturbance rejection. Finally, system performance in the nonlinear mode of operation is difficult to determine from the root loci. (As we shall see in Chapters 9–11, the Bode and Nyquist diagrams enable the designer to deal effectively with common nonlinearities.) Because of these deficiencies, the root locus method is not recommended for control system design.

The root locus method can be valuable for the analysis of the effects of certain parameter variations on stability, and on the nonminimum-phase lag in the link composed of several parallel minimum-phase links (see Sections 3.13 and 4.5). Also, root locus plots make very impressive presentations for high-order systems that have been already designed well using other methods.

8.3 State-space methods and full-state feedback

From the classical control perspective, the linear control system is a block diagram of transfer functions, i.e., Laplace transforms. Of course the system can also be represented by one or several linear differential equations.

The system of equations can be transformed into a set of first-order differential equations by introducing intermediate variables where necessary. The following ***state space*** system description is standard:

$$\dot{x} = Ax + B(u + r)\,, \tag{8.1}$$

$$y = Cx\,, \tag{8.2}$$

where x is a vector (column) of ***state-variables*** (or ***states***), u is the input or ***control vector***, r is the ***reference***, and y is the ***output vector***. The square matrix A is referred to as the ***system matrix***. It describes the dynamics of the system without feedback (i.e., dynamics of the actuator and plant). B is the ***control-input matrix***, and C is the ***output matrix***.

It may be helpful to think about how a SISO system would fit into this format. The system matrix A would be $n \times n$, where n is the order of the combined actuator/plant transfer function. The control-input matrix B would be a column matrix of length n which distributes the scalar control input among the state derivatives. The output matrix C would be a row matrix of length n which reassembles the scalar output (which is a function of time) from the states. Note that the representation is not unique, but depends on the choice of states. It is customary to try to choose states that correspond to some physical variable of the system.

An advantage of the state-space notation is that it is easily generalized to multi-input multi-output systems by changing the matrix dimensions. For example, a two-input three-output system would have a control-input matrix B that is $n \times 2$ and an output matrix C that is $3 \times n$.

The feedback loops are closed when the second component in (8.1) is added to the state vector. In state-space formulation the control u is a linear combination of the states:

$$u = -Kx\,, \tag{8.3}$$

where K is the ***gain matrix***. The closed-loop system is then described by the equations

$$\dot{x} = (A - BK)x + Br\,, \tag{8.4}$$

$$y = Cx\,. \tag{8.5}$$

To be more general, and to conform to the convention adopted by MATLAB, we can allow the control to affect the output directly by introducing the matrix D and rewriting (8.5) as

$$y = Cx + Du\,. \tag{8.6}$$

The state-space block diagram of the feedback system is shown in Fig. 8.4.

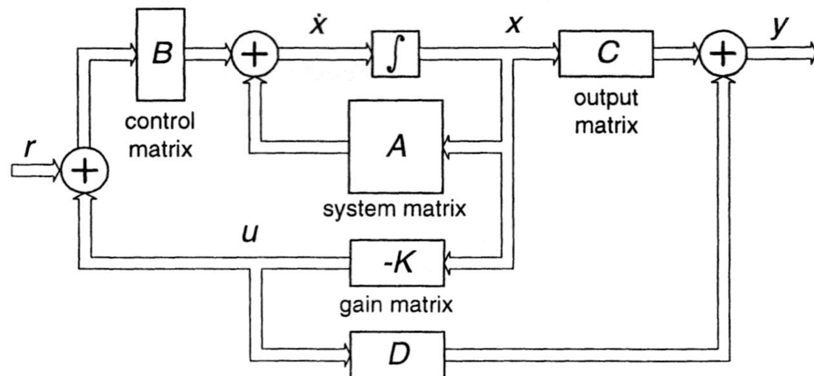

Fig. 8.4 State-space block diagram of a feedback system

The open-loop system corresponds to $u = 0$ in (8.1). The open-loop system becomes

$$\dot{x} = Ax \tag{8.7}$$
$$y = Cx\,. \tag{8.8}$$

Example 1. Suppose that the open-loop plant is the pure double integrator

$$P(s) = \frac{1}{s^2}\,, \tag{8.9}$$

(and that the actuator transfer function is unity). The open-loop system could be represented as follows:

$$y = x_1\,, \tag{8.10}$$
$$\dot{x}_1 = x_2\,, \tag{8.11}$$
$$\dot{x}_2 = u\,. \tag{8.12}$$

Here u is the input and y is the output, both scalars (functions of time). The state vector consists of a position-like and a velocity-like state $x = [x_1\ x_2]^T$. Per our notation, $n = 2$, and A, B, C, and D are as follows:

$$A = \begin{bmatrix} 0 & 1 \\ 0 & 0 \end{bmatrix} \quad B = \begin{bmatrix} 0 \\ 1 \end{bmatrix} \quad C = \begin{bmatrix} 1 & 0 \end{bmatrix} \quad D = 0 \tag{8.13}$$

The nomenclature is important since systems are represented this way in MATLAB. Suppose that we had manually entered the A, B, C, and D matrices for our example into MATLAB. The following command would then produce the open-loop frequency response plot:

```
bode(a,b,c,d,1)
```

The last argument is letting the MATLAB function know that we're interested in the response of the output (all outputs in the general case) to the *first* input (the system in case has only one input and one output). This may seem like a lot of overhead to calculate the frequency response of a double-integrator. Fortunately, the matrices are usually created by other programs. For instance, the block-diagram-oriented SIMULINK has a function `linmod`, which creates the appropriate A, B, C, and D matrices for further analysis:

```
[a b c d] = linmod('model_name')
```

After the gains k_1 and k_2 are chosen, the closed-loop frequency response can be obtained using MATLAB commands to manipulate the system matrices, or by making the appropriate connections in the SIMULINK block diagram and rerunning `linmod`.

The state-space closed-loop design problem is to choose the gain matrix K to obtain the desired closed-loop transient response. (We might already disagree with the practicality of such an approach since obtaining the desired closed-loop response is not the only nor the main purpose of closed-loop control in practical systems.)

Before we discuss the possible strategies for choosing K, some implications of the state-space notation should be noted. An implicit assumption is that the states x are somehow available to be plugged into (8.4) and fed back to the input of the system. For this reason (8.4) is often referred to as ***full-state feedback***. In a typical control system, the order of the actuator/plant combination exceeds the number of sensed outputs, making full-state feedback unrealistic. The missing states must be estimated using the available ones; this is discussed in the next section. Another feature of the state-feedback framework is that it does not allow compensators whose order exceeds the order of the actuator/plant. In our example above, (8.4) restricts the compensator function to consist of a single unrealizable zero: $C(s) = k_1 + k_2 s$. A work-around is to expand the state vector to include some of the compensator dynamics, as is typically done to add integral control in the state-space version of the *PID*.

A more insidious problem with the state-space approach is inherent to the representation of the system by a set of linear matrix differential equations rather than a block diagram of transfer functions. This draws the designer's attention away from the physical elements of the control system, along with their limitations and imperfections, and instead focuses on matrix algebra.

The state-variable approach can be to various degrees mixed with conventional block diagram design methods.

Example 2. The block diagram in Fig. 8.5 has been employed for control of position x (or the attitude angle) in many space systems, especially those having rigid-body plants with small parameter uncertainty. The feedback bandwidth in these systems is typically limited by the sensor quantization noise. The plant is considered rigid, the acceleration is proportional to the actuator force (or torque, for attitude control), the actuator transfer function A is a constant, and the plant P is seen as a double integrator.

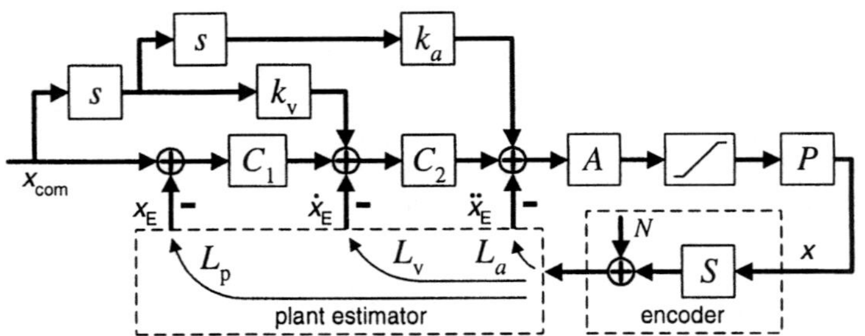

Fig. 8.5 Block diagram of a position control system with position, velocity, and acceleration loops

The position command x_{com}, and the velocity and acceleration commands obtained by differentiation, are forwarded with appropriate gain coefficients to, respectively, position, velocity, and acceleration summing points.

A plant estimator (filter) generates the plant variable estimates x_E, etc. from the noisy readings of the sensor. The transfer functions L_p, L_v, and L_a are m.p., and the phase lag responses are related to the gain responses by Bode integral (3.13). The bandwidth of L_v is wider than the bandwidth of L_p and smaller than that of L_a. The filter cutoff frequencies must be sufficiently low to extensively attenuate high-frequency sensor noise components, but not too low since, first, the filter distorts the output signal and, second, the filter phase lag reduces the available feedback and the disturbance rejection.

The errors in position, velocity, and acceleration are formed by subtraction of the plant variable estimates from the signals arriving to the summing points. The errors are reduced by the three feedback loops. It is seen that when the position error is 0, the output of the compensator C_1 is 0. When the velocity error is also 0, the output of C_2 is 0. When the acceleration error is also 0, the signal at the actuator input is 0. This control scheme can be perceived as multivariable.

Comments on Example 2:

- The three feedback loops are coupled. Still, the design can be made by iterative adjustments, one loop at a time, since, first, the compensators are typically low-order (*PD*), and, second, the three loops substantially differ in bandwidth: the bandwidth of the velocity loop is wider than that of the position loop, and the bandwidth of the acceleration loop is still wider.
- When the plant is flexible, the compensators' order must be much higher than that of the *PD* compensators shown in the block diagram, but the higher order compensators are not easy to fit within this design.
- The differentiators in the feedforward paths are implemented in practice as lead links whose frequency responses approximate the response of the ideal differentiator over the required frequency band. The effects of saturation in the actuator limit the useful bandwidth of the feedforward.
- The torque source actuator (using a driver with high output impedance) simplifies the analysis. On the other hand, a velocity source (a motor driven by a driver with low output impedance) may improve the system accuracy, especially when the plant is flexible with Coulomb friction.

- The controller can be augmented with inclusion of nonlinear links to reduce the windup and to improve the transient response for large-level commands.
- This system is multivariable only formally since it has only one sensor, the plant is rigid and the actuator is a force source. Since the position, rate, and acceleration have unique and simple interrelations, such a system can be equivalently and better described as a single-loop SISO system.
- The complex diagram in Fig. 8.5 can be equivalently transformed into the diagram in Fig. 8.6(a) and further into the diagram in Fig. 8.6(b) which follows the diagram in Fig. 2.1 (the loop transfer function about the plant is the same in these diagrams, and the input-output transfer function without the feedback, i.e., with the sensor transfer function $S = 0$, is the same). The diagram in Fig. 8.6(b) includes only two independent linear links whose transfer functions are defined by the designer: the feedback compensator and the feedforward path.

Therefore, performance of the system shown in Fig. 8.5 cannot be superior to a conventional well-designed system with a prefilter or a feedforward path.

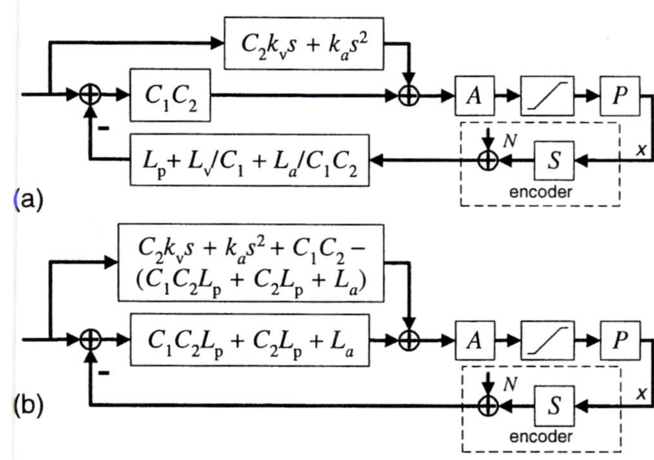

Fig. 8.6 Single-loop equivalents (a) and (b) of the block diagram shown in Fig. 8.5

8.4 LQR and LQG

The general plan of the so called ***modern control theory*** is to take the state-space description of the control system literally, set up some scalar performance index which quantifies the desirable features of the closed-loop system, and then find the gain matrix K which is optimal for this index. One such approach is to minimize a quadratic functional J of the state and control history for the system's step response:

$$J = \int_{t=0}^{\infty} (x^T Q x + u^T R u)\, dt = \min \tag{8.14}$$

where the matrices Q and R are ***weighting matrices***. It is assumed that the desired state is $x = 0$, but the initial condition is non-zero, so the matrix Q penalizes the state error in a mean-square sense. Similarly, the matrix R penalizes the control effort, i.e., limits the control signals' magnitude.

The gain matrix K which minimizes J can be found by solving a matrix Riccati equation. The resulting controller is known as the ***linear quadratic regulator***, and the methodology is referred to as ***LQR***.

Although software is readily available to solve the matrix Riccati equation and thus determine the optimal gain matrix K, it is not advisable to attempt to design a control system using the LQR methodology alone. This is because the features of the control system which constrain the performance are not captured in the LQR framework. There has been no mention of actuator saturation, disturbance rejection, or robustness to plant parameter variations. The only reasonable possibility is to judiciously choose the weighting matrices Q and R, run the LQR software to determine the "optimal" gain matrix K, and then examine the resulting control system using classical frequency-domain analysis. This generally entails several iterations.

The inability to address nonlinearities in the LQR framework is crippling. A common design strategy is to increase the control penalty matrix R until the largest expected transient does not result in saturation of the actuator. This seems wise since actuator saturation can result in windup or even instability for an LQR design. The implications for system design are disastrous, since to achieve the specified performance, the actuator will be oversized to maintain linearity.

As mentioned previously, full-state feedback is not practical. If the LQR framework is to be used for practical problems, the missing states must be continually estimated from the available measurements. Suppose that the available measurements are linear combinations of the state variables. If the measurements are perfect, and the plant model is perfect, the remaining states can usually be reconstructed by repeated differentiation. In fact, the entire future of the state can be predicted with certainty. To make the estimation process non-trivial, the state-space formulation has to be augmented by introducing errors in sensing and modeling. An analytically tractable approach is to assume that the measurements are corrupted by white noise, and that the actual plant differs from the plant model by an additional white noise input. The noise in the measurements is referred to as the ***sensor noise*** and denoted w. The noise added to the plant model is referred to as the ***process noise***, and denoted v. Note that w and v are generally vectors. Let the measurements be z, so that the system description becomes

$$\dot{x} = Ax + Bu + Gw \tag{8.15}$$

$$z = Hx + v\,, \tag{8.16}$$

where G is the ***plant noise distribution matrix*** and H is the ***measurement matrix***.

The ***state estimate*** x_E is to be propagated as

$$\dot{x}_E = Ax_E + Bu + K_E\left(z - Hx_E\right), \tag{8.17}$$

where K_E is the ***estimator gain matrix***. Given the second-order moments of the white processes w and v, the optimal estimator gain can be found which minimizes the mean square error in x_E. This estimator is referred to as ***the linear-quadratic Gaussian***, or ***LQG***. When these estimates are used in conjunction with an LQR controller, the combined approach is referred to as an LQR/LQG regulator.

LQR/LQG regulator theory was intended to resolve the trade-off between the sensor noise and the disturbance rejection. Since this method by itself does not address the robustness issue, it does not provide the best solutions to most practical problems. However, when the plant is known pretty well (say, with 1% accuracy) and the feedback bandwidth is limited by the sensor noise, LQG provides a loop response which is well shaped in the crossover area of the frequency band. This response can be later modified with classical methods for better disturbance rejection at lower frequencies.

The addition of the ***loop transfer recovery*** (LTR) method to the LQG allows addressing the system robustness. The LTR method recalculates the frequency domain loop responses of the system designed in state space, time domain with the LQG method, and allows adjusting the responses to provide the desired stability margins. The process of such design is however not simple, and the quadratic norm is not appropriate for stability analysis on the basis of the closed loop response.

8.5 H_∞, μ-synthesis, and linear matrix inequalities

The state-space approach to control system design and the state-space performance indices are difficult to use during the conceptual design. G. Zames, who initiated the H_∞ method, often said that the processes of approximation in model building and obtaining state-space model do not commute [45], i.e., input-output (black box) formulations is the preferred framework for uncertain (practical) system modeling, and state-space models should came into picture only as internal models at the level of computation and at the level of implementation of control systems.

The computational aspects of control system design have been already advanced to the degree when they cease to be critical for the design of most practical systems. However, building the system model and designing optimal compensators still present a challenge, and are easier to accomplish with the input-output formulations.

In other words, control system engineers should structure the systems as sets of physical blocks interconnected via ports, instead of structuring the systems mathematically in sets of linear matrices. Mathematically, the input-output formulations mean separating the system variables into the sets of local (internal for the blocks) and global (at the blocks' ports) variables, and, typically, the number of the global variables is much less then the number of the local variables.

For the input-output formulations, frequency-domain characterization are in many aspects more convenient than the time-domain ones. As was exemplified in Chapter 7 with the two-ports, linear black boxes can be described by the matrices of their transfer functions and impedances (mobilities), and the entire system, as a conglomerate of linear and nonlinear multiports interconnected via their ports.

H_∞ is an extension of the classical frequency domain design method. It solves in one operation the two problems that are sequentially solved with Bode approach: maximization of the available feedback bandwidth with related shaping of the loop response over the frequency region of crossover frequency and higher, and distribution of the available feedback over the functional bandwidth, as was described in Chapter 4. The method is formulated such that it is directly applicable to multivariable control systems.

The ***H_∞ norm*** is the limit on the magnitude of a vector in the Hilbert space. This norm is an extension of the Chebyshev norm widely used in frequency-domain network synthesis. The H_∞ feedback control design method applies this norm to the closed-loop frequency responses from the disturbance sources to the system output.

With H_∞ method, frequency responses of the disturbance rejection are first specified with ***weight functions***. The weight functions define at which frequencies disturbance rejection should be higher than that at other frequencies, and by how much. The weight functions should be calculated from the known disturbance spectral densities. For the functional feedback bandwidth, the norm on F is nearly the same as the norm on T.

Since it is not easy to properly shape the crossover area of the loop Bode diagram with H_∞ method, the H_∞ method may lead to an overly conservative system. A less conservative solution can been achieved with ***μ-synthesis*** which combines the H_∞ design and ***μ-analysis*** in an iterative procedure. The μ-analysis method introduces into the loop special links that imitate the plant uncertainty. It is required that with these links added, the nominal system should be still stable and perform well.

The H_∞ design method is the method of linear control system design. It optimizes the system performance without paying special attention to the system global stability. Because of this, the H_∞ design often results in Nyquist-stable systems which are not absolutely stable and can burst into oscillation after the actuator becomes overloaded. The solutions to this stability problem are either making several iterations by relaxing disturbance rejection requirements and modifying the weight functions such that the resulting loop response be of absolutely stable type, which is easy to do when the designed system is single-loop, or, better, by using nonlinear controllers that will be studied in Chapters 9-13. The nonlinear controller design methods should be also employed to further improve or optimize the system performance in the nonlinear state of operation when certain commands or disturbances overload the actuators.

The H_∞ control and many other linear control and stability analysis problems can be formulated in terms of ***linear matrix inequalities*** (LMIs). The LMI is the algebraic problem of finding a linear combination of a given set of symmetric matrices that is positive definite. LMIs find applications outside of control, in such diverse areas as combinatorial optimization, estimation, and statistics. Although it has long been recognized that LMIs are important in control, it was only with the advent of the efficient algorithms (based on the interior point methods) that their popularity has increased in the last few years.

Chapter 9

ADAPTIVE SYSTEMS

Large plant uncertainty reduces the available feedback. One remedy is to use an adaptive control law which changes the transfer functions of the compensator, the prefilter, and the feedback path on the basis of accessible information about the plant.

On the basis of the information used for the adaptation, the adaptive controllers can be divided into three types. The first type uses sensor readings of environmental parameters (temperature, pressure, time, etc.) and plant parameter dependencies on these environmental factors to correct the control law. (Obviously the dependencies must be known *a priori*.) The second uses the plant response to the command or disturbances to correct the control law. The third type uses the control loop response to specially generated pilot signals to correct the control law.

The first and the third types of adaptation schemes substantially improve the control when the plant changes at a much slower rate than the control processes. The second method can be useful when the command profile is well known in advance. If this is not the case, the second type may result in a system with rapidly-varying parameters whose stability analysis represents a formidable problem.

It is shown that the plant is easier to identify in the frequency bands where the feedback in the main loop is not large. This identification provides most of the available benefits in the system performance.

A brief description is provided for adaptive systems for flexible plants, for disturbance rejection and noise reduction, and for dithering systems. Examples of adaptive filters are described.

9.1 Benefits of adaptation to the plant parameter variations

Plant parameter uncertainly impairs the available feedback. The uncertainty can be reduced by a ***plant identification*** procedure which gives the plant transfer function estimation as P'. The improved knowledge of the plant should then be used for controller ***adaptation***, i.e., adjustments of the compensator, feedback path, and prefilter to reduce the output error. The rate and dynamics of the adaptation are defined by some ***adaptation law***.

The feedback system output error is contributed to by disturbance sources and by the plant parameter uncertainty. Increasing the feedback reduces both error components, and as explained in Chapter 2, the second component can be additionally reduced by using an appropriate prefilter.

Correspondingly, *the benefit of good plant identification, ($P' \approx P$) is two-fold: the feedback can be increased by appropriate modifications to the compensator C, and the prefilter can be made closer to the ideal*. When plant parameter variations are large, adaptation can significantly improve the system performance.

Adaptation schemes where the plant is identified first are called ***indirect***.

Example 1. Assume the plant gain varies by 20 dB in a loop designed as shown in Fig. 9.1. The feedback bandwidth is limited by sensor noise and/or plant resonances, $f_b < f_{bmax}$. The loop response with maximum available feedback bandwidth is the one when the plant gain is largest. When the plant gain drops, the feedback and the disturbance rejection reduce and the sensitivity of the output to plant parameter variations increase. To alleviate the problem, plant gain can be monitored and the compensator gain coefficient continuously adjusted for the loop gain to remain equal to the loop gain of the case of the maximum plant gain. This is an example of indirect adaptation.

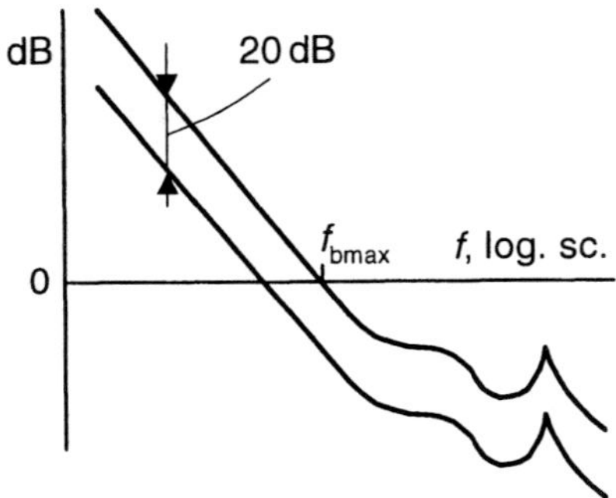

Fig. 9.1 Loop gain responses with maximum and minimum plant gain

Example 2. Assume the loop in the previous example is designed as Nyquist stable to increase the feedback in the functional band. When the plant gain decreases, the system can become unstable. Therefore, the plant gain needs to be monitored and the compensator gain modified not only to maintain the desired level of disturbance rejection, but also just to preserve stability.

Example 3. In the plant, a real pole position can go down by a factor of four from the initial position at $2f_b$. The phase lag increases correspondingly and the system can burst into oscillation. To counteract this pole motion, a zero can be introduced in the compensator transfer function and, after the plant identification, kept close to the pole position to cancel its effect. The accuracy of the pole position identification need not be very high: if the zero-to-pole distance is within 5% of the pole magnitude, the resulting loop gain uncertainty will be less than 0.5 dB.

Example 4. In the previous example, not a real pole but a complex pole related to a plant flexible mode with high Q is uncertain. In this case, if the same remedy is contemplated – compensation of the plant pole with a compensator zero – the plant identification needs to be Q times more accurate than in the previous example which might be difficult to accomplish.

Direct adaptive algorithms do not use plant identification explicitly, but merely tune the compensator to improve the closed-loop performance. For tuning the most popular *PID* controllers, there exist many adaptive algorithms. Time-domain performance adjustment is most often used: the time-response parameters are measured and *P*-, *I*-, and *D*-coefficients of the controllers are adjusted correspondingly. The algorithms typically work well as long as the plant's responses are smooth. The typical goal for such adaptive control is improving the responses to the commands in the system where the plant parameters are varying slowly, and not improving the higher-frequency disturbance rejection or the system tolerance to fast plant parameter variations.

9.2 Static and dynamic adaptation

Plant identification presents no problems and can be performed rapidly when the dependence of the plant transfer function on the environmental variables is well-known and the environmental variables are measured accurately. Such an adaptive control system is shown in Fig. 9.2. The plant is identified, and then the compensator and prefilter transfer functions are adjusted. Such a process is sometimes called ***gain scheduling***. An example of such a system has been given in Section 7.4.2. Commonly, the environment varies much more slowly than the main control loop dynamics, i.e., the adaptation is ***quasi-static***. Therefore, during the system analysis, the main feedback loop dynamics can be assumed to be independent of the process of adaptation, and the prefilter and the main loop links can be considered time-invariant.

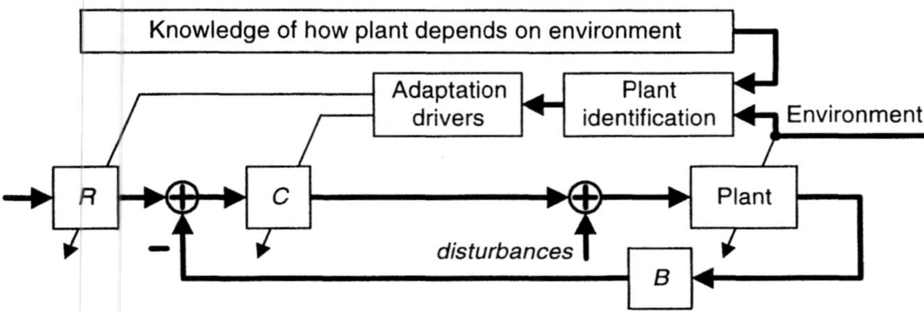

Fig. 9.2 Prefilter and compensator regulation

The rate of the plant identification is limited by effects of noise and disturbances, and by the rate of the information processing. The rate of adaptation – which can be critical when the plant parameters vary rapidly – is limited by the plant identification rate and also by the system stability conditions.

Increasing the speed of the adaptation makes the adaptation ***dynamic*** and leads to LTV systems with rapidly-varying parameters. These systems, generally, have several disagreeable features (as we have seen in an example in Section 3.12). Stability analysis of such systems presents a formidable problem, and ensuring the system stability requires increasing stability margins (and reducing the feedback) in the main control loop. This is why the adaptation laws are commonly designed as quasi-static, i.e., much slower than the processes in the main control loop. In this case, the main loop can be designed as an LTI system.

9.3 Plant transfer function identification

Fig. 9.3 illustrates three methods for the plant transfer function identification that are used with indirect adaptation:

1. Using sensors of environmental variables and the known dependencies of the plant transfer function on these variables.

2. Using the input and output signals of the plant that are generated during normal operation of the control system as the result of commands and disturbances.

3. Using ***pilot signals*** generated and sent to the plant for identification purposes. The pilot signal amplitude should be sufficiently small so as not to introduce substantial error in the system's output.

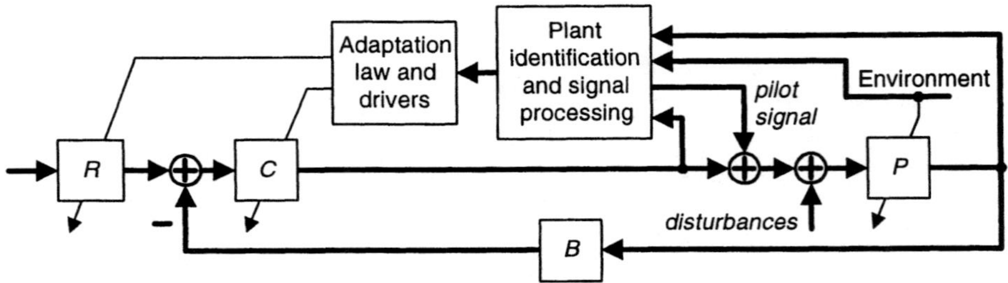

Fig. 9.3 Adaptive feedback system with plant identification

The plant is identified from its measured input and output signals. The measurements are corrupted with the disturbance signals and sensor noise. This limits the accuracy and speed of the adaptation.

Plant identification is ill-conditioned over the bandwidth of large feedback where the error frequency components are small and therefore the pilot signals must be very small. This increases the requirements on the sensors' sensitivity and accuracy, and makes the identification complicated. Also, the frequencies where the feedback is large are relatively low which makes the plant identification process slow. Further reduction of the control error components at these frequencies can be achieved more economically by using nonlinear dynamic compensation, as explained in Chapters 10 and 11.

In contrast, the problem of plant identification is well-conditioned over the frequency ranges of positive feedback and small negative feedback, i.e., from $f_b/4$ up to $4f_b$. This bandwidth is important for feedback bandwidth maximization. The plant identification over this frequency range provides tangible benefits and is relatively easy to implement since at these frequencies the pilot signal amplitudes can be larger and the measurements can be much faster.

9.4 Flexible and n.p. plants

Plant identification might especially improve the control when the plant is flexible and uncertain. For example, when the torque actuator and the angular velocity sensor are collocated in a structural system with flexibility, the plant transfer function is that of a passive two-pole impedance, and the phase variations do not exceed 180°. At higher frequencies, however, due to flexibility of the actuator-to-plant shaft, as shown in Fig. 9.4, the control ceases to be collocated and two poles in a row follow in the plant transfer function. Plant identification allows making a compensator whose transfer function has zeros to compensate for the unwanted plant poles.

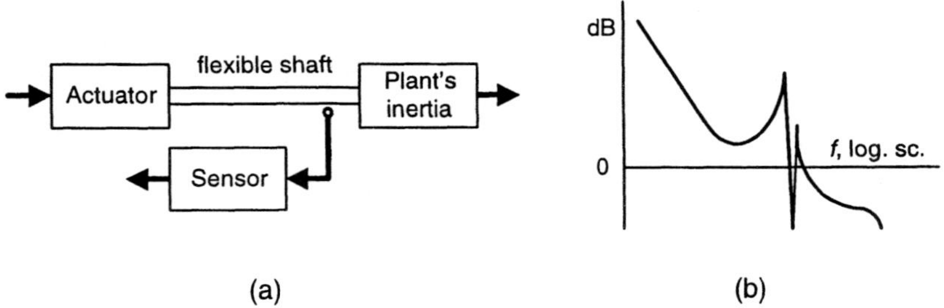

Fig. 9.4 (a) Plant with flexibility and (b) the uncompensated loop response

For this application, using frequency domain identification is also economical since the plant response needs to be identified only at higher frequencies, in the vicinity of the structural modes.

N.p. lag caused by parallel paths of the signal propagation, such as that of rotation and translation (due to the thruster's position and the propellant slosh in the tank of a spacecraft, for example) can be removed from the main feedback loop by the addition of appropriate sensors for the plant identification. The n.p. lag increases with frequency, and the best way to measure it is at higher frequencies, i.e., within the bandwidth of positive feedback.

We conclude that for the flexible plants and n.p. plants, the plant identification is less complicated and most beneficial over the frequency band where the feedback is either positive or small negative.

9.5 Disturbance and noise rejection

The disturbance rejection can be improved by using adaptive loops within the compensator, as diagrammed in Fig. 9.5. The disturbance identification allows modifying the loop gain response to match the disturbance spectral density, under the limitation of Bode integrals (3.7), (3.12). The disturbances can also be compensated in a feedforward manner.

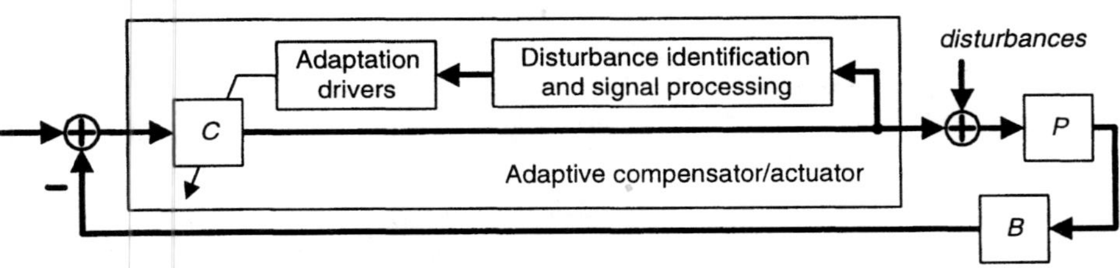

Fig. 9.5 Loop response adaptation for disturbance rejection

Example 1. Using an adaptive ***comb-filter***, i.e., a filter with multiple narrow pass bands, allows the provision of substantial feedback at the harmonics of a periodic disturbance as shown in Fig. 9.6, instead of broad-band feedback with the response shown by the dashed line, the feedback in both of the responses being limited by the Bode integrals. Such a filter tunes by tracking the disturbance frequency f_d using a PLL, a frequency-lock loop (FLL), or some other adaptive algorithm.

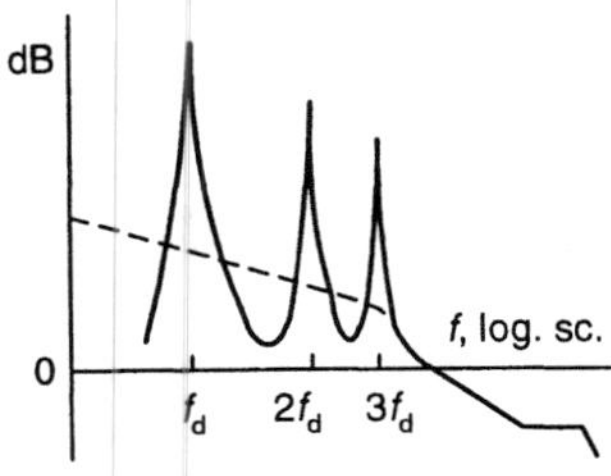

Fig. 9.6 Loop gain Bode diagram with adaptive equalizer

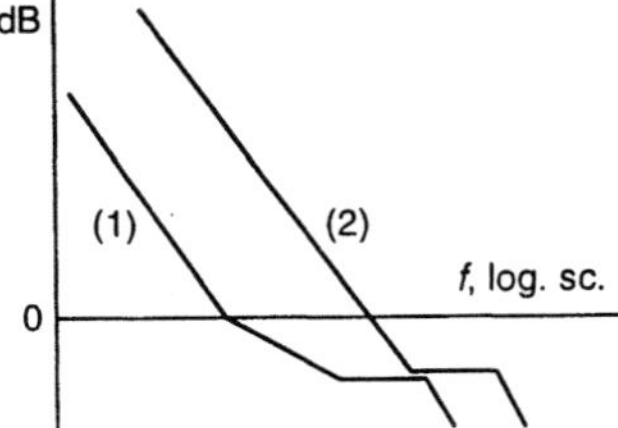

Fig. 9.7 Loop gain Bode diagrams for systems with different bandwidths

Example 2. Diagram (1) in Fig. 9.7 was found optimal for a spacecraft attitude control system which uses reaction wheels in the tracking regime, when the feedback bandwidth is limited by the gyro noise. (This kind of diagram was also used for a spacecraft photocamera temperature control where the feedback bandwidth was limited by the quantizing noise of the temperature sensor.) Diagram (2), on the other hand, is closer to the optimal for the regimes of re-targeting of the attitude control, when the feedback bandwidth needs to be wider and the gyro noise is of small importance. Therefore, for the best performance, the shape of the loop response can be changed from (1) to (2) by an adaptation process.

Example 3. A spacecraft attitude control system using star tracker and gyro attitude sensors is shown in Fig. 9.8. The measurements are corrupted by the noise represented by the sources N_S, N_G. At any moment, the signal from the star tracker depends on the particular stars that happen to be in its field of view. When several bright stars are in view, they are easily recognized when compared with the star map in the computer memory, and the signal-to-noise ratio of the tracker becomes excellent, but when the camera turns in a direction where no bright stars exist, the tracker output is noisy and the attitude control should depend more on the gyro. In the latter case, the bandwidth of the filter in the feedback path from the star tracker must be reduced. An extended ***Kalman filter*** (an algorithm minimizing the mean square error) performs this adaptation task continuously.

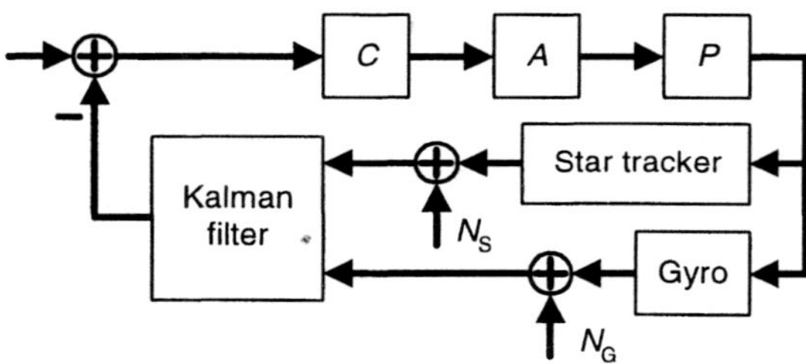

Fig. 9.8 Adaptive control system with a Kalman filter in the feedback path

9.6 Pilot signals and dithering systems

Without pilot signals, ***passive identification*** relies on the feedback system's responses to noise, distortions, and commands, which are all far from optimal test signals for the plant identification. Consequently, passive identification of the plant is slower, less accurate, and less reliable compared with ***active*** plant identification with specially generated pilot signals. When pilot signals are used, there always is a trade-off between the accuracy of the plant identification which increases with the level of the pilot signals, and the error the pilot signals themselves introduce at the system's output.

When the plant changes rapidly, the adaptation must be faster. This requires increased rate of information transmission by the pilot signals, and the pilot signals must have larger amplitudes and/or broader spectrum.

When only the constant multiplier of the gain coefficient of the plant is changing, and this change is slow, a sinusoidal pilot signal with the frequency in the crossover region and very small amplitude suffices to identify the plant accurately. The small-amplitude pilot signal can be superimposed on the command. Or, the pilot signal can be generated in the system itself by self-oscillation. High-frequency oscillation of small amplitude is called ***dither***. A ***dithering system*** block diagram is shown in Fig. 9.9.

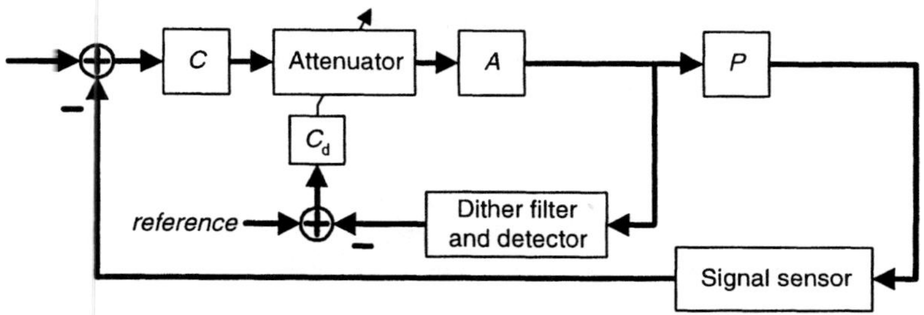

Fig. 9.9 Dithering control system

The Nyquist and Bode diagrams for the system are shown in Fig. 9.10.

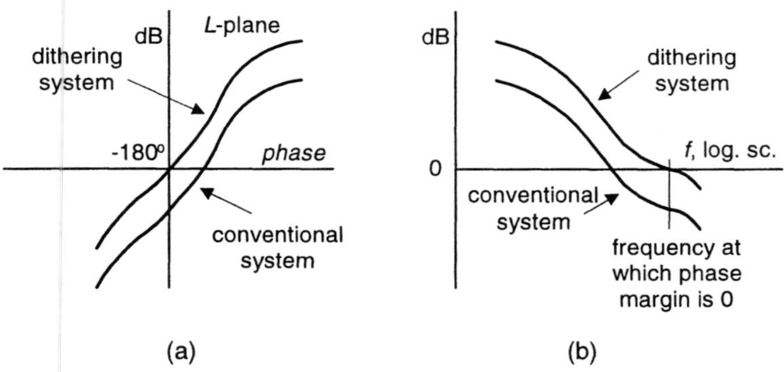

Fig. 9.10 Dithering and conventional control systems, (a) Nyquist diagrams and (b) Bode diagrams

The dithering system oscillates at the frequency where the phase stability margin is zero. In a conventional system, at this frequency the loop gain must be – 8 to –10 dB to provide sufficient gain stability margin. In the dithering system, the loop gain at this frequency is 0 dB which allows an increase in the loop gain by these 8 to 10 dB, i.e., a provision of, approximately, 3 times additional rejection of the disturbances within the functional frequency range.

The gain in the feedback loop needs to be adaptively adjusted in order for the amplitude of the oscillation to remain small. For this purpose, the dither is selected by a band-pass filter, amplified, rectified, smoothed by a low-pass filter, and compared with a reference. The difference between the measured dither and the reference, the error, is amplified and slowly and continuously regulates a variable attenuator in the forward path of the main feedback loop. The attenuator varies the main loop gain so that the dither level becomes as small as the reference. The dither loop compensator C_d implements the desired adaptation law resulting in the desired time-response of the adaptation.

This system has certain similarities with the automatic level control system for AM receivers described in Section 1.10

(In some systems, the dither signal reduces the loop gain by partially saturating the actuator nonlinear link. These systems are simpler since they have neither a dither detector nor a variable attenuator, but their performance is inferior, with smaller range of adaptation, larger error introduced by the dither, and reduced power available from the actuator to drive the plant.)

9.7 Adaptive filters

Adaptive controllers use linear links, also called ***linear filters***, which are varied by some adaptation algorithm. As the filters, symmetrical regulators described in Chapter 7 can be used. Two additional implementation examples are given below.

Example 1. A ***transversal filter*** compensator is shown in Fig. 9.11. It consists of delay links τ and variable gain links, the ***weights*** w_i. With appropriate weights, any desired frequency response of the adaptive filter can be obtained.

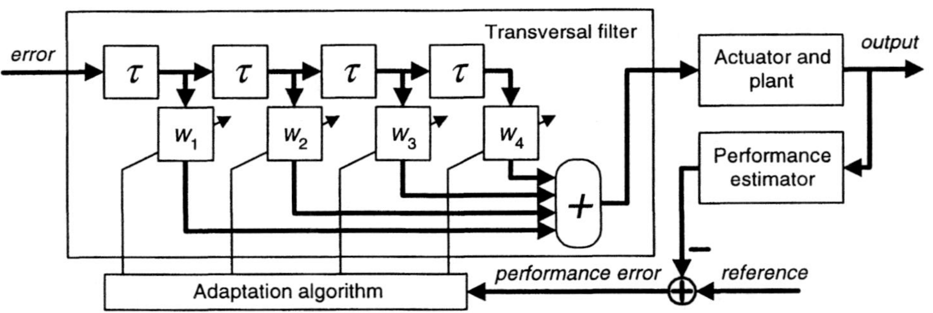

Fig. 9.11 Transversal adaptive filter as a compensator

The weights are adjusted by an adaptation algorithm (defining the dynamics of the adaptation loops) on the basis of the performance error produced by the performance estimator. Such a system can be classified as a self-learning system.

This system works well when the required frequency responses are relatively smooth. Responses with sharper bends and responses with resonance modes would require using too many sections in the transversal filters.

Example 2. Adaptive filters can be made with amplitude modulators, as for example in Fig. 9.12. The first balanced pair of modulators (multipliers) with carriers in quadrature transfers the input signal from baseband to higher frequencies where the signal is filtered by filters $C(s)$, and the second pair of balanced quadrature modulators returns the signal to the baseband. The balanced quadrature modulators are used here to cancel most of the parasitic intermodulation products. The input-output frequency responses can be varied by changing $C(s)$ and using multi-frequency carriers.

These adaptive filters are particularly useful for rejection of periodic non-sinusoidal disturbances.

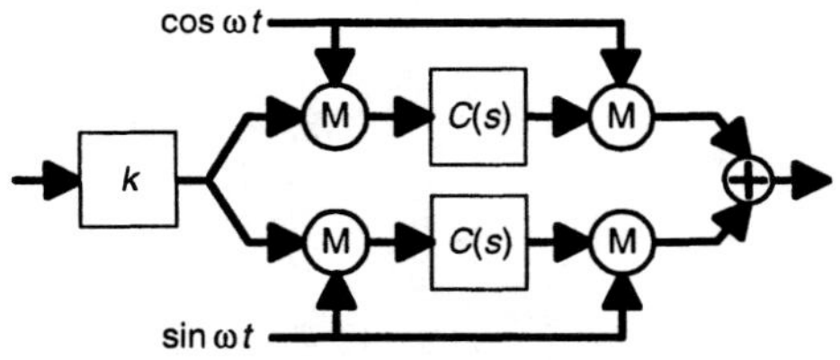

Fig. 9.12 Adaptive filter with modulators

Chapter 10

PROVISION OF GLOBAL STABILITY

The actuator, feedback path and plant nonlinearities are reviewed. Concepts are developed of limit-cycles, stability of linearized systems, conditional stability, global stability, and absolute stability. The Popov criterion is discussed and applied to control system analysis and design. Nonlinear dynamic compensators (NDCs), which ensure absolute stability without penalizing the available feedback, are introduced.

10.1 Nonlinearities of the actuator, feedback path, and plant

In a nondynamic link, the current value of the output variable depends only on the current value of the input variable and not on its previous values. The link has no memory. It is fully characterized by the input-output function. Several examples of the input-output characteristics of nonlinear nondynamic links – hard and smooth saturation, hard and smooth dead zone, dead zone and saturation, and three-position relay – are shown in Fig. 10.1(a),(b),(c), and (d) respectively. These nonlinear links represent properties of typical actuators.

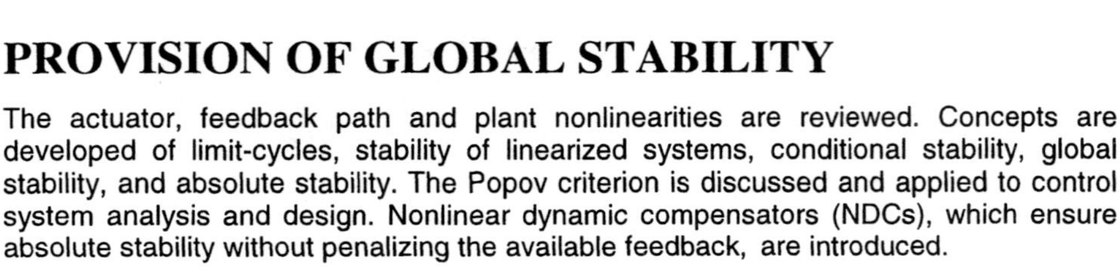

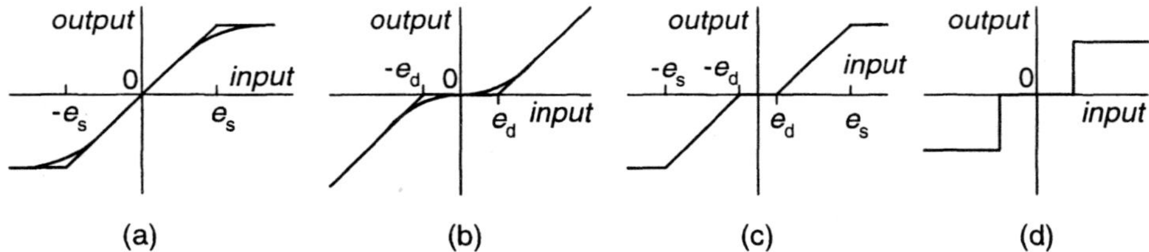

Fig. 10.1 Characteristics of nonlinear nondynamic links: (a) soft and hard saturation, (b) soft and hard dead zone, (c) dead zone and saturation, (d) three-position relay

A nonlinear link $y(x)$ can be placed in the feedback path to implement an inverse nonlinear operator $x(y)$ as shown in Fig. 10.2. When the feedback is large, the error e is small and input $y + e$ approximately equals the signal fed back, $y(x)$.

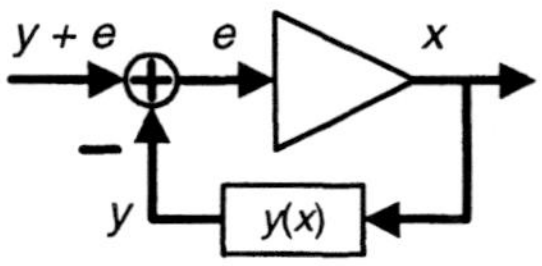

Fig. 10.2 Inverse operator

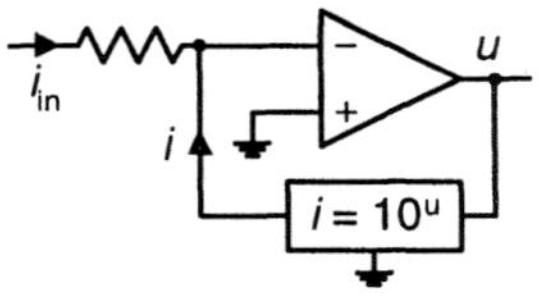

Fig. 10.3 Logarithmic amplifier

Example 1. Putting an exponential link (using a semiconductor *pn* junction) in the feedback path of an op-amp as shown in Fig. 10.3 produces a logarithmic link. If the error is small, the input current $i_{in} = -i$. The range of i_{in} is about 10^4, and the range output voltage u is 4. Therefore, the feedback path gain coefficient increases by a factor of 2500, i.e., 68 dB, when the input signal increases from the maximum to the minimum.

For the error $i_{in} - i$ to be small relative to i_{in}, the feedback must not be smaller than, say, 40 dB even when the y is small and the feedback path coefficient is relatively small. Therefore, the feedback reaches 108 dB when i_{in} is the largest.

The stability conditions limit the feedback bandwidth when the loop gain is largest (i.e., for the largest i_{in}). Therefore, assuming the loop Bode diagram slope is -10 dB/oct, the feedback bandwidth becomes 6.8 octaves smaller for the smallest i_{in}.

To counteract this detrimental effect and to keep the functional feedback bandwidth wide enough when i_{in} varies with time, a nonlinear link with the characteristic approximately inverse to that in the feedback path can be installed in the forward path which will increase the loop gain and the feedback bandwidth for low-level signals.

Two examples of plant nondynamic nonlinearities are the heat radiation and turbulent liquid flow that were mentioned in Section 3.1. These are the ***static nonlinearities*** that can be characterized by the link's input-output function. In a link with ***kinematic nonlinearities***, the relation between the input and the output variables (forces, torques) depends on the output *position*. For example, the force-to-torque ratio in a robotic arm driven by a rotational motor depends on the elbow angle as shown in Fig. 10.4 (a) and (b).

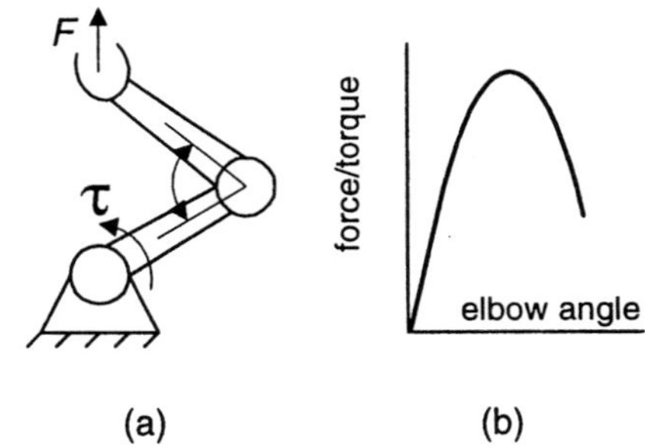

Fig. 10.4 (a) Robotic arm and (b) its force-to-torque ratio as the function of the elbow angle

Dynamic nonlinearities are those where the output nonlinearly depends on the current and the previous values of the input variable. They will be studied in Section 10.7 and Chapters 11, 12, and 13.

10.2 Types of self-oscillation

The most frequently encountered oscillation $x(t)$ is periodic. The ***phase plane*** (x, x'), where $x'(t)$ is the signal time derivative, is shown in Fig. 10.5. The trajectories on the plane reflect the time history of the process. When the oscillation is sinusoidal, the trajectory is a circle, like the oscillation of different amplitudes in a second-order conservative system depicted in Fig. 10.5(a). The unmarket trajectories relate the information about the shape of the oscillation but not about the frequency of the oscillation.

Fig. 10.5(b) depicts a stable second-order system with some damping. Starting with any initial condition, the signal amplitude decreases with time and the trajectories approach the origin, i.e., x and x' approach 0 when time increases indefinitely. In other words, the origin is the ***static attractor*** for the trajectories.

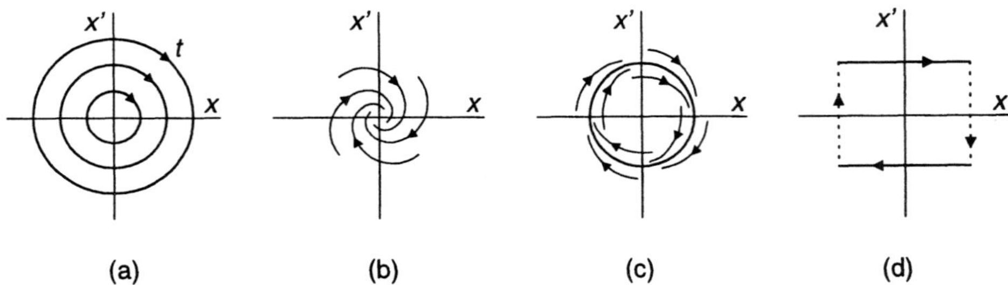

Fig. 10.5 Trajectories on the phase plane, for:
(a) nondissipative second-order system,
(b) dissipative second order system having a static attractor at the origin,
(c) limit cycle for an oscillator of a sinusoidal signal,
(d) limit cycle for an oscillator of a triangular signal

Self-oscillation in physical non-conservative nonlinear systems is initially aperiodic but often asymptotically approaches periodicity. This type of periodic oscillation is called ***limit cycle***. In Fig. 10.5(c), a nonlinear system with a limit cycle is shown. This circular limit cycle describes sinusoidal oscillation.

In many systems, $x(t)$ is rich in high harmonics, as, for example, in the oscillations which are shown in Fig. 10.6. When, for example, the signal is triangular as in Fig. 10.6(a), the limit cycle takes a rectangular shape as in Fig. 10.5(d), with the rate x' changing by instant jumps.

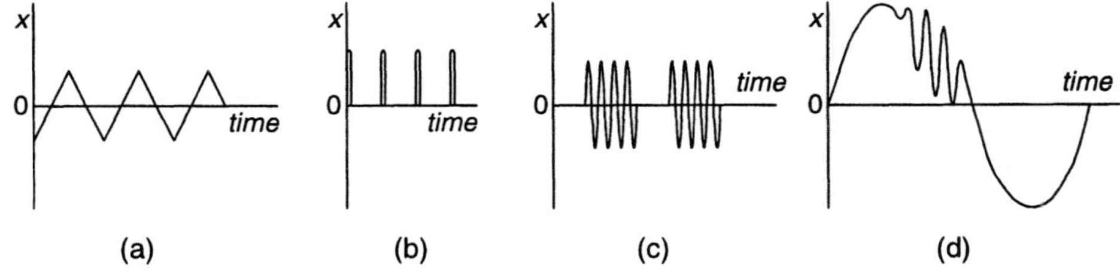

Fig. 10.6 Periodic oscillations with high harmonic content

A nonlinear system might possess several limit cycles, each surrounded with its own ***basin of attraction*** in the phase space. Initial conditions within the basin of attraction lead to this limit cycle. Different initial conditions lead to different limit cycles, and some initial conditions might lead to stability.

The limit cycle and the basin of attraction of a feedback loop which is stand-alone unstable can be modified with linear and nonlinear compensators such as to facilitate stabilizing the system with other feedback loops (see an example in A13.13).

Aperiodic oscillation usually does not arise in control systems of moderate complexity near the border of global stability. Because of this, we will not consider it in this book, although aperiodic oscillations do happen in some engineering feedback systems such as in improperly designed microwave VCOs.

Phase plane can be utilized for the design of low-order relay control systems. For example, when the variable x needs to be constrained by $a < x < a$, the ***switching lines*** $x = a$ and $x = -a$ drawn on the phase-plane define the conditions for the relay to switch, and the x time-history is derived from the trajectories on the phase plane.

10.3 Stability analysis of nonlinear systems

10.3.1 Local linearization

The ***First Lyapunov Method*** for stability analysis is applicable to nonlinear systems with differentiable characteristics. Lyapunov proved that stability of the system equilibrium can be determined on the basis of the system parameters linearized for small increments, i.e., ***locally***.

Example 1. Consider a nonlinear two-pole (an electrical arc, neon lamp, or the output of an op-amp with in-phase current feedback) with an external dc bias voltage source u as shown in Fig. 10.7(a). The differential dc resistance du/di depends on u as shown in Fig. 10.7(b). The resistance is negative on the ***falling branch*** between the two ***bifurcation points***.

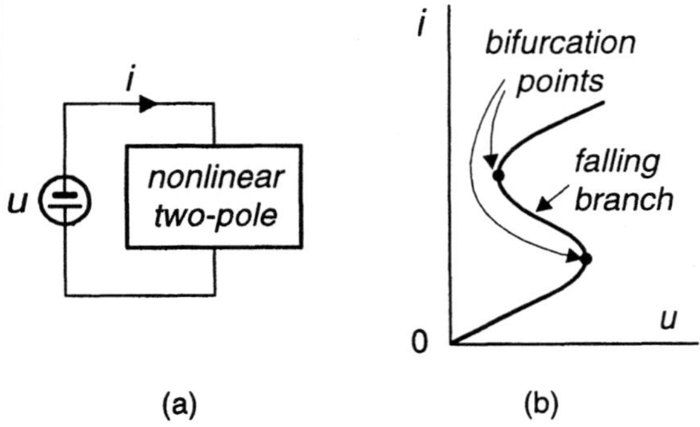

Fig. 10.7 Nonlinear two-pole device (a) with S-type current-on-voltage dependence (b)

According to the First Lyapunov Method, the locally linearized impedance of the two-pole

$$Z(u, s) = \frac{a_m s^m + a_{m-1} s^{m-1} + \ldots + a_o}{b_n s^n + b_{n-1} s^{n-1} + \ldots + b_o},$$

with the polynomial coefficients functions of u, can be used to determine the system local stability. The system is locally unstable when the contour impedance which is $Z(u, s)$ has a zero in the right half-plane of s.

Assume it is known from experiments that outside the falling branch such a device is passive, and therefore the coefficients a_o, b_o, a_m and b_n are positive. Assume that it is also known that on the falling branch the device is stable if connected to a current source, i.e., $Z(s)$ has no poles in the right half-plane.

The differential dc resistance of the two-pole is

$$Z(u,0) = \frac{a_o(u)}{b_o(u)}.$$

As the bias point moves along the characteristic in Fig. 10.7(b) into the falling branch, the resistance and, therefore, coefficient a_o, pass through zero at the bifurcation point and both become negative on the falling branch.

On the other hand, a_m and b_n are determined by the high-frequency behavior of the circuit (by stray inductances and capacitances), and remain positive on the falling branch. With a_m positive and a_o negative, the numerator of Z must have a positive real root. Therefore, when the two-pole is shorted (connected to a voltage source), the contour impedance has a zero in the right half-plane of s, and the system is unstable. We will use this result in Section 12.3.

The First Lyapunov Method justifies using the Nyquist and Bode methods of stability analysis to determine whether a nonlinear system is stable locally, i.e., while the signal deviations from the solution are small. For a system where the deviations can be big, the locally-applied Nyquist and Bode stability conditions are necessary but not sufficient. In particular, a Nyquist-stable system is commonly stable when first switched on, but after being overloaded, can become unstable.

10.3.2 Global stability

Signals that are initially finite and then, when the time increases, remain within an envelope whose upper and lower boundaries asymptotically approach zero, are called ***vanishing***. The system is called ***asymptotically globally stable*** (***AGS***) if its responses are vanishing after any vanishing excitation. Time-exponents $\pm ae^{-kt}$, $k > 1$ are commonly used for the envelope boundaries. Asymptotic stability with such boundaries is called exponential stability.

An AGS system remains stable following all possible initial conditions, as opposed to a ***conditionally stable*** system, which is stable following some initial conditions but in which some other initial conditions trigger instability. Well-designed control systems must be AGS. However, to directly test whether a system is AGS one would be required to try an infinite number of different vanishing signals which is not feasible. This necessitates devising convenient practical global stability criteria.

One such criterion is the ***Second Lyapunov Method***. It uses the so-called ***Lyapunov function*** which is a scalar function of the system coordinates, is equal to zero at the origin, is positive definite (i.e., positive and nonzero) outside of the origin as illustrated in Fig. 10.8 for the case of two coordinate variables, and has negative time derivative. The function V continuously decreases with time and approaches the origin. As the result, each variable approaches zero and the system is AGS.

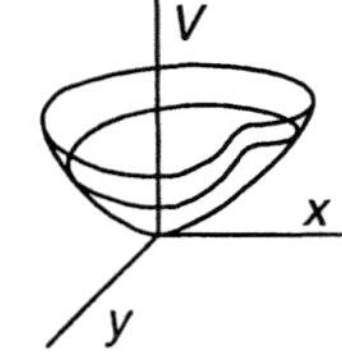

Fig. 10.8 Lyapunov function example

The Lyapunov function is often constructed as a sum of a quadratic form of the system variables and an integral of a nonlinear static function reflecting the system nonlinearity. Finding an appropriate Lyapunov function is simple when the system is low-order and the stability margins are wide. Unfortunately, finding the function for a practical nonlinear system with a high-performance controller is, typically, difficult.

10.4 Absolute stability

Many practical feedback systems consist of a linear link $-T(s)$ and a nonlinear link (actuator) that can be well approximated by a memoriless (i.e., nondynamic) nonlinear link $v(e)$, as shown in Fig. 10.9.

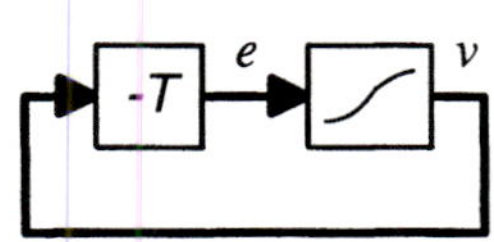

Fig. 10.9 Feedback system with a nonlinear link

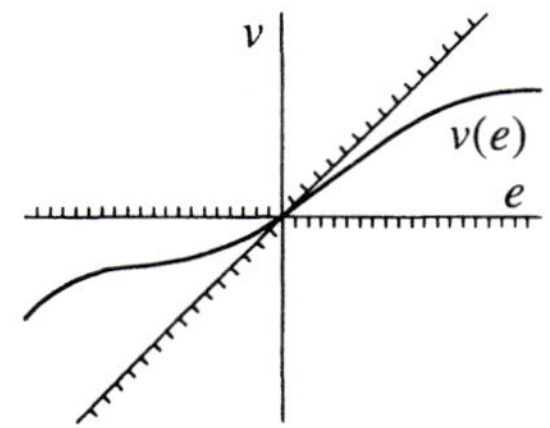

Fig. 10.10 Characteristic of the nonlinear link in Fig. 10.9

This system is said to be ***absolutely stable*** (***AS***) if it is AGS with any characteristic $v(e)$ constrained by

$$0 < v(e)/e < 1 \tag{10.1}$$

as illustrated in Fig. 10.10. Hard and soft saturation, dead zone, and the three-level relay belong to the class of nonlinear characteristics defined by (10.1).

10.5 Popov criterion

10.5.1 Analogy to passive two-poles' connection

The criterion due to V. M. Popov is based on the Parseval's theorem and is equivalent to the absolute stability criterion that has been previously obtained by A. I. Lurie with the Second Lyapunov Method. The Popov criterion can be readily applied to the systems defined by plots of their frequency responses, and is instrumental in developing controllers with improved performance.

The Popov criterion can be understood through the mathematical analogy between the feedback system and the connection of electrical two-poles (refer to Section 2.10).

Consider a nonlinear inductor which in response to current i creates magnetic flux $q\phi(i)$, where q is some positive coefficient and $\phi(i)/i > 1$ as shown in Fig. 3(a). Since the flux has the same sign as the current, the energy $qi\phi(i)/2$ stored in the inductor is positive. Consider also a nonlinear resistor with the dependence of voltage on current $u = \phi(i) - i$. The power dissipated in the resistor $ui = i(\phi(i) - i)$ is positive for all i.

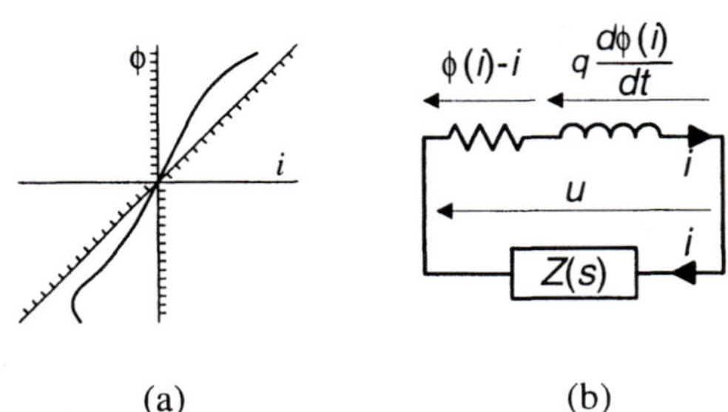

Fig. 10.11 (a) Function $\phi(i)$ and (b) an AGS circuit

Let us connect this inductor, this resistor, and a passive linear two-pole with impedance $Z(s)$ (that is positive real) in the contour shown in Fig. 10.11(b). The resistor and the two-pole $Z(s)$ draw and dissipate the energy stored in the inductor. Since this energy cannot become negative, i.e. the energy cannot be overdrawn from the magnetic field, the current i decays with time and approaches 0. The system is AGS.

The voltage u in the circuit in Fig. 10.11(b) is, according to the equations related to the upper branch,

$$u = \phi(i) - i + q\frac{d\phi(i)}{dt}.$$

Using Laplace transforms $U = \mathsf{L}\,u$ and $I = \mathsf{L}\,i$, the voltage

$$U = \mathsf{L}\phi(i) - I + qs\,\mathsf{L}\phi(i) = (1 + qs)\,\mathsf{L}\phi(i) - I.$$

The current,according to the Ohm's equation for the the lower branch, is

$$I = -\frac{1}{Z(s)}U.$$

These equations describe the block diagram (a) in Fig. 10.12.

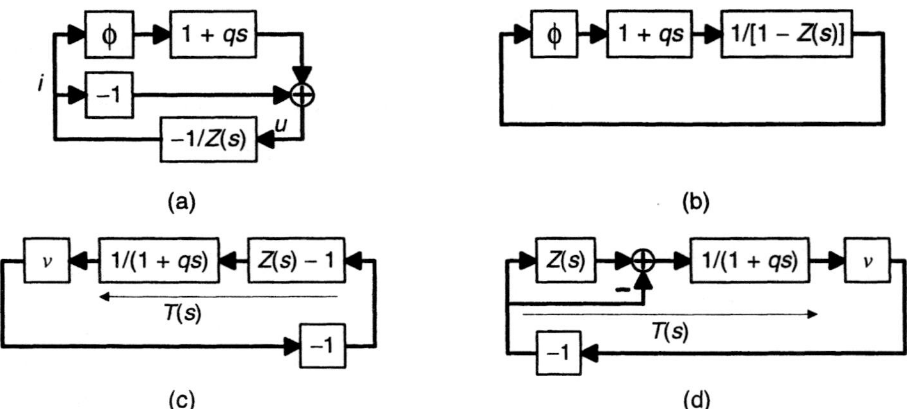

Fig. 10.12 AGS equivalent feedback systems

Since the link −1 can be viewed as the feedback path for the link $-1/Z(s)$, the diagram can be converted to that in (b). Next, reversing the direction of the signal propagation generates the diagram in (c) where the function v is the inverse of ϕ. Notice that such v satisfy condition (10.1). The diagram in (c) is equivalently redrawn with conventional clockwise signal transmission in (d).

The feedback system shown in Fig. 10.12(d) is described by the same equations as the passive circuit in Fig. 10.11 and is therefore AGS. The return ratio of the linear links of the loop is

$$T(s) = [Z(s) - 1]/(1 + qs). \tag{10.3}$$

Therefore, a feedback system consisting of a nonlinear link v and the linear link $T(s)$ is AGS if $T(s)$ is representable in the form (10.3) with any p.r. $Z(s)$. In other words, <u>*the system is AS if there exists a positive q such that the expression*</u>

$$Z(s) = (1 + qs)T(s) + 1 \tag{10.4}$$

<u>*is positive real (p.r.)*</u>. This means that $T(s)$ should have no poles in the right half-plane, and at all frequencies

$$\text{Re}\,[(1 + jq\omega)T(j\omega)] > -1. \tag{10.5}$$

The ***Popov absolute stability criterion*** follows: *if the system is stable open-loop and there exists a positive q such that (10.5) is satisfied at all frequencies, the system is AS.*

To check whether a system satisfies the Popov criterion, one might plot the Nyquist diagram for $(1 + qs)T(s)$. *If a positive q can be found such that the Nyquist diagram for $(1 + qs)T(s)$ stays to the right of the vertical line −1, the system is AS.*

The Popov criterion (which is sufficient but not necessary for AS) is more restrictive than the Nyquist criterion (which is necessary but not sufficient for AS).

Instead of the Nyquist diagram, equivalent Bode diagrams can be used.

Example 1. Consider a system with a loop gain response which is flat at lower frequencies and has a high-frequency cut-off with −10 dB/oct slope. The phase shift in this system varies from 0 to −150° as shown in Fig. 11.13(a). By multiplying the loop transfer function by the Popov factor $1 + qs$ with large q, so that the Popov factor's angle is nearly +90°, we obtain the function $(1 + qs)T(s)$ with the angle from +90° to −60°. The real part of this function is positive at all frequencies, and the absolute stability condition (10.5) is satisfied.

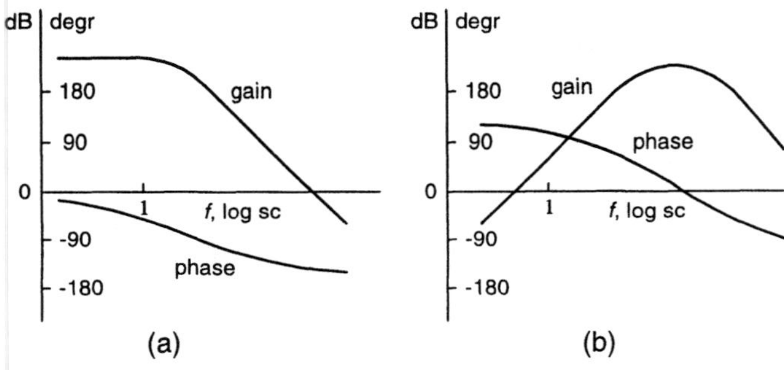

Fig. 10.13 Loop responses for (a) an AS system and (b) for a system for which AS cannot be proved with the Popov criterion

Example 2. Consider a band-pass system with a low-frequency roll-off slope of 10 dB/oct and with the associated phase shift of 150° as shown in Fig. 10.14(b). The phase shift of the Popov factor $(1 + jq\omega)$ is within the 0 to 90° limits. Therefore, for the frequencies on the roll-off, the phase shift of the expression in the brackets in (10.5) is within the 150° to 240° limits, and (10.5) is not satisfied. This system therefore falls into the gap between the Nyquist and the Popov criteria, and no judgment can be passed on whether the system is AS.

Example 3. In a Nyquist-stable system, T must be real and less than −1 at some frequency as shown in Fig. 11.14. At this frequency, since T is real, q does not affect the left side of (10.5), and since T is less than −1, the inequality is not satisfied. Hence, Nyquist-stable systems do not satisfy the Popov criterion.

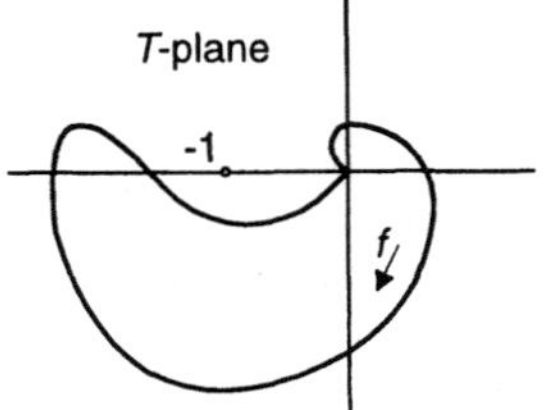

Fig. 10.14 Nyquist-stable system

10.5.2 Different forms of the Popov criterion

Condition (10.5) uses the real part of a function. The condition can be changed into an equivalent form which makes use of Bode diagrams, which is more convenient. Adding qs to $Z(s)$ does not change whether $Z(s)$ is positive real since $qj\omega$ is purely imaginary. From this follows the *second form* of the criterion:

the system is AGS if a real positive q exists such that $(1 + qs)T(s) + 1 + qs$ *is p.r., i.e.,*

$$(1 + qs)F(s) \quad \text{is p.r.} \tag{10.6}$$

i.e., at all frequencies

$$|\arg[(1 + jq\omega)F(j\omega)]| < \pi/2 . \tag{10.7}$$

Therefore AGS can be established by plotting the Bode diagram for $(1 + jq\omega)F(j\omega)$ and, using the phase-gain relations, deciding whether (10.7) is satisfied.

The *third form* of the Popov criterion uses the ***modified Nyquist plane*** (ReT, ωImT). When drawn on this plane, the Nyquist diagram is vertically compressed at lower frequencies and expanded at higher frequencies.

Condition (10.5) can be rewritten as $\text{Re}\,T(j\omega) - q\omega\text{Im}\,T(j\omega) > -1$ or

$$\omega\,\text{Im}\,T(j\omega) < q^{-1}[\text{Re}\,T(j\omega) + 1]. \tag{10.8}$$

With the sign " < " replaced by " = ," (10.8) represents the ***Popov line*** shown in Fig. 10.15. It passes through the point (–1,0) with slope $1/q$.

Therefore, inequality (10.8) is satisfied and *the system is AS if a Popov line can be drawn entirely to the left of the Nyquist diagram on the modified plane*.

For example, the system characterized by the plot shown in Fig. 10.15 is AS.

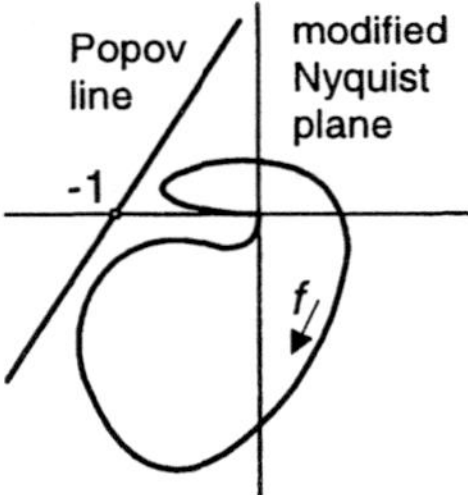

Fig. 10.15 Popov line on modified Nyquist plane

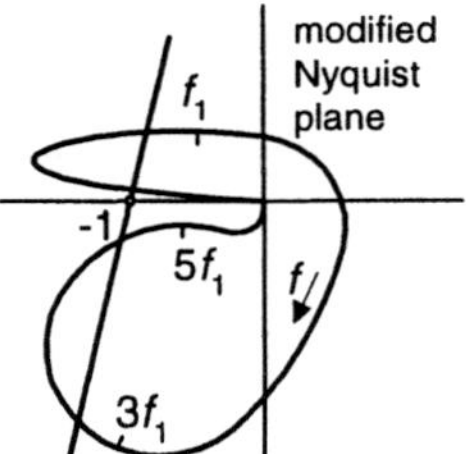

Fig. 10.16 On the basis of this diagram a judgment can be passed that a limit cycle with fundamental f_1 is not possible, but no judgment can be made about whether the system is AS.

The plot shown in Fig. 10.16 neither indicates that the system is stable nor that the system is unstable. This plot only guarantees the absence of certain types of oscillation. It has been proven that *no periodic self-oscillation can take place if a Popov line can be drawn to the left of all the points of the modified Nyquist diagram that relate to the Fourier components of the oscillation*. For instance, periodic oscillation with fundamental f_1 cannot exist in the system represented in Fig. 10.16.

The Popov criterion can be extended using more complex circuits than that in Fig. 10.12, e.g. circuits including several various nonlinear two-poles and several passive two-poles. It was proven, in particular, that the criterion is valid with the function $(1 + j\omega q_1)/(1 + j\omega q_2)$ replacing the Popov factor $(1 + qs)$.

10.6 Applications of Popov criterion

10.6.1 Low-pass system with maximum feedback

The Popov criterion is not a tool in everyday use like the Nyquist criterion or Bode diagrams. Its major application is to prove that the Nyquist and Bode criteria established for linear systems also guarantee absolute stability of most practical nonlinear systems, including those with maximized feedback.

The low-pass system with Bode optimal cut-off is AS. This can be established with the second form of the Popov criterion and employing $(1 + q_1 s)/(1 + q_2 s)$ instead of the Popov factor. With $1/q_1 << 1$ and $1/q_2 = 4\omega_c$, the inequality (10.7), expressed as

$$|\arg F + \arg(1 + j\omega q_1)/(1 + j\omega q_2)| < \pi/2,$$

is satisfied, as seen in Fig. 10.17.

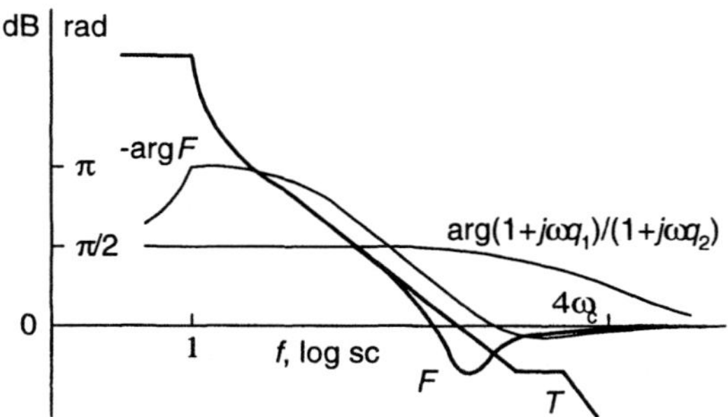

Fig. 10.17 Application of Popov criterion to Bode optimal low-pass cut-off

10.6.2. Band-pass system with maximum feedback

Consider a band-pass system with the feedback maximized over a large relative frequency band, with phase stability margin $y\pi$ in the high-frequency cut-off region and $y_l\pi$ in the low-frequency roll-up region, as shown in Fig. 10.18(a) and (b). It is seen that the combination of $y = 0$ and $y_l = 0.5$ satisfies condition (10.7), i.e., $\arg(1 + jq\omega)T$ varies only from $-\pi/2$ to $\pi/2$, if $1/q$ is chosen close to the mean square frequency of the passband.

Therefore, stability can be proved with the Popov criterion only when $y_l \geq 1/2$. However, the value of $y_l = 1/6$ does well in practical design. Making y_l excessively large to satisfy the criterion imposes little impairment to the available feedback but requires an extension of the loop gain to lower frequencies.

In the narrow-band band-pass system, the Popov sufficient stability criterion requires that the roll-up and cut-off phase stability margins, y_l180° and $y180^\circ$, exceed 90°. Satisfying this requirement dramatically reduces the maximum available feedback compared with using the typical stability margin of $y/6$ in the band-pass optimal

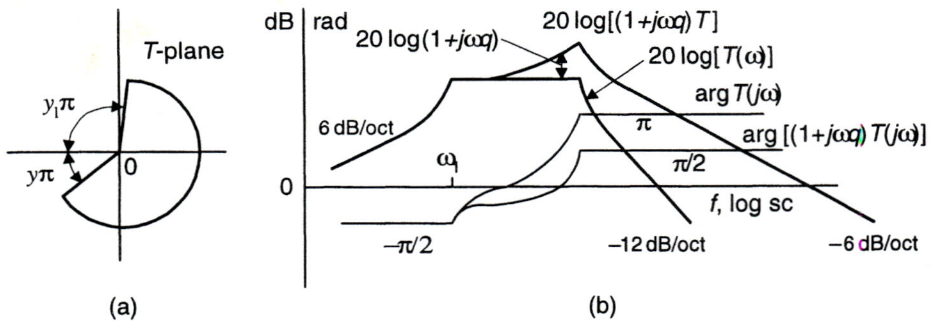

Fig. 10.18 Band-pass system analysis with Popov criterion

cut-off. The smaller the relative bandwidth, the smaller the feedback limited by the Popov criterion, because of the deficient filtering properties of the Popov factor $(1 + jq\omega)$.

For narrow-band band-pass systems, the Popov criterion is excessively restrictive. As mentioned, phase stability margins of $\pi/6$ both at higher and lower frequencies suffice for AGS of a system with a saturation link, regardless of the relative bandwidth – although this has not yet been proven theoretically.

10.7 Absolutely stable systems with nonlinear dynamic compensation

10.7.1 Nonlinear dynamic compensator

Larger feedback is available in Nyquist-stable systems. Such systems are not AS when the compensators are linear. Still, they can be made AS by using ***nonlinear dynamic compensators*** (***NDCs***). The NDC can be built of linear and non-dynamic nonlinear links.

In the system shown in Fig. 10.19, the nonlinear link $1 - v(e)$ in the local feedback of the NDC uses the same nonlinear function $v(e)$ as the nonlinear link of the actuator. Typically, $v(e)$ is a saturation link so that $1 - v(e)$ represents a dead zone. The rest of the links in the block diagram are linear. For the AS analysis, it can be assumed the command is 0.

We denote by T_P the return ratio for the plant measured when the link v is replaced by 1 (and the link $1 - v$, by 0). Then, when $v(e)$ is saturation, the compensator transfer function for small level signals is expressed as T_P/P. When the signal level is very large, the return ratio in the NDCs local loop becomes G.

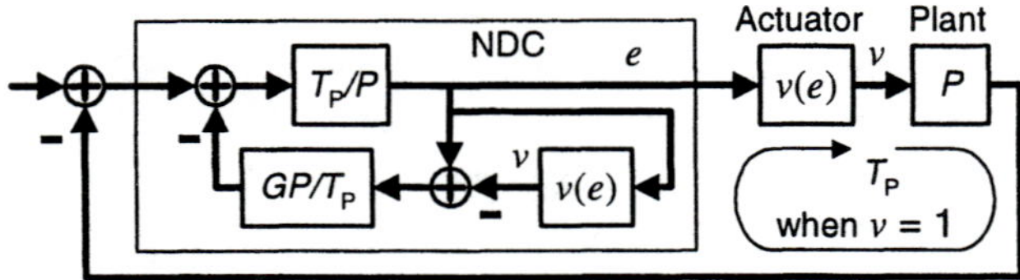

Fig. 10.19 Feedback system with nonlinear link v in the actuator and link $1 - v$ in the feedback path of the NDC

10.7.2 Reduction to equivalent system

The diagram shown in Fig. 10.19 depicts a system that has two identical nonlinear links $v(e)$, with the same input signal e and, therefore, the same output signal v. For the sake of stability analysis, the system can be modified equivalently into the one shown in Fig. 10.20 which contains only one nonlinear link v. The linear links within the dashed envelope form a composite linear link. We denote the negative of its transfer function as T_E (equivalent return ratio). If T_E satisfies the Popov criterion, the system must be globally stable.

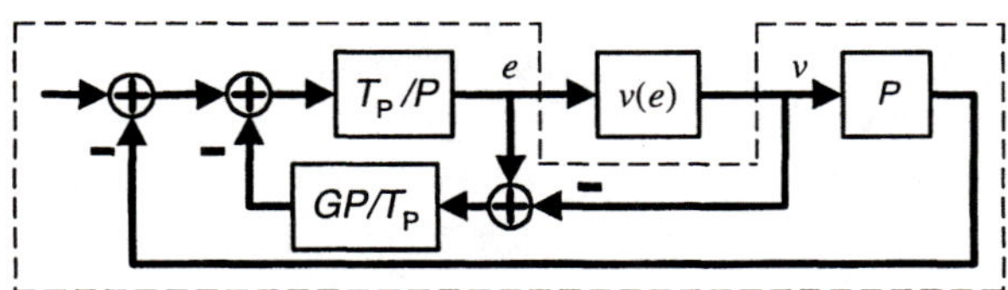

Fig. 10.20 Equivalently transformed system containing a single nonlinear link v

To find the expression for T_E, the diagram in Fig. 10.20 is further redrawn as shown in Fig. 10.21.

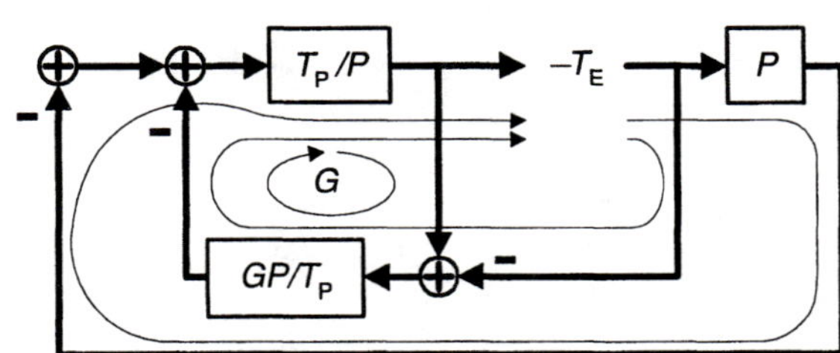

Fig. 10.21 Calculation of T_E using two parallel paths and a loop tangent to both paths

From this diagram, using Mason's rule, the negative of the transfer function from the output of the nonlinear link to its input is

$$T_E = \frac{T_P - G}{1+G}. \tag{10.9}$$

Given T_E and T_P, the NDC linear link transfer function is

$$G = \frac{T_P - T_E}{1+T_E}. \tag{10.10}$$

From (10.10), the plant feedback is

$$1 + T_P = (1 + G)(1 + T_E). \tag{10.11}$$

Eq. (10.11) states that *the plant feedback is the product of the local feedback in the NDC for large-level signals and the feedback in the equivalent system*. Therefore, the areas of substantial positive feedback in the G and T_E loops should not overlap or else the positive feedback in the plant will be excessive and the phase stability margin in the plant loop, correspondingly, small. If positive feedback in each loop is substantial, *the crossover frequency of $T_E(j\omega)$ must be either much smaller or much larger than the crossover frequency of $G(j\omega)$*.

We can see the advantage of a system with an NDC as follows: in a conventional system, T_P must satisfy the Popov criterion, but in a system with an NDC, the only requirements are that T_P satisfies the weaker Nyquist criterion, and that T_E, defined by (10.9), satisfies the Popov criterion. Exploiting this extra design flexibility leads to better performance as will be shown in the next section.

Commonly, there is no need to maximize to the full extent the disturbance rejection while keeping specified stability margins, or maximizing the stability margins while providing specified disturbance rejection response. In these cases, there still remains some freedom in choosing the responses for G and T_E. This freedom can be utilized for the provision of desired transient responses for large level signals. For a large-level signal, the dead zone in the feedback path of the NDC can be replaced approximately by a unity link, and the system becomes a tracking system with the loop transfer function

$$\frac{T_P}{G+1} \tag{10.12}$$

and saturation in the actuator. Such a system has response without an overshoot if the guard point phase stability margin is 90° or so. To achieve this, the Bode diagram for (10.12) must be rather shallow over a rather wide frequency interval including the crossover frequency. Examples with such responses will be demonstrated in the next section.

The equations derived for the block diagram in Fig. 10.19 are also valid for the system shown in Fig. 10.22 where the actuator exhibits the same nonlinearity as the function $v(e)$ in the NDC feedback path but for the k-times larger signal.

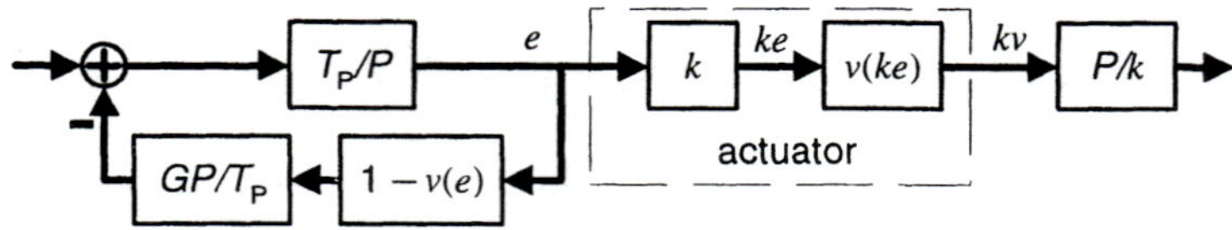

Fig. 10.22 Scaling down the nonlinear link in the NDC by a factor of k

10.7.3 Design examples

With lower-order compensation, the disturbance rejection and the benefits of NDC application decrease. Still, they remain substantial, and in the following Examples 1–3 we will use systems with low-order links to illustrate such benefits of nonlinear dynamic compensation as making AS the system with rather steep Bode diagram, and, in some cases, improving the transient responses to large commands.

Example 1. In the homing system diagrammed in Fig. 10.23, $v(e)$ is a saturation, and the plant P is a single integrator. The NDC forward path transfer function T_P/P (a lead-with-integration compensator) and the plant return ratio for small-level signals T_P are

$$\frac{T_P}{P} = \frac{2(s+0.5)}{s(s+2)},$$

$$T_P = \frac{2(s+0.5)}{s^2(s+2)}.$$

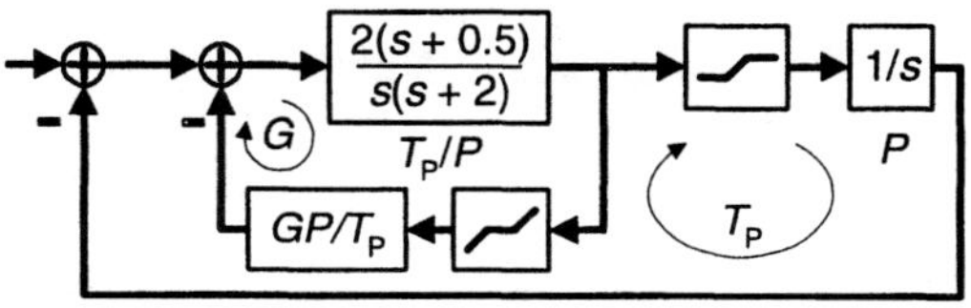

Fig. 10.23 Block diagram of a system with an NDC

The asymptotic Bode diagram for T_P shown in Fig. 10.24(a) is typical for simple lead compensation. The logarithmic Nyquist diagram for T_P is plotted in Fig. 10.24(b) with nyqlog function from Bodestep toolbox described in Appendix A14:

```
np = [2 1]; dp = [1 2 0 0]; nyqlog(1,np,dp)
```

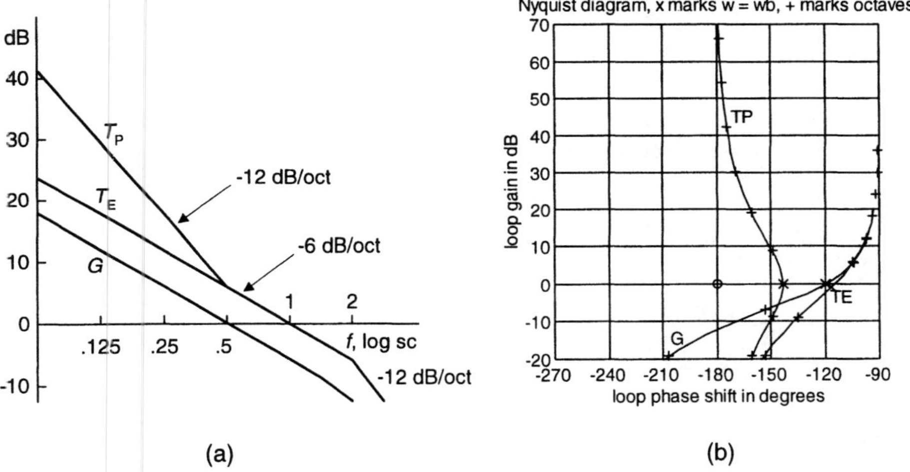

Fig. 10.24 (a) Asymptotic Bode diagrams and (b) L-plane plots for Example 1(A)

In the linear mode of operation (for small signals), the guard-point phase stability margins is 37° and the closed-loop transient response has 43% overshoot.

The general rules for selecting T_E and G have been formulated in Section 10.7.2. Consider two choices of low-order T_E and G.

(A) We will start the design by a guessed response for T_E:

$$T_E = \frac{2}{s(s+2)}.$$

As shown in Fig. 10.24 his response merges with T_P at higher frequencies but has a shallower slope at lower frequencies and, correspondingly, lesser phase lag. The system with such a T_E is AS (and even the *processes* in this system are stable; process stability is discussed in the next chapter). With such a T_E, from (10.10),

$$G = \frac{1}{s(s^2+2s+2)}.$$

The asymptotic Bode diagrams for these T_E and G are shown in Fig. 10.24(a). The L-plane Nyquist diagrams are plotted in Fig. 10.24(b) with the following script (which includes the previously shown line for plotting T_P):

```
np = [2 1]; dp = [1 2 0 0]; nyqlog(1,np,dp); hold on
ne = 2; de = [1 2 0]; nyqlog(1,ne,de); hold on
ng = 1; dg = [1 2 2 0]; nyqlog(0.5,ng,dg); hold off
gtext('TP'); gtext('TE'); gtext('G') %place labels with mouse
```

The transfer function of the link in the feedback path of the NDC is

$$\frac{GP}{T_P} = \frac{s+2}{(s+0.5)(s^2+2s+2)}.$$

The large-amplitude loop transfer function (10.12)

$$\frac{T_P}{G+1} = \frac{2(s+0.5)}{s^2(s+2)}\frac{s^3+2s^2+2s}{s^3+2s^2+2s+1} = \frac{2s^3+5s^2+6s+2}{s^5+4s^4+6s^3+5s^2+2s}.$$

The large-amplitude closed-loop response (not shown here) is dissimilar to the response of a Bessel filter; therefore, closed-loop transient responses to large commands are not expected to be acceptable.

(B) If, while retaining the NDC forward path link, we interchange the functions for G and T_E making them

$$T_E = \frac{1}{s(s^2+2s+2)}, \quad \text{and} \quad G = \frac{2}{s(s+2)},$$

the equations (10.9) – (10.11) are still satisfied since G and T_E enter these equations symmetrically. The transfer function of the link in the feedback path of the NDC

$$\frac{GP}{T_P} = \frac{2}{2s+1}.$$

The stability margins for this T_E (i.e., for G in Fig. 10.24) are smaller than in version (A), but sufficient for the system to remain AS (although not process-stable).

The large-signal loop transfer function (10.12) is

$$\frac{T_P}{G+1} = \frac{2(s+0.5)}{s(s^2+2s+2)}.$$

This frequency response and the related closed-loop response are plotted in Fig. 10.25(a). The closed-loop step-response is shown in Fig. 10.25(b); it has no overshoot. These linear system responses indicates that the overshoot in the nonlinear regimes in response to large commands will be small or non-existent.

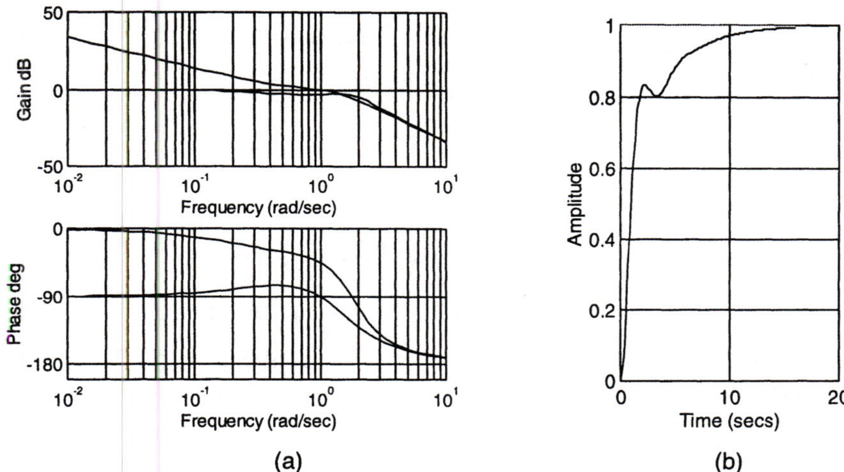

(a) (b)

Fig. 10.25 (a) Open-loop and closed-loop responses for large signals and (b) transient response to large step commands in Example 1

Indeed, the system transient responses to step commands obtained with SIMULINK and shown in Fig. 10.26 demonstrate that NDC improves the responses.

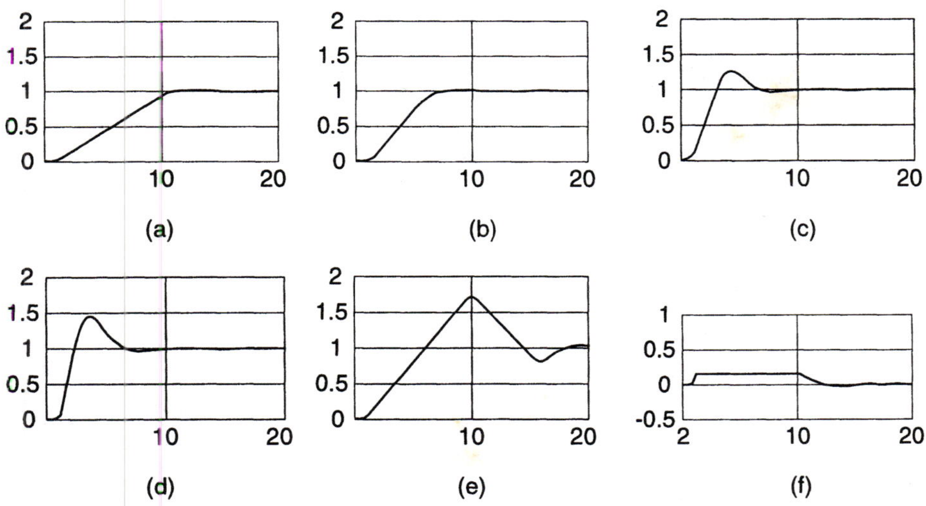

(a) (b) (c)

(d) (e) (f)

Fig. 10.26 Transient response to unity step command in Example 1, case (A): (a),(b),(c) – systems with threshold of saturation and dead zone set to, respectively, 0.1; 0.2; 0.5; (d) system without NDC; (e) windup in the system where threshold is set to 0.2, and the NDC feedback path disconnected; (f) actuator output for (a)

In the linear mode of operation (d), the overshoot is large, approximately 45%. In (a),(b),(c) the responses are shown for the systems with the unity step command and different saturation thresholds. The larger the ratio of the command to the saturation threshold, the smaller the percentage of overshoot.

Fig. 10.26(e) shows the windup response which occurs when the feedback path in the NDC is disconnected and the actuator overloaded by a relatively large disturbance or command. The effects of the NDCs on the transient response for large-level signals will also be considered in Chapter 13.

Fig. 10.26(f) shows that the actuator output is constant over some time range and then rapidly drops to nearly zero. This indicates good utilization of the actuator output power capability.

The system exhibits performance which is desirable for many homing systems: the absolute value of the overshoot is limited to a small value. Large (percentwise) overshoots occur in the responses to small disturbances, but not in the responses to large disturbances.

The performance can be further improved with higher-order compensation.

Example 2. The system block diagram is shown in Fig. 10.27. The transfer function of the NDC forward path and the plant small-signal return ratio are

$$\frac{T_P}{P} = \frac{2s^2 + 1.2s + 0.1}{s^3 + 2s^2},$$

$$T_P = \frac{2s^2 + 1.2s + 0.1}{s^4 + 2s^3},$$

i.e.,

$$T_P = \frac{2(s+0.5)(s+0.1)}{s^3(s+2)}.$$

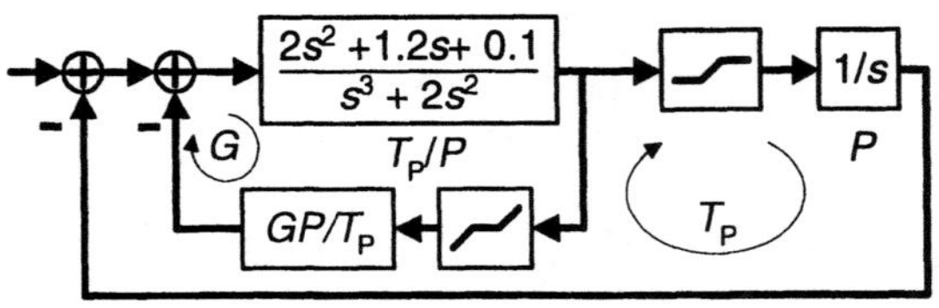

Fig. 10.27 Block diagram of a system with an NDC

As seen on the asymptotic Bode diagram in Fig. 10.28(a), $|T_P|$ is larger at lower frequencies than that in Example 1 and therefore provides better disturbance rejection. The system is Type 3. The Nyquist diagram for T_P is shown in Fig. 10.28(b), the system is Nyquist-stable.

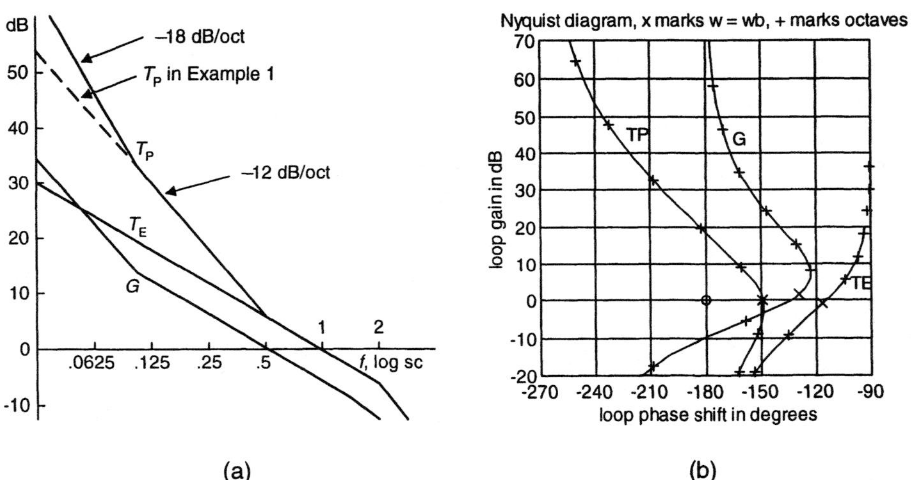

Fig. 10.28 (a) Asymptotic Bode diagram and (b) L-plane plots for Example 2

Let us consider three choices of T_E and G, keeping them low-order.

(A) With the same

$$T_E = \frac{2}{s^2 + 2s}$$

as in case A of Example 1, G is found from (10.10) as

$$G = \frac{1.2s + 0.1}{s^2(s^2 + 2s + 2)}.$$

The asymptotic Bode diagrams for the T_E and G are shown in Fig. 10.28(a). The L-plane plots are shown in Fig. 10.28(b). For large-level signals, the system is Type 3 – 2 = 1 since the NDC local loop is Type 2. The NDC feedback path transfer function is

$$\frac{GP}{T_P} = \frac{(0.6s + 0.05)(s + 2)}{(s^2 + 2s + 2)(s + 0.5)(s + 0.1)}.$$

The system is globally stable and the stability margins in the equivalent system are large, but it does not qualify as a good homing system because the transient responses to large-amplitude signals (not shown here) are far from the best possible.

(B) With the functions $T_E(s)$ and $G(s)$ interchanged, i.e.,

$$T_E = \frac{1.2s + 0.1}{s^2(s^2 + 2s + 2)} \quad \text{and} \quad G = \frac{2}{s(s + 2)},$$

the NDC feedback path becomes

$$\frac{GP}{T_P} = \frac{2}{s(s + 2)} \frac{s^2(s + 2)}{2(s + 0.5)(s + 0.1)} = \frac{s}{2(s + 0.5)(s + 0.1)} = \frac{s}{2s^2 + 1.2s + 0.1}.$$

The type of the system for large-level signals is 3 – 1 = 2. The transient responses for large magnitude step commands (not shown here) still deserve improvement.

(C) The numerator of the NDC feedback path transfer function is further adjusted by trial and error, in order to improve the transient response to a large-amplitude step command. With NDC feedback path transfer function

$$\frac{GP}{T_P} = \frac{1.6s + 0.16}{2s^2 + 1.2s + 0.1}, \qquad G = \frac{GP}{T_P}\frac{T_P}{P} = \frac{1.6s + 0.16}{s^3 + 2s^2},$$

$$G + 1 = \frac{s^3 + 2s^2 + 1.6s + 0.16}{s^3 + 2s^2}, \qquad \frac{T_P}{G + 1} = \frac{2s^2 + 1.2s + 0.1}{s(s^3 + 2s^2 + 1.6s + 0.16)},$$

$$T_P + 1 = \frac{s^4 + 2s^3 + 2s^2 + 1.2s + 0.1}{s^4 + 2s^3}, \qquad T_E = \frac{0.4s^2 + 1.04s + 0.1}{s^4 + 2s^3 + 1.6s^2 + 0.16s}.$$

The large signal open-loop and closed-loop Bode diagrams are shown in Fig. 10.29(a). With such diagrams, the transient response shown in Fig. 10.29(b) has no overshoot, and the nonlinear system transient response shown in Fig. 10.30 becomes satisfactory. The Nyquist diagram for T_E is shown in Fig. 10.31, where the crossover frequency $\omega_b = 0.65$. The system is AS.

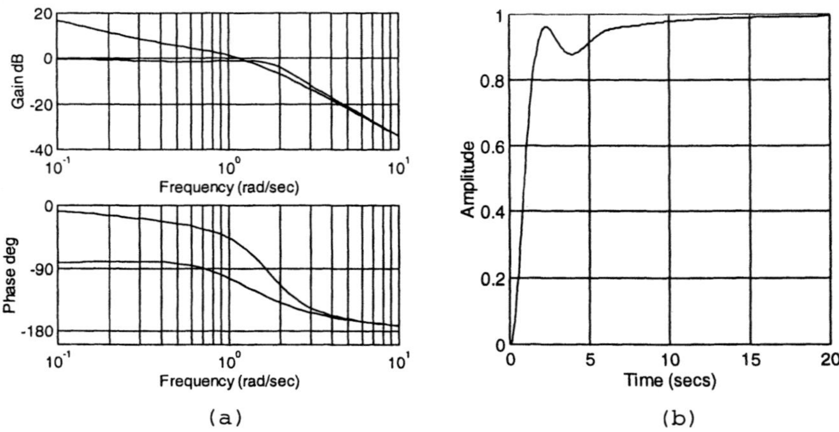

Fig. 10.29 (a) Bode diagrams and (b) transient response for $T_P/(G+1)[1+T_P/(G+1)$

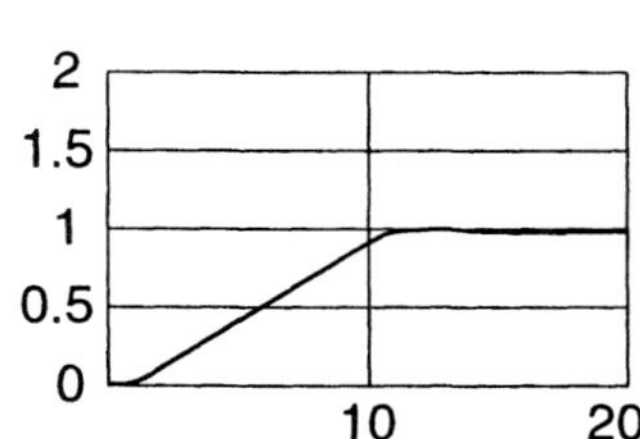

Fig. 10.30 Transient response of the nonlinear system to large step commands in Example 2, case (C), obtained with SIMULINK.

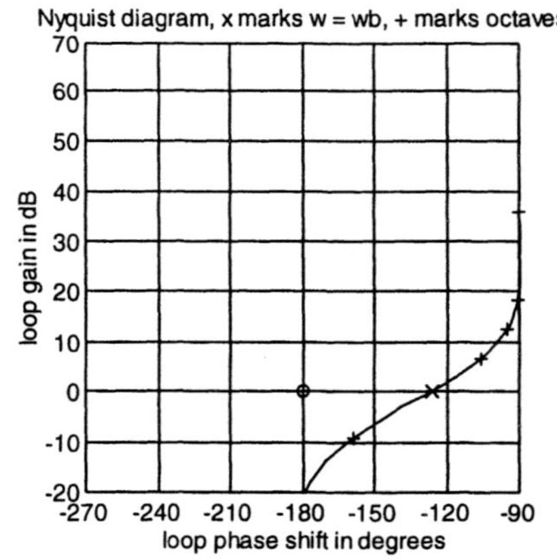

Fig. 10.31 Nyquist diagram for T_E

Example 3. In Examples 1 and 2, the plant is $1/s$. Let us next consider the system with a double-integrator plant.

To retain the plant loop return ratio the same as before, let us multiply the NDC forward path transfer functions in Examples 1 and 2 by s, and to retain the same internal feedback in the NDC, let us multiply the feedback path by $1/s$. The system with the $1/s^2$ plant and the modified NDC has the same G and T_E, and is therefore AS.

However, since the NDC reduces the system's type for large signal amplitudes by the number of poles of G at dc, the output transient responses to large-magnitude step commands undershoot. When good large-signal transient responses must be assured in the nonlinear system, the Bode diagram for T_E should be made steeper. Acceptable performance can be obtained with

$$P = \frac{1}{s^2}, \quad \frac{T_P}{P} = \frac{2s^3 + 1.2s^2 + 0.1s}{s^3 + 2s^2}, \quad \frac{GP}{T_P} = \frac{s}{1.5s^3 + 1.5s^2 + 0.15s + 0.002}.$$

Better transient responses can be achieved by making the dead zone in the NDC somewhat smaller than the actuator's threshold of saturation (although in this case the system analysis with Popov criterion is no more valid, and the disturbance rejection for large signals will decrease, which might or might not be of importance).

The presented Examples 1, 2, and 3 use explicit algebraic low-order transfer functions. The following three examples use graphical representation for the gain responses. The related phase responses can be calculated using either graphical methods given in Chapter 4, or the programs from Appendices 4 or 14. After this, the responses for G and for the compensator T_P/P can be approximated by rational functions.

Example 4. Fig. 10.32 shows the diagrams for $G + 1$ as Bode optimal cut-off with $f_b = 8$ Hz, 30° of phase stability margin, and 10 dB of amplitude stability margin resulting in 40 dB of feedback over the bandwidth [0,1]. T_E is chosen with only 10 dB over the functional bandwidth, and having the phase stability margin more than 90°. The attained feedback $|T_P + 1|$ of 50 dB exceeds the 40 dB available in the single-loop AS system. The system is Nyquist-stable and AS.

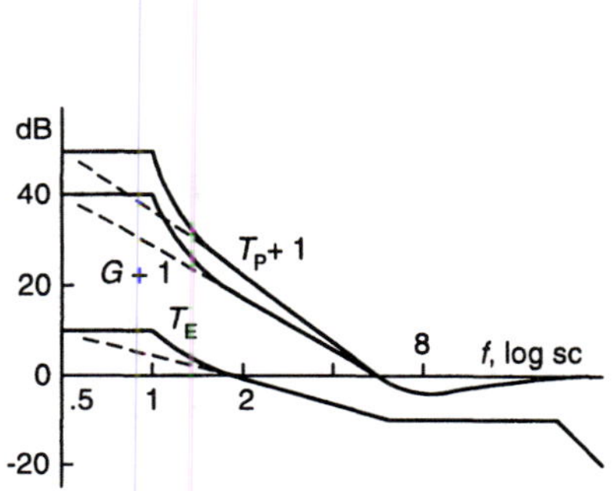

Fig. 10.32 Bode diagrams for the AS system of Example 4

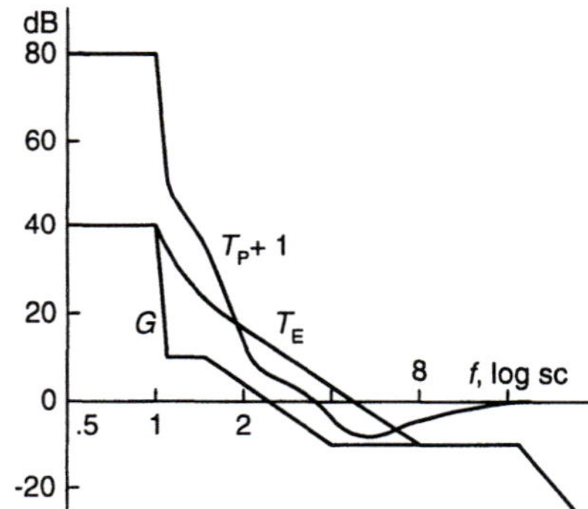

Fig. 10.33 Bode diagrams for the AS system of Example 6

At lower frequencies the responses can be reshaped, for example as shown by the dashed lines. The rational function approximation for the obtained responses can be found as described in Chapter 6.

Example 5. As indicated in Section 10.6, a band-pass transform of the low-pass Bode optimal cut-off with a phase stability margin smaller that 90° does not satisfy the Popov criterion. Application of an NDC resolves this problem. The band-pass transform of the Bode diagrams shown in Fig. 10.32 produces an AS band-pass system with 50 dB of feedback.

Example 6. The Bode plot for T_E displayed in Fig. 10.33 is chosen as the Bode optimal cut-off with 40 dB of feedback and a 30° phase stability margin. The available feedback in the AS system without the NDC is 40 dB.

A Nyquist-stable local loop in the NDC is chosen to reduce the loop crossover frequency in order for the area of positive feedback not to fall upon the area of positive feedback in the T_E loop. The Nyquist-stable plant feedback is 80 dB in the operational frequency range, much larger than 40 dB allowable in the AS system without an NDC.

In Examples 4 – 6, no attention was paid to the closed-loop transient responses. These responses, if desired, can be improved with the command feedforward technique or with additional nonlinear feedback loops.

We can conclude that NDCs improve performance of AS systems.

Another approach to designing feedback systems with nonlinear dynamic compensation will be discussed in the next chapter.

10.8 Problems

1 (a) Make an amplifier equalizing the attenuation of a span of a coaxial cable (the cable attenuation in dB is proportional to the square root of the frequency) using a passive two-port imitating the cable; (b) make an A/D converter from a D/A converter; (c) make a voltage-to-frequency converter from a frequency-to-voltage converter.

2 What is the input-output characteristic of a feedback system with a high-gain forward path and a feedback path where (a) saturation (b) dead zone is placed?

3 Figure out the spectral density plots (Fourier transforms) for the signals in Fig. 10.6.

4 Find the point-to-point correspondence between the limit cycles in Fig. 10.5 and the time responses for x and x'.

5 Draw phase plane limit cycle curves for (a) a triangular symmetrical signal; (b) a triangular signal with different front and back slopes.

6 What is the current-to-voltage characteristic of the output of an amplifier with in-phase voltage feedback, if the amplifier gain coefficient is 20, the feedback voltage divider consists of two 10 kΩ resistors, and the power supply voltage is ±10 V? What external load resistance makes the circuit stable?

7 Is the system with saturation AS if
(a) $T(s) = 100(s + 3)(s + 6)/[(s + 10)(s + 200)s]$;
(b) $T(s) = 100(s + 2)(s + 16)/[(s + 20)(s + 200)s]$;
(c) $T(s) = 300(s + 1)(s + 5)/[(s + 100)(s + 200)s]$;
(d) $T(s) = 300s(s + 1)(s + 5)(s + 5000)/[(s + 100)(s + 120)(s + 130)]$.
(Draw asymptotic Bode diagrams or use computer generated Bode diagrams, and apply the Popov criterion.)

8 Are the systems from the previous problem AS when the saturation link is replaced by a dead zone link? By a three-position relay?

9 A system is AS. Is it globally stable (yes, no, or no judgment can be passed) if the nonlinear link $v(e)$ is replaced by a non-dynamic link described by the following equations: (a) $v = e^2/2$; (b) $e = v^2/2$; (c) $v = e^3/2$; (d) $e = v^3/2$; (e) $v = e - e^2/2$; (f) $e = v - e^2/2$; (g) $e = 5v$; (h) $v = 5e$; (I) $e = v + 0.25v^3$; (j) $e = v + v^{25}$; (k) $e = v/(v - 1)$ over the range $|v| < 1$, no solution outside this range.

10 Draw the characteristic of the nonlinear link obtained by connecting in parallel a unity link and the negative of a three-position relay link.

11 Design an AS system with an NDC, with $f_b = 1$ kHz, using the design prototype in Example 1, Fig. 10.24(a). Make MATLAB or SPICE simulations. Plot the output time responses to step-function input of small and large amplitudes.

12 Design an AS system with an NDC, with $f_b = 140$ Hz, using the design prototype in Example 2, Fig. 10.24(b). Make MATLAB or SPICE simulations. Plot the output time responses to step-function input of small and large amplitudes.

13 Using the block diagram shown in Fig. 10.22, scale down the saturation link for the NDC, when the plant saturation threshold is 20 V, and the maximum output signal of the driver is 2 V.

14 What is the meaning of the word "asymptotic" in the terms: asymptotic Bode diagram, AS, AGS. Can an asymptotic Bode diagram be used in determining AS?

15 Research project:

- Design a Nyquist-stable system for a plant specified by the instructor, with the feedback bandwidth specified by the instructor, with saturation in the actuator. Verify the design with SIMULINK. Apply a large-amplitude vanishing signal to the system's input, and observe the following self-oscillation.
- Introduce a link with gain coefficient 100 in the loop after the compensator.
- Introduce local feedback about this link that makes the main loop response AS-type. Introduce a dead zone into the local feedback path.
- Transform equivalently this system to a single-loop system. Modify the linear links if necessary for the system to satisfy the Popov criterion.
- Study the responses to commands of different shapes and amplitudes. Modify the linear links if necessary to improve the responses.
- Compare the disturbance rejection of the designed system with the disturbance rejection of an AS system without an NDC and with (a) Bode step response or (b) PID response.

16 Research project: Do the design similar to that of the previous problem for a system with a three-position relay in the actuator and an appropriate nonlinear link in the compensator.

17 Research project: Do the same as in Problem 13 for the system with the characteristic of the nonlinear link in the actuator described by the equation $e = -v - 0.25v^3$ and an appropriate nonlinear link in the compensator.

18 Research area: Design of nonlinear prefilters and feedforward paths complementing nonlinear feedback loops.

19 Research area: Design of NDCs with multiple nonlinear links.

20 All links in the system diagrammed in Fig. 10.34 are linear except the two identical links v. Which of the following conditions allows stability analysis with the Popov criterion: (a) $L_1 = L_2$; (b) $L_2 = L_3$; (c) $L_3 = L_4$; (d) $L_2 = L_4$; (e) $L_1 = 2.7L_4$

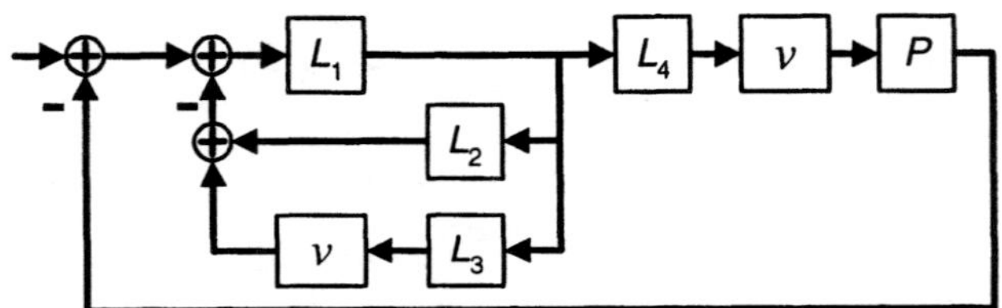

Fig. 10.34 Feedback system

Answers to selected problems

3 The sketches of the spectral density for the oscillations are shown in Fig. 10.35.

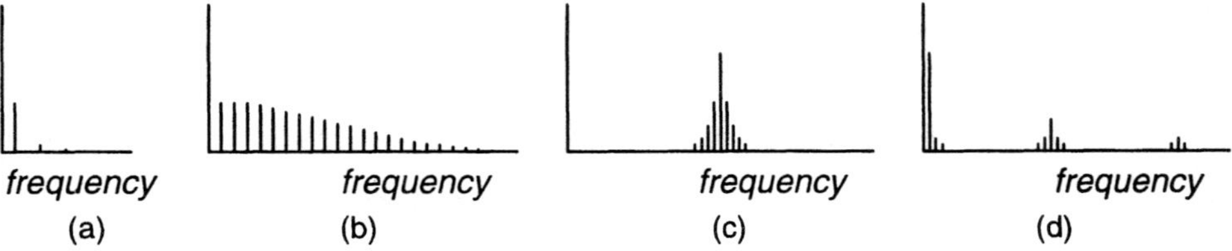

Fig. 10.35 Spectral density plots for oscillation shapes in Fig. 10.6

6 The circuit is depicted in Fig. 10.36(a), and the characteristic is shown in Fig. 10.36(b). The output impedance in the linear mode of operation can be calculated with Blackman's formula as $R = R_o F(\infty) = (1 - 10)20\ \text{k}\Omega = -8\ \text{k}\Omega$. When the voltage exceeds VCC which is 10 V, the amplifier is saturated, and the output resistance is $R_o = 20\ \text{k}\Omega$ (we assume the amplifier output impedance remains infinite in the linear mode of operation and while it is saturated). The system is certainly stable when the load impedance is 0 since this disconnects the feedback. The negative impedance on the falling branch contains a positive real zero.

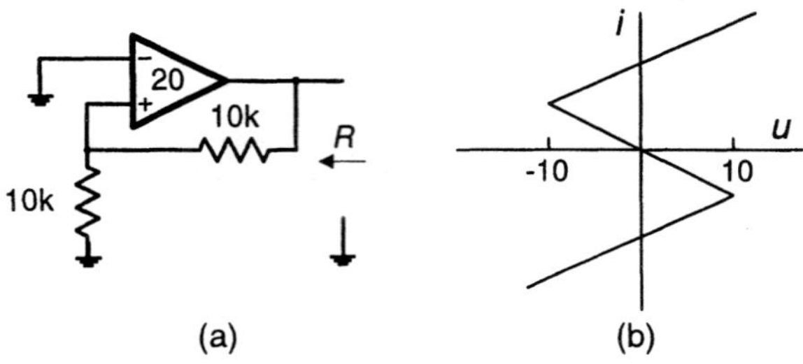

Fig. 10.36 (a) Amplifier with in-phase voltage feedback and (b) its output current-to-voltage characteristic

7 (a) The system is AS, as can be proved with $q = 0.1$, using MATLAB to verify (10.7) or any of the other equivalent conditions.

20 The answer is (c) since in this case the input to both identical nonlinear links is the same and the outputs are, therefore, the same. Therefore, for the purpose of stability analysis, one of the nonlinear link can be removed.

Chapter 11

DESCRIBING FUNCTIONS

Stability analysis based on harmonic balance provides sufficient accuracy when applied to well-designed control feedback loops having low-pass filter properties.

Describing function (DF) stability analysis is simple and convenient. In this chapter, DFs are derived for most common nonlinear links: saturation, dead zone, three-position relay, and hysteresis. Simple approximate formulas are derived and used in the design of nonlinear controllers.

A bang-bang controller with a hysteresis link is described.

Iso-*f* and iso-*E* responses are introduced. The responses characterize the dependence of DF on the signal amplitude and frequency.

Nonlinear dynamic compensators are the links that provide phase advance for large-level signals. A loop is analyzed which includes an NDC and an additional nonlinear link (actuator). Stability margins for such systems are defined.

An NDC can be built of one nonlinear and several linear links. Conditions are discussed for maximizing the phase advance for large-level signals. NDCs are described with parallel nonlinear channels and with nonlinear local feedback paths.

It is known that SISO and MIMO Nyquist-stable systems provide larger feedback thus enhancing the disturbance rejection. However, their stability is conditional when the only nonlinear links in the systems are the actuators' saturation. NDCs eliminate limit cycles in such systems. NDCs also improve transient responses for large command amplitudes and reduce the effects of process instability. For all these purposes, NDCs can be built in the same configurations as the NDCs designed for provision of global stability. NDCs are neither expensive nor complex – an analog NDC can be made with an op-amp and 5 to 10 passive elements, resistors and capacitors. A digital NDC is trivial to program. So, there is little excuse for lowering the system performance by using only linear controllers.

Nonlinear interaction between local and common loops is reviewed, and the reader is warned about practical cases where such interaction can lead to a limit cycle.

The effects of harmonics and intermodulation on the system stability and accuracy are discussed. The chapter ends by describing the procedure for testing whether the system is AGS.

11.1 Harmonic balance

11.1.1 Harmonic balance analysis

The condition for periodic self-oscillation (limit cycle) is unity transmission about the feedback loop, i.e., after passing about the loop, the same signal must return to any initially chosen cross section. Since periodic signals can be presented in Fourier form, each harmonic of the return signal is the same in amplitude and phase as the harmonic of the signal at the beginning of the feedback loop. This condition is called ***harmonic balance***.

Since the harmonics interact in the nonlinear links, the superposition principle cannot be used here. We cannot consider a single harmonic in isolation from the others. For each of the harmonics, the transfer function's equality to 1 should be verified in the presence of all other harmonics. Although generally this procedure could be cumbersome, harmonic balance analysis is simplified when the Fourier series can be justifiably truncated to a few terms. To establish whether this is possible, let us consider a typical feedback system example.

Example 1. The feedback system diagrammed in Fig. 11.1 contains a common type of plant with an integrator and two real poles. The compensator has a real gain coefficient k (the compensator is not optimally designed). The loop phase lag reaches π at frequency 0.5 Hz. A self-oscillation takes place when the loop gain coefficient exceeds 1 at this frequency, which happens when $k > 11$. The self-oscillation initially grows exponentially until, due to the saturation link in the loop, the signal stabilizes with some specific amplitude and shape.

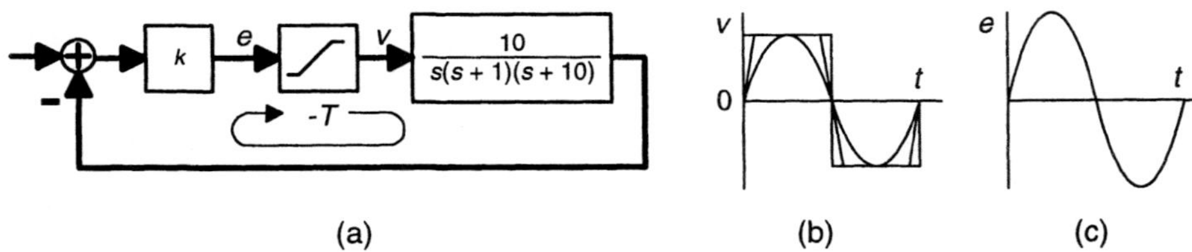

Fig. 11.1 (a) Block diagram of a feedback loop and the shapes of self-oscillation at (b) the output and (c) the input of the saturation link

The shapes of the signal $v(t)$ at the output of the saturation link are illustrated in Fig. 11.1(b) for three particular values of k. When the gain is barely sufficient for the self-oscillation to occur, $v(t)$ is sinusoidal. With larger k, the signal $e(t)$ is clipped in the saturation link and $v(t)$ becomes nearly trapezoidal. When k is large enough to make the loop gain coefficient 20 or more, $v(t)$ becomes nearly Π-shaped.

Because of the saturation symmetry, $v(t)$ is symmetrical and, therefore, contains only odd harmonics. The amplitudes of the harmonics increase when the shape of $v(t)$ approaches rectangular. In this case, the Fourier series for $v(\omega t)$ is

$$v(\omega t) = (4/\pi)\left[\sin \omega t + (1/3)\sin 3\omega t + (1/5) \sin 5\omega t + (1/7)\sin 7\omega t + \ \ldots\right].$$

At the same time, as seen in Fig. 11.1(c), the signal $e(t)$ at the input to the nonlinear link looks smooth. This happens because $v(t)$ is nearly doubly integrated by the loop linear links. As the result, $e(t)$ and its first and second derivatives are continuous. Another explanation is that higher harmonics of $v(t)$ are effectively filtered out by the low-pass properties of the loop's linear links. In fact, $e(t)$ does not differ much from the sinusoid. Because of this, *the cross section at the input to the nonlinear link is, generally, the one simplifying the stability analysis.*

11.1.2 Harmonic balance accuracy

During self-oscillation, the signal $e(t)$ at the input to the nonlinear link is not exactly sinusoidal. The interference of its harmonics in the nonlinear element contributes to the fundamental V of $v(t)$. However, we generally expect the effects of the higher harmonics of $e(t)$ on V to be small because of the following features of feedback control systems' typical nonlinear and linear links:

(a) The nonlinear link characteristic $v(e)$ is such that *small-amplitude harmonics of $e(t)$ have negligible effect on V.* This assertion is fair for common types of control system nonlinear links.

(b) *T(jω) possesses properties of a filter* attenuating the return signal harmonics. This statement is called the ***conjecture of a filter***.

The filter conjecture is typically satisfied for well-designed control systems with relatively smooth response and small n.p.s. If a periodic self-oscillation appears in such a system, then, due to Bode phase-gain relations, the phase condition of the oscillation arg $T(j\omega) = -\pi$ occurs at some frequency where the average steepness of the Bode diagram is −12 dB/oct. Therefore, the third harmonic is attenuated by $12\log_2 3 \approx 20$ dB relative to the fundamental by the linear links of the loop. Since in the Π-shaped $v(t)$ itself, the third harmonic is 10 dB below the fundamental, we conclude that in $e(t)$, the amplitude of the third harmonic is 30 dB (i.e., 30 times) lower than the fundamental. The higher harmonics in $v(t)$ are even smaller, and they are attenuated even more by the loop linear links as indicated in Fig. 11.2.

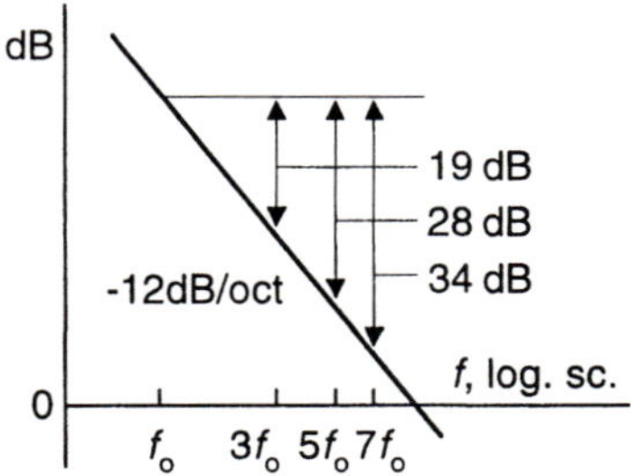

Fig. 11.2 Filtering higher harmonics by loop linear links

With conditions (a) and (b) satisfied, looking at the fundamental only and neglecting the harmonics gives a good estimate for whether the system is stable, and for the value of the stability margins. When, however, the slope of the gain response of the loop links is not steep and the phase condition of oscillation is met due to nonminimum phase lag or due to phase lag in nonlinear links, then harmonic analysis must involve not only the fundamental but also several harmonics of the signal $e(t)$.

11.2 Describing function

Using harmonic stability analysis for nonlinear control systems was suggested almost simultaneously by several scientists – in 1947 by L. Goldfarb in Russia and by A. Tustin in the U.K., and in 1948, by R. Kochenburger in the USA (who introduced the term "describing function") and several other scientists (in Germany and France).

The ratio of the fundamental's complex amplitude V to the amplitude E of the sinusoidal signal applied to the link input is known as the ***describing function*** (***DF***):

$$H(E, j\omega) = V/E \tag{11.1}$$

In general, the describing function can be different at different frequencies, and it is a complex number, describing the signal attenuation and phase shift. The real and imaginary parts of the describing function are defined by Fourier formulas as

$$\mathrm{Re}\, H = \frac{1}{\pi E}\int_0^{2\pi} v \sin\omega t \, d\omega t \tag{11.2}$$

and

$$\mathrm{Im}\, H = \frac{1}{\pi E}\int_0^{2\pi} v \cos\omega t \, d\omega t \,. \tag{11.3}$$

When the characteristic of the nonlinear link is symmetrical, the dc component of its output is zero. When, further, the harmonics at the input to the nonlinear links are negligible compared to the fundamental, the oscillation condition in a feedback system is, approximately,

$$T(j\omega)\, H(E, j\omega) = 1 \tag{11.4}$$

or

$$T(j\omega) = 1/H(E, j\omega)\,. \tag{11.5}$$

(When the characteristic is asymmetric, the dc signal component needs to be taken into account.)

Fig. 11.3 illustrates two ways to do stability analysis. The trajectories formed by the DF $H(E)$ and ***inverse DF*** $1/H(E)$ at a specific frequency and varying E are shown by dashed lines. *The oscillation conditions* (11.4, 11.5) *become satisfied when the DF trajectory passes through the critical point* in Fig. 11.3(a), or, equivalently, *when the inverse DF line* in Fig. 11.3(b) *intersects the Nyquist plot*.

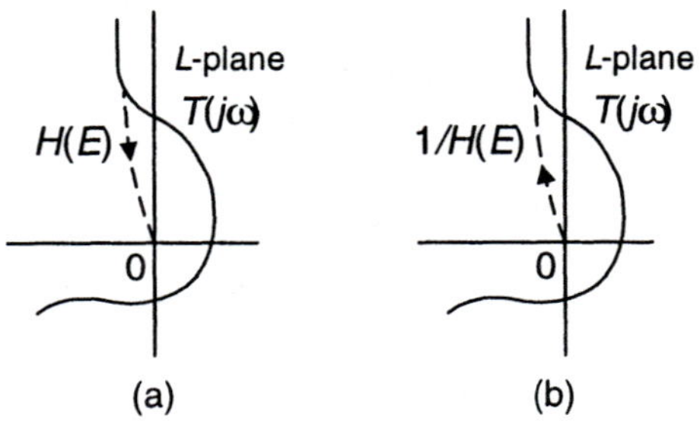

Fig. 11.3 Using (a) DF and (b) inverse DF

When H does not depend on frequency, analysis with inverse DF appears to be easier. However, when the DF is different at different frequencies which is typical for multiloop nonlinear systems and systems with nonlinear dynamic compensation, the direct DF method is more convenient.

DF analysis replaces the nonlinear link by an equivalent linear link with transmission function $H(j\omega)$. The difference is that H depends on E, and generally, the superposition principle and Bode integral relationships do not apply. The integrals can be applied only in a modified form, as shown in Appendix 4.

11.3 Describing functions for symmetrical piece-linear characteristics

11.3.1 Exact expressions

Characteristics of hard saturation, of dead zone, of saturation with a dead zone, and of a three-position relay are shown in Fig. 11.4. The saturation threshold is e_s, and the dead zone is e_d.

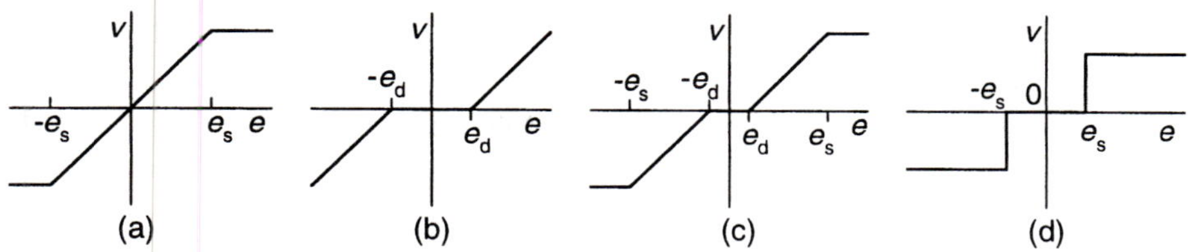

Fig. 11.4 Characteristics of (a) saturation, (b) dead zone, (c) saturation with dead zone, and (d) three-position relay

The characteristics in Fig. 11.4 are symmetrical. Therefore, when the inputs to the links are sinusoidal, the outputs are symmetrical, i.e., do not contain dc components. The output is not shifted in time relative to the input. Hence, the links' DFs are real (the DFs have no phase shift). These DFs do not depend on frequency.

Let us derive the DF for a dead-zone link. The input and the output signals for the dead-zone link are shown in Fig. 11.5. When e is positive, the output is 0 as long as $e < e_d$, i.e., when the angle $\omega t < \arcsin(e_d/E)$. Therefore, the output is $\sin\omega t - e_d/E$ up to the angle $\pi - \arcsin(e_d/E)$.

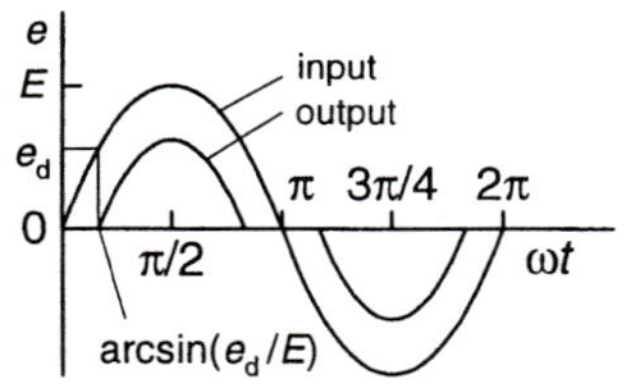

Fig. 11.5 Input and output signals of the dead-zone link

Since the output is not shifted in time relative to the input, the phase of the DF and the imaginary part of the DF are 0. (In other words, since the integrand in (11.3) is an odd function, (11.3) becomes 0.) Therefore, DF can be found with only (11.2). Because the integrand is an even function and the function is symmetrical, we can take the integral from $\arcsin(e_d/E)$ to $\pi/2$ and multiply the result by 4:

$$H = \frac{4}{\pi}\int_{\arcsin(e_d/E)}^{\pi/2}\left(\sin\omega t - \frac{e_d}{E}\right)\sin\omega t\, d\omega t =$$

$$\frac{4}{\pi}\left[\frac{\omega t - \sin\omega t\cos\omega t}{2} + \frac{e_d}{E}\cos\omega t\right]_{\arcsin(e_d/E)}^{\pi/2}$$

or

$$H = \frac{4}{\pi}\left[\frac{\pi}{2} - \frac{1}{2}\arcsin\frac{e_d}{E} + \frac{e_d}{2E}\cos\left(\arcsin\frac{e_d}{E}\right) - \frac{e_d}{E}\cos\left(\arcsin\frac{e_d}{E}\right)\right]$$

i.e.,

$$H = \frac{2}{\pi}\left[\pi - \arcsin\frac{e_d}{E} - \frac{e_d}{E}\sqrt{1 - \left(\frac{e_d}{E}\right)^2}\right].$$

This expression for the dead zone DF is valid for the signal with the amplitude $E > e_d$. When $E < e_d$, the DF is 0.

Any piece-linear characteristic can be obtained by connecting a link k_o in parallel with several links with dead zones $e_{d1}, e_{d2}, e_{d3}, \ldots$ followed by linear links $k_1, k_2, k_3, \ldots$ as with several links with dead zones e_{d1}, e_{d2}, e_{d3}, ... followed by linear links k_1, k_2, k_3, ... as shown in Fig. 11.6; here, a, b, c, ... are certain constants. The DF of the total link can be obtained as the sum of the DFs of the parallel paths, since the integral in (11.2) is a linear function.

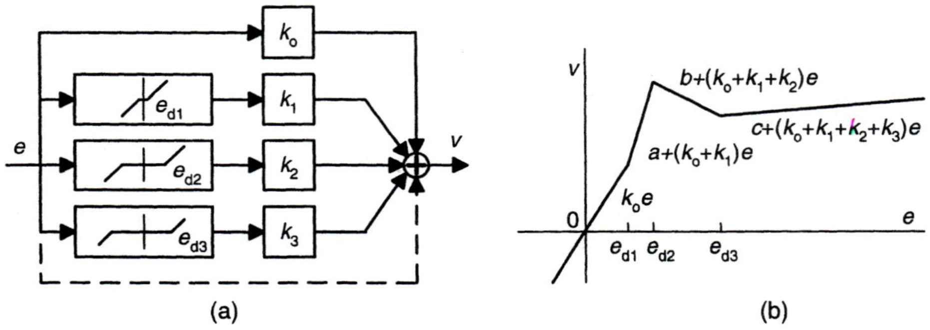

Fig. 11.6 (a) Parallel connection of links with dead zones to implement piece-linear characteristic (b)

For example, the *saturation* link DF is the DF for the link with the dead zone e_s subtracted from 1. Therefore, the saturation DF equals 1 for $E < e_s$, and for $E > e_s$ is given by

$$H = \frac{2}{\pi}\arcsin\frac{e_s}{E} + \frac{e_d}{\pi E}\sqrt{1-\left(\frac{e_s}{E}\right)^2}\ . \tag{11.6}$$

For a ***three-position relay*** actuator ("forward," "stop," "reverse") with the characteristic shown in Fig. 11.4(d), the output $v(t)$ is two-polarity pulses, shown in Fig. 11.7.

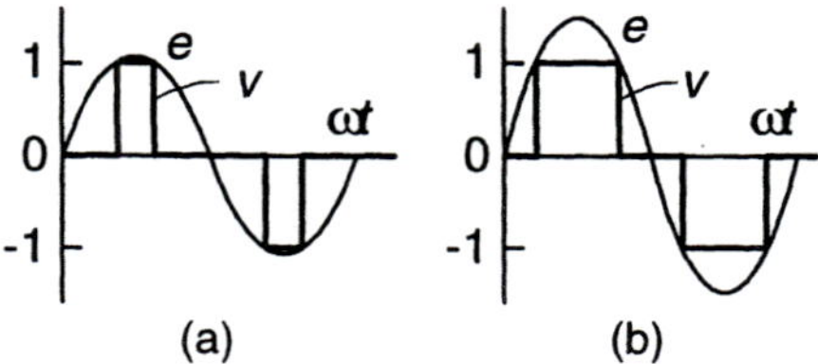

Fig. 11.7 Output signal of a three-position relay for (a) $E = 1.05$ and (b) $E = 1.5$

The amplitude of the pulses is 1. Therefore, from (11.2) the describing function is

$$H = \frac{4}{\pi}\int_{\arcsin e_d/E}^{\pi/2} \sin\omega t\, d\omega t = \frac{4}{\pi}\cos\left(\arcsin\frac{e_d}{E}\right),$$

or

$$H = \frac{4e_d}{\pi E}\sqrt{1-\left(\frac{e_d}{E}\right)^2}. \tag{11.7}$$

When E/e_d is large, the second component under the square root can be neglected and $H \approx (4/\pi)e_d/E$.

Describing functions for saturation, dead zone, and three-position relay are plotted in Fig. 11.8.

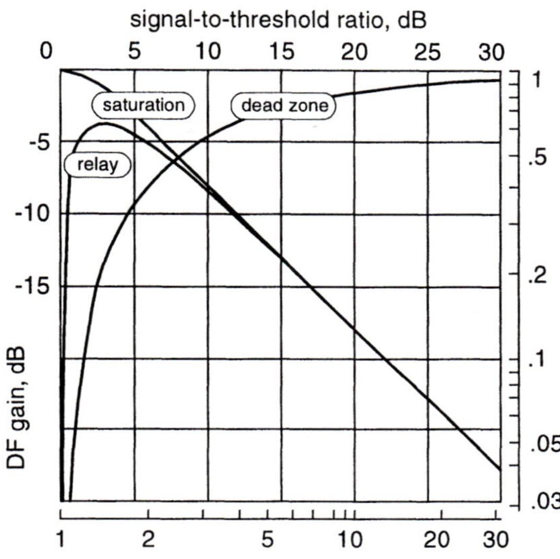

Fig. 11.8 Describing function characteristic for saturation, dead zone, and three-position relay

When E is only slightly larger than 1, then the pulses are short, the fundamental of $v(t)$ is small, the loop gain for the fundamental is small, and oscillation cannot take place. The oscillation is more likely to take place in practical systems when DF is close to the maximum. As seen in Fig. 11.8, the relay DF is maximum when $E \approx 1.5$. In this case, the pulses are rather wide as shown in Fig. 11.7(b), the harmonics in $v(t)$ are therefore relatively small, and the DF analysis is sufficiently accurate.

Example 1. The stability analysis of a system with a three-position relay and the loop transfer function

$$T = \frac{50{,}000(s+500)}{s(s+20)(s+100)^2}$$

is shown in Fig. 11.9. The Nyquist diagram and the inverse DF plot are shown in Fig. 11.9(a). The direct DF analysis is presented in Fig. 11.9(b). The frequency at which the return ratio phase lag is 180° is 5.1 Hz. The loop gain of the linear links at this frequency is 5.4 dB. From the diagram in Fig. 11.8, for the DF of –5.4 dB, the signal-to-threshold ratio is 2, i.e., 6 dB. The conclusions of the inverse and the direct DF analyses are, certainly, the same: the system can oscillate with the signal amplitude about twice

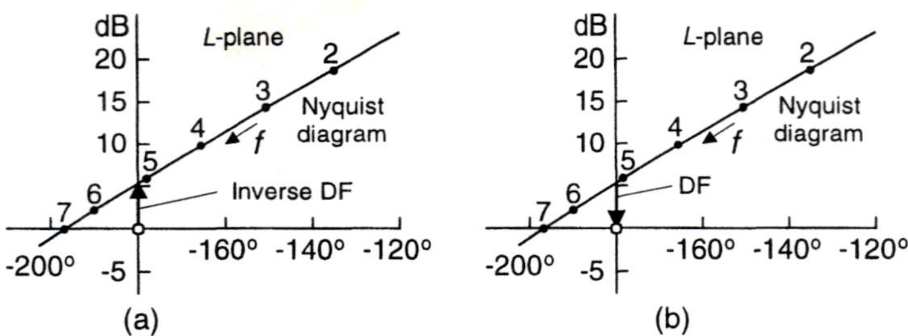

Fig. 11.9 Nyquist diagram and the (a) inverse and (b) direct DF plots for a system with a three-position relay

11.3.2 Approximate formulas

Expression (11.6) for *saturation* DF is conveniently approximated by

$$H \approx (4/\pi)(E/e_s)^{-1} - (4/\pi - 1)0.27(E/e_s)^{-4}$$

or

$$H \approx 1.27(E/e_s)^{-1} - 0.27(E/e_s)^{-4} \tag{11.8}$$

with the error smaller than 0.1 dB. Calculations can be further simplified by omitting the second term in (11.8) which rapidly vanishes for large E. It contributes less than 2 dB; 0.6 dB; 0.35 dB correspondingly for E larger than, respectively, e_s; $1.5e_s$; $2e_s$.

The link with the *dead zone* characteristic of Fig. 11.4(b) can be replaced by a parallel connection of a unity link and an inverting saturation link. Hence, the DF for the dead zone link for $E > e_d$ is

$$H \approx 1 - 1.27(E/e_d)^{-1} + 0.27(E/e_d)^{-4}. \tag{11.9}$$

A nonlinear link with characteristic including both *dead zone and saturation* is shown in Fig. 11.4(c). It can be represented by parallel connection of a saturation link with the threshold e_s and an inverting saturation link with the threshold e_d. Then, for $E \in [e_d, e_s]$, H is as in (11.9), and for $E > e_s$

$$H \approx 1.27[E/(e_s - e_d)]^{-1} - 0.27[(E/e_s)^{-4} - (E/e_d)^{-4}]$$

or

$$H \approx \{1.27/[1 - (e_d/e_s)]\}(E/e_s)^{-1} - \{0.27/[1 - (e_d/e_s)^4]\}(E/e_s)^{-4}$$

A *saturation link with frequency-dependent threshold* can be made by cascading a linear link L, a saturation link, and a link $1/L$ as shown in Fig. 11.10. The threshold is $e_s|L(j\omega)|$. For example, if $L(s)$ approximates s over the frequency band of interest, this is a ***rate limiter***. In a similar manner, a link in which the width of the dead zone is frequency-dependent can be made.

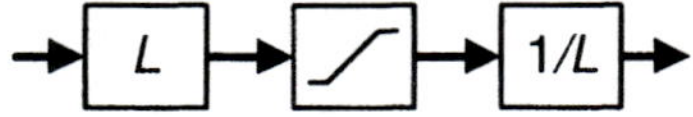

Fig. 11.10 Nonlinear link with frequency-dependent threshold

11.4 Hysteresis

Fig. 11.11(a) shows the output/input characteristic of smooth saturation with hysteresis.

11.4 Hysteresis

Fig. 11.11(a) shows the output/input characteristic of smooth saturation with hysteresis. The output $v(t)$ in Fig. 11.11(b) is found by using branch v_1 while the input $e(t)$ is rising, and branch v_2 while $e(t)$ is decreasing. The time delay of the output relative to the input indicates that the DF must have a negative imaginary component.

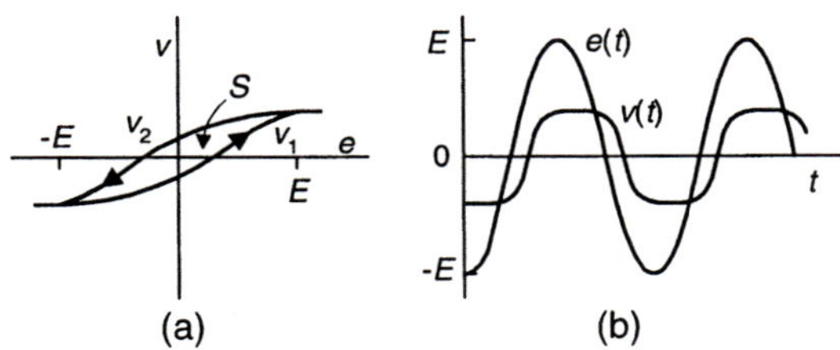

Fig. 11.11 (a) Output/input characteristic of smooth saturation with hysteresis and (b) time-history of the input and the output

The DF phase shift is

$$\arg H = \arcsin\frac{\mathrm{Im}H}{H}. \tag{11.10}$$

From (11.3),

$$\mathrm{Im}H = \frac{1}{\pi E}\int_0^{2\pi} v\, d\sin\omega t.$$

After replacing $d\sin\omega t$ by $E^{-1}de(t)$ and taking the integral of the output from $-E$ to E using $v_1(e)$, and back to $-E$ using $v_2(e)$, we have

$$\mathrm{Im}H = \frac{1}{\pi E^2}\int_{-E}^{E} v_1\, de + \frac{1}{\pi E^2}\int_{E}^{-E} v_2\, de = \frac{1}{\pi E^2}\int_{-E}^{E}(v_1 - v_2)\, de = -\frac{S}{\pi E^2} \tag{11.11}$$

where S stands for the area within the hysteresis loop. By substituting (11.11) and $H = V/E$ into the right side of (11.10), we have

$$\arg H = -\arcsin\frac{S}{\pi EV}. \tag{11.12}$$

Example 1. In Fig. 11.11, $V \approx 0.7E$ and $S \approx 0.3$. Therefore,

$$\arg H \approx -\arcsin\frac{0.3}{0.7\pi} \approx 0.14\mathrm{rad}.$$

Example 2. The Schmitt trigger has a rectangular hysteresis characteristic, shown in Fig. 11.12(a). The characteristic for the ***backlash*** shown in Fig. 11.12(b) is typically caused by air gaps in gears. For the backlash, the area of the hysteresis depends on the signal amplitude. The time-history for the backlash link output when the input signal is sinusoidal is shown in Fig. 11.13.

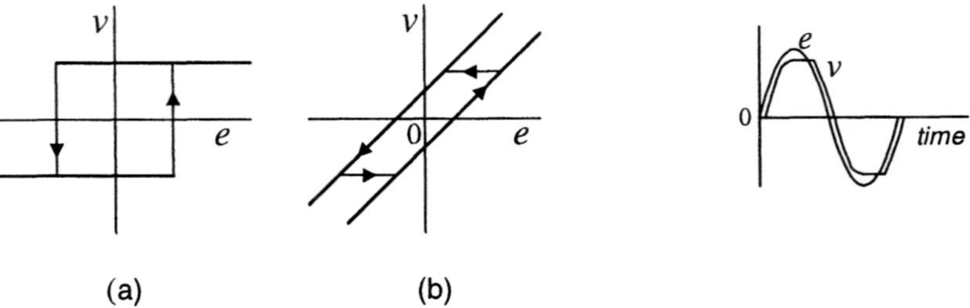

Fig. 11.12 Characteristics of (a) Schmitt trigger and (b) backlash

Fig. 11.13 Time response of the backlash link output

Example 3. If the width of the backlash is 2, and $e = 20 \sin\omega t$, then the output signal amplitude is nearly the same as that of the input, and the phase lag is $\arcsin [2\times20/(400\pi)] = 0.064$ rad, i.e., 3.6°.

Because the phase lag reduces the available feedback, efforts are always made to eliminate or decrease backlash in gears and machinery. Reducing the backlash is important even for manually operated tools and equipment (like lathes) because of the man-machine feedback loop via the operator's tactile and visual sensors and the operator's brain (compensator). The backlash necessitates slowing the actions of the operator, or else the overshoots in the feedback system become large or the system becomes unstable.

Example 4. Hysteresis links are frequently used in oscillating feedback loops of which the saw-tooth signal generator shown in Fig. 11.14 is representative. The feedback loop is composed of an inverting integrator with transfer function $-1/(R_3Cs)$ and a non-inverting Schmitt trigger. The input thresholds of the Schmitt trigger shown by dashed lines in Fig. 11.14(c) are $u_{th} = \pm [R_1/(R_1 + R_2)]VCC$, where VCC is the power supply voltage. The integrator output is constrained within the ***dead beat*** (or ***dead band***) $[-u_{th}, u_{th}]$.

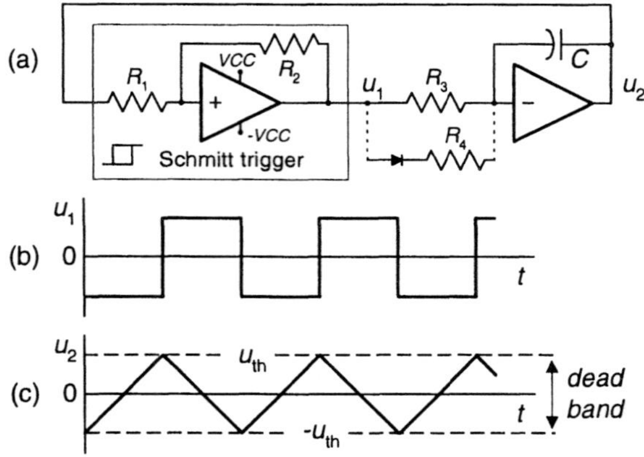

Fig. 11.14 (a) Saw-tooth signal generator and signal histories at (b) the Schmitt-trigger output and (c) the integrator output

When the output voltage of the integrator arrives at u_{th}, the trigger output switches from $-VCC$ to VCC. After the switching, the integrator output begins decreasing with constant rate $VCC/(R_3C)$ (V/sec) until the next switching occurs. The period is therefore

$$T = 4(u_{th}/VCC)R_3C = 4R_1R_3C/(R_1 + R_2).$$

The output signals' rates of rising and falling are the same in Fig. 11.14(b). They can be made different by adding in parallel to R_3 an additional resistor R_4 in series with a diode, so that in one direction the resistance will be smaller and the integrator gain coefficient larger, thus increasing the rate.

It is seen in Fig. 11.14 that the linear link and the nonlinear link each lags the signal by 90° so that the signal comes back in phase after passing about the loop. It is also seen that *during the analysis of such a system the harmonics cannot be neglected, i.e., DF analysis cannot be used*: if the sawtooth signal at the input to the Schmitt trigger is replaced by its fundamental, the phase lag in the Schmitt trigger will be less than 90° and the condition of oscillation will be not satisfied.

Example 5. An ***on-off*** (or ***bang-bang***) oscillating temperature controller is shown in Fig. 11.15(a). The actuator is a ***two-position relay with hysteresis*** which can be implemented either as an electromechanical device or electronically, for example, employing a Schmitt trigger. The temperature of the single integrator plant oscillates within the dead band. The positive temperature rate depends on the power supplied by the heater, and the negative rate (the cooling rate) depends on the heat transfer and radiation conditions. The narrower the dead band is, the higher is the control accuracy but also the higher is the oscillation frequency. The oscillation frequency cannot be chosen to be excessively high since each switching in the physical actuators consumes some energy and/or wears out the contact mechanisms.

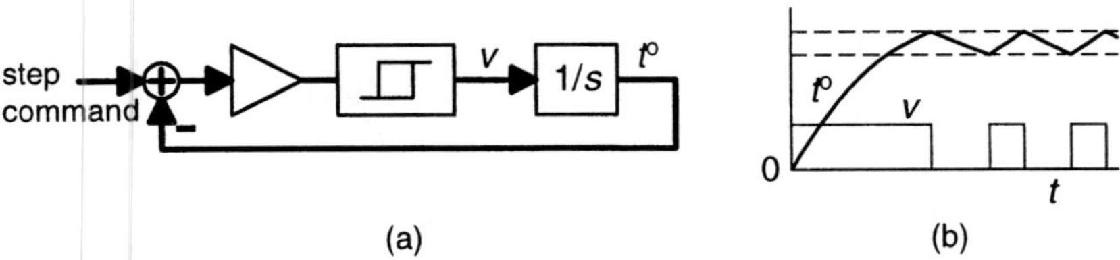

Fig. 11.15 (a) Bang-bang temperature controller and (b) its output time-response

When the plant in this feedback system contains extra high-frequency poles, the control law is often augmented by additional rules. For example, switching can be done not in the instant the output approaches the end of the dead band, but somewhat earlier in time or position, to counteract additional plant inertia. The accuracy of such an on-off controller, although sufficient for many applications, is typically inferior to the well-designed controllers using pulse width modulated (PWM) drivers. When the actuator is electrical, switching on and off with sufficiently high frequency does not present a problem, and using PWM is common. However, when the switching cannot be fast and causes noticeable power losses (or propellant losses, as when switching gas thrusters on and off, since the thrust of such thrusters does not develop instantly), bang-bang control may be preferred.

11.5 Nonlinear links yielding phase advance for large-amplitude signals

As mentioned before, DF is a function of E and ω (or f). The DF ***iso-f lines*** (or ***f-lines***) are shown in Fig. 11.16 by the dashed lines. The ***iso-E lines*** (or ***E-lines***) are shown by the solid lines. Either set of lines can be used for the stability analysis. If no line passes over the critical point, the system is considered stable. For the iso-lines shown in Fig. 11.16, arg DF increases with E. The system is stable because the loop includes a link that provides a certain phase advance for large-amplitude signals. We will call such links ***nonlinear dynamic compensators*** (***NDCs***). The AS analysis and design of the NDCs has been already introduced in Chapter 10. In this chapter we will employ DF methods. These methods allow designing certain NDCs which cannot be analyzed or designed with AS methodology.

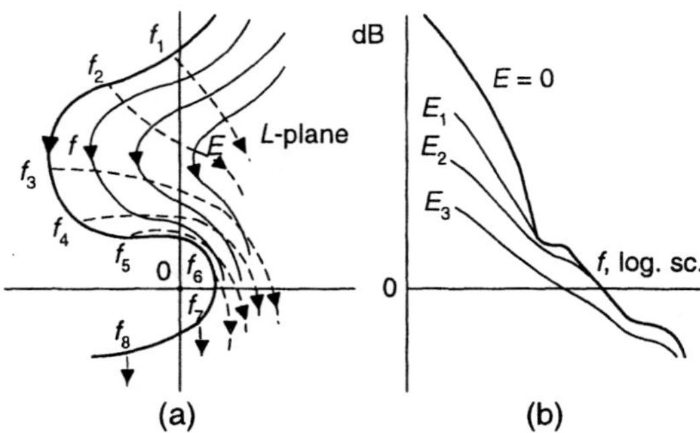

Fig. 11.16 (a) iso-f (solid) lines and iso-E (dashed) lines, (b) iso-E Bode diagrams

From the iso-E lines, ***iso-E Bode diagrams*** can be drawn as shown in Fig. 11.16(b) (MATLAB functions described in Appendix 14 can be used to plot the iso-E diagrams for some typical NDCs). We should keep in mind, however, that these Bode diagrams do not uniquely define the phase shift. The Bode relations can only be used when constant E causes the DF of nonlinear links to be constant at all frequencies so that the DF can be equivalently replaced by a constant-gain linear link.

Example 1. Fig. 11.17(a) shows a simple proportional-integral (PI) NDC using parallel connection of a linear link with a nonlinear link. Because of the saturation, the upper channel contributes relatively less to the output when E increases. The compensator iso-E Bode diagrams are shown in Fig. 11.17(b). Consequently, the NDC lag decreases as E increases. An iso-E Bode diagram of the loop gain of a feedback system with such an NDC is shown in Fig. 11.17(c).

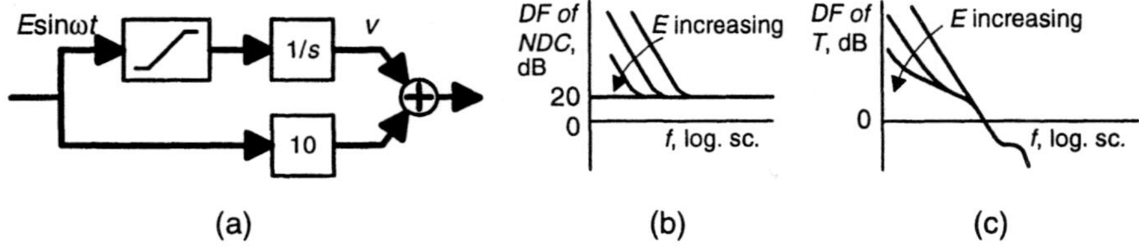

Fig. 11.17 (a) NDC with parallel channels, and its (b) iso-E Bode diagrams and (c) the loop iso-E Bode diagrams; phase shift can be calculated from these diagrams

This example shows that an effective NDC can be implemented rather simply, and that using iso-E Bode diagrams is convenient for certain practical classes of NDCs.

Before concentrating on the NDC design, we first consider the NDC performance in the loop which also includes another nonlinear link, the actuator.

11.6 Two nonlinear links in the feedback loop

A feedback system including an NDC and a nonlinear actuator is shown in Fig. 11.18. The NDC and the actuator are separated by some linear link L. For the purpose of stability analysis the loop can be cross-sectioned at the input to either of the nonlinear links as shown in the figure. The loop return ratio DF can be measured by applying some test-signal either to the input to the NDC, with some amplitude E_1, or to the input of the actuator with some amplitude E_2.

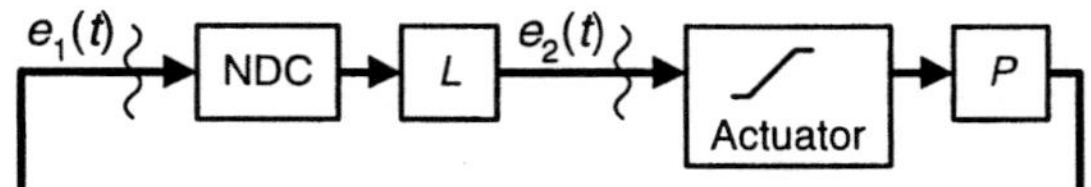

Fig. 11.18 Feedback loop with two nonlinear links separated by linear links

Although different, the iso-f lines for E_1 and E_2 have a common point when $|T| = 1$, as shown in Fig. 11.19. This can be seen by considering that the condition $|T| = 1$ is satisfied with some specific values of $E_1 = E_{1C}$ and $E_2 = E_{2C}$. Then, if the loop is broken at the input to the NDC and E_{1C} is applied to the input of the loop, the value E_{1C} returns to the cross-section, and the value E_{2C} appears at the input to the actuator. Also, when the loop is broken at the input to the actuator and E_{2C} is applied to the input of the loop, the value E_{2C} returns to this cross-section, and the value E_{1C} appears at the input of the NDC. In both cases, the same signals appear at the inputs of both nonlinear links, and therefore, the loop DF is the same and the loop phase shift is the same. Hence, these two iso-E lines intersect when $|T| = 1$.

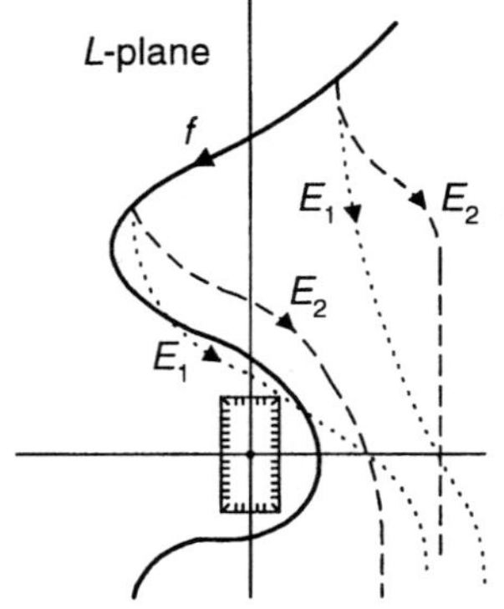

Fig. 11.19 Iso-f lines in a system with two nonlinear links

The nominal guard-point phase stability margin, therefore, does not depend on the position of the cross section. However, some confusion can be seen with the gain stability margin: this stability margin seemingly depends on the cross-section chosen for examining the loop. The proper cross section is that at the input to the NDC, since in this case, uncertainty in the plant gain directly affects the loop return ratio: when plant gain reduces by a dB, then the Nyquist diagram and the iso-f lines simply sink down by a dB. Therefore, in Fig. 11.19, the stability margin is just exactly satisfied.

The accuracy of DF analysis suffers when the loop incorporates two nonlinear links since in this case the linear links separating them, although, commonly, still low-pass filters, are not as frequency selective as the linear link in the loop with a single nonlinear link. Nonetheless, when the feedback in the loop is large and the Bode diagrams of the linear links are reasonably steep (they are quite steep when the system is Nyquist stable), the phase uncertainty caused by the harmonics typically does not exceed ±20°. This range seems large: generally, such uncertainty can make the entire difference between an unstable and a high performance system. However, using DF enables an easy design of a simple NDC providing phase advance of more than 120°. Therefore, even with the ±20° phase error, the NDC gives the phase advance no less than 100° – which is certainly better than no phase advance at all, and is adequate for most practical applications.

11.7 NDC with a single nonlinear nondynamic link

For an NDC composed of several linear links and a single nonlinear non-dynamic link with DF w, the normalized NDC transfer DF depends on w as a ratio of two linear functions, i.e., ***bilinearly***:

$$H(w) = \frac{w+M}{w+N} \tag{11.13}$$

where M and N are some functions of s.

When the signal amplitude changes, w changes and $H(w)$ changes. When w changes from 0 to ∞, H changes from M/N to 1 and therefore $\arg H$ changes by $\arg(M/N)$. To maximize the phase advance produced by the NDC, $\arg N$ and $-\arg M$ must be made as large as possible. This angle is, however, bounded by the continuity considerations.

During the gradual change in the signal amplitude and in w, the transitions in H are desired to be monotonic and smooth. The most critical values of w, from this point of view, are those equal in modulus to either $|M|$ or $|N|$, as shown in Fig. 11.20. For these values of w, if the angle of M (or, correspondingly, N) is π, the vector $w + M$ (or, correspondingly, $w + N$) vanishes, and in the neighborhood of this point, H becomes excessively sensitive to w. To avoid this situation with a $\pi/6$ ***safety margin***, neither $|\arg M|$ nor $|\arg N|$ should be allowed to exceed $2\pi/3$. This requirement limits the phase shift of the NDC to $4\pi/3$, i.e., 240°, which is more than enough for all practical purposes.

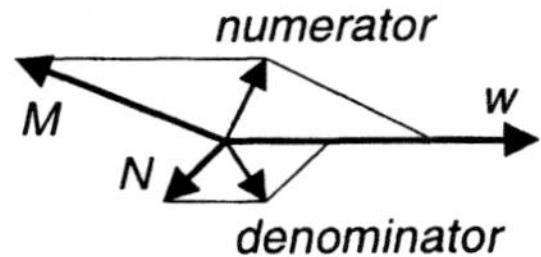

Fig. 11.20 Vectors ($w + M$) and ($w + N$) with minimum moduli

A flowchart implementation of (11.13) is shown in Fig. 11.21(a). (Notice that when w is very large, the error signal at the beginning of the upper branch decreases and the branch output signal becomes negligible.) Fig. 11.21(b) exemplifies the case where w is a high-gain amplifier with the gain coefficient k and a dead zone, and $M = 1/s$, $N = s$. This NDC reduces the phase delay by 180° for large-amplitude signals. Fig. 11.21(c) exemplifies the case when a saturation link is used, with $M = s$, $N = 1/s$ over the frequency range of interest, resulting in a 180° phase advance for large-amplitude signals.

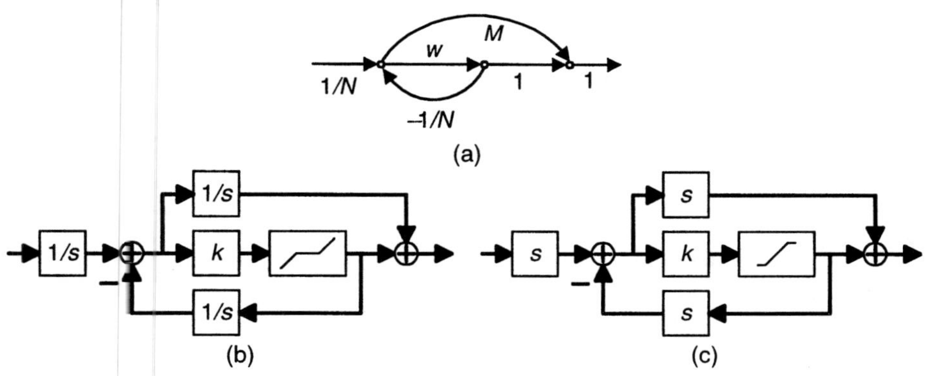

Fig. 11.21 NDC flowchart (a) and block diagrams (b), (c)

To shift the phase over the full available range of 240°, the NDC includes a forward path with w, a path in parallel with the forward path, and a feedback path. When the NDC includes only a parallel path or only a feedback path, i.e., when either M or N is zero, the available phase shift change must not exceed 120° – which still suffices for most applications.

In addition to iso-f and iso-E responses, ***iso-w Bode diagrams*** can be used. These are the frequency responses measured while maintaining constant the value of the signal amplitude at the input to the nonlinear element. Such a response can be calculated after replacing the nonlinear element with a constant linear element. The iso-f and iso-E responses can be further calculated using the set of iso-w responses, DF for the nonlinear element, and the signal amplitude at the nonlinear element.

Example 1. In the NDC shown in Fig. 11.21(b), $k = 10$. Calculate and plot the iso-w Bode diagrams, for three values of $k\times$(DF of the dead zone link): 0.1, 1, and 10. The SPICE simulation input file is shown below, and uses the schematic diagram in Fig. 11.22. The iso-w Bode diagrams are shown in Fig. 11.23.

```
**** ch9ex1.cir for iso-w simulation of NDC Figs. 11.21(b),
**** 11.22
*** input integrator
G2 2 0 0 1 1
C2 2 0 1
R2 2 0 1MEG
*** feedback summer
G3 3 0 7 2 1
R3 3 0 1
***  kDF path:  G5 =.1, 1, or 10
G5 5 0 0 3 10
R5 5 0 1
*** forward path, inverting
G4 4 0 3 0 1
C4 4 0 1
R4 4 0 1MEG
*** forward summer, the output is VDB(6)
G6 6 0 4 5 1
R6 6 0 1
*** feedback path
```

```
G7 7 0 0 5 1
C7 7 0 1
R7 7 0 1MEG
***
VIN 1 0 AC 1
RIN 1 0 1MEG
 .AC DEC 20 .001
10
 .PROBE
 .END
```

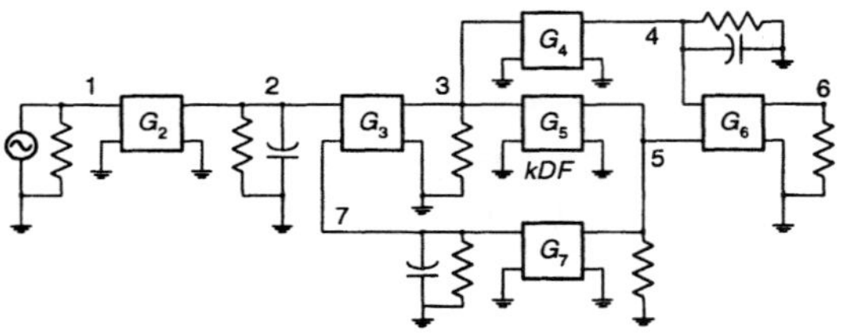

Fig. 11.22 SPICE model for iso-*w* response

With the increase of *w*, i.e., with the increase in the signal at the input to the NDC, the plot gradually changes from that of a double integrator to a constant gain response. Correspondingly, the phase lag decreases from π to 0.

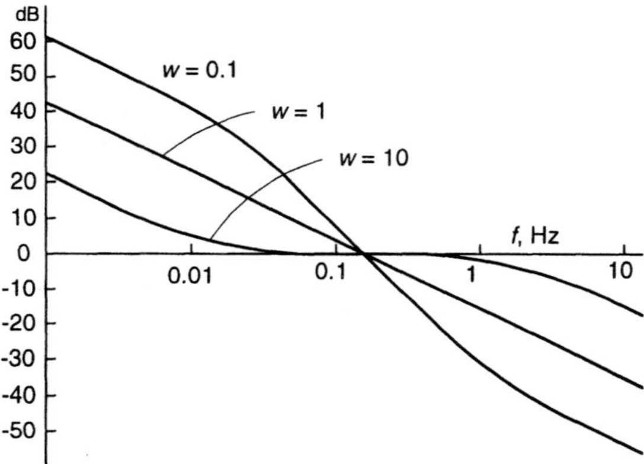

Fig. 11.23 Iso-*w* Bode diagrams

When the system is linear and the loop gain is large, the signal amplitude after the summer is nearly constant at all frequencies where the loop gain is large, since the input block is an integrator, and also, the feedback decreases with frequency.

11.8 NDC with parallel channels

Simple examples of NDCs with parallel paths are the PID controller with saturation in the *I*-channel (or in the low-frequency channel, as described in Chapter 6) and the *PI* controller shown in Fig. 11.17.

The feedback system in Fig. 11.24(a) includes an actuator with saturation and an NDC with two parallel channels. Nonlinear non-dynamic links are placed here in both channels. The first channel starts with a saturation link with unity threshold, and the second, with unity dead zone. When the signal amplitude is low, the loop transfer function is $T_{01} = C_1AP$. The second channel is off when the input signal amplitude is less than 1, and takes over when the first channel becomes overloaded. In this case $T_{02} = C_2AP$. The NDC can be equivalently implemented with a saturation link as shown in Fig. 11.24(b).

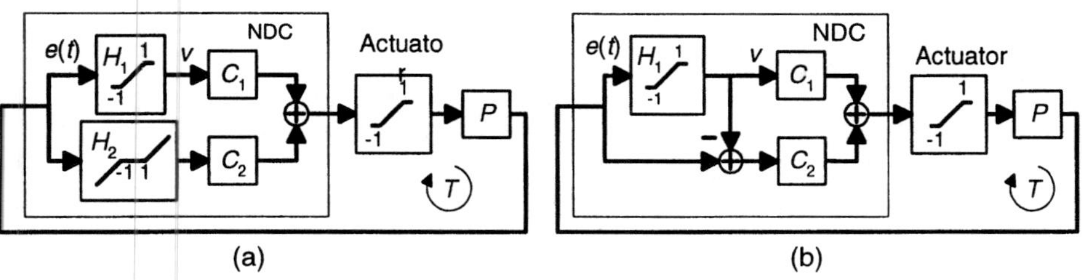

Fig. 11.24 (a) Feedback system with two nonlinear elements in the NDC, and (b) an equivalent diagram with a single nonlinear element in the NDC

Three more equivalent block diagrams for the same NDC are shown in Fig. 11.25. The versions with saturation links shown in Fig. 11.24(b) and in Fig. 11.25(a) are typically easier to implement.

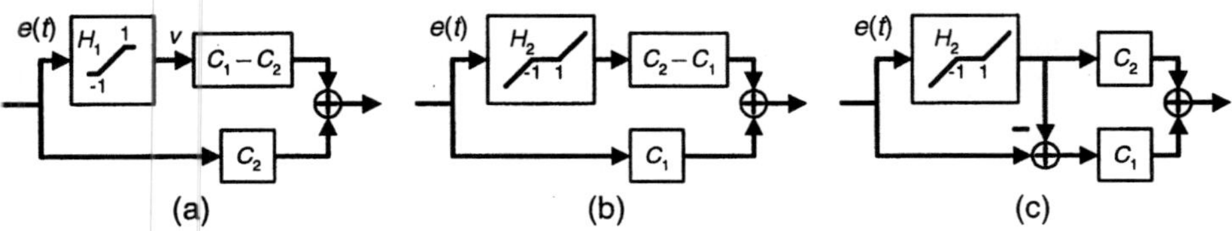

Fig. 11.25 Three more NDC configurations

The analysis and iterative design of the compensator can be performed by calculating and plotting iso-E Bode plots, or by plotting f-lines on the T-plane as exemplified in Fig. 11.26.

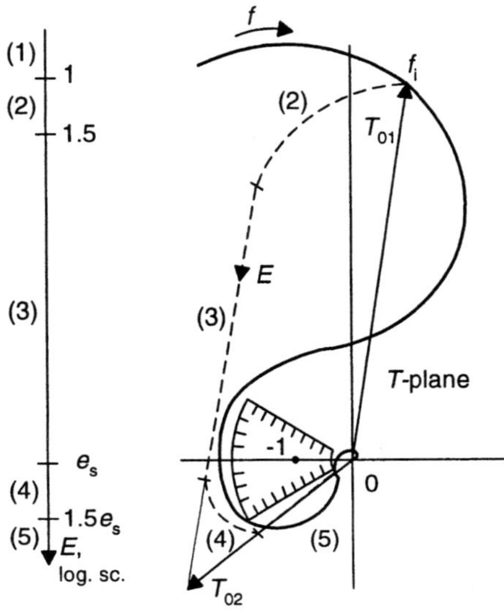

Fig. 11.26 Intervals on E-axis and on the iso-f line for a system with two parallel channels

An iso-f line for

$$T = T_{01} H_1(E) + T_{02} H_2(E) \tag{11.14}$$

is displayed in Fig. 11.26. We can examine it piece by piece using simplified formulas (11.7) and (11.8) for the intervals of E shown in the left of the figure.

On the *first interval*, $E < 1$ and $T = T_{01}$.

The *second interval*, where $E \in [1, 1.5]$, is comparatively short. Over this interval, H_1 reduces from 1 to 0.8, and H_2 increases from 0 to 0.25 as can be verified with the plots in Fig. 11.7. This segment of the iso-f line is curvilinear, and its exact shape is not important for the stability analysis.

On the *third interval* where $E \in [1.5, e_s]$ the expression (11.13) reduces to

$$T = T_{01} 1.27/E + T_{02} - 1.27 T_{02}/E = T_{01} + T_{02}(1 - 1.27/E) \tag{11.15}$$

This piece of the iso-f line presents a segment of a straight line aimed at the end of the vector T_{02}. In the vicinity of the stability margin boundary, the iso-f line can be approximated by the side of the parallelogram shown in Fig. 11.26.

On the *fourth interval* $E \in [e_s, 1.5e_s]$. This section is curvilinear, short, and does not deserve detailed analysis.

On the *fifth interval* $E > 1.5e_s$ and

$$T = 1.27 T_{01}/E + 1.27 T_{02}/E = 1.27(T_{01} + T_{02})/E \tag{11.16}$$

so that the iso-f line is straight and directed toward the origin. From the plot Fig. 11.24 we can conclude that at this frequency, limit cycling is avoided with sufficient stability margins.

Example 1. Fig. 11.27 shows the L-plane Nyquist diagram and iso-f lines measured in an experimental system designed with the method presented in this section. The system has no limit cycles.

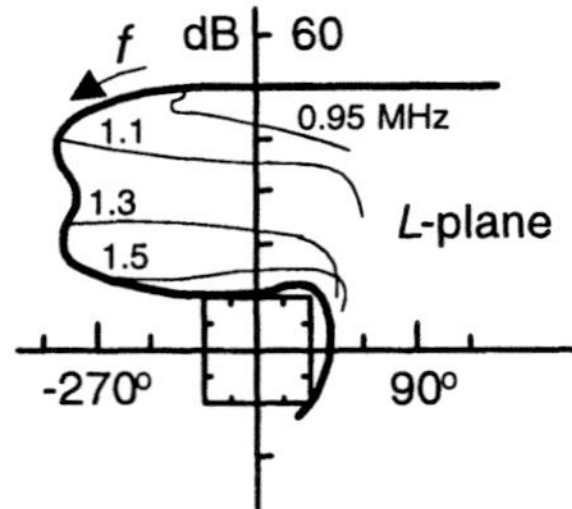

Fig. 11.27 Nyquist diagram and iso-f lines for an experimental system

11.9 NDC made with local feedback

In many practical links the DFs modulus is the same as that for saturation, but the phase is nonzero and changes with the signal level. We will call such links ***dynamic saturation*** links. A typical case is presented in Fig. 11.28(a). The phase shift in the nonlinear dynamic link varies with the signal level, while the output signal amplitude and, therefore, the magnitude of the DF $|H|$ of the composite link, do not vary after the saturation threshold is exceeded.

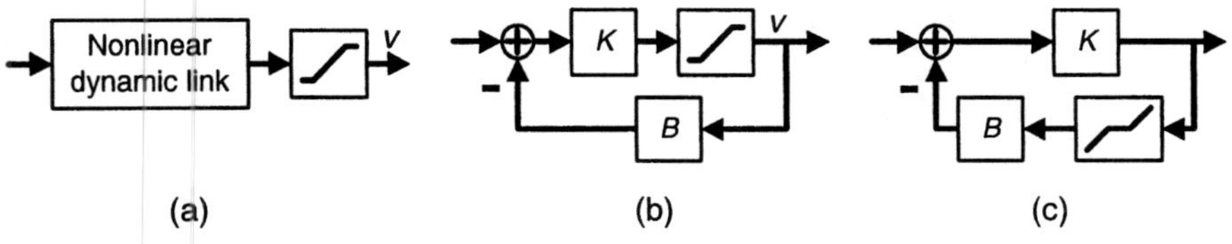

Fig. 11.28 Dynamic saturation

Fig. 11.28(b) gives another implementation of dynamic saturation. Here, $|K(j\omega)|$ is large although decreasing with frequency, and B is real. When the signal is small, the closed-loop transfer function is $-1/B$. When the signal is large, the describing function of the forward path decreases, the feedback becomes negligible, and the phase lag becomes that of $K(j\omega)$.That is, the phase lag increases with the signal level.

Still another version is shown in Fig. 11.28(c) using similar links $K(j\omega)$ and B. For small signal amplitudes, the link transfer function is $K(j\omega)$. As the signal amplitude gets much larger than the dead zone, the dead zone DF approaches 1, large feedback is introduced, and the output signal is limited as if by saturation. The phase shift of the link will then be determined by the feedback path. That is, the phase lag reduces with the signal level. This link is especially suitable for nonlinear dynamic compensation.

When the linear link loop gain is high, and the dead zone link DF w changes from 0 to 1, the DF of the link changes from $-1/B$ to K, i.e., the phase shift changes by $\arg(KB)$, i.e., by the angle of the local loop phase shift. As indicated in Section 11.7, this value must be limited to 120°. This value is sufficient for most applications.

Example 1. Consider an NDC with dead zone in the local feedback path as shown in Fig. 11.29, with $B = s^{-4/3}$ (i.e., the Bode diagram for B has constant slope -8 dB/oct, and the phase shift is $-120°$), and $K = 200$. Since V and the dead zone link DF w are uniquely related and this relation does not depend on frequency, the full set of iso-V Bode diagrams is the same as the full set of iso-w diagrams, and either one can be employed for the stability analysis.

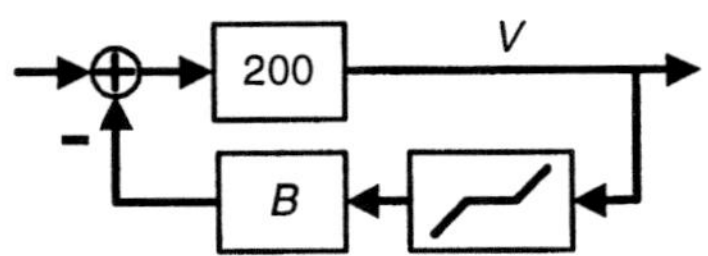

Fig. 11.29 Example of an NDC with dead zone in the feedback path

The iso-V Bode diagrams for the NDC are shown in Fig. 11.30(a), and the iso-V lines of the main loop on the L-plane might look as shown in Fig. 11.30(b). The system has no limit cycle.

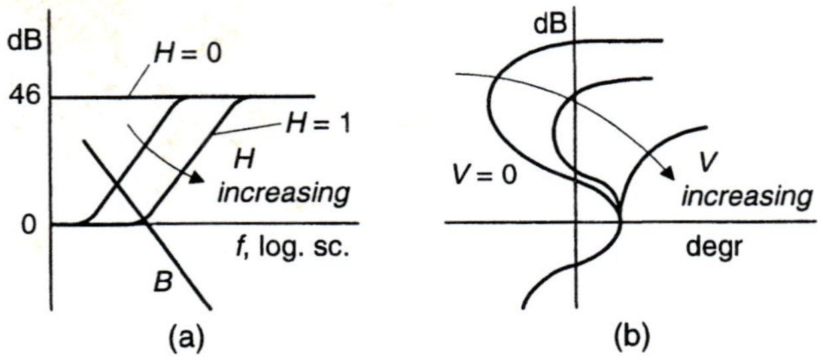

Fig. 11.30 (a) Bode diagrams of the NDC shown in Fig. 11.29 with the dead zone replaced by equivalent linear links, and (b) iso-V Nyquist diagrams for the main loop

Example 2. A tunnel-effect accelerometer is shown in Fig. 11.31(a). The proof mass and the soft springs the mass is suspended on are etched of silicon. The position of the proof mass is regulated by electrostatic forces between the proof mass and the upper and lower plates. The accelerometer uses a tunnel effect sensor to measure the proof mass position. The tunnel current flows between the proof mass and the sharp tunnel effect tip when the distance between them is sufficiently small. The proof mass is gradually brought into the vicinity of the tunnel effect tip by an additional feedback loop using capacitive sensors, not shown in the picture. The voltage on the lower plate equals the voltage on the upper plate plus some bias. It can be shown that with proper bias voltage, the upper plate voltage is proportional to the measured acceleration.

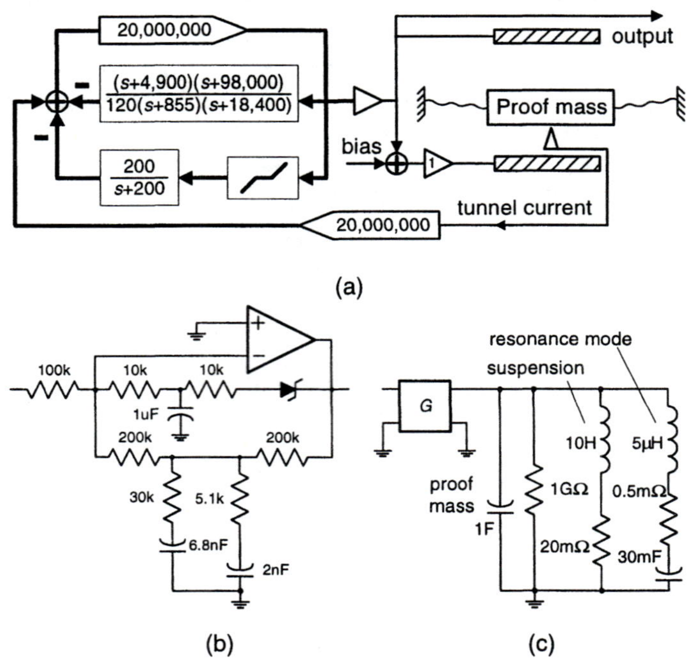

Fig. 11.31 (a) Silicon accelerometer block diagram, (b) compensator, and (c) SPICE plant model

To achieve the desired accuracy, the feedback in the proof mass control loop must be made larger than 100 dB at frequencies up to 5 Hz. The feedback crossover frequency f_b is limited by the dynamics (structural resonances) of the proof mass and suspension system to less than 3 kHz.

The tunnel current is the exponent of the inverse of the tunnel sensor gap. The normal value of the gap is approximately 6 angstroms, but when the gap is smaller, the tunnel current is exponentially larger. Since in this case the derivative of the current to the gap width (the tunnel sensor gain coefficient) increases, the loop gain becomes bigger than nominal. Without an NDC, such a system would not be globally stable. Global stability is provided by an NDC with a dead-zone in the local feedback path.

The mechanical plant might have some resonance modes with uncertain frequencies over 500 Hz. The quality factor of the resonances is not higher than 20, i.e., 26 dB. The SPICE model for the plant with such a resonance is shown in Fig. 11.31(c). The 1GΩ resistor is for the SPICE algorithm to converge.

The compensator is shown in Fig. 11.31(b). The dead-zone element was chosen to be non-symmetrical (a Zener diode) since the characteristic of the tunnel effect sensor is also non-symmetrical. For low-level signals the Zener does not conduct, and the compensator response is determined by the lower feedback path. Two series *RC* circuits shunting the feedback path provide two leads giving sufficient phase stability margins over the range 200 to 3000 Hz. The Bode diagram and the Nyquist plot for signals of small amplitudes simulated in SPICE are shown in Fig. 11.32.

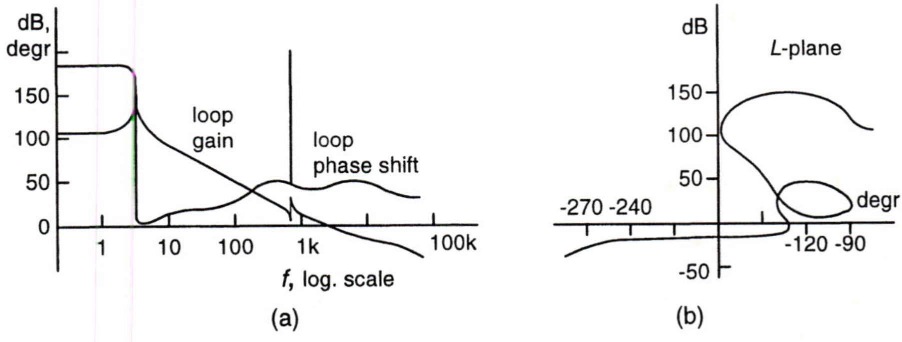

Fig. 11.32 Accelerometer (a) Bode diagram and (b) Nyquist diagram

When the signal exceeds the Zener threshold, the diode opens and the upper feedback path, which is an *RC* low-pass, reduces the compensator gain at lower frequencies by approximately 30 dB. This gain reduction reduces the slope of the Bode diagram, substantially increases the phase stability margin at frequencies below 200 Hz, and improves the transient response of the closed loop which is important since the acquisition range of the tunnel effect sensor is very narrow, only about 15 angstroms. In experiments, the system locks rapidly into the tracking mode and remains stable whatever the initial conditions are.

Two additional examples of applications of NDCs with local feedback incorporating a dead zone link are given in Appendix 13.

11.10 Negative hysteresis and the Clegg Integrator

Let us consider two alternatives to NDCs.

Negative hysteresis is a link with characteristics as in Fig. 11.12(a) but with reversed directions of the arrows on the branches of the characteristic. The negative hysteresis effect can be achieved by switching the output at specific levels of the incident signal as shown in Fig. 11.33. Such a link introduces phase lead up to 90° for signals of certain amplitudes. However, the link does not pass signals with small amplitudes, and it is very sensitive to the signal amplitude and shape, and to the noise. Negative hysteresis links are rarely used since NDCs are simpler, more robust, and able to provide much larger phase lead.

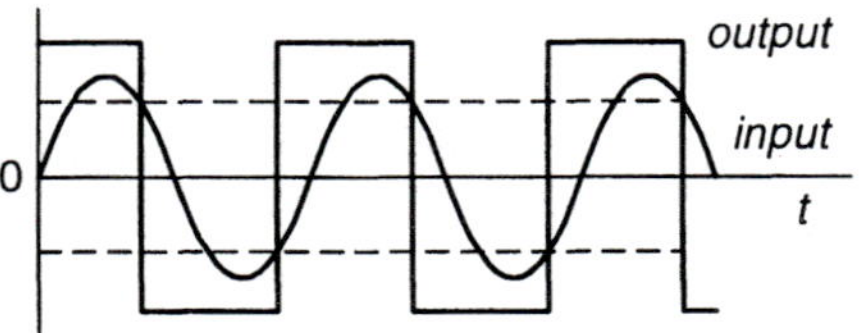

Fig. 11.33 Negative hysteresis

A generalization of the ***Clegg Integrator*** shown in Fig. 11.34 consists of a splitter, two different linear links L_1 and L_2, a full-wave rectifier (i.e., absolute value link), a high-gain link with saturation realizing the *sign* operator, and a multiplier M. The instantaneous amplitude of the output signal is determined by the upper channel output $v_1(t)$ shown in Fig. 11.33(b). The sign of the output signal is defined by the sign of the lower channel output. Therefore, the composite link output signal is $v_{out}(t) = |v_1(t)|$ sign $v_2(t)$ as seen in Fig. 11.34(c).

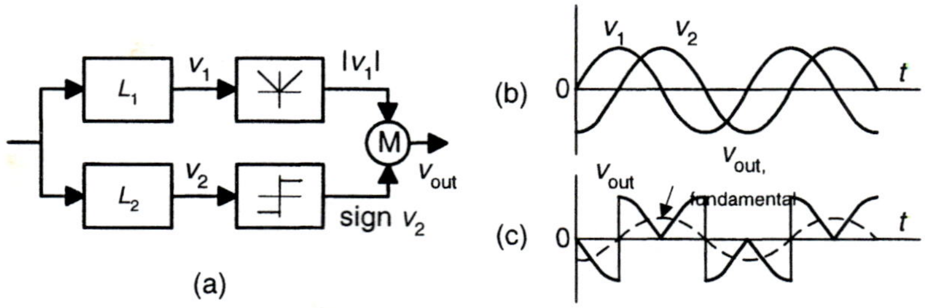

Fig. 11.34 (a) Clegg Integrator and (b) its signal histories

In particular, when $L_1 = k/s$ and $L_2 = 1$, the integrator gain decreases with frequency without introducing the 90° phase lag of a conventional linear integrator, as seen in Fig. 11.34(b). When this idea was first introduced, the hope was expressed that this method would work well for small-amplitude signals, thus allowing circumvention of Bode limitations and the causality principle. This, however, is not possible. A large-amplitude second harmonic is present at the output of the Clegg Integrator. This harmonic's interference with the fundamental produces large phase lag for the fundamental. This and high sensitivity of H to the shape of the input signal (also the disadvantage of negative hysteresis) prevent the circuit from being used in practice. In fact, the Clegg Integrator and negative hysteresis are described here only to inform the reader that such ideas have already been explored and found not particularly useful.

11.11 Nonlinear interaction between the local and the common feedback loops

Saturation links are located in the local and common loop actuators. In addition to the common loop, a local loop is often employed in the ultimate stage of a linear amplifier to linearize its characteristic. Local feedback is employed widely in electrical and electro-mechanical actuators to make their characteristics more linear and stable in time, thus benefiting the main loop.

In control feedback systems, the optimal characteristic for the actuator is hard saturation, since it makes the loop gain constant up to the maximum of the output amplitude. Hence, when the feedback is limited by stability conditions, the best actuator transfer characteristic is saturation. When it is not, a predistortion memoriless nonlinear link could be installed at the input to the actuator to make the total nonlinearity the saturation, or a local feedback about the actuator could be introduced.

The value of the local feedback varies with the signal level. Interference of the local loop with the main loop is important to understand.

Consider an example of local feedback about the actuator in the block diagram in Fig. 11.35.

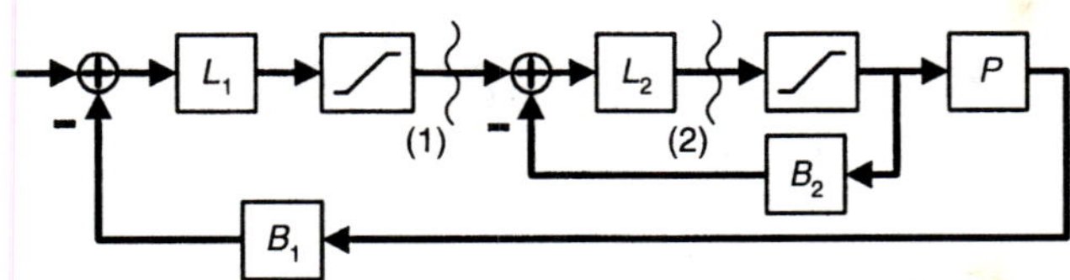

Fig. 11.35 Local (actuator) and common loops with saturation

When the main loop is disconnected at either of the cross sections (1) or (2), the system is stable (when the actuator loop is properly designed). Either of the cross sections can be used for the stability analysis (the Bode-Nyquist criterion for successive loop closure has been described in Section 3.4). In both cases, the resulting loop includes a nonlinear dynamic loop, in one case, with parallel channels, in the other case, with local feedback. The nonlinear dynamic links can be analyzed as we did with NDCs. These links may introduce phase lag in the main loop that can result in a limit cycle.

For example, when L_2B_2 is $s^{-4/3}$ and B is real, the actuator's overload results in the reduction of the local feedback and in the introduction of a 120° phase lag in the main loop. To prevent an oscillation, an extra NDC with a 120° phase lead can be introduced in the loops, at either of the cross sections (1) or (2).

Another common problem is variations in the output impedance of the driver. This impedance depends on the signal level since local current or voltage feedback loops are often employed in the driver amplifier to make its output impedance correspondingly high or low, and the feedback in these loops is affected by the actuator or driver saturation.

Example 1. A typical example is using current drivers for magnetic windings of the actuators: voice coils, reaction wheels of spacecraft attitude control systems, flux winding for electrical power generators, solenoids, etc., as shown in Fig. 11.36(a). The output current of the driver is proportional to the driver's input voltage due to large local current feedback; without this feedback, the driver amplifier would be a voltage amplifier. Since the output force (torque) is proportional to the coil current, the transfer function of the actuator is a real number and the phase lag is 0. (Back emf does not affect the output when the signal source impedance is high.) However, when the driver is overloaded and the gain in the local feedback loop vanishes, the driver's output impedance drops, its output becomes a *voltage*, and the actuator output force becomes an integral of this voltage. In other words, overload introduces an extra integrator into the main loop, and can trigger a limit cycle. To prevent this from happening, an extra saturation link can be placed at the input of the driver, as shown in Fig. 11.36(b), to limit the signal amplitude at the driver's input.

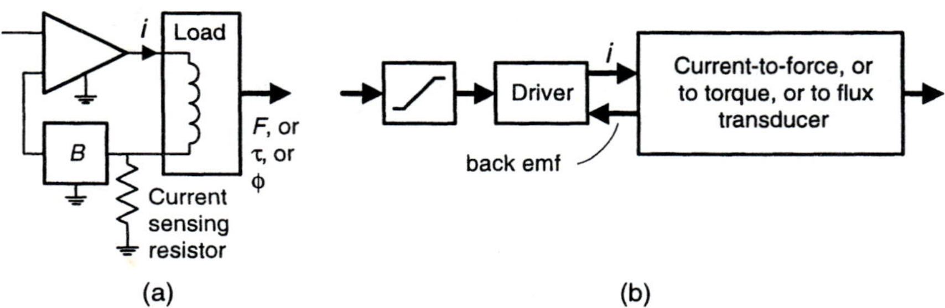

Fig. 11.36 (a) Current feedback driver for inductive loads and (b) using extra saturation link at its input

11.12 NDC in multiloop systems

As was discussed in Chapter 2, MIMO control systems most often have individual actuators for each dimension of the system output, and the plant is to a large extent decoupled, i.e., the diagonal terms are much larger than the others. Still, the coupling exists and might cause instability, especially the coupling due to the plant structural resonances and due to the plant nonlinearity.

Let us consider the two-input, two-output feedback system shown in Fig. 11.37. This system can be, for example, an x-y positioner, with some coupling between the x and y directions because of, say, a resonant mode of the payload that is orthogonal neither to x nor to y. Although the plant gain coefficient in the main direction is substantially larger than the coupling coefficient, the coupling is still not negligible. In this example, let us consider the main direction gain coefficients to be 1, and the coupling coefficients, k_c.

Assume that each loop is stable and robust in the absence of coupling. Then, consider the effect of the coupling on the y-loop. The effect will be easier to see if we redraw the system block diagram as shown in Fig. 11.38.

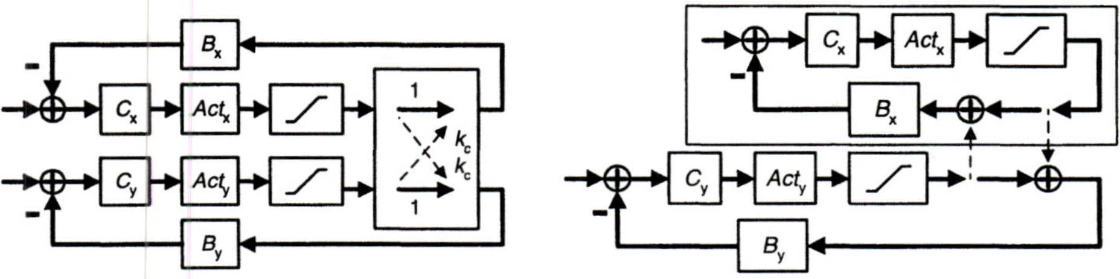

Fig. 11.37 Two-input, two-output system **Fig. 11.38** Equivalent block diagram

The coupling results in a composite link with the transfer function

$$k_c^2 \frac{B_x C_x A_x}{B_x C_x A_x + 1}$$

in parallel with the plant transfer coefficient.

When the feedback in the x-loop is large, the composite link gain coefficient is close to k_c^2. But if the feedback is positive, the gain correspondingly increases. In the nonlinear mode of operation, when the x-actuator is overloaded, the feedback can become positive at any frequency below the crossover. This factor introduces additional uncertainty in the y-loop. To ensure the necessary robustness, either the stability margin must be increased with the resulting reduction in performance, or an NDC must be introduced in front of each actuator – even if the NDCs are not required for the individual loop operation (but they will certainly do no harm to the individual loops, quite the opposite, they will improve the individual loop performance!). When the x-actuator becomes overloaded, the NDC introduces some phase lead in the x-loop; this eliminates positive feedback in this loop and reduces the composite link gain. Hence, with the NDC, stability margins can be reduced in the y-loop and performance can be improved without sacrificing robustness.

11.13 Harmonics and intermodulation

11.13.1 Harmonics

DF analysis neglects the effects of harmonics on the output fundamental amplitude. Let us consider the resulting error.

Fig. 11.39 shows the practically important case of the incident signal amplitude far exceeding the saturation levels, which are indicated by the dashed lines. The output $v(t)$ is clipped most of the time. The third harmonic in $e(t)$ delays the 0-line crossing of $e(t)$ and, therefore, of $v(t)$. This results in an extra phase lag for the fundamental of $v(t)$. The lag could reach 12° in control systems with conventional loop gain response. Harmonics higher than the third also contribute to this effect, although their contributions are smaller. Therefore, phase stability margins for the iso-f lines should be larger by approximately 15° than the phase stability margin accepted for the linear state of operation, i.e., for the Nyquist diagram.

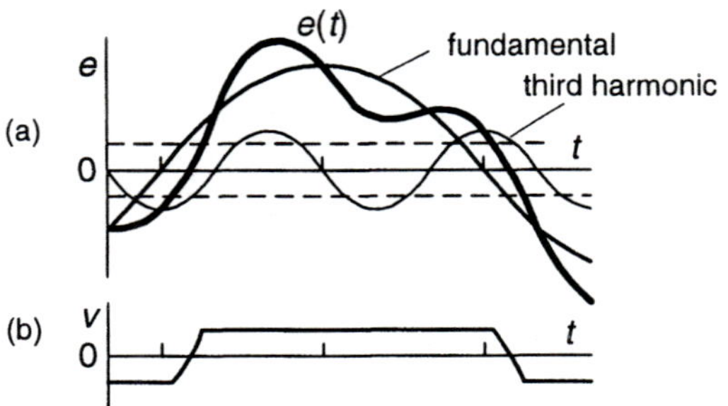

Fig. 11.39 Effect of the third harmonic on the fundamental time delay

Accounting for the harmonics in $e(t)$ might be important while studying systems with resonant peaks and/or valleys on the loop Bode diagram and systems with large pure delay in which the loop phase lag can reach π without the loop Bode diagram being steep.

When the Bode diagram for the loop gain is not monotonic, as in feedback systems with resonance modes in the plant exemplified in Fig. 11.40, DF analysis is still quite satisfactory at all frequencies except for those whose third harmonic is relatively large. At these frequencies, the effect of the third harmonic must be calculated since it can produce a change of up to 30° or 40° in the phase shift for the fundamental f_o. Accordingly, some extra phase stability margin must be provided by reducing the steepness of the Bode diagram in the region close to f_o as shown in Fig. 11.40.

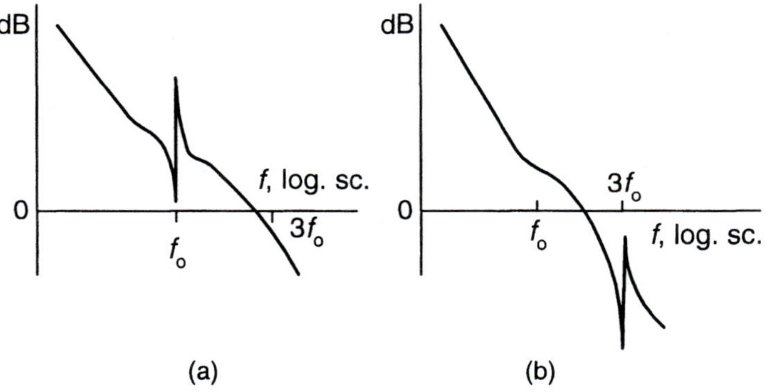

Fig. 11.40 Open-loop Bode diagram of a system with a flexible mode (a) reducing the gain of the fundamental and (b) increasing the gain at the third harmonic frequency

After the DF analysis and design, the system robustness must be verified with computer simulation.

11.13.2 Intermodulation

Application of a signal having two Fourier components with different frequencies, f_1 and f_2, to the input of a nonlinear link results in the output signal $v(t)$ containing not only the harmonics of f_1 and f_2, but also intermodulation products with frequencies $mf_1 \pm nf_2$.

For example, the signal shown in Fig. 11.41 with large-amplitude low-frequency components and small-amplitude high-frequency components is very typical for audio signals. When the signal is clipped, the information contained in the high-frequencies components over the time of clipping is lost. The information would be preserved if the lower-frequency and the higher-frequency components are first separated by a fork of low-pass-high-pass filters, and then combined at the output after amplification by separate amplifiers. In this case, nonlinear distortions of the low-frequency components will not affect the high-frequency components, i.e., will not produce intermodulation. (The intermodulation in the speakers can be reduced by using separate speakers for higher and lower frequencies.)

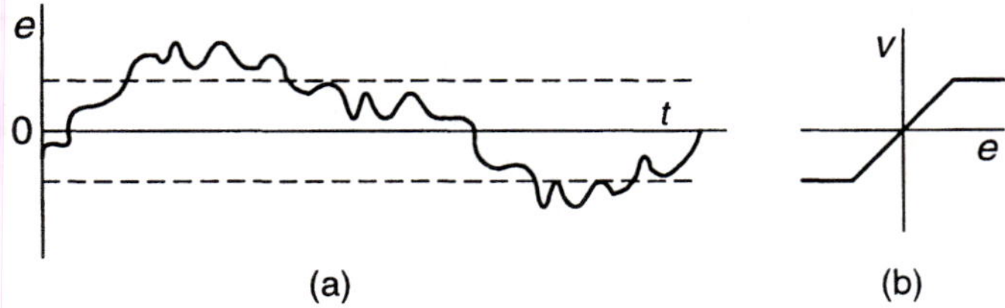

Fig. 11.41 (a) Clipping a multicomponent signal by a link with saturation (b)

For example, antenna pointing can be affected by wind disturbances, the latter having both lower-frequency, higher-amplitude components and higher-frequency, lower-amplitude components. Similar kinds of disturbances occur in vibration isolation actuators. For large-amplitude, low-frequency vibrations not to cause the actuator, after clipping, to stop rejecting high-frequency disturbances, two different actuators for two separate bands of disturbances can be employed.

While audio recording, it is important to hide the high-frequency noise of the magnetic head and the amplifiers. When high-frequency components of the signal are large, the noise is not noticeable, and the high-frequency gain should be large for better signal reproduction. When the signal high-frequency components are small, the amplifier bandwidth should be reduced to reduce the noise that otherwise would be clearly heard. This function is performed by specially designed IC's.

In control systems, similar problems can occur. The IC's designed for audio signal processing can serve the control system purposes as well.

11.14 Verification of global stability

The condition of harmonic balance at some frequency does not always lead to periodic self-oscillation at this frequency. For such self-oscillation to persist, the process of self-oscillation must be stable. The process is stable, by definition, if vanishing disturbances in parameters of the oscillation (in the amplitude, for example) cause deviations from the solution which exponentially decay in time. This condition signifies a limit cycle and is illustrated in Fig. 11.42(a). If the deviations from the periodic solution grow exponentially, as shown in Fig. 11. 42(b), the process is not stable and the limit cycle does not take place.

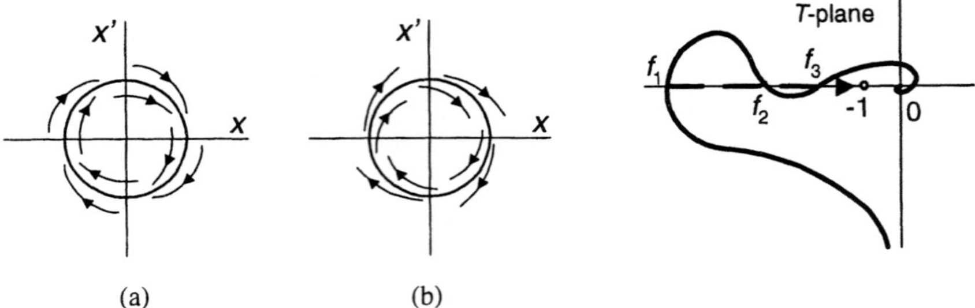

Fig. 11.42 (a) Limit cycle and (b) unstable periodic solution

Fig. 11.43 Nyquist diagram and iso-f lines causing two limit cycles

For example, in a system with saturation and the Nyquist diagram shown in Fig. 11.43, three iso-f lines related to the frequencies f_1, f_2, and f_3 cross the critical point. From these three, the limit cycles are associated with frequencies f_1 and f_3, and the solution at frequency f_2 is unstable. This can be shown as follows.

Consider the limit cycle with frequency f_3. In this case, illustrated in Fig. 11.44(a), the saturation DF reduces the loop gain such that the equivalent Nyquist diagram shrinks and passes through the point 1. An extra increase of the signal level reduces the DF, and the Nyquist diagram shrinks further. The critical point -1 then occurs outside of the Nyquist diagram which is the mapping of the left half-plane of s. Then, the exponent corresponding to the frequency of oscillation has a negative real component, and the signal amplitude reduces. On the other hand, if the signal level gets smaller, the DF increases, the Nyquist diagram expands, the critical point appears inside the Nyquist diagram which is the mapping of the right half-plane of s, and the signal starts rising. Therefore, deviations of the signal amplitude from the equilibrium amplitude starts the process of exponential adjustment of the amplitude toward the equilibrium.

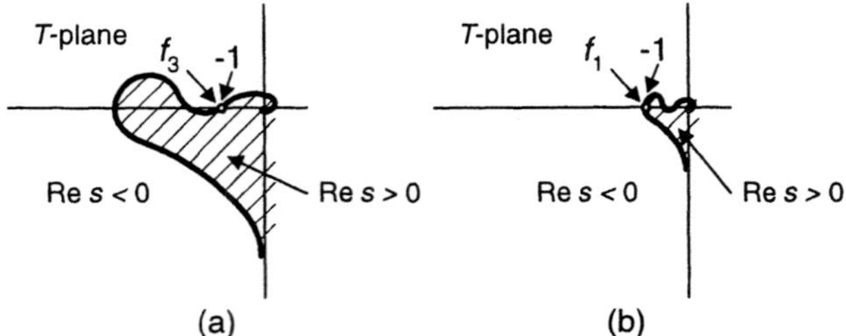

Fig. 11.44 Nyquist diagram when the system oscillates with frequency (a) f_3, (b) f_1

The analysis for the limit cycle with frequency f_1 gives similar results, but the analysis corresponding to oscillation with frequency f_2 leads to the opposite conclusion, pointing at instability of this process. Therefore, if the oscillation at frequency f_2 is created, the signal deflects from the solution exponentially (and very rapidly, i.e., practically, by a jump), gets into the basin of attraction of some of the limit cycles, and this limit cycle will proceed. Process instability will be further studied in the next chapter.

The simple test for global stability involves application of a step-function or a large-amplitude pulse to the command input. Such a test discovers most of the hidden limit cycles, but not always all of them.

To discover the suspected limit cycle, the initial conditions must belong to the basin of attraction of the limit cycle. In other words, the initial conditions should be chosen close to the conditions that will exist during the limit cycle. Such a test uses bursts of oscillation, of large amplitude and of various frequencies, such as those illustrated in Fig. 11.45. Although, theoretically, even this test might not discover all limit cycles, and a counter-example system can be imagined where a limit cycle is triggered by only a special key-signal, using a set of bursts of oscillation is good enough for testing stability of practical control systems.

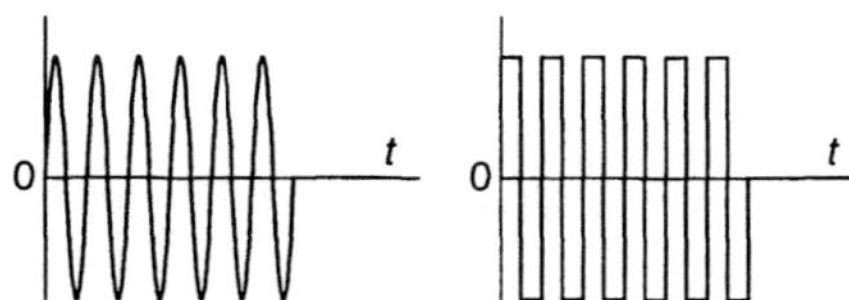

Fig. 11.45 Signals to excite limit cycles during a global stability check

To estimate the stability margins, extra phase lag and/or extra gain is gradually added to the linear links of the loop until self-oscillation starts, and these extra phase and gain values are considered to be the phase and gain stability margins.

11.15 Problems

1 The loop transfer function is
(a) $T = 200(s + 300)(s + 600)/[s(s + 20)(s + 50)(s + 100)]$;
(b) $T = 1000(s + 400)(s + 500)/[s(s + 10)(s + 40)(s + 80)]$;
(c) $T = 200{,}000(s + 100)(s + 400)/[s(s + 10)(s + 30)(s + 600)]$
(d) $T = 1400(s + 600)(s + 800)/[s(s + 15)(s + 20)(s + 1200)]$;
(e) $T = 180(s + 272)(s + 550)/[s(s + 12)(s + 30)(s + 80)]$;
(f) $T = 2500(s + 1000)(s + 1200)/[s(s + 30)(s + 80)(s + 180)]$;
(g) $T = 5{,}000{,}000(s + 500)(s + 550)/[s(s + 30)(s + 32)(s + 3000)]$.

What is the frequency of oscillation (approximately), if there is a saturation link in the loop?

If there is a dead-zone link in the loop, is the system stable, unstable, or conditionally stable? If it is conditionally stable, how to trigger self-oscillation?

2 What is the relative amplitude of the 3th, 5th, and 7th harmonics in the Π-shaped symmetrical periodic signal?

3 During oscillation in a feedback control system where the loop phase lag is 180° and the slope of the Bode diagram is constant, how much smaller are the (a) 5th, (b) 7th harmonics at the input to the saturation link relative to the fundamental?

4 Invent a nonlinear link whose output's fundamental is very sensitive to the shape (i.e., to the harmonics) of the input signal, and use it as a counterexample to the (incorrect) statement that describing function analysis gives sufficient accuracy in all situations. Are the outputs of nonlinear links of common control systems very sensitive to the signal shape?

5 DF analysis might fail to accurately estimate the frequency of oscillation in systems where stability margins are small over broad frequency bands. Why is this failure of small importance for control system designers?

6 Find the value of DF for saturation when the ratio of the signal amplitude to the threshold is (a) 5 dB (b) 15 dB (c) 25 dB (d) 4 times (e) 8 times, using the chart in Fig. 11.8.

7 Find the value of DF for a dead zone when the ratio of the signal amplitude to the threshold is (a) 5 dB (b) 15 dB (c) 25 dB (d) 3 times (e) 5 times, using the chart in Fig. 11.8.

8 Find the value of DF for a three-level relay when the ratio of the signal amplitude to the threshold is (a) 5 dB (b)15 dB (c) 25 dB (d) 5 times (e) 10 times, using the chart in Fig. 11.8.

9 Do Problems 6–8 using approximate formulas for describing functions.

10 The signal amplitude at the input to a saturation link with the threshold 0.12345 increases (a) from 123 to 1230; (b) from 314 to 628. How many times does the DF change? (Make an engineering judgment about what should be the accuracy of the answers.)

11 The signal amplitude at the input to a link with 0.12345 dead zone increases (a) from 123 to 1230; (b) from 2.72 to 7.4. How many times does the DF change?

12 Write an approximate expression for DF which is valid for input sinusoidal signals with amplitudes more than the saturation threshold, for the link with
(a) dead zone 0.2 and saturation threshold 1.3;
(b) dead zone 0.3 and saturation threshold 2.3;
(c) dead zone 0.4 and saturation threshold 3.3;
(d) dead zone 0.8 and saturation threshold 5.

13 The linear loop links' gain is (a) 25 dB; (b) 10 dB; (c) 15 dB, and the phase is 180° at some frequency. What is the ratio of the signal amplitude to the threshold in the saturation link during self-oscillation? If the loop contains a dead-zone link, what is the ratio of the signal amplitude at the input to the link to the dead zone?

14 Plot the transfer function on the *L*-plane with MATLAB and make inverse and direct DF stability analysis for a system with saturation, with threshold 1, for:
(a) $T = 200(s+300)(s+650)/[s(s+20)(s+45)(s+90)]$;
(b) $T = 10000(s+400)(s+550)/[s(s+10)(s+35)(s+80)]$;
(c) $T = 20000(s+100)(s+400)/[s(s+10)(s+30)(s+600)]$;
(d) $T = 140000(s+600)(s+800)/[s(s+15)(s+20)(s+120)]$;
(e) $T = 1800(s+272)(s+700)/[s(s+12)(s+35)(s+75)]$;
(f) $T = 2500(s+1200)(s+1200)/[s(s+22)(s+80)(s+160)]$;
(g) $T = 600000(s+650)(s+550)/[s(s+16)(s+32)(s+300)]$.

15 The area of the hysteresis loop is one half of the product of the sinusoidal input signal amplitude and the amplitude of the output signal fundamental. What is the phase lag of the link DF?

16 Find the current-voltage characteristic of the Schmitt trigger's (a) transfer function (b) input. Discuss the stability conditions.

17 Calculate the dead beat zone and the period of oscillation in the system in

Fig. 11.14 where VCC = 12 V, the Schmitt trigger resistors are R_1 = 100 kΩ and R_2 = 200 kΩ, and the integrator elements are R_3 = 1 MΩ and C = 10 nF.

18 Calculate the oscillation period in the on-off control system shown in Fig. 11.15 if the pre-amplifier gain is 1, the dead beat zone is 10°C, the heater power is 2 kW, the cooling losses are 200 W, and the payload thermal capacitance is 2 Btu/deg. Is the oscillation symmetrical?

19 Calculate the corner frequency (the frequency of the zero) in the plot in Fig. 11.17(b) when the DF of the saturation link at all frequencies is (a) 0.5; (b) 0.1; (c) 0.05. To what values of the signal amplitudes do these DFs correspond?

20 In the nonlinear link in Fig. 11.17, what is the phase shift at frequency 10 mHz if the signal-to-threshold ratio in the saturation link is (a) 1; (b) 10; (c) 100?

21 Will the saturation in the forward path of the NDC in Fig. 11.25(c) substantially affect the NDCs performance?

22 Plot with MATLAB the iso-E Bode gain diagrams and phase diagrams for the link in Fig. 11.17(a) for the range of $1 < E < 100$, with the threshold of the saturation link equal to 2. Mark the plots with the values of E calculated with approximate formulas for DF.

23 Produce the iso-w Bode plot shown in Fig. 11.23 and the related phase responses with MATLAB or SIMULINK.

24 In the NDC shown in the block diagram in Fig. 11.29, $B = 20/(s + 20)$, and the dead zone (for each polarity) is 1.5. Plot iso-w and iso-V Bode diagrams using MATLAB and the approximate expression for the DF of the dead zone link.

25 (a) With SIMULINK, plot iso-w Bode diagrams for the NDC shown in Fig. 11.31(a). (b) Using SPICE, plot iso-w Bode diagrams for the NDC shown in Fig. 11.31(b).

26 The composite link is a cascade connection of an integrator and a saturation link. The local feedback about this composite link makes the link gain response flat. When the saturation link becomes overloaded by a large amplitude signal, what extra phase shift will be introduced in the main loop? Is this situation dangerous from the point of view of making the system not globally stable?

27 A link, whose gain response is flat and whose phase shift is zero over the band of interest, is enlooped by local feedback, the feedback path being an integrator. When the link is overloaded by a large-amplitude signal, what extra phase shift will be introduced in the main loop? Is this situation dangerous from the point of view of making the system not globally stable?

28 A Nyquist-stable system is made globally stable by an NDC. What initial conditions must be used to verify by computer simulations that the system is globally stable?

29 Is the system with return ratio $T = 20{,}000{,}000(s + 500)(s + 550)(s + 600)/[(s + 10)(s + 30)(s + 32)(s + 33)(s + 3000)]$ stable in the linear mode of operation? Does the system have a limit cycle if there is a saturation link in the loop? At what frequencies is the phase stability margin equal to zero (read these frequencies from frequency responses obtained with MATLAB)? What is the loop gain at these frequencies? What is the frequency of the limit cycle? What is the value of the saturation DF if the system oscillates? What is the shape of the signal at the output of the nonlinear link if the system oscillates? What (approximately) is the shape of the signal at the input to the saturation link? What kind of signal should be applied to the system to discover the limit cycle?

30 Research project: Design a Nyquist-stable system with an NDC with parallel channels. Compare versions with different sequences of the links in the channels.

31 Research project: Design a Nyquist-stable system with an NDC with local feedback. Compare versions with different sequences of the links in the local feedback paths.

32 Research project: Design a Nyquist-stable globally-stable system with the AS approach and with the DF approach, and compare the performance.

Answers to selected problems

1 (a) Since the phase shift of saturation DF is zero, oscillation will occur at the frequency where $\arg T \approx -\pi$, if the loop gain at this frequency is more than 0 dB. This happens at $f \approx 4.2$ Hz, while the loop gain is 17 dB. If a dead zone link is included in the loop, the system is certainly conditionally stable.

The loop response can be obtained by simulation with MATLAB, or by simulation with SPICE.

When employing SPICE, use a chain of voltage-controlled current sources. *Each source should be loaded at a series RL two-pole to obtain a function zero, with L = 1 and R equal to the zero value, or, to implement a function pole, at a parallel RC two-pole with C = 1 and R equal to the inverse value of the pole.* The SPICE schematic diagram is shown in Fig. 11.46; the input file for the open-loop frequency response calculation is

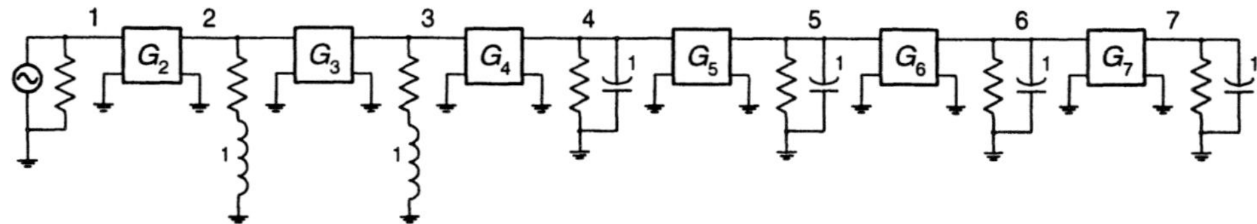

Fig. 11.46 Schematic diagram for SPICE open loop simulation

```
*ch9oc1.cir for determining oscillation condition
*T = 200(s+300)(s+600)/[s(s+20)(s+50)(s+100)], open loop
G2 2 0 0 1 200    ; gain, zero at 300
L2 2 8 1
G3 3 0 0 2 1      ; zero at 600
L3 3 9 1
R3 9 0 600
G4 4 0 0 3 1      ; pole at 20
R4 4 0 0.05
C4 4 0 1
G5 5 0 0 4 1      ; pole at 50
R5 5 0 0.02
C5 5 0 1
G6 6 0 0 5 1      ; pole at 100
R6 6 0 0.01
C6 6 0 1
G7 7 0 0 6 1      ; integrator
```

```
R7 7 0 1MEG       ; de-floating resistor
C7 7 0 1
VIN 1 0 AC 1
RIN 1 0 1MEG      ; source loading resistor
 .AC DEC 20 1 100
 .PROBE
 .END
```

Use a cursor to find the frequency at which the phase shift VP(7) equals –180°. Check whether the loop gain VDB(7) is positive at this frequency.

With a dead zone in the loop, the system is conditionally stable, the limit cycle being at the frequency where the phase margin is zero. To excite the self-oscillation, a single pulse or a burst of oscillation at frequency 4.2 Hz should be applied to the closed system input. (If it is desired to perform this simulation, a dead-zone link and a summer should be added to the input file, and the loop should be closed.)

6 (a) –3 dB

7 (a) –9 dB

8 (a) – 4.5 dB

Chapter 12

PROCESS INSTABILITY

In asymptotically process-stable (APS) systems, any infinitesimal increment to the input process causes only an infinitesimal increment of the output process. The necessary and sufficient APS conditions require the stability margins to be much larger than those commonly used.

The APS requirement contradicts feedback maximization, and the majority of practical control systems are not APS – the effects of the output process instability need only to be bounded. The acceptable values of process instability can be reflected into certain boundaries on the Nyquist diagram, so they can be easily accounted for in a necessary and sufficient way by appropriate sizing of the stability boundaries. This is an important advantage of the frequency domain approach to nonlinear system design

The bounds for process instability are found for sinusoidal test signals. The relationship between the amplitude of the input sinusoidal signal and the amplitude of the fundamental of the error is hysteretic. As the input amplitude is gradually increased, the error amplitude increases smoothly until the error jumps up and then continues to increase very slowly. If the input amplitude is then gradually reduced, the error and the output remain almost the same until the input amplitude is reduced to a certain value, at which point the error decreases (again, by a jump). This phenomenon is called jump resonance. Boundaries on the *T*-plane specify the values and the thresholds of the jumps.

The odd subharmonics may originate in control systems with saturation by a jump, when a parameter of the system is varied continuously. The second subharmonic only occurs in systems comprising nonlinear elements with asymmetric characteristics, and the subharmonic originates smoothly. With stability margins of common values, subharmonics do not show up.

12.1 Process instability

The output process is considered unstable if an infinitesimal increment in the input triggers a finite or exponentially growing increment in the output. ***Process instability*** manifests itself in sudden bursts of oscillation or jumps in the output signal. These phenomena contribute to the output error and therefore need to be limited.

A system is said to be ***process-stable*** when the output processes are stable for all conceivable inputs. In accordance with the First Lyapunov Method, process stability implies stability of the locally linearized (generally, time-varied) system.

In most cases, process instability is not a critical design issue. However, during experiments with, testing, and troubleshooting feedback systems, process instability and the associated nonlinear phenomena like jump-resonance and subharmonic oscillation may occur and can be very confusing. For this reason, the process instability conditions and manifestations need to be well understood.

12.2 Absolute stability of the output process

The system in Fig. 12.1(a) is said to be ***absolutely process-stable*** (***APS***) if it is process-stable with any characteristic of the nonlinear memoriless link $v(e)$ whose differential gain coefficient is limited by

$$0 < \frac{dv}{de} < 1 . \qquad (12.1)$$

For example, the monotonic characteristic illustrated in Fig. 12.2 satisfies the condition (12.1).

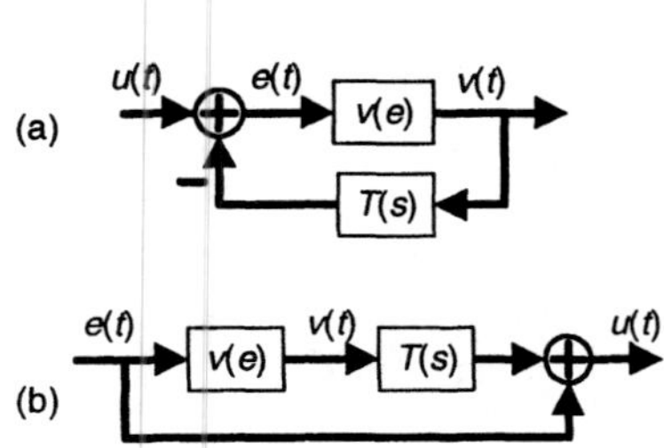

Fig. 12.1 Feedback system (a) and its equivalent (b)

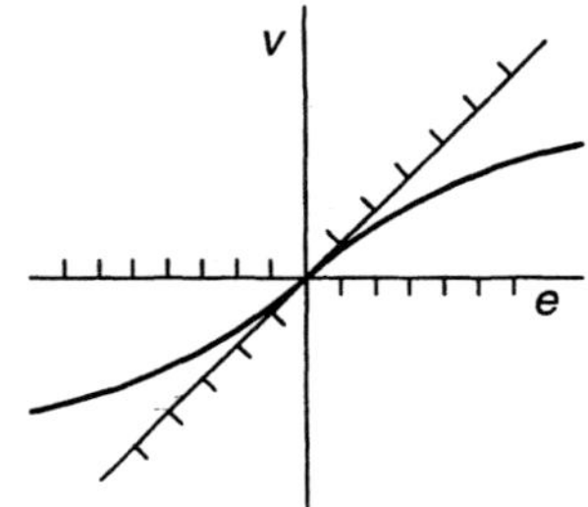

Fig. 12.2 Characteristic of the nonlinear link with limited differential gain

The system is APS if at all frequencies

$$\text{Re}\, T(j\omega) > -1 . \qquad (12.2)$$

To prove this statement we need to show that the feedback system linearized for deviations shown in Fig. 12.3(a) is stable. Here, the nonlinear link $v(e)$ is replaced by a linear time-variable link with the gain coefficient

$$g(t) = \frac{dv}{de} .$$

In accordance with (12.1), the coefficient $g(t) \in (0,1)$, or

$$1/g(t) > 1. \qquad (12.3)$$

The analogy between the feedback system and the two-poles connection discussed in Section 3.10.2 yields the equivalent electrical circuits, shown in Fig. 12.3(b), (c). Since $F(s) = T(s) + 1$ is positive real, the two-pole in the lower part of the circuit in Fig. 12.3(c) is passive. The two-pole $1/g(t) - 1$, although time-variable, is resistive and positive, i.e., energy-dissipative. Hence, no self-oscillation can arise in this system. Therefore, the system in Fig. 12.1 is APS if it satisfies conditions (12.1) and (12.2). (Notice that the analysis is valid even if the linear links in Fig. 12.3 are time-variable. Therefore, (12.2) gives a sufficient stability criterion for LTV feedback systems.)

Condition (12.2) is not only sufficient but in some sense necessary for APS. If (12.2) is not satisfied, i.e., Re $T < -1$ at some frequency ω, then there can be found (a) some function $v(e)$ satisfying (12.1) and (b) some periodical input signal with the fundamental ω, that together bring forth an unstable output process. Particularly, it can be shown that when $v(e)$ is saturation, and the input is a П-shaped periodical signal with frequency ω, gradual change in the input signal amplitude renders a jump in the output signal amplitude [9].

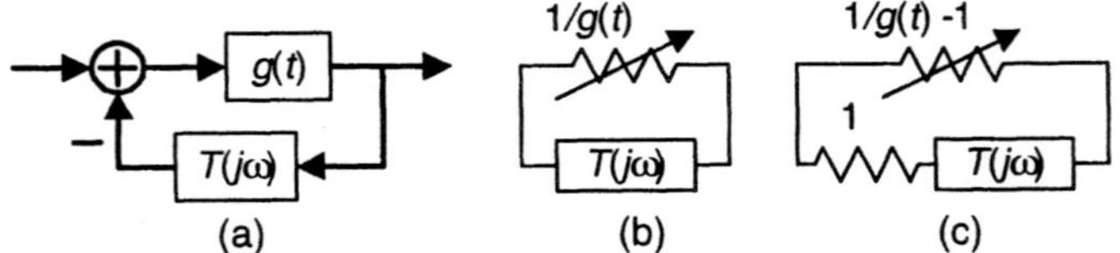

Fig. 12.3 Feedback system with time-variable real gain coefficient $g(t)$ (a) and the schematic diagrams (b),(c) described by the same equations

The condition (12.2) restricts the position of the Nyquist diagram as shown in Fig. 12.4 and requires the phase lag at large loop gains not to exceed 90°. This limits the slope of the Bode diagram to –6 dB/oct and thereby reduces the available feedback.

In practical systems not satisfying the APS condition, typically process instability does not contribute much to the output error, and then only when the actuator is saturated and a rather unusual command is applied to the input. On the other hand, excessive process instability is not acceptable. The required feedback system parameters should therefore lie somewhere between these two extremes. To evaluate and certify the systems, appropriate test-signals should be selected among those typical for the practical system inputs. For the systems where the inputs can be periodical with large amplitudes, the relative jump values of the ***jump-resonance*** (described in the following section) are employed as the figure of merit.

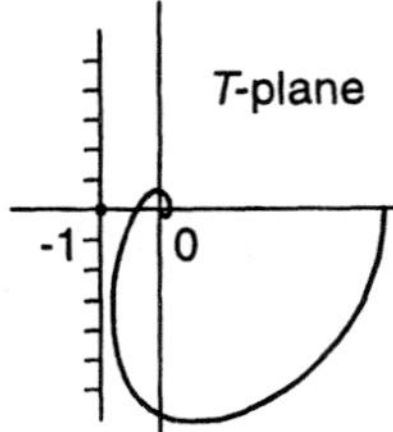

Fig. 12.4 Restriction of the Nyquist diagram by the APS condition

12.3 Jump-resonance

The amplitude and the shape of the output periodical signal generally are multivalued functions of the input periodical signal parameters, and depend on the input signal prehistory. Particularly, the output might depend on whether the current value of the input signal was arrived at by gradually increasing or gradually decreasing the input signal amplitude.

Jump-resonance can be observed in the nonlinear feedback system in Fig. 12.1(a) excited by sinusoidal input $u = U \sin \omega t$. While the input signal amplitude U is gradually increased, at a certain amplitude $U = U''$ after the saturation level, an infinitesimal increase in the amplitude of the input, from U'' to $U'' + 0$, causes the output signal $v(t)$ to change by a jump from the time-response (a) to the time-response (b) shown in Fig. 12.5. As the result, the amplitude of the fundamental of the output V increases by a jump.

Next, gradual reducing the amplitude of the input starting at a value bigger than U'' down to a certain value U', causes a jump down from the time-response (c) to the time-response (d) when the input amplitude is reduced by an infinitesimal increment, to $U' - 0$.

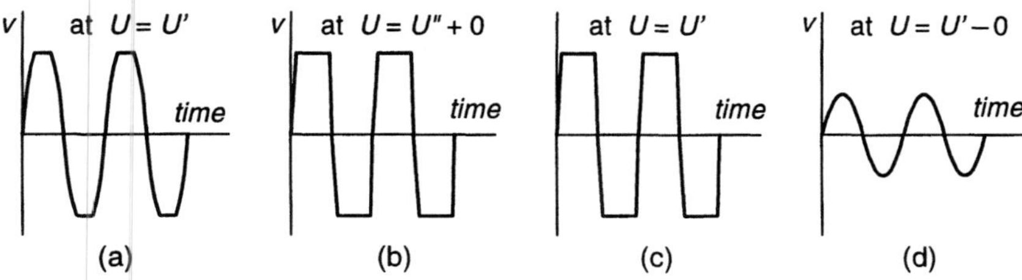

Fig. 12.5 Shapes of the output signal $v(t)$ before the jump up (a) and after the jump up (b); before the jump down (c), and after the jump down (d)

The conditions for the jumps can be found by initially analyzing the system in Fig. 12.1(b), which implements the same relationships as the feedback system in Fig. 12.1(a) (as can be easily verified). Using the DF approach, we assume that $e(t) \approx E \sin \omega t$.

When the phase stability margin is rather small, say 15°, i.e., arg T is −165°, vector addition of the outputs of the two parallel channels generates a collapse of $U = 2E\sin(y\pi/2)$ in the region where $V|T| \approx E$. This is exemplified in Fig. 12.6.

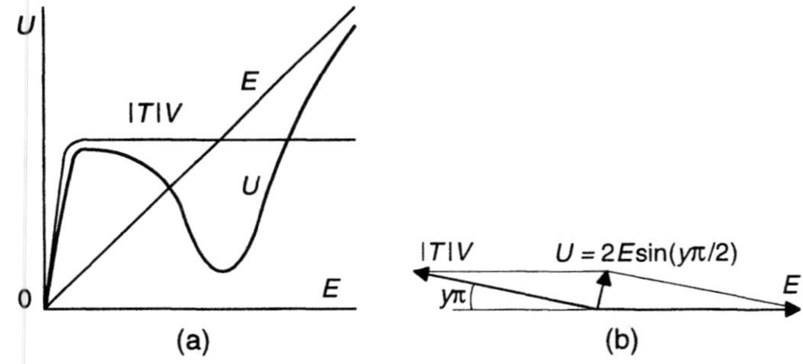

Fig. 12.6 Output signal amplitude for equivalent parallel channel system, (a) dependence of U on E, and (b) vectors' addition at the collapse

The dependence $U(E)$ in Fig. 12.6 is single-valued. The inverse operator $E(U)$ redrawn in Fig. 12.7 is three-valued. This plot comprises a ***falling branch*** over the interval (U', U'') between the ***points of bifurcation*** marked dark. When U passes U'' while being gradually increased, E jumps from before-the-jump value E_b'' to the after-the-jump value E_a'', and when U is reduced to U', then E jumps down to E_a'.

The proof that the solution on the falling branch is unstable with a real positive pole was given in Section 10.3.1. Since the system is unstable, the signal rises exponentially – jumps – between the two stable solutions as shown by the arrows in Fig. 12.7.

In the system with saturation, only the jumps down can be large. From the curves in Figs. 12.6 and 12.7 it is seen that the after-the-jump amplitude E_a' depends on the loop gain and phase. The calculated dependencies are plotted in Fig. 12.8. These curves allow one to specify the required stability margins from the values of the allowed jumps, and also, while experimenting, to calculate the stability margins from the observed jumps.

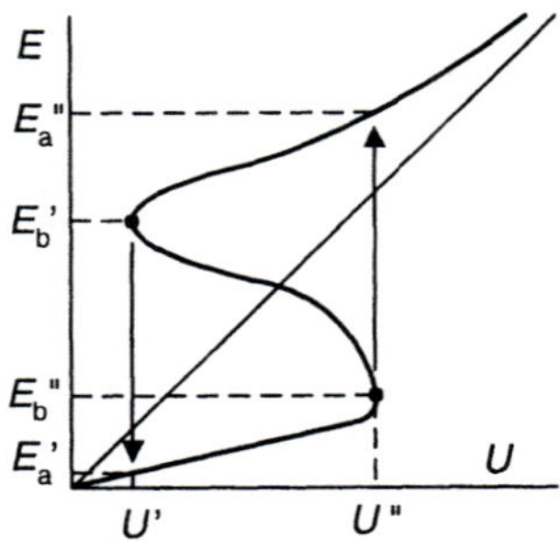

Fig. 12.7 Three-valued dependence of the error amplitude on the input signal amplitude

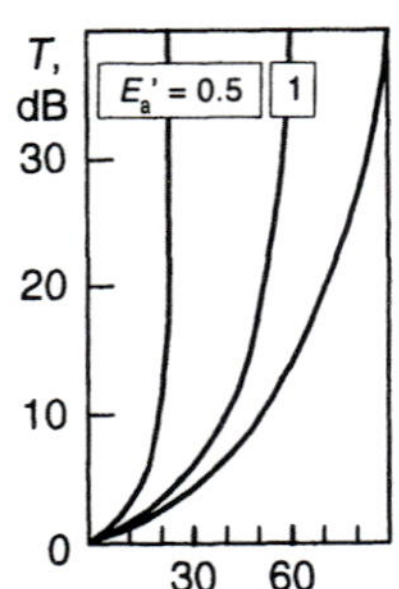

Fig. 12.8 Lines of equal values of jumps down in a system with saturation

The jumps occur in feedback systems when $\arg T$ is such that the curve $U(E)$ in Fig. 12.6 possesses a minimum. This takes place within certain frequency intervals. On the plane (U, f) in Fig. 12.9, the area is traced where the dependence $E(U)$ has three solutions. At the lowest and the highest frequencies, the limit case occurs of $U' = U''$ and $dU/dE = 0$ with zero-length falling branch. The jumps can be caused by varying either U or f or both.

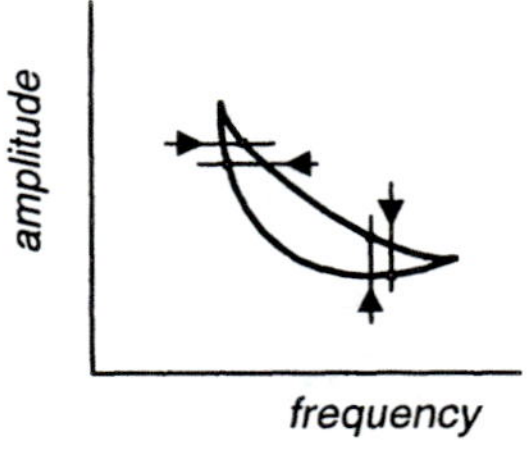

Fig. 12.9 Amplitude-frequency areas of three-valued amplitude U

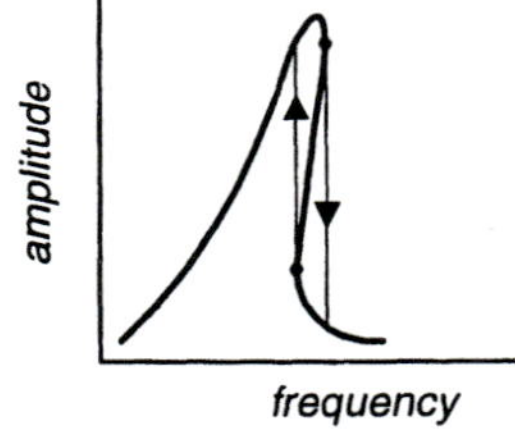

Fig. 12.10 Jump-resonance when the frequency of the input signal is being changed

Hysteretic frequency responses of the output signal amplitude similar to that shown in Fig. 12.10 can be recorded while the input signal amplitude is kept constant. Such responses were first observed during the study of resonance tanks with nonlinear inductors having ferromagnetic cores. Because of the analogy described in Section 3.10.2, a parallel connection of an inductor and a capacitor is described by the same equations as a feedback system with the return ratio $1/(LCs^2)$. The phase stability margin is small (although it is nonzero due to the inductor losses), and therefore the jump-resonance is prominent.

For feedback systems with dynamic saturation (including those with NDCs), it can be shown [9] that the after-the-jump amplitude

$$E_a' = \frac{1.27}{\sup |M|_{E>1}}. \tag{12.4}$$

The denominator can be readily found by plotting the iso-f line on the Nichols chart. This is exemplified in Fig. 12.11. On the iso-f line, $|M|$ of 3.5 dB is maximum, i.e., $|M| = 1.5$, hence $E_a' = 0.85$.

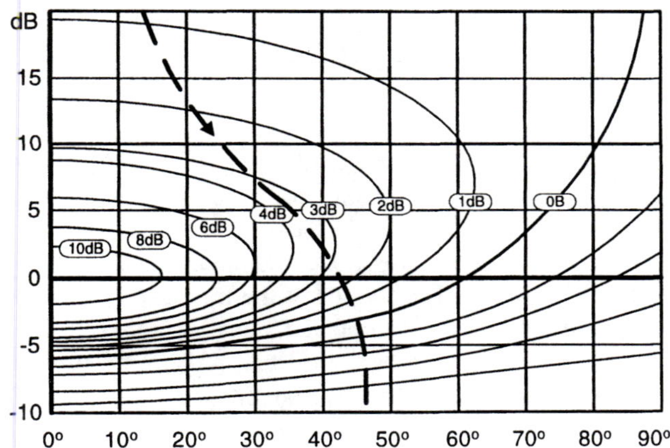

Fig. 12.11 The jump value corresponds to the maximum of $|M|$ on the iso-f line.

Example 1. It is difficult to measure the loop responses in RF and MW feedback amplifiers since these measurements require loading the port of the open feedback loop on a two-pole with impedance imitating the impedance of the disconnected port as shown in Fig. A6.3. Instead, phase stability margins at various frequencies can be found by observing the jump-resonance in the closed-loop configuration and using the plot in Fig. 12.8.

Example 2. Jump-resonance has been observed in the attitude control loop on the Mariner 10 spacecraft. Nitrogen thrusters of the solar panels' attitude control excited large-amplitude periodic oscillation in the panels which have high-Q structural mode. The gyro bearing nonlinearity lead to jump-resonance in the gyro loop. This in turn lead to the panel attitude control thrusters consuming an abnormally large amount of the propellant. The problem was rectified by reducing the gain in the control loop.

12.4 Subharmonics

12.4.1 Odd subharmonics

As will be shown below, the subharmonics in low-pass systems with saturation could only become excited if the phase stability margin at the subharmonic frequency ω/n is rather small. This implies a steep cut-off of the Bode loop gain diagram. Therefore, the signal at the input to the nonlinear link mainly consists of the lower frequency components: $E \sin \omega t + E_{sb} \sin(\omega t/n)$.

Let us examine whether odd subharmonic oscillation is possible with either small or big amplitudes of E and E_{sb}. We assume the saturation threshold is 1.

Evidently, with both E and E_{sb} being so small that $E + E_{sb} < 1$, the subharmonics are not observed since in this case the system behaves linearly.

If $E > 1$ and $E_{sb} << 1$, part of the time the nonlinear link is saturated by the signal.

For small signal increments given to its input, the link can be seen as an equivalent linear time-varying one, a pulse element with the sampling frequency ω. Considering the subharmonic as the input increment, the output increment is the product of $E_{sb}\sin(\omega t/n)$ and the Fourier expansion with the fundamental frequency ω which characterizes the pulse element. In this product, the component with the frequency ω/n can be generated only due to the constant component of the Fourier series. Since this component is real, the nonlinear link does not introduce in the loop any phase shift for the subharmonic.

Therefore, when the parameters of the input signal are manipulated slowly and continuously, the odd subharmonics may only originate by a jump, with nonzero steady-state amplitude E (this is called ***hard oscillation***).

On the other hand, if $E_{sb} >> 1$ and $E < E_{sb}$, the input to the nonlinear link can be as shown in Fig. 12.12(a). This signal is clipped at the levels shown by dashed lines, and the output of the saturation link $v(t)$ accepts the shape of trapezoidal pulses, shown in Fig. 12.12(b). These subharmonic pulses are shifted in phase by some angle due to the signal $E\sin\omega t$. If this angle exceeds the phase stability margin at the frequency of the subharmonic, and the loop gain at this frequency is more than 1, a steady-state subharmonic oscillation may be excited by creating appropriate initial conditions.

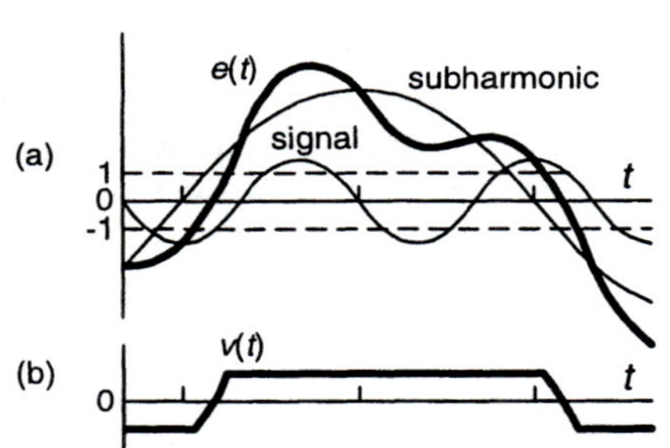

Fig. 12.12 Third subharmonic mechanism

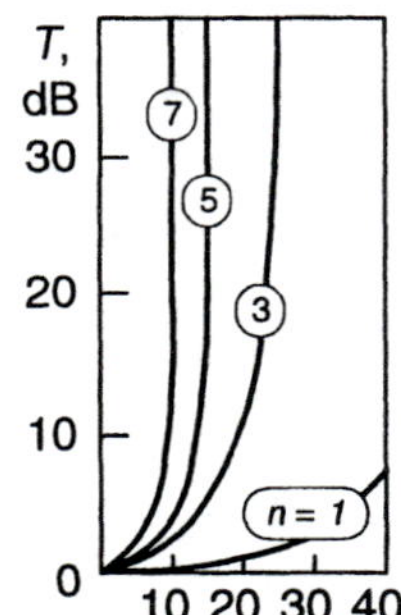

Fig. 12.13 *L*-plane boundaries for odd subharmonics in a system with saturation

It is seen from Fig. 12.12 that the phase shift is less than $\pi/2n$. The boundaries for the third- and for higher-order subharmonics are displayed in Fig. 12.13 [9]. In practical control systems the phase stability margin is always bigger than $\pi/6$, and odd subharmonics are not observed.

12.4.2 Second subharmonic

The second subharmonic can be observed only if the characteristic of the loop nonlinear link is asymmetric, and the greater the asymmetry is, the wider are the areas in which the subharmonic exist. For a system with one-sided saturation, the boundaries of the subharmonic's existence that correspond to different values of the signal amplitude are shown in Fig. 12.14 [9]. As seen from the figure, the second subharmonic is not generated in systems where the phase stability margins are greater than 30°.

Unlike odd subharmonic self-excitation, second subharmonic self-excitation can be

soft. Soft excitation means that when either U or ω or both are gradually changed along any trajectory entering the domain where the subharmonic can be observed, the amplitude of the subharmonic increases steadily from zero, without jumps.

We may conclude that in single-loop systems with saturation neither the odd nor the even subharmonics present real danger, since meeting the mandatory requirement for eliminating windup and substantial jump-resonance automatically excludes the subharmonics.

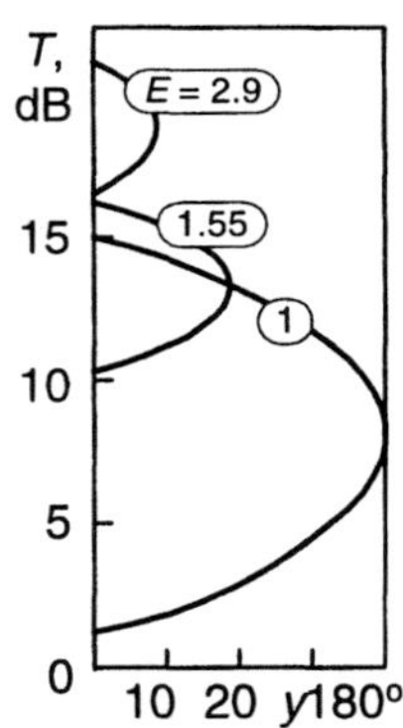

Fig. 12.14 *L*-plane boundaries for second subharmonic existence in a system with single polarity saturation

12.5 Nonlinear dynamic compensation

Nonlinear dynamic compensation can make the system process-stable without sacrificing the available feedback. The conversion of the system with two nonlinear links to an equivalent system with one nonlinear link that was described in Section 10.7 is as well applied to designing NDCs to satisfy the process stability criteria. Example 3 in Section 10.7.2 describes a process-stable system with large feedback.

12.6 Problems

1 Prove that the systems in the block diagrams in Fig. 12.1(a),(b) are described by the same equations.

2 In a system with saturation, with crossover frequency $f_b = 1$, the phase stability margin is 30° and the slope of the Bode diagram below f_b is −10 dB/oct. Is this system process-stable? If not, what must be the slope for the system to be process-stable? Approximately, what will be the change in the feedback at 0.03 Hz?

3 Is a system with saturation APS if
(a) $T = 100/(s + 0.01)$?
(b) $T = 1000/[(s + 0.01)(s + 2)]$?
(c) $T = 123/(s + 0.21)$?
(d) $T = 500/[(s + 0.02)(s + 3)]$?
(e) $T = 272/[(s + 2.72)(s + 27.2)]$?
(f) $T = 5000/[(s + 0.08)(s + 0.2)(s + 0.3)]$?

4 Prove the validity of the following equivalent forms for condition (12.2) of APS:

$$\text{Re } F(j\omega) > 0 \tag{12.5}$$

$$\text{Re } 1/F(j\omega) > 0 \tag{12.6}$$

$$\cos \arg T(j\omega) > 1/|T(j\omega)| \tag{12.7}$$

and

$$|M(j\omega)| = \frac{T(j\omega)}{F(j\omega)} > \frac{1}{\sin \arg T(j\omega)} \tag{12.8}$$

5 Consider jump-resonance in a system with a dead zone. Draw responses analogous to those shown in Fig. 12.6 for a system with saturation.

6 Show that in a system with dead zone and saturation, the output-input amplitude dependence can be 5-valued, by drawing a response analogous to that in Fig. 12.6.

7 On a Nichols chart, the minimum value of $|M|$ for a system with saturation is (a) 6 dB, (b) 8 dB, (c) 4 dB. What is the value of E after a jump down from the state of saturated output?

8 The loop gain is 20 dB in a system with saturation at some frequency. The jump down is (a) 2 times, (b) 1.5 times. What is the phase stability margin (approximately) at this frequency?

9 In a system with saturation, what must be the stability margin for the fifth subharmonic to be observable?

Chapter 13

MULTI-WINDOW CONTROLLERS

Since, generally, nonlinear controllers perform better than linear controllers, NDC design methods are of profound interest. We already considered the design of NDCs using AS and DF approaches. In this chapter we consider the NDC design from yet another angle, as multi-window composite controllers.

Composite nonlinear controllers consist of linear high-order controllers and the means of transition between the linear controllers according to some participation rules. Each of the linear controllers is a local linear approximation to the optimal nonlinear controller. The size of the regions where a single linear controller is operational and the complexity of the elementary control laws are discussed.

Multi-window control makes transitions between the elementary linear controllers on the basis of the amplitude of the error. It is relatively simple to implement and provides much better performance than linear controllers.

During the transition between the elementary controllers, it makes a big difference whether the controller to be put on is "cold" or "hot." Hot controllers are those with the input connected to the signal. This eliminates large transients caused by the initial switching on of the controller.

Windup is the large and/or long overshoot in nonlinear systems. It often occurs in the systems with a large integrating component in the compensator. The widely used anti-windup controllers include nonlinear links directing the signal into different paths, depending on the signal level.

The acquisition and tracking problem is that of, first, finding and locking into the target, and then, precisely tracking it. To perform each of these two tasks optimally, the control law cannot be the same: for acquisition, the control bandwidth must be larger but the feedback can be much smaller. The transition between the two laws must be gradual in order not to de-acquire the target. When combining linear compensators via multi-window nonlinear summer, it is important to guarantee that the combined transfer function remains m.p. This can necessitate using more than two parallel branches and windows.

Another typical application for the multi-window controllers is time-optimal control. This problem is related to the acquisition and tracking problem, and the solutions are similar. Several examples of nonlinear controller applications are presented.

Still another related problem is the problem of switching between the existing controller to a new controller in an industrial environment when the manufacturing process cannot be stopped. The switching should not generate disruptive transients.

Many issues in multi-window controller design have not yet been investigated, and many questions remain as to how to make the best use of such controllers.

The chapter ends with a brief discussion of command feedforward in a system with multi-window controllers.

13.1 Composite nonlinear controllers

It was demonstrated in Chapters 10 and 11 that some nonlinear controllers perform much better than any linear controller. The optimal controller is, in general, nonlinear.

For a small region in the variable space, the nonlinear optimal controller can be approximated well by a linear controller. For an adjacent region, another linear controller can be designed that would be optimal over this region, and so on. The design problem, therefore, arises of integrating these locally-optimal linear controllers into a composite nonlinear controller and providing smooth transitions between these regional linear laws.

Transitions between the control modes can be characterized by ***participation rules*** defining the set of ***participation functions*** illustrated in Fig. 13.1. Over some transition interval of a variable or a condition, the controller action is the sum of the actions of the adjacent regional control modes, and at the ends of the interval only one of the modes (or conditions, or actions) takes place.

The transition rule can be linear and expressed as

$$action = (1 - k) \times action_1 + k \times action_2, \tag{13.1}$$

where the scalar k changes from 0 to 1 over the transition interval. With these rules and when the actions are scalars or collinear vectors, the transitions between the control laws are as illustrated in Fig. 13.1(a). A smoother rule is illustrated in Fig. 13.1(b). Commonly, the precise shape of the participation functions does not matter much as long as it is monotonic, not too steep, and not too shallow.

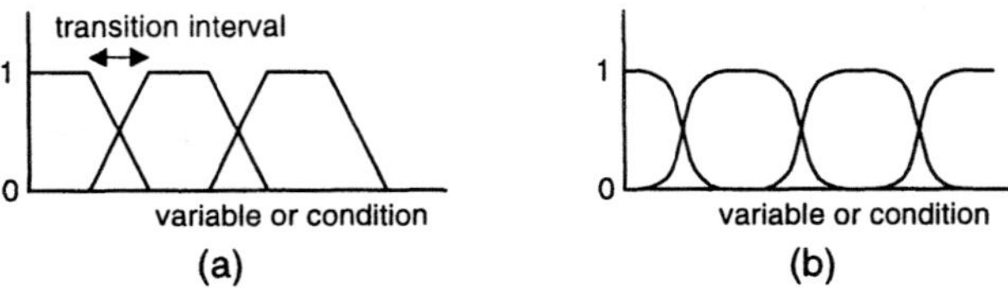

Fig. 13.1 Participation functions of control laws in composite controllers

In general, a monotonic shape of the participation functions does not yet guarantee the smoothness of the transition between different actions. It is also required that the adjacent control laws mix well, i.e., in particular, that the combined action of the adjacent controllers exceeds that of each individual controller. This is not always the case. For example, even if some residue that needs to be cleaned out can be removed by either an acid or a base, an acid and a base should not be used as a mixture with gradually changing content. For regulation of a reactive electrical current, a variable capacitor or a variable inductor can be used, but these elements should not be combined in series or in parallel since they might produce resonances. A low-pass link should not be carelessly mixed in parallel with a high-pass link or else notches and n.m.p. shift might result.

Fuzzy logic controllers break each smooth transition into several discrete steps. This increases the total number of regions with different control laws. Since these regions become very small, fuzzy logic control can use low-order regional control laws. Hence, fuzzy control design can be based on phase-plane partitions, and on passivity theory expressed in state variables. In fuzzy logic controllers, many variables need to be sensed and processed to define the boundaries of the regions.

What region size is optimal for composite controllers?

There are two advantages to making the regions small. The first is that the control laws in the adjacent regions might become very similar, which enables smooth transitions between them without taking special precautions. The second advantage is that the linear controller can be of low order, and the phase plane can be used for the controller analysis and design. As claimed by some fuzzy logic advocates, such controllers can be designed even by those ignorant of control theory. However, when the number of the regions is large, the number of boundaries between them becomes very large. Correspondingly, the number of decision-making algorithms and instruments for changing between the control modes becomes very large. This complicates both the controller design and the designed controller.

On the other hand, *higher-order linear control laws can be made to remain nearly optimal over a much broader region than low-order laws. This reduces the number of the regions and the number of the boundaries between them*.

For the design of higher-order regional control laws, not the phase plane but frequency domain methods should be used. The partition between the regions should be also defined in the frequency domain. This approach requires caution and application of certain rules discussed in Chapter 4 to provide good blending of the regional control laws at the boundaries between the regions. Nevertheless, this approach is not difficult and leads to economical and nearly optimal controllers.

For controllers designed in frequency domain, summation (13.1) may cause nonminimum phase lag. In this case it is worth considering logarithmic transition rule

$$log(action) = (1 - k)\ \log(action_1) + k\ \log(action_2), \tag{13.2}$$

i.e., multiplication

$$action = action_1 \times (\, action_2/action_1)^k. \tag{13.3}$$

13.2 Multi-window control

In the following, we will consider single-input nonlinear controller. The input of the controller is the output of the feedback summer, i.e., the feedback error. Unlike fuzzy-logic controllers, no other sensors and variables are employed to modify the control low.

As shown by N. Weiner, the output of a nonlinear operator can be approximated by applying the input signal to a bank of linear operators, and then combining various products of the linear operators' outputs. In other words, nonlinear dynamic links can be approximated by interconnections of linear dynamic links and nonlinear static links. The multi-window controllers make a small subclass of such links. Although relatively simple, this subclass allows for much richer varieties of the input-output relationships than those of the linear system. Multi-window controllers perform significantly better than linear controllers, do not require high accuracy of design and implementation (i.e., are robust), and therefore allow developing some rules of thumb suitable for the conceptual design and the design trade-offs. In spite of the subclass simplicity, rigorous strightforward methods for the synthesis of such controllers are yet to be developed, and the initial design is further optimized in practice by repeating the computer analysis. As we concentrate on the applications, the presentations in this chapter are not rigorous and rely largely on examples and simulations.

The signal components of the error can be divided into several sets bounded by two-dimensional ***windows*** shown in Fig. 13.2. The windows divide the frequency spectrum (or, equivalently, the time-response behavior) and the amplitude range. Within each window, a regional linear controller (compensator) is employed.

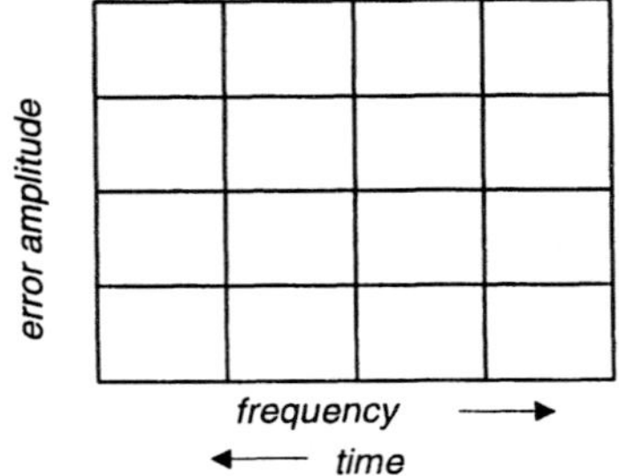

Fig. 13.2 The multi-window control concept: the choice of the linear controller defined by the error amplitude and frequency content

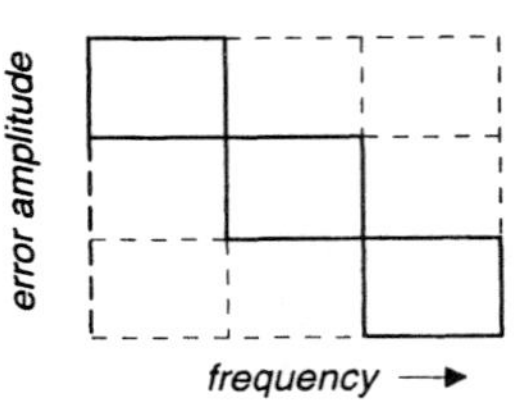

Fig. 13.3 Diagonal windows

The multi-window nonlinear controller can be implemented as follows: the error is partitioned into components falling into different windows, the components processed by the linear operators of the windows, and the results added up or combined by nonlinear functions. This architecture is referred to as ***multi-window***, and a great number of useful nonlinear control schemes can be cast in this form.

The regional linear controllers are optimized using Bode integrals as the performance bounds. The criterion for minimum phase behavior is employed as the condition for smooth blending with the adjacent regions differing in frequency. Due to this smooth blending, the exact shape of the participation rules is not critical. The static nonlinearities used to implement the transition between the control modes can all be chosen to be of hard saturation type, and the threshold of the saturation need not be exact.

A strong correlation exists in many systems between the error's amplitude and the error frequency content so that the errors fall into a set of diagonal windows as shown in Fig. 13.3. This is the case, for example, when the disturbance is a stochastic force with flat spectrum density, applied to a rigid body and causing the body displacement with the spectrum density inversely proportional to the square of the frequency, and when a displacement command profile is chosen in consideration of the limited force available from the actuator. Due to this correlation, the signal components that should go to a specific window can be selected either by the amplitude or by the frequency (the order of the selection is further discussed in Section 13.5). The controllers for these systems can be composed of linear operators of the windows and non-dynamic (static) nonlinear functions for pre- or post-processing and for splitting and/or combining the signals. Such controllers provide good performance for a wide variety of practical problems.

The simplest multi-window compensators are two-window compensators. The large amplitude components are processed by the regional compensator with low low-frequency gain; the low-amplitude signals are processed with higher low-frequency gain. The two-window controllers are widely employed, in particular, in anti-windup schemes, in acquisition and tracking systems, and in Nyquist-stable systems for provision of global and process stability.

The multi-window compensator is nonlinear dynamic. And conversely, the NDCs discussed in Chapters 10 and 11, whether they are made as a combination of parallel channels or as links with nonlinear local feedback, can be viewed as multi-window controllers.

13.3 Switching between hot controllers and to a cold controller

The sharpest participation rules are the instant switching between the controllers. Even with the switching, the transition between the regional laws can be made smooth.

Assume the inputs of several linear controllers are connected to the output of the feedback summer as shown in Fig. 13.4(a). The controllers are "hot," i.e., they process the error all the time. A simple switching or some nonlinear law can be used to choose one of the outputs or a nonlinear function of all or some outputs to send further to the actuator. Since the controllers are hot, and if the difference between the controllers is not excessive, the output signals of the controllers are to some extent similar, and switching between them will not create large transients.

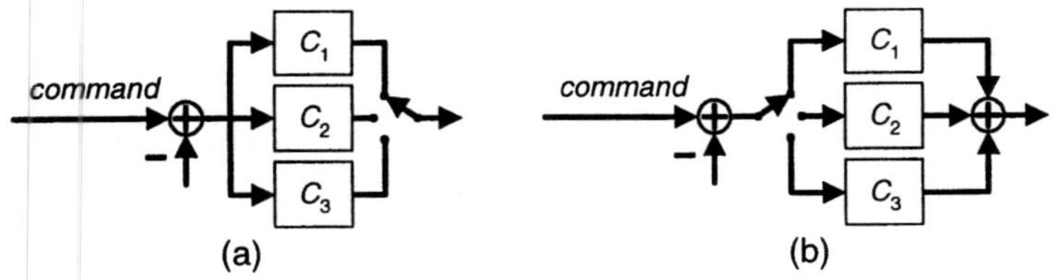

Fig. 13.4 Switching (a) between "hot" controllers and (b) to a "cold" controller

On the other hand, when a nonlinear law allocates the inputs, but the outputs are both permanently connected to the actuator input as shown in Fig. 13.4(b), the controller which has been off for a long time is "cold," and its output signal is zero or some constant. When the actuator input is switched to this controller from even a similar but "hot" other controller, large transients can result. The following example illustrates the problem.

Example 1. Consider the system with a single-integrator plant and two switchable linear compensators shown in Fig. 13.5(a).

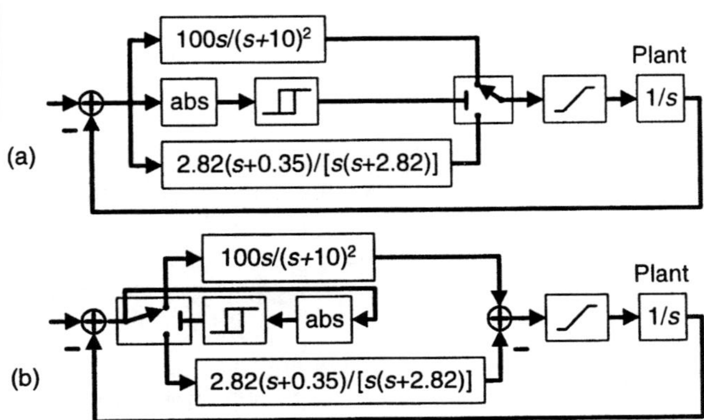

Fig. 13.5 Switching between controllers, (a) at the input, (b) at the output

The lower path compensator is a lead with an integrator. The upper path provides much larger feedback bandwidth but smaller feedback at lower frequencies (these responses will be discussed in Section 13.5). The compensators are permanently connected to the output of the feedback summer so that the compensators are "hot." Switching between the compensators' outputs occurs when the error magnitude achieves a certain threshold value, with some hysteresis to avoid frequent switching back and forth. In Fig. 13(b), a similar system is shown but here the switching occurs from a hot to a cold compensator. The saturation thresholds are –100, 100; for the hysteresis link, the thresholds are 0.1, 0.2, and the output switches between 1 and 0. The switch engages the upper path when 1 is applied to the switch input, and the lower path when 0 is applied to the switch input.

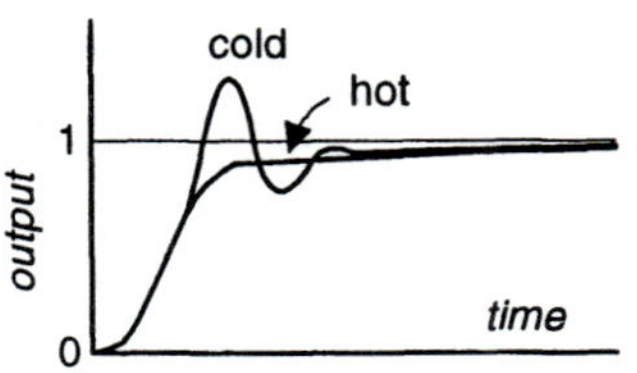

Fig. 13.6 Transient responses with switching to "hot" and "cold" compensators

The output transient responses for both systems are shown in Fig. 13.6. It is seen that in the system with hot compensators the transients caused by the switching are small while the transients caused by switching to the cold compensator are large.

13.4 Windup and anti-windup controllers

The time-responses of a system with saturation to small- and large-amplitude steps can be as shown in Fig. 13.7. The overshoot for the large-amplitude input step is excessive and persistent – this phenomenon is called ***windup***. The windup can be many times longer than the overshoot in the linear mode of operation (i.e., the overshoot for a step command of smaller amplitude).

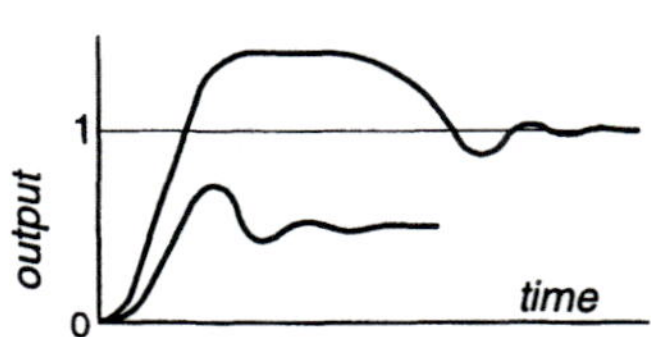

Fig. 13.7 Transient response in linear mode (lower curve) and windup (upper curve)

Windup is typically caused by a combination of two factors: the error integration in the compensator and the actuator saturation. The saturation limits the return signal and therefore prevents the error signal accumulated in the compensator integrator from being compensated. During the initial period after the step command is applied, when the output is still low and the error is large, the compensator integrates the error. When the time comes at which this integrated error would be compensated by the return signal in the linear mode of operation, this does not happen for large commands, since the return signal is reduced by the actuator saturation. Therefore, it might take a long time for the feedback to compensate the integrated error. The error "hangs up," and only after some time does the output signal drop to the steady state value.

Example 1. We will illustrate the windup with an example of a simple system with a single-integrator plant, shown in Fig. 13.8(a). The asymptotic loop response is shown in Fig. 13.8(b).

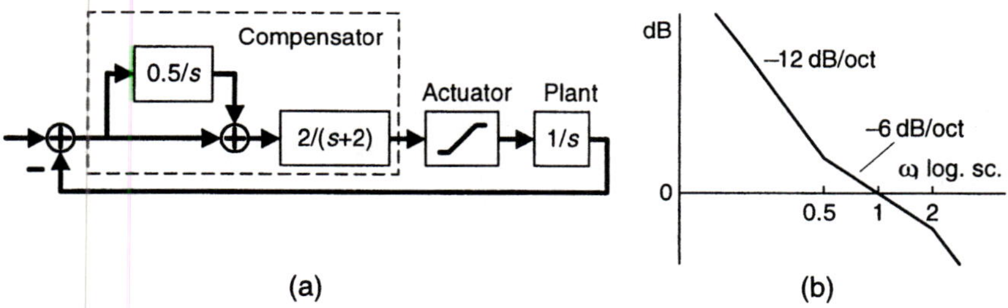

Fig. 13.8 (a) System with saturation and (b) the loop asymptotic Bode diagram

The compensator includes a parallel connection of an integrator and a unity gain path, and a following link with a pole at 2 rad/sec. The crossover frequency is 1 rad/sec. As seen from the asymptotic Bode diagram, the system must be stable. The loop transfer function (return ratio), the closed-loop transfer function, the transfer function from the input to the error, and the transfer function to the input of the actuator are, respectively:

$$T = \frac{2s+1}{s^3+2s^2} = \frac{m}{n}$$

$$M = \frac{m}{n+m} = \frac{2s+1}{s^3+2s^2+2s+1}$$

$$\frac{error}{command} = \frac{1}{F} = \frac{n}{n+m} = \frac{s^3+2s^2}{s^3+2s^2+2s+1}$$

$$\frac{actuator\ input}{command} = Ms = \frac{2s^2+s}{s^3+2s^2+2s+1}$$

The time-diagrams obtained with the MATLAB `step` command for this system without saturation are shown in Fig. 13.9. It is seen that the signal at the actuator input has a large peak. The output response overshoot follows this peak with about 90° delay because of the plant integration.

Clipping the signal peak at the actuator by the actuator saturation causes windup. Fig. 13.10 shows the output response to the step-function command when the saturation level in the actuator is ±0.25. The windup increases the height and the length of the overshoot.

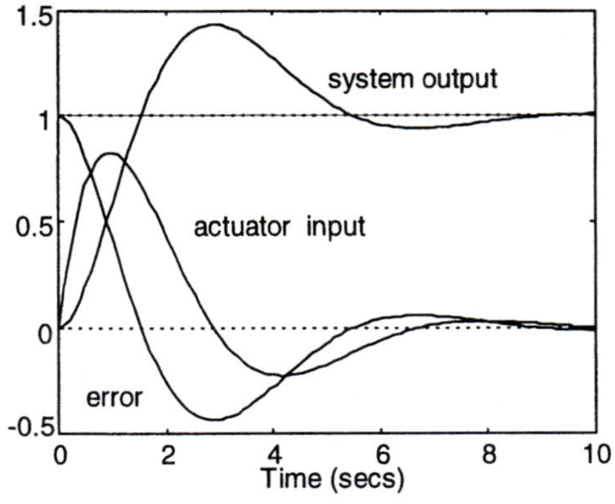

Fig. 13.9 Time-responses of a linear system

Using the DF concept, the qualitative explanation for the windup phenomenon goes as follows: Actuator saturation reduces the describing function loop gain thus shifting the equivalent crossover frequency down. The resulting overshoot is long, corresponding to this low crossover frequency. The value of the windup depends on the loop phase lag. When the phase stability margin is more than 70°, the windup is practically nonexistent, but it is large when the stability margin is 30° or smaller.

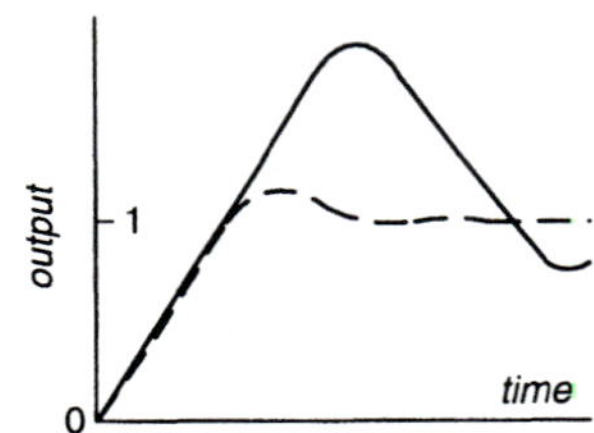

Fig. 13.10 Time-responses for the system in Fig. 13.8, solid line; with added saturation with threshold 0.1 in front of the integrator in the compensator, dashed line

The windup can be reduced or eliminated by employing nonlinear dynamic compensation. For example, placing an extra saturation link with saturation level 0.1 in front of the integrator in the compensator produces the response shown in Fig. 13.10 by the dashed line.

The explanation of the wind-up phenomenon can be also based on the idea of intermodulation: large amplitude low-frequency component overload the actuator and prevent it from passing higher-frequency components; the remedy suggested by this analysis is using separate channels for processing lower-frequency and higher-frequency signal components.

Windup in a *PID* system is commonly reduced or eliminated by placing a saturation in front of or after the integrator, or by ***resetting*** the integrator, i.e., by changing its output signal to zero at the rise-time.

In Fig. 13.11, two measures are shown which are widely used in analog compensators to reduce or prevent the windup. In Fig. 13.11(a), the diodes (or Zener diodes) in parallel to the capacitor in the integrator limit the maximum voltage on the capacitor and, therefore, the maximum charge in it that can be accumulated during the transients to a step command. In Fig. 13.11(b), the reset option is shown: simultaneously with the application of the step-command, the capacitor is shorted for certain time (close to the rise-time) to prevent the built-up charge.

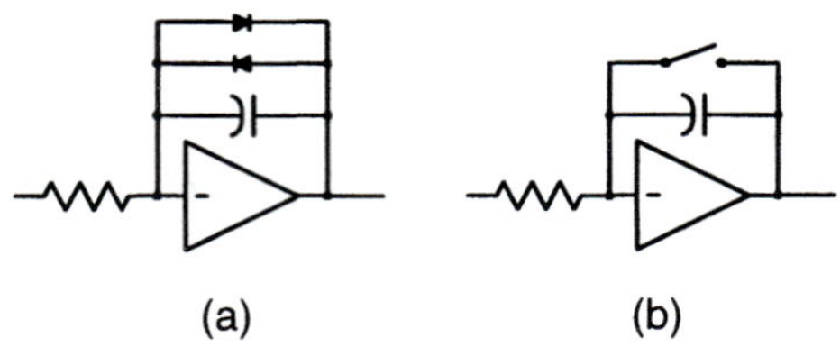

Fig. 13.11 Analog integrators with (a) parallel diodes and (b) with reset switch

Another method of the windup elimination is placing a rate limiter like that shown in Fig. 6.19 in the command path.

A saturation link is sometimes placed as well in front of the *P*-term as shown in Fig. 13.12, with a larger threshold than that of the saturation in front of the *I*-term. This makes the controller a three-window one.

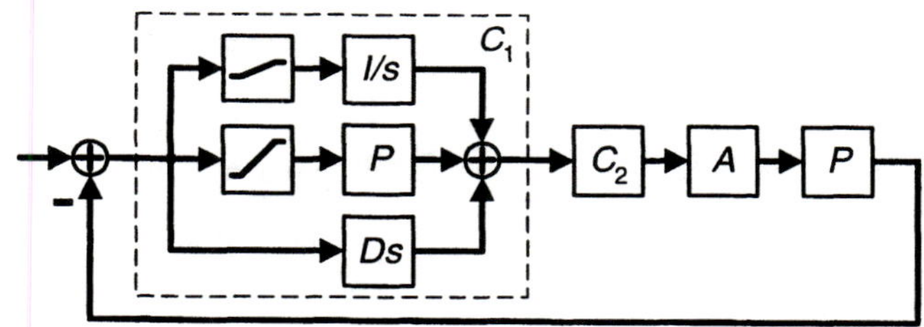

Fig. 13.12 Saturation links in front of the *I*- and *P*-paths

13.5 Selection order

The diagrams in Figs. 13.2 and 13.3 are somewhat ambiguous since they do not indicate whether the frequency selection or the amplitude selection is performed first. Often a particular order is required. This order is different in the block diagrams in Fig. 13.13 which exemplifies two types of architectures for multi-window compensators.

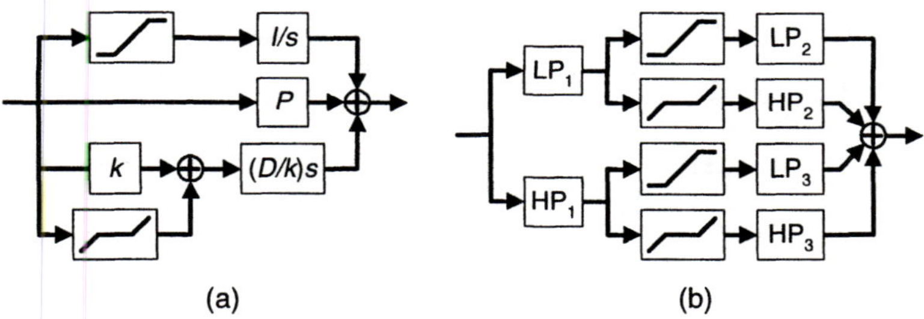

Fig. 13.13 Multi-window compensators with parallel channels

Fig. 13.13(a) shows compensators with parallel channels. In such compensators, saturation links are commonly placed in the low-frequency channel since this channel's gain response dominates at lower frequencies. At large signal levels, the saturation link reduces the low-frequency gain, and the phase lag of the compensator decreases thus reducing or eliminating the windup. Placing a saturation link with an appropriate threshold in front of the *I*-path commonly reduces the length and the height of the overshoot. The value of the threshold is not critical. Placing a dead-zone element in front of the high-frequency channel with $k < 1$ reduces the phase lag at large signal amplitudes, which helps to eliminate windup and to improve the transient response.

When the saturation link follows the linear ***filter fork*** (a filter fork is a combination of low-pass and high-pass filters for splitting or combining low-frequency and high-frequency signals, in this case low-pass LP_1 and high-pass HP_1) as in the left side of the block diagram in Fig. 13.13(b), the amplitude of the overshoot can be regulated by adjustments of the saturation threshold. However, in some cases, while the value of the overshoot reduces nicely, the time of the overshoot remains. The overshoot can be reduced to say, only 1%, but the hang-off can last a long time. (In a version of the temperature controller described further in this section, the overshoot of 1% lasted 1 hour.) In some applications, small amplitude windup is acceptable, but not in all applications.

The block diagram in Fig. 13.13(b) shows an architecture of mixed order of nonlinear windows and filter forks.

The best architecture of a *PID* controller with anti-windup depends on the sort of command the controller must respond to and also on the nature of the disturbances. When the only command is a step, placing the saturation first will prevent the accumulation of integrated error that causes windup. However, if there is large-amplitude, short pulse-type random noise, and the low-frequency components of this noise are substantial, the integral term needs to be functional to reduce the static error – but the integration is not effective if the peaks are being clipped by the saturation preceding the integration. In this case it would be better to lower the amplitude of the pulses by placing a low-pass filter (or an integrator) in front of the saturation link.

When the error signal can be arbitrary, the best performance (in a minimax sense) might be provided by a combination of the two strategies, i.e., the saturation link could be sandwiched between two low-pass filters. Here, two options exist: (a) half of an integration can be placed before the saturation, and the other half after; or (b) using two first-order links – one before the nonlinearity, to cut off the higher frequencies using a single pole, and one after, cutting off low-frequencies, with a zero compensating the pole of the first link. These implementations are indicated in Fig. 13.14. The pole/zero frequency can be adjusted by a single knob. But still, counting the saturation level, two extra knobs are required compared with the linear *PID* controller. (For the half-integrator version, there is only one additional knob.)

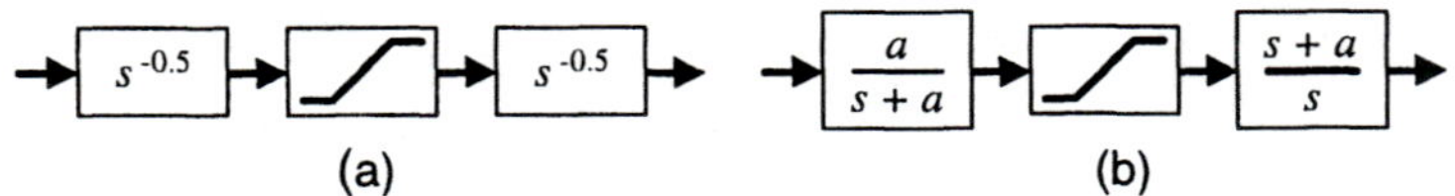

Fig. 13.14 Possible *I*-path implementations for anti-windup

These NDCs and the used for the same purpose NDC with local nonlinear feedback shown in Fig. 13.15 can be designed with the describing function approach as described in Chapter 11. The NDC shown in Fig. 13.15 can also be designed as described in Chapters 10 and 12 with the absolute stability and process stability approaches.

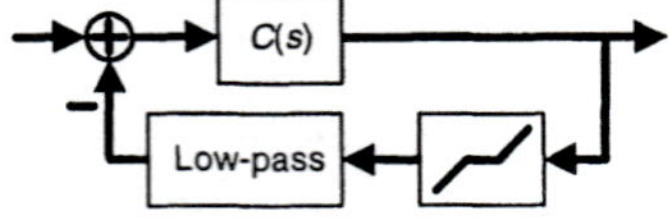

Fig. 13.15 Anti-windup NDC with local feedback

13.6 Acquisition and tracking

Acquisition and tracking systems, like those used in homing missiles, are designed to operate in two modes: ***acquisition mode*** when the error is large, and ***tracking mode*** when the error is small. An example of the acquisition/tracking type is a pointing control system for a spacecraft-mounted camera, in which a rapid retargeting maneuver is followed by a slow precise scanning pattern to form a mosaic image of the object. Another example is clock acquisition in the phase-locked loops of telecommunication systems and frequency synthesizers.

When the error signal is large, the system is in the acquisition regime, as indicated in Fig. 13.16(a), and the controller should respond rapidly, i.e., the feedback bandwidth should be wide, as shown in Fig. 13.16(b). In the acquisition mode it is not necessary however that the loop gain be very large, since the error is big and even with a small gain in the compensator, the actuator applies maximum available power to the plant. In contrast, in the tracking regime, the feedback bandwidth needs to be reduced to reduce the output effects of the sensor noise, but the value of the feedback at lower frequencies should be made rather large to minimize the jitter and the tracking error.

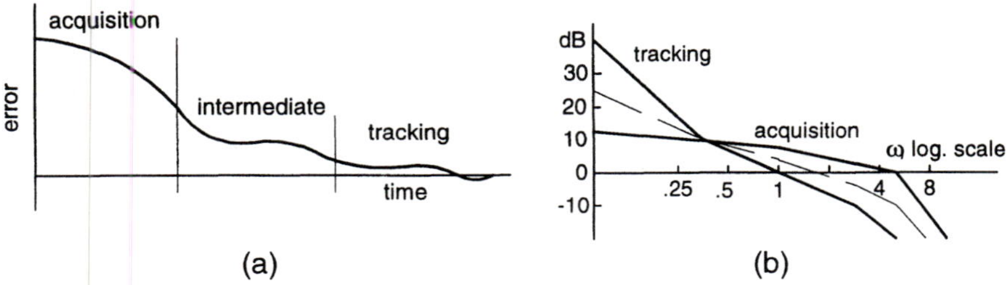

Fig. 13.16 (a) Error time-history and (b) acquisition/ tracking loop responses

While the determination of the optimal frequency responses for the acquisition mode and for the tracking mode is straightforward, guaranteeing smooth transient responses during transition from acquisition to tracking is not trivial. The transition can generate large transients in the output and in the error signals. If the transients are excessively large, the target can be de-acquired.

The transition between the responses can be made by switching as shown in Fig. 13.5, or by using nonlinear windows: the small errors are directed to the tracking compensator, and the large errors directed into the acquisition compensator. A special care must be taken to ensure that all intermediate frequency responses of the combined channel are acceptable. An improper intermediate response might result in an unstable system, or in a system with small stability margins and, therefore, would produce large-amplitude transient responses.

As an example showing the importance of paying attention to the intermediate responses, let the total loop response be the weighted sum (13.1) of the acquisition and the tracking responses shown in Fig. 13.16(b):

$$T = (1-k)T_{\text{acq}} + kT_{\text{tr}}\,, \tag{13.4}$$

and suppose that k smoothly varies from 0 to 1. As the transition from acquisition mode to track mode occurs, the acquisition response gradually sinks, the tracking response rises, and the frequency at which the moduli of the two components are equal, increases. When this frequency is still low, the difference in phase between the two transfer functions at this frequency exceeds 180°; therefore the total transfer function T has a zero in the right half-plane of s. (The conditions for the transfer function $W_1 + W_2$ to become n.m.p. when each of the channels W_1 and W_2 is m.p. were given in Chapters 4 and 5.) The nonminimum phase lag reduces the phase stability margin and may result in self-oscillation. The transients generated while the system passes these values of k can be big and disruptive, even causing the target to be lost.

For example, in a modification of the system in Fig. 13.5 where the compensators' outputs are combined via nonlinear windows, the overshoot reached 500%.

To reliably avoid the nonminimum phase lag and excessive transients, the slopes of the Bode diagrams of regional responses should not different more than by 9 dB/oct in two-window controllers. Hence, the two-window controller, although substantially better than a linear controller, still does not allow implementation of the best possible responses for acquisition and for tracking. This can be done with a three-window controller using an intermediate frequency response like that shown by the dashed line in Fig. 16(b).

The appropriate responses can be also obtained by changing the response continuously. One method for this was described in Sections 6.7.2, 11.7.

Example 1. With the regulation function

$$Q(s) = \frac{1.1(s+0.25)(s+2)(s+3)(s^2+5s+64)}{s(s+0.5)(s+7)(s^2+10s+100)}$$

and the intermediate return ratio

$$T_o(s) = \frac{250(s+0.3)}{s(s+0.005)(s^2+10s+100)}$$

the responses of (6.17) for the return ratio T are shown in Fig. 13.17.

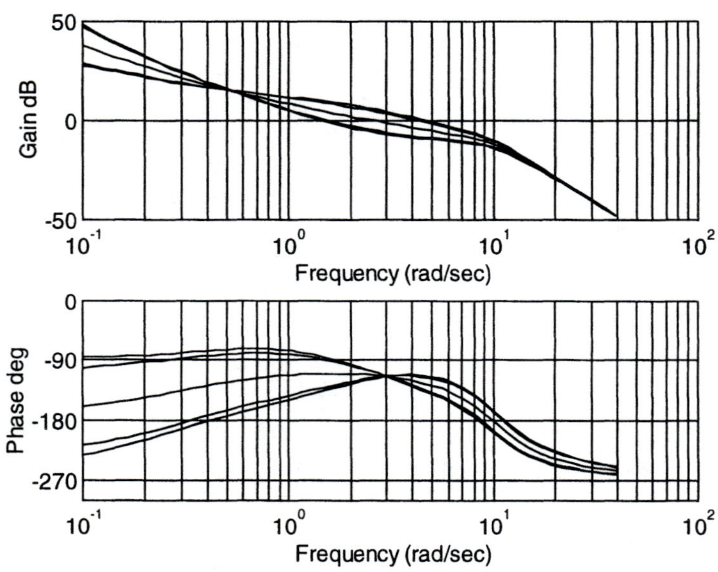

Fig. 13.17 Responses for smooth transition from acquisition to tracking

13.7 Time-optimal control

Time-optimal control changes the output variable between the commanded levels in minimum time, using an actuator with limited force or power. It has been proven that for the control to be time-optimal, the magnitude of the actuator action must be maximum all the time during the transition of the output variable between the limits. For example, shifting a rigid body with a force actuator in minimum time requires the force profile shown in Fig. 13.18.

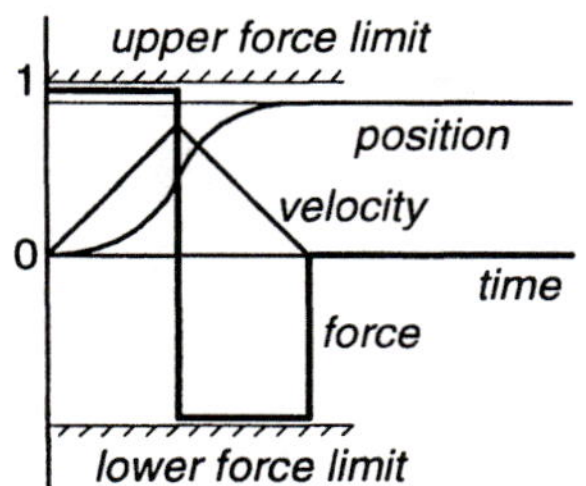

Fig. 13.18 Time-optimal control of the position of a rigid body

Time-optimal controller is a relay controller. It switches the action on, off, and between the positive and the negative values, at appropriate instants depending on the plant dynamics. When the plant is uncertain, the timing for the switching cannot be exactly calculated in advance, and the open-loop control entails considerable errors.

A stable closed-loop large-feedback control of an uncertain plant cannot employ a switch actuator since this actuator is equivalent to a saturation element with a preceding *infinite* gain linear link, and in practical systems the loop gain cannot be infinite at *all* frequencies. A good approximation to a closed-loop time-optimal controller in a system with saturation requires using *large* loop gain over the *wide bandwidth*. In order to obtain the best practical results, a proper balance must be kept between the achieved loop gain and the achieved feedback bandwidth.

Whatever the chosen feedback bandwidth, the feedback must be maximized under the limitation of keeping the system globally stable and without a windup. This requires using an NDC which can be implemented as a multi-window controller as was shown in Chapter 10. Using a multi-window controller can also help reduce the settling time.

For most common practical problems of time-optimal control, a two-window controller suffices. More windows may be necessary when the required settling error is very small, like in beam pointing of space optical telecommunication systems.

13.8 Examples

Example 1: Despin Control for S/C Booster Separation. The spacecraft booster is stabilized by spinning at 85 RPM. After separation from the booster, the spacecraft shown in Fig. 13.19(a) is despun by a yo-yo to about 2 RPM. (The yo-yo is a weight at the end of a cable wrapped several times around the spacecraft. When the spacecraft is released from the booster rocket, the weight is also released and begins unwrapping the cable. When all the cable length is unwrapped, the cable is separated from the spacecraft, and the yo-yo takes away most of the rotation momentum.) The remaining spacecraft spin needs to be removed by firing thrusters. Spinning of the spacecraft about the z-axis is unstable since the spacecraft is prolate and not absolutely symmetrical, and when left for some time uncontrolled, the spacecraft will tumble. Therefore, the despin should be fast. Because of large uncertainty in the initial conditions after the separation, with various positions and spin rates and different types of coupling between the axes, the controller design for the despin function must be made very robust, and at the same time, it must perform in a nearly time-optimal fashion. After the despin is complete, the controller must be changed to provide better control accuracy in the cruise mode.

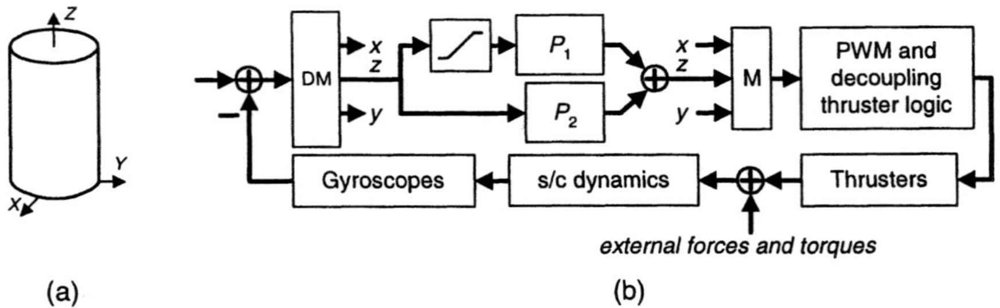

Fig. 13.19 Spacecraft (a) local frame coordinates, (b) attitude control block diagram

The controller shown in Fig. 13.19(b) uses pulse width modulated (PWM) thrusters. Since each thruster produces x-, y-, and z-torques, they are combined in pairs and decoupled by the thruster logic matrix. This renders the control of each axis independent to a certain extent. The problem is, however, complicated by coupling between the x-, y-, and z-rotations due to the spacecraft dynamics, including spinning of fuel and oxidizer, initially at the rate of the booster. Due to large plant uncertainty, the despin was chosen to be proportional, providing a large phase stability margin over the entire frequency range of possible plant uncertainty and x-, y-, and z-controllers coupling.

In the block diagram, the demultiplexer DM separates the error vector into its components. The multiplexer M does the opposite. The compensators are independent for the x-, y-, and z-rotations, i.e., the controller matrix is diagonal.

When the controllers' gains were chosen such as to despin the s/c without substantial overshoot, the z-axis response was as shown in Fig. 13.20(a). It is seen that the control is not time-optimal.

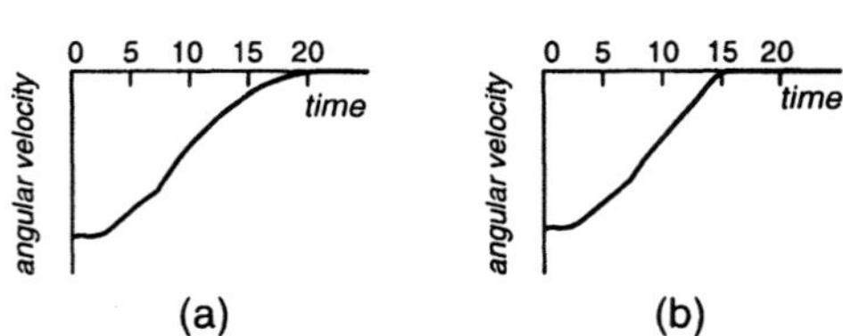

Fig. 13.20 Time-response of z-axis despinning: (a) linear controller, (b) two-window controller

A better controller is a two-window nonlinear controller which changes the control law on the basis of the absolute value of the error in each channel. This was done by passing the errors via saturation/dead zone windows with smooth transition between the control laws. The resulting control law is nearly perfect for the despin function and as well for the cruise mode. The transient response for this controller is shown in Fig. 13.20(b). The despin time was reduced by 20%.

The two-window controller performs better and is at the same time more robust than the original linear controller, with larger stability margins for the large error mode when the cross-axis coupling is the largest. (This two-window controller was, however, designed too late and has not been incorporated into the Mars Global Surveyor software. The initial simple *PID* controller was employed and provided sufficiently good despin and cruise control.)

This example shows that even for complicated plants with multi-channel coupled nonlinear feedback loops, a nonlinear two-window controller using only the error in individual channels for changing the control law provides nearly time-optimal performance, substantially better than that of linear controllers.

Example 2: Cassini spacecraft attitude control with thrusters (without PWM). The plant is close to a pure double integrator, although there are flexible modes at high frequencies. The thrusters are not throttled and not modulated, and the torque is some fixed positive or negative value, or zero (similarly to a 3-position relay.) These controllers often do not include an *I*-channel (low-frequency disturbances are almost nonexistent), and only include a *P*-channel and a high-frequency channel. To avoid windup, they commonly use saturation in the *P*-channel, which is then considered to be the low-frequency channel.

Example 3: Temperature controller for the mirrors of Cassini spacecraft's Narrow Angle Camera. The camera represents a small telescope. The primary and the secondary mirrors of the telescope must be kept at approximately the same temperature in order for the mirror surfaces to match each other, and the image in the focal plane to be clear.

Fig. 13.21 shows an electrical analogy to a thermal control system for a spacecraft-mounted telescope (recall Section 3.1). The plant is highly nonlinear because of the nonlinear law of heat radiation into free space. The heater H_2 is used to keep the temperature of the primary and the secondary mirrors within 1.6° K of each other. (Another loop, which is not discussed here, drives the primary mirror heater H_1 which maintains the absolute temperature within the 263°–298° K range). The heaters are pulse-width modulated with the modulation period of 6 sec, and the pulse-width timing resolution of 125 ms. The total heater power cannot exceed 6 W. The frequency response of the plant transfer function from the heater to the temperature differential is basically that of an integrator; however there are also radiative losses *G*, which are nonlinear.

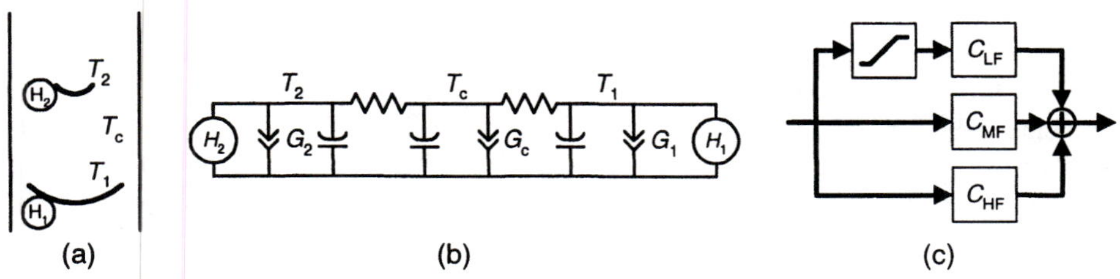

Fig. 13.21 Narrow Angle Camera (a), its thermal controller electrical analogy (b), and the controller configuration (c)

The compensator is implemented in three parallel channels. The compensation for the high-frequency (HF) channel is a complex pole pair:

$$C_{\text{HF}} = \frac{0.1}{s^2 + 0.125s + 0.1}. \tag{13.5}$$

The medium-frequency (MF) and low-frequency (LF) channels are first-order:

$$C_{\text{MF}} = \frac{0.5}{s + 0.1} \quad \text{and} \quad C_{\text{LF}} = \frac{0.5}{s + 0.035}, \tag{13.6}$$

and a saturation element precedes the LF compensation. The separate and combined frequency responses of the compensator channels (ignoring the nonlinearity) are shown in Fig. 13.22, and the loop frequency response is shown in Fig. 13.23. The parallel connection of the MF and HF channels forms a Bode step on the Bode diagram near 30 mHz. The controller was implemented as digital, and the feedback bandwidth was ultimately limited by sampling effects.

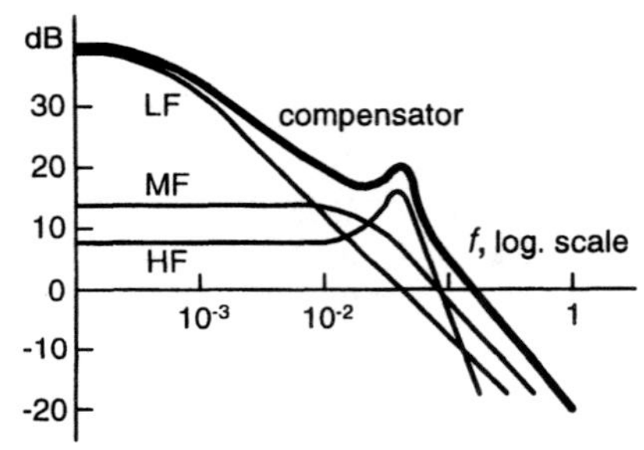

Fig. 13.22 Parallel-channel compensator responses for thermal controller

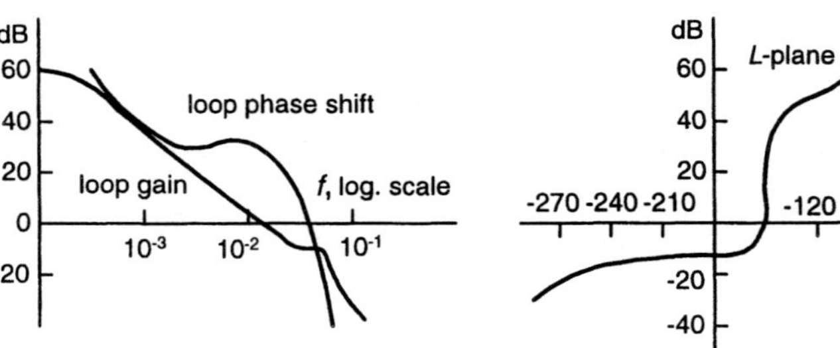

Fig. 13.23 Loop frequency response for thermal controller

The LF compensation steepens the response in the 1–10 mHz range, providing larger feedback at low frequencies. This would result in windup, i.e., excessive overshooting, for transients in which the heater saturates, unless an anti-windup device is provided. (For the typical power-on transient, the heater saturates immediately.) The anti-windup device used here is a saturation element preceding the LF path. This prevents the LF path from "integrating up" excessively when the actuator is saturated and the error is large. After a few (5 – 10) step response simulations were observed, the saturation threshold in the LF path was chosen to be 0.8° K. The closed-loop system transient response is notably insensitive to variations in this level, which makes a good level easy to determine. Note that placement of the saturation element *after* the LF compensation results in a transient with a windup error that is small but takes an excessive amount of time to decay. Industrial controllers, which often place the saturation link after the *I*-path, frequently use integrator reset features to overcome this problem.

The power-on step response for the controller is shown in Fig. 13.24. The heater power is maximum most of the time while the mirror is heated up. The controller is nearly time-optimal, and the overshoot is insignificant.

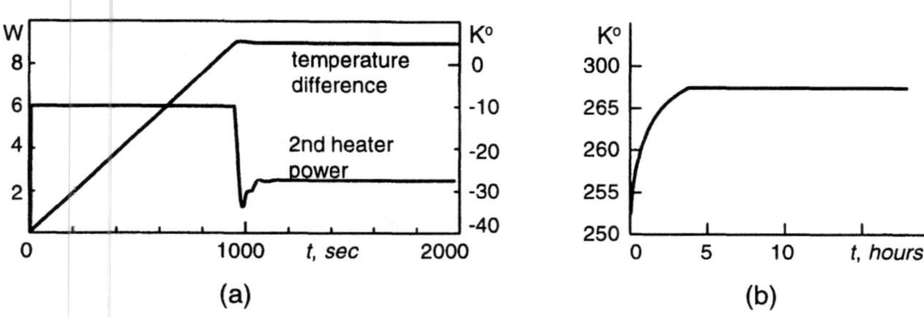

Fig. 13.24 Step response for thermal controller

Example 4: The microgravity accelerometer that was described in Section 11.9 is another example of a two-window controller. This controller not only provides global stability with loop phase shift of π at frequencies where the loop gain is large, but also eliminates windup, reduces the overshoot, and increases the acquisition band of the tunneling condition. The tunnel effect is an exponential function, and if the feedback loop were initialized when the distance in the tunnel sensor gap was much smaller than normal, then the loop gain would be much larger, and the system would become unstable if it were not for the gain reduction by the NDC.

Example 5. The antenna pointing controller described in Section 6.10.5 is another example of a two-window controller.

13.9 Problems

1 In a system with saturation, a double integrator plant, and a *PID* controller, study the effect on the windup of the saturation links placed in front of the *I* and *P* paths. Use SIMULINK or SPICE, and make simulations with different saturation thresholds.

2 In the previous problem, use dead zone feedback paths about the integral and proportional paths.

3 Make SIMULINK simulations of the system shown in Fig. 13.5.

4 Make a simulation for the acquisition and tracking problem with switching between two responses, similar in shape (of the *PI*-type) but one shifted relative to the other by an octave along the frequency axis.

5 The nominal plant is $P_o = 1/[(s + 10)(s + 100)]$ with up to ± 2 dB variations in gain. The actuator is a saturation link with unity threshold. The feedback bandwidth is 200 Hz (it is limited by sensor noise or the plant uncertainties). Design a good controller.

6 In Example 1 in Section 13.8, assume the plant gain is uncertain within 3 dB. Plot the transient response for plant gain deviations up and down. What is the conclusion?

7 Study command feedforward with different frequency responses of the command feedforward link in nonlinear modes of operation.

8 Study a command feedforward system for
(a) large-level commands and
(b) large deviations of the plant response from the nominal.

9 Study multi-window controllers with bounded internal variables in the plant and the actuator.

10 Study a system with multi-window compensators, command feedforward, prefilter, and feedback path.

Appendices

Appendix 1 Feedback control, elementary treatment

A1.1 Introduction

The easiest way to comprehend complex systems is to break them into building blocks. The block diagram method became standard in computer aided analysis and design. Understanding systems relies on understanding the blocks' interaction and, particularly, on understanding the feedback.

The term ***feedback*** was applied to closed-loop system engineering by Harold Black of The Bell Laboratories in the 1920s. It describes regulation processes in engineering and in biological, economical, social and political systems – everywhere where information about the results comes back and influences the input.

The fundamentals of feedback can be expressed in simple terms. Feedback systems can and, in our opinion, should be taught as a part of a science course in high school, preceding and facilitating the teaching of physics, chemistry, biology, and social sciences. It is important to demonstrate not only how automatic control works, but also how the system dynamics limit the speed and accuracy of control, and why and how feedback systems fail.

The word ***feedback*** is often employed in a much simplified sense, denoting merely the obtaining of information on the results of one's action. There is much more, however, to the quantitative meaning and methods of feedback. In the modern world, feedback is employed in spacecraft and missile control, in cars, and in TV sets, and is widely used to explain and quantify the processes studied in biology, economics, and social sciences. We hope the following material will provide a better perception of how the systems of this world operate.

A1.2 Feedback control, elementary treatment

A1.2.1 Feedback block diagram

For the purposes of analysis and design of a complicated system, the system can be presented as an interconnection of smaller parts of the systems called subsystems or ***blocks***.

Pictures like Fig. A1.1 are called ***block diagrams***. Here, in response to the ***input*** value of 3, the ***output*** value of 25 is produced. The diagram describes this arithmetic:

$$3 \times 10 = 30; \quad 30 - 5 = 25.$$

Fig. A1.1 A block diagram with a summer

The factor by which the block's output is larger than the input (10 in this case) is called the block's ***gain coefficient***.

In the block diagram in Fig. A1.2, the output of 5 is fed back to the input summer forming a ***feedback loop***. The arithmetic is the following:

$6 - 5 = 1;\ 1 \times 5 = 5.$

Fig. A1.2 Block diagram of a feedback loop

An inscription in the block (or close to it) often gives the name of a physical device represented by the block.

A1.2.2 Feedback control

We start with examples.

Example 1. While a rifle is being aimed, an eye looks at the target through the rifle sights. In Fig. A1.3, the rifle points down and to the left of the target. The pointing ***error*** is the difference between the direction to the target and the rifle pointing direction.

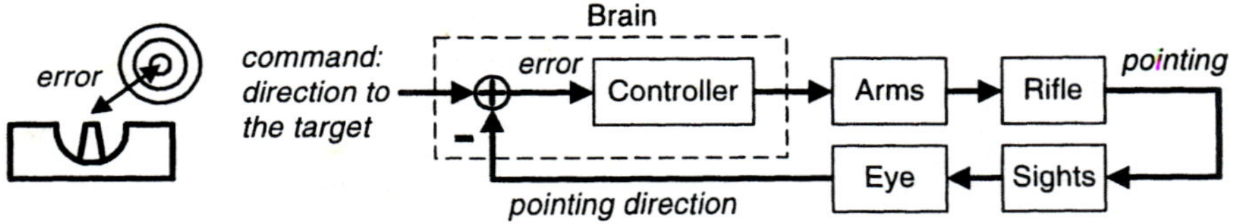

Fig. A1.3 Block diagram describing aiming of a rifle

Using the sights, the eye registers the rifle pointing direction and communicates this information to the brain as indicated in Fig. A1.3. The brain (a) calculates the error by subtracting the pointing direction from the command, and (b) acting also as a ***controller***, issues appropriate orders to the arms' muscles to correct the aim.

Example 2. While steering a four-wheel-drive car in a desert, the "command" can be: "drive west." The block diagram in Fig. A1.4 shows the process of the control system operation. The eyes estimate the actual direction of the car's motion, and the brain calculates the error and gives orders to the hands.

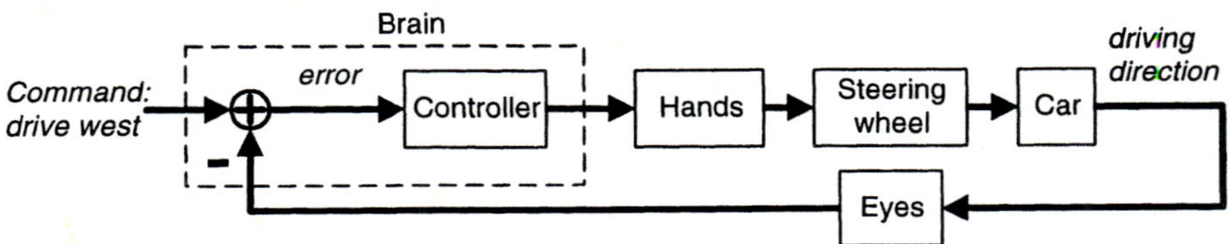

Fig. A1.4 Block diagram describing driving a car

Example 3. While pointing a small spacecraft with a hard-mounted telescope to take a good picture of a planet, the flight computer calculates the direction to the planet and sends this data as a command to the control system shown in Fig. A1.5. The pointing angle sensor here can be, for example, a camera with a wider angle than the telescope; the steering means can be the jets rotating the spacecraft. The signal at the summer output is the difference between the command and the actual readings of the sensor, i.e., the error.

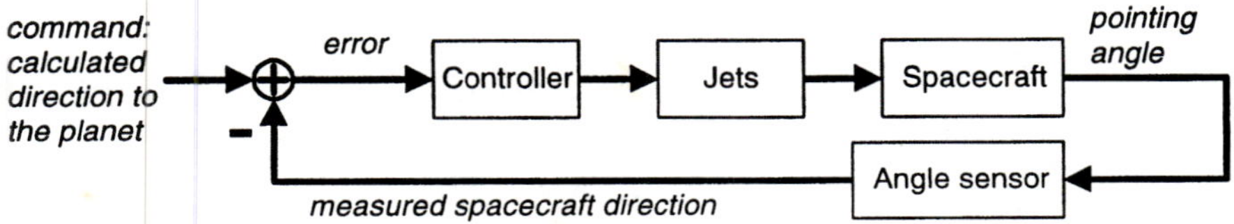

Fig. A1.5 Block diagram describing pointing a spacecraft

We may now generalize the ***feedback control system*** which is also called the ***closed-loop system*** to the form shown in Fig. A1.6. The ***actuator*** drives the object of control, called the ***plant***.

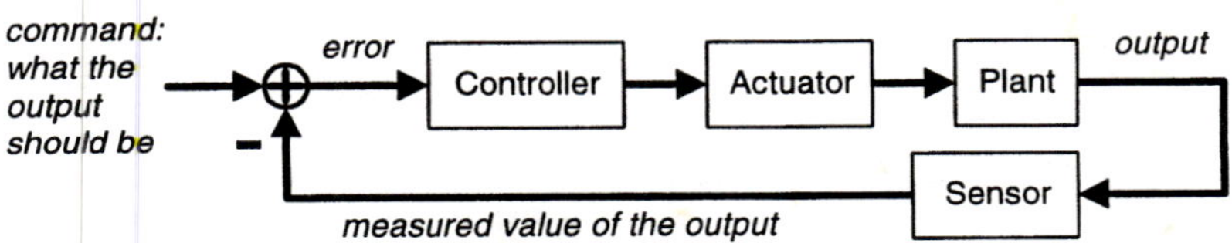

Fig. A1.6 General block diagram describing control of a plant

If the error is 0, no action is taken.

The controller's gain coefficient is large. It senses even a small error and aggressively orders the actuator to compensate for the error.

In a typical control system, the actuator is powerful, but not as accurate as the sensor. The sensor is accurate, but not powerful. The feedback control integrates the best features of both the actuator and the sensor. It is widely employed in biological and engineering systems.

We now know enough to start *designing* control systems.

Example 4. Design a system to maintain the temperature of 1206° C within an industrial furnace.

We use the general diagram of feedback control in Fig. A1.6. The command here is: "1206° C." The actuator is now an electrical heater. The plant is the furnace with the payload. The sensor is an electrical thermometer. The resulting block diagram is shown in Fig. A1.7.

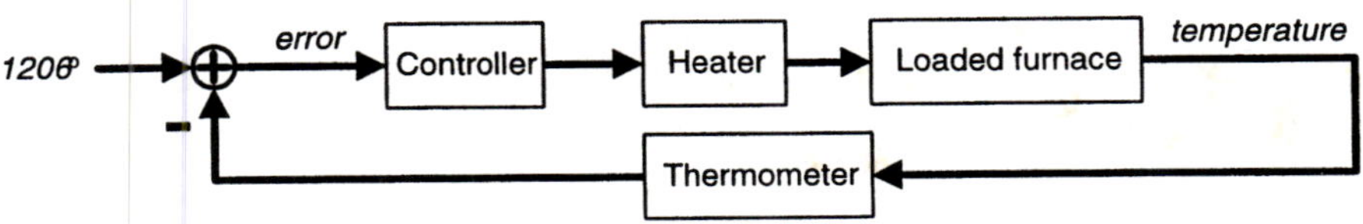

Fig. A1.7 Block diagram of temperature control

Example 5. Design a system to maintain a pressure of 2.2 atmospheres in a chamber. Now, the command is 2.2 atm, the actuator is a pump, the plant is the chamber, and the sensor is a pressure gauge, as shown in Fig. A1.8.

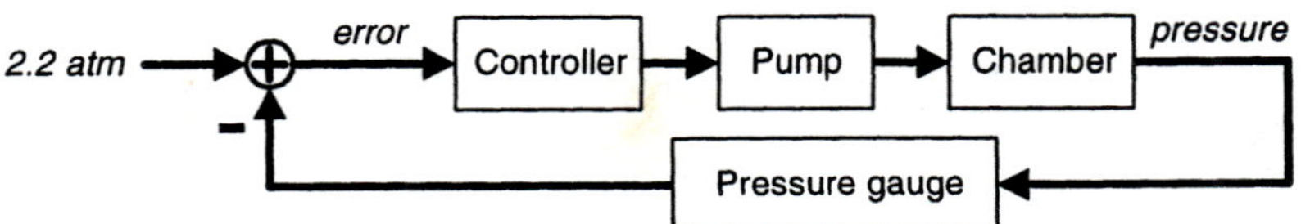

Fig. A1.8 Block diagram of pressure control

Assume that the pressure gauge reading is 2.15 atm. This means that the command is not performed perfectly, and the error is 0.05 atm.

Example 6. Design a block diagram of a biological system to produce a certain amount of a specific tissue.

The block diagram Fig. A1.9 will do it. (What is inside the blocks, i.e., how the tissue is manufactured and measured, is not considered here.)

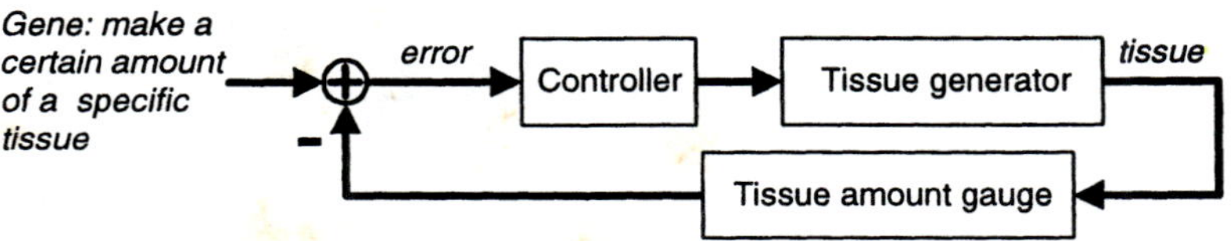

Fig. A1.9 Block diagram of tissue manufacturing control

When the feedback mechanism fails, the tissue continues to be manufactured even when there is more than enough of it. This may cause a serious health problem.

A1.2.3 Links

Feedback systems are composed of links.

An electronic thermometer, for example, produces electrical voltage proportional to the temperature. This link speaks two languages: its input understands degrees Fahrenheit, and its output speaks in volts. When the temperature is 1°F, the output is 0.01 V. At 100°F, the thermometer output is 1 V. That is, this particular thermometer generates 0.01 V per each degree as indicated in Fig. A1.10.

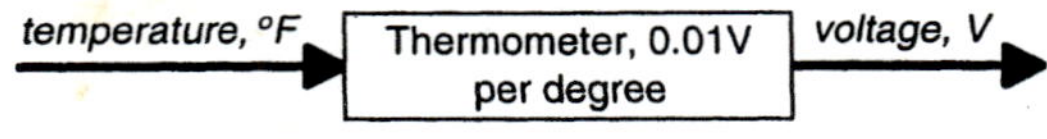

Fig. A1.10 Thermometer link

The electronic pressure gauge displayed in Fig. A1.11 produces 1 V output for each atmosphere of pressure. In other words, its transmission coefficient is 1 V/atm.

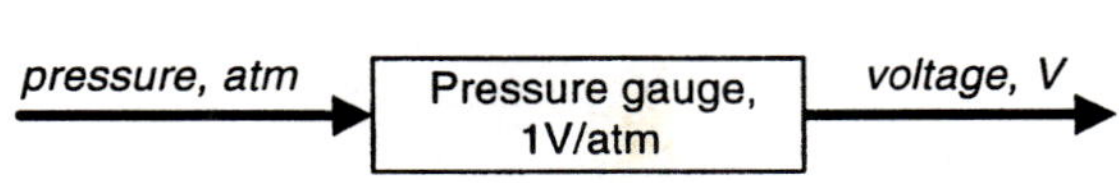

Fig. A1.11 Pressure gauge link

Fig. A1.12 shows a connection of two links making a composite link. An electrical heater consumes electrical power from the input, in watts, and produces heat, 0.24 calories each second per each watt. The heat raises the temperature of the payload in the furnace chamber by an amount depending on the size of the payload.

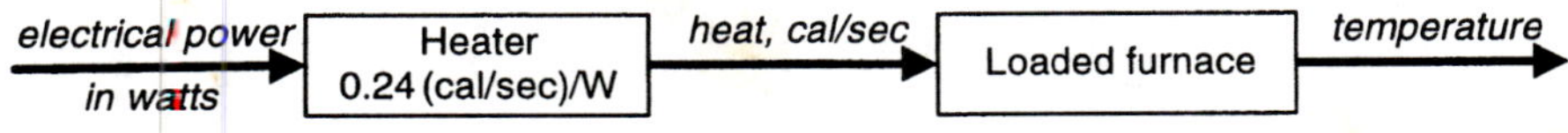

Fig. A1.12 Composite link

Can we connect two arbitrary links: a thermometer and a pressure gauge, for example, as in Fig. A1.13? No, this will not work, and not only because 13 is an unlucky number: the links speak different languages and do not understand each other.

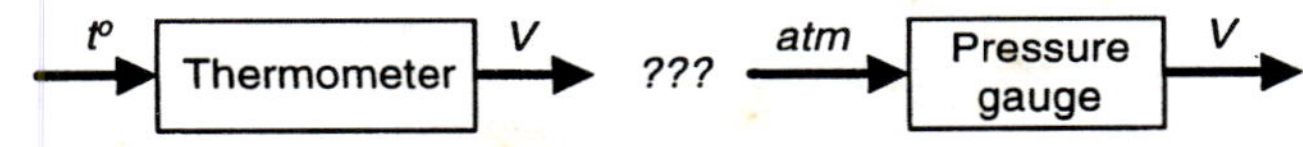

Fig. A1.13 Links that cannot be connected

At the link joint, the language must be common to the links. We can, for example, connect an electrical thermometer to the output of the links of Fig. A1.12, as shown in Fig. A1.14.

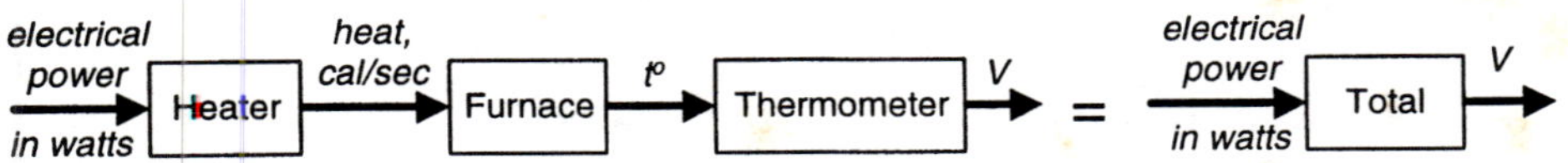

Fig. A1.14 Equivalent composite link

When several links are correctly connected in a chain, the resulting gain coefficient is the product of these links' gain coefficients.

A1.3 Why control cannot be perfect

A1.3.1 Dynamic links

We assumed before that thermometers measure the temperature instantly. This is only approximately correct. A mercury thermometer has to be kept in the mouth for several minutes for the readings to approach the mouth temperature, as shown in Fig. 15(a). Electronic thermometers settle faster, but still not instantly. The thermometer readings depend not only on the instant temperature, but also on what the temperature was seconds and minutes ago. Thus, the thermometer has *memory*.

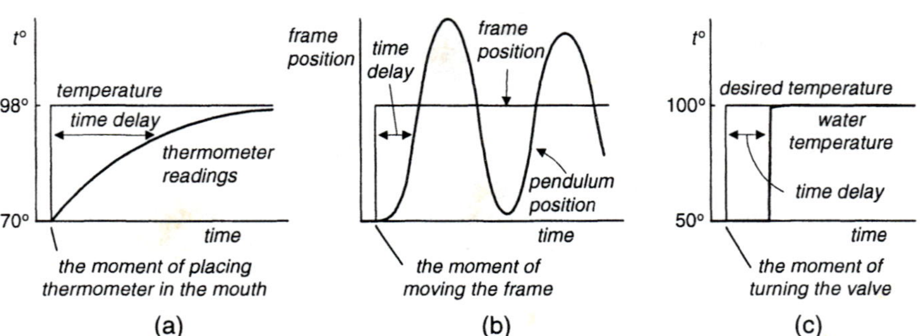

Fig. A1.15 Dynamic links time histories: (a) time history of thermometer readings; (b) pendulum position history; (c) temperature history after a delay

Consider now a pendulum suspended from a frame, initially at rest, as shown in Fig. A1.16 by dashed lines. Let us push the frame by some distance, considering this distance the input of the link. The output of the link is the pendulum position. It depends not only on the frame position at the current moment, but also on the *previous* frame position and when the position changed. The plot for the pendulum position after the frame was pushed is shown in Fig. A1.15(b). The output depends on what happened in the past. The pendulum has memory.

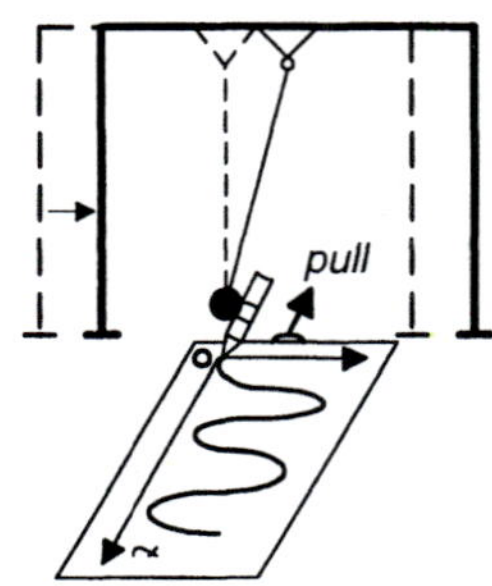

Fig. A1.16 Pendulum

The plot can be plotted by pulling the paper as shown.

Another link with memory is the shower. The input is the hot water valve position, the output is the shower water temperature. The input and output time histories are shown in Fig. A1.15(c). The output is delayed by the time it takes the water to flow through the pipe.

Devices or processes with memory are called ***dynamic***. When the input is changed instantly by a certain amount, the change in the output is delayed, the output can grow gradually, and can be oscillatory.

Certainly, a feedback system composed of dynamic links itself becomes dynamic.

A1.3.2 Control accuracy limitations

After a step command (s) is issued, it takes some time for the output of a dynamic control system to change, as seen in Fig. A1.17, curve (a). The error decreases with time but does not completely vanish.

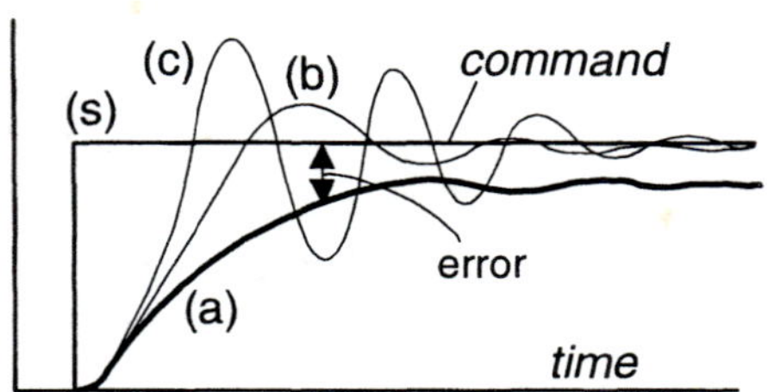

Fig. A1.17 Control system's output time history

A more aggressive control, with larger controller gain, reduces the error as exemplified in Fig. A1.17, curve (b). However, with bigger controller gain, the problem arises of stopping the actuator action immediately after the error is reduced to zero. Since the links in the feedback loop are dynamic, the information from the sensor that the error is already zero comes back to the actuator with some delay, and the actuator action proceeds for some time after the moment it should be terminated. Then, an error of the opposite sign will appear at the output. The process then repeats itself, and the output oscillates as seen in the Fig. A1.17, curve (b). If the controller gain is even larger, the oscillation amplitude increases and output will look more like Fig. A1.17(c); with further increase of the controller gain, the oscillation becomes periodical and with large amplitude like that in Fig. A1.18. A simple way to explore this process experimentally is by trying to regulate the shower temperature while being very impatient.

The larger the total delay of all dynamic links in the feedback loop, the smaller must be the controller gain for the system to remain stable, and the less accurate and the more sluggish will be the control. Thus, while a feedback control system is being designed, major attention should be paid to reduction of delays in the loop.

A1.4 More about feedback

A1.4.1 Self-oscillation

The time history of an oscillation is shown in Fig. A1.18(a). It can be drawn by a pen bound to the pendulum on Fig. A1.16 while a sheet of paper is being dragged in the direction perpendicular to oscillation. This curve is called a ***sinusoid***. The number of oscillations per second is called the oscillation ***frequency*** in Hertz (Hz). A single oscillation is a ***cycle***.

The cycle is 360° long. Oscillation (b) is ***in phase*** with oscillation (a). Notice that if shifted by 360°, the oscillation remains in phase. Oscillation (c) is 90° delayed compared with oscillation (a). Oscillation (d) is delayed by 180° which is equivalent to changing the sign of the oscillation.

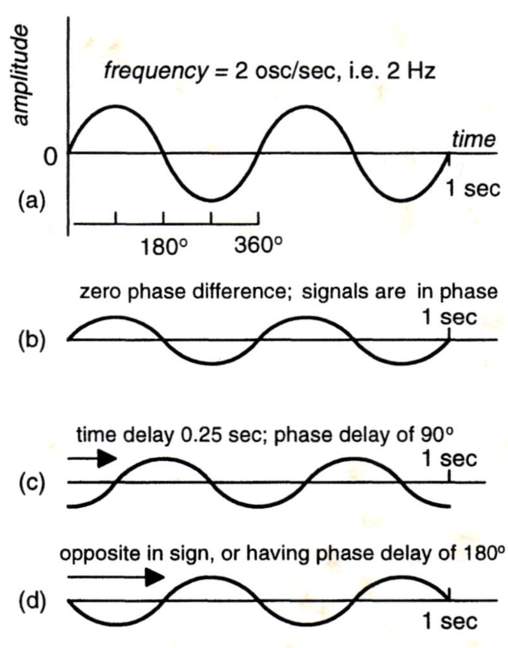

Fig. A1.18 Time histories of motion

Rigorously speaking, oscillation of the pendulum is not exactly sinusoidal, and the oscillation gradually dies. The oscillation amplitude can be maintained constant if the friction-caused losses of energy are compensated by some mechanism injecting energy in the system by some actuator.

The operation of a swing shown in Fig. A1.19 can be explained with the help of the block diagram in Fig. A1.20. The actuator is the kid's muscles. The kid jerks his body to produce extra tension in the rope, to sustain the oscillation. The sensors he uses are in his vestibular apparatus. He detects the proper timing for his movements by feeling zero velocity in the rightmost and leftmost positions. To sustain the self-oscillation, the return signal in the feedback system must be in phase with the swing motion *and* must be strong enough.

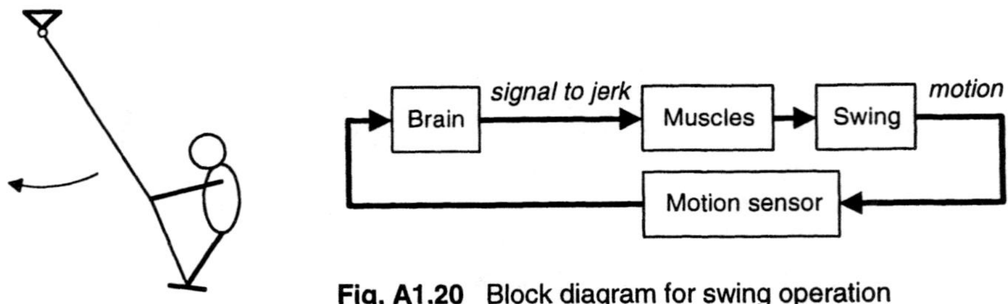

Fig. A1.20 Block diagram for swing operation

Fig. A1.19 Swing

Similar feedback systems are employed to generate radio and TV signals and in the dynamos generating electricity at power stations.

A1.4.2 Loop frequency response

Links and entire systems can be tested with a set of sinusoidal inputs with different frequencies. This method is used, for example, in testing audio recording systems like that illustrated in Fig. A1.21. This system contains a CD player, a power amplifier, and speakers. The input to output gain coefficient expressed in decibels (dB) is the system gain. The gain frequency responses are exemplified in Fig. A1.22 for a good quality system with nearly equal gain at all frequencies from the lower frequencies of 25 Hz to the higher frequencies of 18,000 Hz, and for a portable boombox where the lower and higher frequencies are not well reproduced, thus making the sound different from the original.

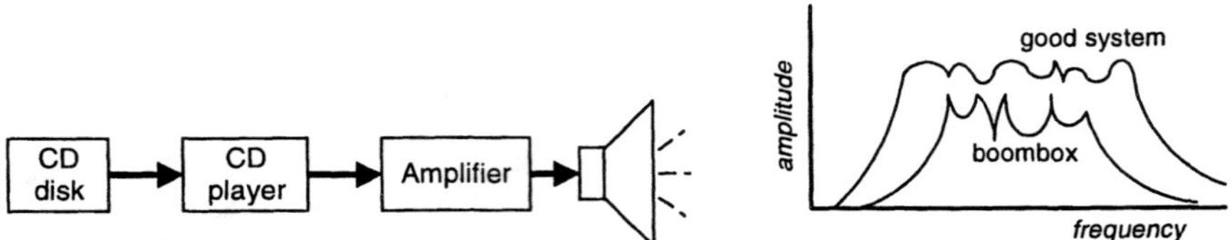

Fig. A1.21 CD player block diagram

Fig. A1.22 Frequency responses of a CD player

The gain frequency response is not flat (as would be desired) because the speakers resonate at many frequencies with various amplitudes. Better and more expensive speakers (bigger, with better magnets, with larger and firmer enclosures, with some special filling inside the enclosures) have wider and flatter gain responses.

Audio systems are typically characterized by only the sound amplitude responses since our ears are to a large extent insensitive to the phase of the sound. For feedback systems, however, the phase shift in the loop is important as well – as we already know from the analysis of the swing.

A1.4.3 Control system design using frequency responses

Self-oscillation in control systems is potentially disastrous. A control system must be ***stable***, i.e., self-oscillation must not occur at any frequency. To prevent self-oscillation, feedback control systems are designed such that at frequencies where the return signal is big, its phase is not such that supports oscillation, and at those frequencies where the return signal is in a phase that supports oscillation, the return signal is sufficiently small.

During control system design, the gain and the phase shift frequency responses about the feedback loop are first calculated with computers and then measured experimentally and displayed with a signal analyzer as is shown in Fig. A1.23.

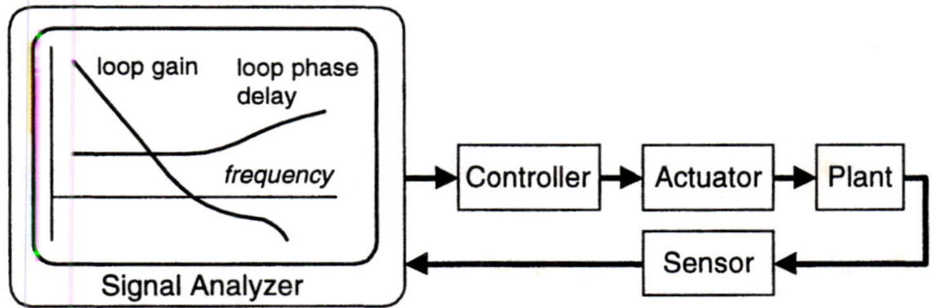

Fig. A1.23 Measuring loop transmission frequency responses

A1.4.4 Some algebra

We already know qualitatively that when the controller gain is large, the error is small. Using some algebra we will find how small the error is, and how many times it is reduced by the feedback.

As mentioned, in a series link connection, the equivalent gain coefficient is the product of the gain coefficients of all the elementary links. Then, the feedback system on Fig. A1.24 implements the following equations:

$$output = error \times CAP \quad \text{(A1.1)}$$

$$error = command - fbs \quad \text{(A1.2)}$$

$$fbs = error \times CAPS \quad \text{(A1.3)}$$

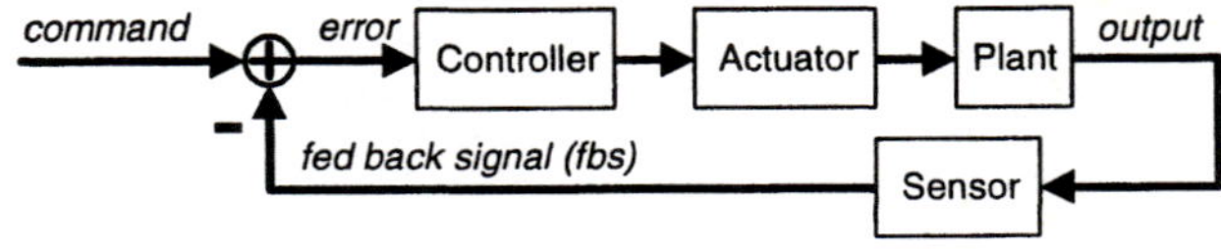

Fig. A1.24 Feedback control system

By substituting (A1.3) into (A1.2) we get

$$error = command - error \times CAPS,$$

wherefrom

$error = command/(1 + CAPS)$.

This expression shows that the *error* is (1 + *CAPS*) times smaller than the command. The expression 1 + *CAPS* is, numerically, the ***feedback***. The larger the feedback, the smaller the error. And, we already have concluded that the feedback cannot be arbitrarily large.

After substituting this expression for the error into (A1.1), we find that

$$output = command \frac{-1}{S} \frac{CAPS}{1 + CAPS} \tag{A1.4}$$

When the product *CAPS* is large, much more than 1, then 1 in the denominator can be neglected, and

$$output \approx command \times \frac{-1}{S}. \tag{A1.5}$$

For example, if $S = -2$, $CA = 20$, and $P = 1$, or 10, or 100, the *output* from (A1.4) is correspondingly 0.488, 0.4988, 0.49988, i.e., very close to 0.5 from (A1.5).

A1.4.5 Disturbance rejection

The actuator's inaccuracy and the environmental ***disturbances*** *D* shown in Fig. A1.25 add some unwanted components to the plant's output. The value of these unwanted components at the system's output can be calculated with (A1.4) when using the disturbance as the command:

$$disturbances\ at\ the\ output = D \frac{P}{1 + CAPS}$$

Without the feedback, the effect of disturbances at the system's output would be *DP*. Therefore, *the feedback reduces the output effect of disturbances (1 + CAPS) times*. The feedback can be therefore used to reduce the effects of mechanical vibrations on some precision instruments and machinery.

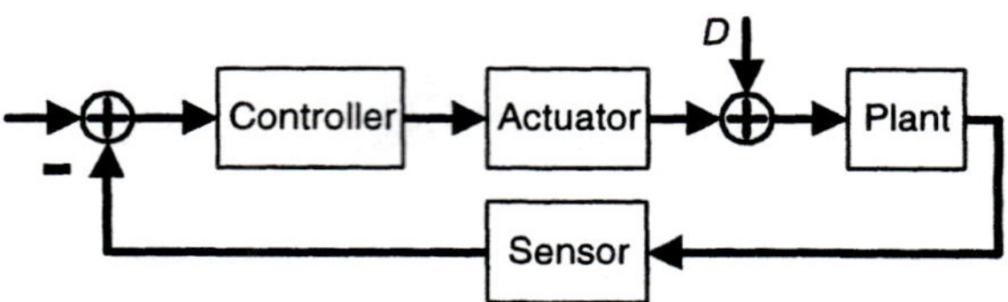

Fig. A1.25 Disturbances at the plant input

For example, floor vibrations caused by passing cars, by the air-conditioner motors, and by people walking around, disturb precision optical systems mounted on a desk. By using position sensors and piezoelectric motors to move the desktop, the amplitude of the optics' vibrations can be reduced many times.

A1.4.6 Conclusion

In this short introduction to feedback systems we considered single-loop control with a single actuator and a single sensor. Sometimes, several sensors are employed. While operating our hands, for example, we use position and stress sensors in the muscles, tactile and temperature sensors in the skin, and the eyes. In complex systems, many loops coexist to regulate various parameters: heartbeat frequency and strength, amount of enzymes in the stomach, body temperature, and many others. It would be difficult to count all feedback loops in a TV set – there are hundreds of them.

A1.5 New words

The following list was composed when the material of this Appendix was used by one of the authors to teach feedback control to his then eleven-year-old daughter Helen. The list might be useful for those readers who will attempt a similar task of explaining feedback control to their children.

actuator - a device like a motor, power amplifier, muscle
closed-loop control, feedback control - using data from a sensor to correct actuator actions
disturbance - a source of error in the system
dynamics - description of physical system motion when forces are applied
feedback - return of signal or of information from the output to the input
frequency in Hertz (Hz) - a number of oscillation periods per second
frequency response - a plot of gain or phase dependence on frequency
gain coefficient - a number by which the signal is amplified
in phase - without phase delay
link - representation of one variable dependence on another
payload - a useful or paid for load to be moved or heated
phase shift - a change in phase
plant - an object to be controlled
positive feedback - feedback supporting oscillation
quantitative - expressed in numbers
sensor - a measuring device
sinusoid - a curve describing periodic oscillation
summer - a device whose output equals the sum of its inputs
variable - a numerical description of a feature of a system (such as temperature or distance)

Appendix 2 Frequency responses

A2.1 Frequency responses

Linear systems have the property that *when they are driven by a sinusoidal signal, the output variable* – and, in fact, any variable of the system – *is also sinusoidal with the same frequency*. Thus, when signal $u_1(t) = U_{m1}\sin \omega t$ is applied to the input of a link, the output signal of the link is $u_2(t) = U_{m2}\sin(\omega t + \varphi)$, where U_{m1} and U_{m2} are the signal amplitudes, ω is the signal frequency in rad/sec, and φ is the ***phase shift*** between the output and the input signals. The ratio U_{m2}/U_{m1} is called the ***gain coefficient*** of the link, and $20\log_{10}(U_{m2}/U_{m1})$ is called the ***gain*** of the link in dB.

Example 1. If a sinusoidal signal with unity amplitude is the input, and the output sinusoidal signal is delayed in phase with respect to the input by $\pi/3$ (i.e., by 60°), and the output signal amplitude is 2, then the phase shift is $-\pi/3$, or -60°, the gain coefficient is 2, and the gain is approximately 6 dB.

Fig. A2.1 shows how to measure the phase shift and the gain of an electrical link: a signal generator is connected to the link's input, the voltmeters read the amplitudes of the input and the output, and the phase difference between the signals can be seen with a two-beam oscilloscope or another phase difference meter.

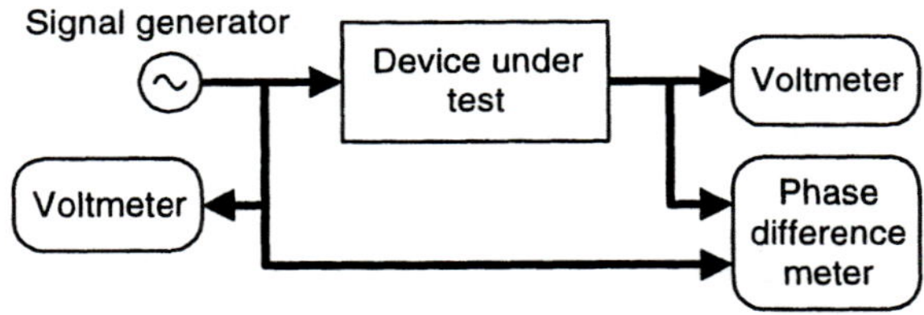

Fig. A2.1 Measurement of a link's gain and phase shift

The gain coefficient and the phase shift both are, generally, functions of the frequency. The plots of these functions are called gain and phase shift ***frequency responses***. The frequency axis is commonly logarithmic.

Fig. A2.2 depicts a truss structure (which can be a model of a stellar interferometer or something else). Some relatively noisy (vibrating) equipment is placed on platform 1 (pumps, motors, reaction wheels, tape recorders) while platform 2 is the place where some sensitive optics operates. It needs to be measured to what extent the vibrations from the upper platform propagate through the truss structure to the second platform which is supposed to be quiet. The transfer function is defined as the ratio of the measured acceleration of a specified point on platform 2 to the force applied at a specific point to platform 1. The measurements are performed in the frequency domain, and the signals are sinusoidal. For the purpose of the measurements, the force is applied by a ***shaker***. Inside of the shaker there is an electromagnet that moves up and down a body called a ***proof mass***. This motion generates reacting force F applied by the case of the shaker to the platform. An accelerometer (a small proof mass placed on a piezoelement, or a magnetic core of a coil suspended on a spring in the field of a permanent magnet) produces an electrical signal proportional to the acceleration a.

Instead of two voltmeters and a signal generator, a ***signal analyzer*** is used that incorporates these three devices together with a display and a computer that sends signals of appropriate amplitude and frequency to the input of the power amplifier driving the shaker.

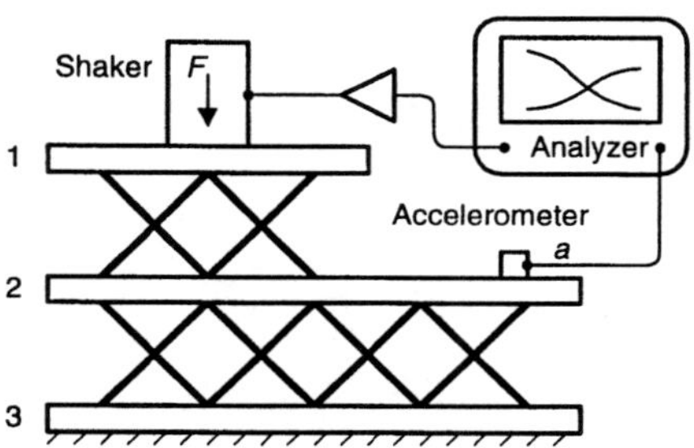

Fig. A2.2 Measurements of a link's gain and phase with a signal analyzer

Example 2. The application of force $F = 5\sin \omega t$ to a body with mass M as shown in Fig. A2.3(a), and the calculation of the resulting acceleration, velocity, and position is reflected in the three-link block diagram shown in Fig. A2.3(b).

The input to the first link is the force, and the link's output is the acceleration

$a = (1/M)5\sin \omega t.$

This link's gain coefficient is $|a/F| = 1/M$, and the phase shift is 0. The gain in dB is plotted in Fig. A2.3(c).

The second link is an ***integrator***: its input is the acceleration and its output is the velocity

$v = \int a\,dt = [5/(M\omega)]\cos \omega t.$

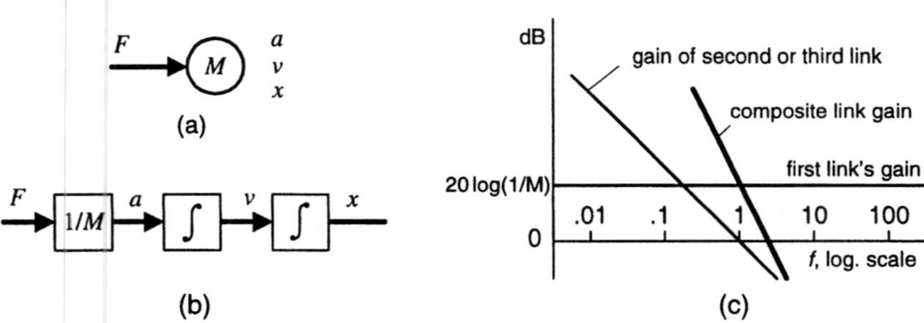

Fig. A2.3 (a) Force acting on a rigid body, (b) block diagram that relates the force to the body's position, and (c) the links' frequency responses

The second link's gain coefficient is $|v/a| = 1/\omega$, and the phase is $-90°$.

The third link is also an integrator: its input is the velocity and its output is the position $x = \int v\,dt$. All three links can be integrated to form a single composite link with the gain coefficient $1/(M\omega^2)$ and the phase shift $-180°$.

Since the second and the third links' gain coefficients are inversely proportional to the frequency, the gain of these links decreases with frequency with a constant slope of −20 dB per decade. The slope of the composite link gain is therefore −40 dB/dec, or, equivalently, −12 dB/oct.

A2.2 Complex transfer function

The two scalar variables, the gain coefficient and the phase shift, can be seen as the magnitude and the phase of a complex ***transfer function***.

Example 1. Saying that a transfer function is $5\exp(j\pi/6)$, or $5\angle 30^\circ$, means that the gain coefficient is 5, and the phase shift is 30°.

Example 2. In Section A2.1's example of a mass driven by a force, the total transfer function from force to position is $-1/(M\omega^2)$. Notice that it can be written as $1/[(j\omega)^2 M]$.

A2.3 Laplace transform and the *s*-plane

The Laplace transform

$$F(s) = \int_0^\infty f(t)\, e^{-st}\, dt$$

or

$$F(s) = \mathrm{L}\,[f(t)]$$

renders a unique correspondence between a function of time $f(t)$ and a function $F(s)$ of ***Laplace complex variable*** $s = \sigma + j\omega$. The transform is linear and exists for all practical stable functions $f(t)$. Some of the transforms are shown in Table A2.1. Here, $\delta(t)$ is the delta-function (infinitely narrow pulse whose area is 1), and $1(t)$ is the step-function.

Example 1. For the Laplace transform $(s+a)^{-1}$, the function $f(t) = e^{-at}$. In the region of small times: $f(t) \approx 1 - at$;
for transform $(s+a)^{-2}$, $f(t) \approx t - at^2$;
for transform $(s+a)^{-3}$, $f(t) \approx t^2/2 - at^3/2$.

The ***initial-value*** and ***final-value theorems*** are the following:

$$f(0) = \lim_{s\to\infty} sF(s)$$

and

$$\lim_{t\to\infty} f(t) = \lim_{s\to 0} sF(s).$$

Table A2.1 Laplace transforms

$F(s)$	$f(t)$
1	$\delta(t)$
$1/s$	$1(t)$
$1/s^2$	t
$1/s^3$	$t^2/2$
$(s+a)^{-1}$	e^{-at}
$(s+a)^{-2}$	te^{-at}
$(s+a)^{-3}$	$(t^2/2)e^{-at}$
$a/[s(s+a)]$	$1 - e^{-at}$
$a/[s^2(s+a)]$	$(at - 1 + e^{-at})/a$
$a^2/[s(s+a)^2]$	$1 - e^{-at}(1 + at)$
$a/(s^2 + a^2)$	$\sin at$
$s/(s^2 + a^2)$	$\cos at$
$\dfrac{a}{(s+b)^2 + a^2}$	$e^{-bt}\sin at$
$\dfrac{s+b}{(s+b)^2 + a^2}$	$e^{-bt}\cos at$

When $F(s)$ is rational, the related function $f(t)$ can be found by, first, presenting $F(s)$ as a sum of partial fractions, and then, summing the time-functions that correspond to the fractions.

Fig. A2.4 shows various poles of $F(s)$ in the s-plane and the related functions of time. As seen in Table A2.1, real poles make exponential signals; purely imaginary poles make sinusoidal signals; complex poles make oscillatory time-responses. As long as the pole is in the left half-plane, $\sigma < 0$ and the envelope of the signal is exponentially narrowing with time.

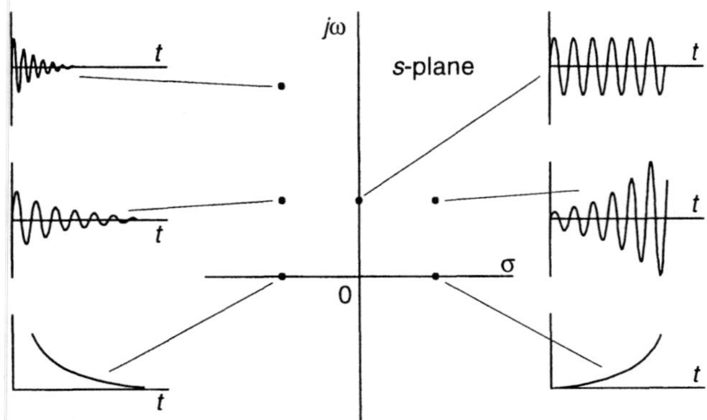

Fig. A2.4 Poles in the s-plane and related time exponents

A2.4 Laplace transfer function

The ***Laplace transfer function*** is the ratio of the Laplace transform of a link's output to that of its input.

Example 1. Since the step-function is an integral of the delta-function, and the ratio of the step-function Laplace transform to the delta-function Laplace transform according to Table A2.1 is $1/s$, evidently, *1/s is the Laplace transfer function of an integrator*. Conversely, *multiplying a Laplace transform by s is analogous to differentiating the corresponding time-function*.

Since the Laplace transform is linear, the transfer function of several links connected in parallel equals to the sum of the links' transfer functions.

According to the convolution (Borel) theorem, *the Laplace transfer function of a chain connection of links is the product of the Laplace transfer functions of the individual links*.

Example 2. The step-function $1(t)$ is applied to the input of a link composed of two cascaded links with transfer functions $5/(s + 1)$ and $(s + 1.4)/(s + 2)$. Find the output time-function.

Since the step-function's Laplace transform is $1/s$, the Laplace transform of the output is

$$\frac{5(s+1.4)}{(s+1)(s+2)s} = \frac{5s+7}{s^3+3s^2+2s}.$$

The partial fraction expansion of the Laplace transform of the output

$$\frac{-2}{s+1}+\frac{-1.5}{s+2}+\frac{3.5}{s}$$

can be found with the MATLAB function `residue`, as is illustrated in Section 7.2.1.

The time-function of the output can be obtained by summing the time-functions from Table A2.1 that correspond to the partial fractions:

$$-2e^{-t}-1.5e^{-2t}+3.5\times 1(t)\,.$$

The output time-function can be plotted using MATLAB either with

```
num = [5 7];
den = [1 3 2 0];
impulse(num, den)
```

or with

```
num = [5 7];
den = [1 3 2];
step(num, den)
```

With the `impulse` command, the numerator and the denominator of the Laplace transform of the output are used (as the response to the δ-function whose Laplace transform is 1); with the `step` command, the numerator and the denominator of the transfer function are used.

Laplace transforms and inverse Laplace transforms (i.e., finding $f(t)$ from $F(s)$) can be directly found with MATLAB commands `laplace` and `invlaplace`.

When the input signal is an exponent $\mathrm{Re}(e^{-s_o t})$ and s_o is a zero of the transfer function, the output signal is zero.

For the input signal $\mathrm{Re}(e^{-s_p t})$ where s_p is a pole of the transfer function, the transfer function is infinite, i.e., the signal is infinitely amplified and becomes infinitely large at the output.

When a pole of a transfer function is in the right half-plane of s, the output signal in response to δ-function input is an exponentially growing signal. Such a system is considered unstable since there always exist δ-function components in the input noise. In practical systems, the output grows exponentially until it becomes so large that, because of the power limitation in active devices, the output signal becomes limited, and the system no longer can be viewed as linear.

To determine whether a system with a given rational transfer function is stable, one might calculate the roots of the denominator polynomial and check whether there are roots with positive real parts. This can be done with MATLAB and many other popular software packages.

The Routh-Hurwitz criterion indicates the presence of right-sided polynomial roots when certain inequalities hold; verifying these inequalities is simpler than actually finding the roots. The criterion is not described here since software packages calculate the roots in no time.

To calculate the response to sinusoidal inputs, s should be replaced by $j\omega$. This results in a complex transfer function.

Example 3. If the input signal is Re[exp($j\omega_o$)], and at this ω_o the Laplace transfer function is $A\angle\varphi$, then the output signal of the link will be ARe[exp($j\omega_o + j\varphi$)]. In other words, the signal is amplified A times and the phase is shifted (advanced) by φ radians.

Example 4. The transfer function is $100/(j\omega + 5)$. At frequency $\omega = 5$ rad/sec, the function equals $14.1\angle{-45^\circ}$, i.e., the output amplitude is 14.1 times larger than that of the input, and the output is delayed by 45°.

A2.5 Poles and zeros of transfer functions

The locations of the transfer function's poles and zeros manifest themselves in two important aspects:

- They show which signals are amplified infinitely (exponential signals with the exponents equal to the poles), and which are not transmitted at all (exponential signals with the exponents equal to the zeros). In Chapter 4, the former issue is related to system stability, and the second one, to nonminimum phase shift.
- They affect the transmission of purely sinusoidal test-signals, i.e., the frequency responses.

Example 1. In the example shown in Fig. A2.3, the Laplace transfer function is $1/(Ms^2)$. This is a rational function with a double pole at the origin. *It is seen that s replaces jω.*

Example 2. Consider the transfer function $L(s) = 50/(s + 5)$. The pole is $s = -5$. The gain frequency response is the plot of the function

$20 \log|L(j\omega)| = 20 \log[50/(j\omega + 5)]$.

The vector $j\omega + 5$ is shown in Fig. A2.5(a). It is evident that this vector is lowest in magnitude at lower frequencies where therefore the transfer function is largest. At zero frequency, the gain is 20 dB and the phase is 0. When the frequency increases, the vector becomes longer and eventually the gain coefficient decreases in proportion to the frequency. That is, when the frequency doubles, i.e., increases by an octave, the gain coefficient is halved, i.e., decreases by 6 dB. Therefore, the gain decreases with the constant slope of −6 dB/oct.

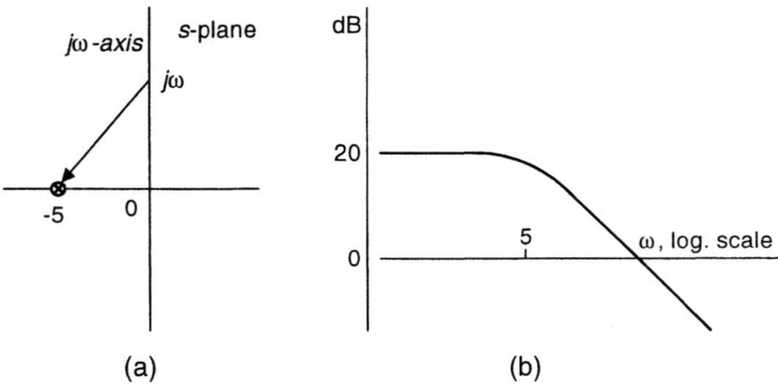

Fig. A2.5 Gain frequency response (b) corresponding to the pole −5 on the s-plane (a)

When ω changes from 0 to ∞, the phase changes from zero to $-\pi/2$.

Laplace rational transform functions for physical systems always have only real coefficients. Therefore, their poles and zeros are either real or come in complex-conjugate pairs. The transfer function multipliers for complex poles (or zeros) are commonly written as

$$[s-(\sigma_o+j\omega_o)]\,[s-(\sigma_o-j\omega_o)]=s^2-2\sigma_o\,s+\omega_o{}^2=s^2+2\zeta\omega_o\,s+\omega_o{}^2.$$

Here, σ_o is negative and $\zeta=-\sigma_o/\omega_o$ represents the damping coefficient for the pole.

Example 3. The transfer function

$$L(s)=\frac{c}{s^2+2\zeta\omega_o s+\omega_o^2}=\frac{c}{[s-(\sigma_o-j\omega_o)][(s-(\sigma_o+j\omega_o)]}$$

has two complex-conjugate poles, $\sigma_o\pm j\omega_o$. Fig. A2.6(a) shows the poles on the s-plane where σ_o is negative and small; the two vectors represent the multipliers in the denominator of the function. The frequency response is obtained by substituting $j\omega$ for s in the function. Then, when ω approaches ω_o, the first multiplier becomes $-\sigma_o$, the denominator becomes $2\sigma_o\omega_o$, and $|L(j\omega)|$ becomes large. With a further increase of the frequency the denominator gets bigger and the modulus of the function decreases. Fig. A2.6(b) shows the gain frequency response for the function.

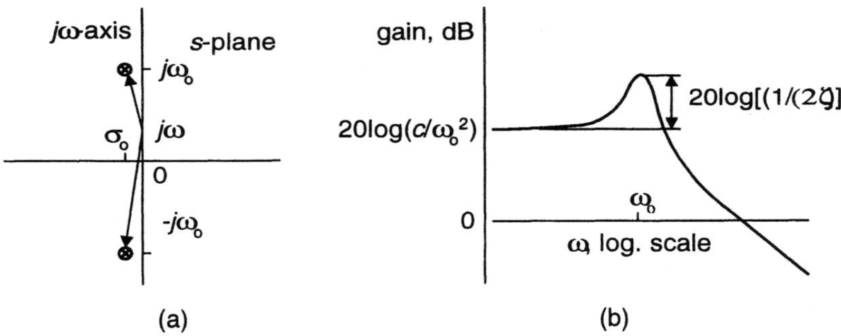

Fig. A2.6 (a) Complex poles and (b) the related frequency response

The smaller the damping coefficient ζ (and therefore $|\sigma_o|$) is, the higher is the peak on the gain-frequency response. At higher frequencies $\omega >> \omega_o$, $|L(j\omega)|$ decreases as the square of the frequency, and the gain decreases with constant slope –12 dB/oct.

A2.6 Pole-zero cancellation, dominant poles and zeros

Next, consider a pole-zero pair with little distance between the pole and the zero, as shown in Fig. A2.7. Since the vectors corresponding to the pole and to the zero are nearly the same, they compensate each other and insignificantly affect the frequency response. Adding such a pair to any transfer function will certainly change the order of the system but will have negligible effect on the frequency response. And, since the frequency response fully characterizes the performance of a linear link, there will be no substantial difference in these links' performance in any application.

For example, a cluster of two poles and one zero can be replaced by one pole. If the cluster is far from the frequency range of interest on the $j\omega$-axis, the replacement is adequate even when the cluster is not very tight. But if the cluster is very close to the frequencies of interest, then the distance between the singularities must be very small for the replacement to be adequate.

In many systems, several poles and zeros mutually compensate or have small effect on the transfer function within the frequency range of interest, and only one or a few poles and zeros have prominent effect on the frequency response. Such poles and zeros are called dominant.

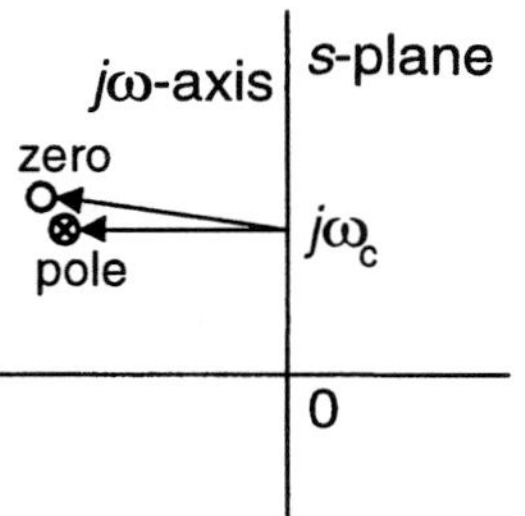

Fig. A2.7 Pole-zero pair

Much more about frequency responses' relation to the poles and zeros is explained in Chapter 6.

A2.7 Time-responses

In the calculation of time-responses, the poles of the transfer function characterize the exponents of the solution to a system of certain linear differential equations. The closer the poles are to the $j\omega$-axis, the more oscillatory the solution is. Therefore, the closed-loop responses should not have poles too close to the $j\omega$-axis in order that the output time history not differ substantially from the command.

A2.8 Problems

1 Can the system be linear if the input signal is $3\sin(34t + 5)$ and the output is: (a) $3\sin(5t + 34)$; (b) $3\sin(34t + 34)$; (c) $5\sin(5t - 5)$; (d) $-13\sin(34t + 5)$; (e) $2.72\sin(2.72t + 2.72)$; (f) $3\sin(340t - 5)$; (g) $5t\sin(34t - 5)$.

2 What is the phasor for the function (express the phase in radians): (a) $22\sin(\omega t + 12)$; (b) $-2\sin(\omega t + 2)$ (c) $3\sin(\omega t + 12)$; (d) $-5\cos(\omega t + 2)$; (e) $-2.72\cos(\omega t + 2)$.

3 What is the phasor of the sum:
(a) $22\sin(\omega t + 12) - 12\sin(\omega t + 12)$;
(b) $12\sin(\omega t + 12) - 10\sin(\omega t + 12)$;
(c) $4\angle 30^\circ - 40\angle 60^\circ$;
(d) $4\angle 30^\circ + 40\angle 60^\circ$;
(e) $4\angle 60^\circ - 40\angle 30^\circ$;
(f) $4\angle 60^\circ + 40\angle 30^\circ - 20\angle 0^\circ$.

4 What is the transfer function if the input signal phasor is $4\angle 49^\circ$ and the output is: (a) $4\angle 49^\circ$; (b) $40\angle -59^\circ$; (c) $40\angle 229^\circ$; (d) $1000\angle -30^\circ$; (e) $2.72\angle -2.72^\circ$.

5 Find the originals of the Laplace transforms: (a) $1/s$; (b) s; (c) $2/(s + 3)$; (d) $4/[(s + 5)(s + 10)]$ (Do it using Table A2.1, and using the MATLAB function `invlaplace`.)

6 Find the Laplace transforms of the signals: (a) δ-function, (b) step-function with the value of the step equal to 8, (c) $5t^2$, (d) $10te^{-2t}$, (e) $13(1 - e^{-4t})$, (f) $6\sin 2t$,

(g) $4e^{-4t}\cos 5t$, (h) $-4\cos 2t$ (Use both the Table A2.1 and the MATLAB function `laplace`.)

7 Find transfer function of the linear operators: (a) $df(t)/dt$, (b) integral of $f(t)$, (c) double integral of $f(t)$ (Use Table A2.1.)

8 The input signal is the δ-function, the output signal Laplace transform is
(a) $(s^2 + 2s + 4)/(2s^3 + 4s^2 + 30)$;
(b) $(s^2 + 3s + 24)/(2s^3 + 12s^2 + 60)$;
(c) $(s^2 + 4s + 42)/(2s^3 + 12s^2 + 80)$;
(d) $(s^2 + 40s + 46)/(2s^3 + 2s^2 + 90)$;
(e) $(s^2 + 50s + 40)/(2s^3 + 22s^2 + 120)$;
(f) $(s^2 + 60s + 100)/(2s^3 + 20s^2 + 300)$;
Find the link's transfer function.

9 Which of the following expressions can be transfer functions of stable systems:
(a) $(20s^2 + 30s + 40)/[(s + 43)(s + 85)(s + 250)(s + 2500)]$;
(b) $(s^2 + 30s + 4)/(2s^3 - 2s^2 + 30)$;
(c) $-(10s^2 + 10s + 40)/(2s^4 - 2s^3 + 3s^2 + 80)$;
(d) $(s^4 + 5s^3 + s^2 + 20s + 5)/(s^3 + 23s^2 + 200s + 300)$;
(e) $-60(s + 3)(s - 16)/[(s + 33)(s + 75)(s + 200)(s + 2000)]$;
(f) $10(s + 2)(s - 22)/[(s + 40)(s + 65)(s + 150)]$;
(g) $-20(s + 2)(s + 26)/[(s + 43)(s + 85)(s + 250)(s + 2500)]$.

10 Use the MATLAB command `root` to calculate the poles and zeros of the function:
(a) $(20s^2 + 30s + 40)/(2s^4 + s^3 + 5s^2 + 36)$;
(b) $(s^2 + 30s + 4)/(s^4 + 2s^3 + 2s^2 + 36)$;
(c) $(10s^2 + 10s + 40)/(2s^4 + 2s^3 + s^2 + 3)$;
(d) $(s^2 + 20s + 5)/(s^4 + 5s^3 + s^2 + 3)$;
(e) $(2.72s^2 + 27.2s + 20)/(s^4 + 2.72s^3 + 7s^2 + 2.72)$;
(f) $(s^2 + 10s + 8)/(s^4 + 12s^3 + 12s^2 + 33)$.

11 Use the MATLAB command `poly` to convert the following functions to ratios of polynomials:
(a) $50(s + 6)(s + 12)/[(s + 50)(s + 85)(s + 110)(s + 1200)]$;
(b) $-60(s + 7)(s + 15)/[(s + 53)(s + 95)(s + 210)(s + 2300)]$;
(c) $10(s + 8)(s + 62)/[(s + 50)(s + 65)(s + 150)]$;
(d) $-20(s + 9)(s + 66)/[(s + 63)(s + 95)(s + 240)(s + 2700)]$;
(e) $2.72(s + 7)(s + 10)/[(s + 70)(s + 130)(s + 1200)]$.

12 Write the frequency response function by replacing s by $j\omega$ for:
(a) $(20s^2 + 30s + 40)/(2s^4 + s^3 + s^2 + 3)$;
(b) $(s^2 + 30s + 4)/(s^4 + 2s^3 + 2s^2 + 3)$;
(c) $(10s^2 + 10s + 40)/(2s^4 + 2s^3 + s^2 + 3)$;
(d) $(s^2 + 20s + 5)/(s^4 + 5s^3 + s^2 + 3)$;
(e) $(2.72s^2 + 27.2s + 20)/(s^4 + 2.72s^3 + 7s^2 + 2.72)$;
(f) $(s^2 + 10s + 8)/(s^4 + 12s^3 + 12s^2 + 130)$.

13 Three links with transfer functions, respectively,
$-20(s + 2)(s + 26)/[(s + 85)(s + 250)]$,
$(s^2 + 30s + 4)/(s^3 + 2s^2 + 30)$, and
$(10s^2 + 10s + 40)/(s^4 + 2s^3 + 30s^2 + 1000)$
are connected (a) in series (cascaded), (b) in parallel, (c) with the first and the second in parallel and the third in cascade, (d) with the first and the second in

series and with the third in parallel to this composite link. Find the transfer functions of the resulting composite links.

14 Use MATLAB to plot the frequency response for the first-, second-, and third-order functions:
(a) $10/(s+10)$;
(b) $100/(s+10)^2$;
(c) $1000/(s+10)^3$.
Describe the correlation between the slope of the gain-frequency response and the phase shift.

15 Use MATLAB to plot the frequency response and the step time-response for the first- and second-order functions:
(a) $10/(s+10)$;
(b) $-10/(s+10)$;
(c) $100/(s^2+4s+100)$;
(d) $-100/(s^2+4s+100)$;
(e) $100/(s^2+2s+100)$;
(f) $100/(s^2+s+100)$.
Describe the correlation between the shapes of the frequency responses and those of the step-responses.

16 Use MATLAB to convert the function to a ratio of polynomials and plot the frequency response for the function:
(a) $50(s+3)(s+12)/[(s+30)(s+55)(s+100)(s+1000)]$;
(b) $60(s+3)(s+16)/[(s+33)(s+75)(s+200)(s+2000)]$;
(c) $10(s+2)(s+22)/[(s+40)(s+65)(s+150)]$;
(d) $-20(s+2)(s+26)/[(s+43)(s+85)(s+250)(s+2500)]$;
(e) $2.72(s+7)(s+20)/[(s+10)(s+100)(s+1000)]$;
(f) $-25(s+2)(s+44)/[(s+55)(s+66)(s+77)(s+8800)]$.
What is the value of the function at dc (i.e., when $s=0$)? What does this function degenerate into at very high frequencies?

17 Compare the frequency responses of:
(a) $(s+2)/(s+10)$ and $(s+2)(s+5)/[(s+10)(s+5.1)]$;
(b) $(s+2)/(s+10)^2$ and $(s+2)/[(s+9)(s+11]$;
(c) $(s+2)/(s+10)^2$ and $(s+2)/[(s+7)(s+14)]$;
(d) $(2.4s^2+25s+20)/(s^4+2s^3+8s^2+3)$.
Draw a conclusion.

18 Plot time-responses to a step command for the functions in the previous example. Draw a conclusion.

19 Plot and compare frequency responses of the functions with complex poles and zeros:
(a) $(s+2)/(s+10)$ and $(s+2)(s+5)/[(s+10)(s+5.1)]$;
(b) $(s+2)/(s+10)^2$ and $(s+2)/[(s+9)(s+11)]$;
(c) $(s+2)/(s+10)^2$ and $(s+2)/[(s+7)(s+14)]$;
(d) $(2s^2+22s+20)/(s^4+2.5s^3+7.5s^2+2.5)$.
Draw a conclusion.

20 Plot the frequency responses (on the same plot, using `hold on` feature) of the functions:
(a) $(s+1)/(s+10)$;

(b) $(s + 1)/(s + 20)$;
(c) $(s + 1)/(s + 1000)$;
(d) $(s + 10)/(s + 1000)$.
Over what frequency range do the functions approximate the differentiator s?

21 Make a transfer function to implement the following frequency response:
(a) 10 at lower frequencies, rolling down at higher frequencies inversely proportionally to the frequency;
(b) 10 at high frequencies, rolling up at low frequencies proportionally to the frequency;
(c) 10 at low frequencies, 100 at high frequencies;
(d) 100 at low frequencies, 10 at high frequencies;
(e) 10 at medium frequencies, rolling up proportionally to the frequency at lower frequencies, rolling down inversely proportionally to the frequency at higher frequencies;
(f) resonance peak response, rolling up proportionally to the frequency at lower frequencies, rolling down inversely proportionally to the frequency at higher frequencies;
(g) resonance peak response, flat at low frequencies and rolling down inversely proportionally to the square of the frequency at higher frequencies;
(h) resonance peak response, flat at high frequencies and rolling up proportionally to the square of the frequency at higher frequencies;
(i) notch responses of three different kinds, with different behavior at lower and higher frequencies.

22 Which of the following functions are positive real?
(a) $(s + 2)(s + 22)/[(s + 40)(s + 65)(s + 150)]$;
(b) $(s + 2)(s - 22)/[(s + 40)(s + 65)(s + 150)]$;
(c) $(s + 2)(s + 22)/[(s + 40)(s - 65)(s + 150)]$;
(d) $(s + 2)(s + 22)/[(s + 40)(s + 65)(s + 150)(s + 250)]$;
(e) $(s + 2)(s + 22)/[(s + 40)(s + 65)]$;
(f) $(s + 2)(s + 22)(s + 150)/[(s + 40)(s + 150)]$;
(g) $-(s + 2)(s + 22)/[(s + 40)(s + 65)(s + 150)]$.

Appendix 3 Causal systems, passive systems, and positive real functions

When s is equal to a pole, the transfer function is ∞. A transfer function pole in the right half-plane means that the response to a finite input signal is an exponentially rising output signal. This cannot be a property of a ***passive system***, i.e., a system without sources of energy. Therefore, passive systems must have no transfer function poles in the right half-plane of s.

Causal systems are those systems whose output value at any t_1 does not depend on the input signal at $t > t_1$. All stable systems are causal.

Positive realness of a function $\theta(s)$ means that for all s in the right half-plane, $\mathrm{Re}\theta(s) > 0$. All passive driving point impedances (a driving point impedance is the ratio of the voltage to the current at the same port) are positive real (p.r.). Such an impedance has non-negative resistance for all sinusoidal signals and positive resistance for all rising exponential signals (whether oscillating or not). A p.r. impedance does not generate but only dissipates power.

The driving point admittances of passive systems are also p.r.

The transfer impedances and admittances of passive systems can be p.r. but are not necessarily so.

The driving point impedances of active systems can be p.r. but are not necessarily so.

Collocated control is a feedback control of a passive plant where the actuator and the sensor are collocated, and the ratio of the sensor readings to the actuator action is the driving point impedance or admittance, or a derivative or an integral of the impedance. This limits the range of the plant phase variations to 180°.

Any p.r. function can be ***realized*** (i.e., implemented) as a driving point impedance of a passive system, i.e., it is always possible to make a system composed of passive elements whose driving point impedance is the prescribed p.r. function. In some cases, the system is an arrangement of resistances, capacitances, and inductors connected in series and in parallel. For some p.r. functions, however, realization requires bridge-type circuits or transformers. A p.r. function can also be realized as a driving point impedance of an active *RC* circuit.

The function is p.r. if the following three conditions are satisfied:

A. There are no poles or zeros in the right half-plane of s.

B. Poles and zeros on the $j\omega$-axis are single.

C. $\mathrm{Re}\,\theta(j\omega) > 0$ at all frequencies.

Appendix 4 Derivation of Bode integrals

A4.1 Integral of the real part

The Laurent expansion at $s \to \infty$ for an m.p.f. $\theta(s) = A(s) + B(s)$, where $A(s)$ is an even function and $B(s)$ is an odd function, is

$$\theta(s) = A_\infty - \frac{B_1}{s} - \frac{A_2}{s^2} - \frac{B_3}{s^3} - \frac{A_4}{s^4} - \dots$$

The expansion converges over the entire right half-plane of s. On the $j\omega$-axis, it accepts the form

$$\theta(j\omega) = A_\infty + j\frac{B_1}{\omega} + \frac{A_2}{\omega^2} + \dots \tag{A4.1}$$

The function $\theta - A_\infty$ is m.p. as well. Therefore, the contour integral of $\theta - A_\infty$ around the right half-plane of s equals 0. The contour of integration may be viewed as composed of the $j\omega$-axis completed by a π-radian arc of infinite radius R as shown in Fig. A4.1. The integral along the arc equals πB_1; the integral along the whole $j\omega$-axis equals twice the integral of the even part of the integrand, i.e. of $A - A_\infty$, along the positive semiaxis. Therefore

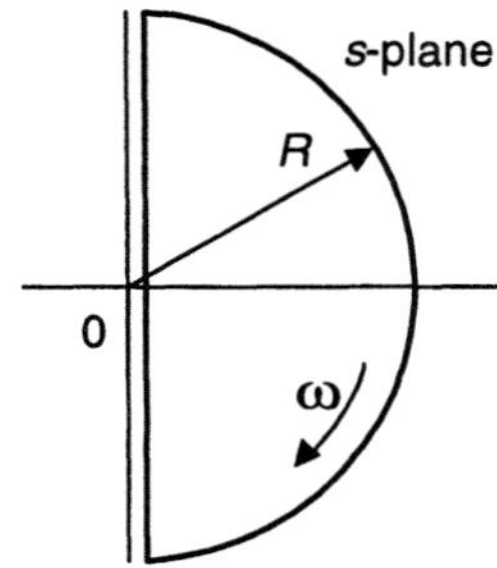

Fig. A4.1 Contour on the s-plane

$$\int_0^\infty (A - A_\infty)d\omega = -\frac{B_1}{2\pi}. \tag{A4.2}$$

A4.2 Integral of the imaginary part

If θ is m.p., then θ/s is analytical in the right half-plane of s and on the $j\omega$-axis, excepting the origin. So if the origin is avoided along the π-radian arc of infinitesimal radius as shown in Fig. A4.2, the integral of θ/s around the contour enclosing the right half-plane equals 0. The integral along the small arc equals πA_o where A_o is the value of θ at zero frequency. The integral along the large arc equals $-\pi A_\infty$ where, as follows from (A4.1), A_∞ is the value of θ at very high frequencies. Therefore, the integral of the even part of θ/s along the $j\omega$-axis equals

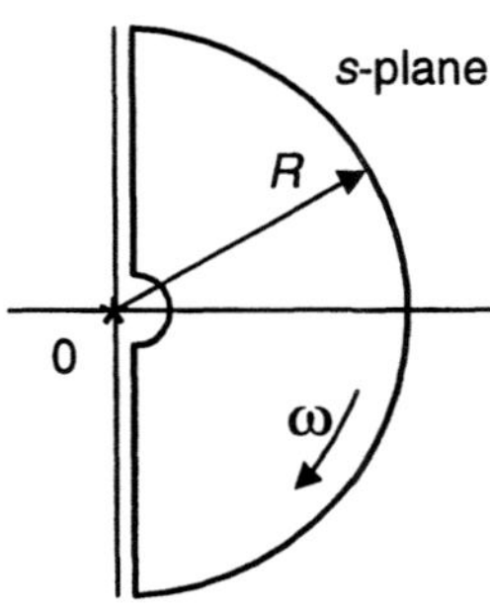

Fig. A4.2 Contour on the s-plane avoiding a pole at the origin

$$\int_{-\infty}^{\infty} \frac{B}{\omega} d\omega = \frac{\pi(A_\infty - A_0)}{2}$$

i.e.,

$$\int_{-\infty}^{\infty} B(u)\, du = \frac{A_\infty - A_0}{2} \tag{A4.3}$$

where $u = \ln(\omega/\omega_c)$. This relation is known as the ***phase integral***.

Another important relation between the real and imaginary components results from setting to zero the integral of $\theta - A_\infty\sqrt{sW}$ around the same contour; here, W is a ***reactance function***, i.e., an impedance function of a reactance two-pole. On the $j\omega$ axis, W is purely imaginary, either positive or negative. The function sW is purely real, positive or negative, on the $j\omega$-axis.

The function $\sqrt{sW}$ therefore alternates between being purely real or purely imaginary on adjoining sections over the $j\omega$-axis. It has branch points at the joints of these sections. The sign of the radix at the sections must be chosen so that the whole contour of integration belongs to only one of the Riemann folds. On this contour, the function Re $W(j\omega)$ must be even and Im $W(j\omega)$, odd.

For those W reducing at higher frequencies to s, the integrand decreases with s at least as s^{-2}, and the integral along the large arc vanishes. Since the total contour integral is zero, the integral along the $j\omega$-axis equals zero as well. Its real part is certainly zero:

$$\int_{-\infty}^{\infty} \mathrm{Re}\frac{\theta - A_\infty}{sW} d\omega = 2\int_{0}^{\infty} \mathrm{Re}\frac{\theta - A_\infty}{sW} d\omega = 0. \tag{A4.4}$$

If, in particular, $W = (1 + s^2)/s$, then $\sqrt{sW} = \sqrt{1-\omega^2}$ is real for $|\omega| < 1$ and imaginary otherwise, and (A4.4) yields

$$\int_{0}^{1} \frac{(A - A_\infty)}{\sqrt{1-\omega^2}} d\omega + \int_{1}^{\infty} \frac{B}{\sqrt{1-\omega^2}} d\omega = 0,$$

i.e.

$$\int_{\omega=0}^{\omega=1} (A - A_\infty)\, d\arcsin\omega = -\int_{1}^{\infty} \frac{B}{\sqrt{1-\omega^2}} d\omega. \tag{A4.5}$$

A4.3 General relation

Let's define the frequency at which the phase shift is of interest as ω_c, and also define $\theta_c \equiv A_c + jB_c \equiv \theta(j\omega_c)$, where A_c and B_c are real. Consider the function of s

$$(\theta - A_c)\left(\frac{1}{s/j - \omega_c} - \frac{1}{s/j + \omega_c}\right) = (\theta - A_c)\frac{2\omega_c}{-s^2 - \omega_c^2}. \tag{A4.6}$$

This function is analytical in the right half-plane of s and on the $j\omega$-axis except at the points $-j\omega_c$, $j\omega_c$. Therefore, the integral of the function taken around the contour shown in Fig. A4.3 equals zero. The contour consists of several pieces; the sum of the integrals along these pieces is 0.

The integral along the arc of infinite radius R equals 0 because of the term s^2 in the denominator of the integrand.

As s approaches $j\omega_c$, $\theta - A_c$ approaches B_c, and the second multiplier in the left side of (A4.6) tends to $1/(s/j - \omega_c)$. Then the integral along the infinitesimal arc centered at $j\omega_c$ equals

$$\int \frac{B\,ds}{s/j - \omega_c} = -\pi B_c .$$

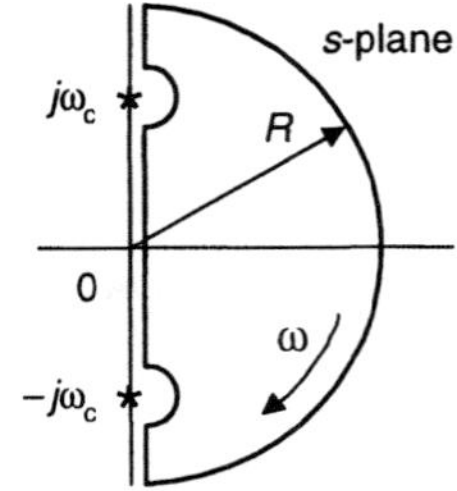

Fig. A4.3 Contour on *s*-plane avoiding imaginary poles

The integral along the small arc centered at $-j\omega_c$ equals $-\pi B_c$ as well.

Next, equating the sum of all components of the contour integral to 0, we see that the integral along the $j\omega$-axis equals $2\pi B_c$. Neglecting the odd component of the integrand whose integral is annihilated within these symmetrical boundaries, we have

$$\int_{-\infty}^{\infty} (A - A_c) \frac{2\omega_c}{\omega^2 - \omega_c^2}\, d\omega = 2\pi B_c$$

and finally

$$B_c = \frac{2\omega_c}{\pi} \int_0^{\infty} \frac{d\omega}{\omega^2 - \omega_c^2} (A - A_c)\, d\omega . \tag{A4.7}$$

Since

$$\int \frac{A - A_c}{\pi} \frac{2\omega_c}{\omega^2 - \omega_c^2}\, d\omega = \int U\, dV = UV - \int V du \tag{A4.8}$$

where we denote:

$$U = \frac{A - A_c}{\pi},$$

$$V = \int \frac{2\omega_c\, d\omega}{\omega^2 - \omega_c^2} = -\ln\coth\frac{u}{2},$$

$$u = \ln(\omega/\omega_c),$$

(A4.7) equals

$$B_c = \frac{-1}{\pi}\left[(A - A_c)\ln\coth\frac{u}{2} \right]_{-\infty}^{\infty} + \frac{1}{\pi}\int_{-\infty}^{\infty} \frac{dA}{du} \ln\coth\frac{u}{2}\, du. \tag{A4.9}$$

The left side of the equation is real. Hence, the imaginary components on the right side are annihilated after summing. We can therefore count the real components only. Since

$$\ln\coth\frac{-u}{2} = j\pi + \ln\coth\frac{u}{2},$$

replacing ln coth(u/2) by ln coth|u/2| does not change the real components of (A4.9) and is therefore permitted. After the replacement, the function in square brackets becomes even and is annihilated because of symmetrical limits. Thus,

$$B(\omega_c) = \frac{1}{\pi}\int_{-\infty}^{\infty}\frac{dA}{du}\ln\coth\frac{|u|}{2}\,du. \tag{A4.10}$$

Appendix 5 Program for phase calculation

The integral (A4.10) can be calculated with MATLAB using numerical routines `find_phase2.m`, `table_maker.m`, and `integral_u`. These three are among several routines for calculation of different Bode integrals written by Michael Kantner from Caltech. The routines can be used to calculate the phase shift response of a physical m.p. plant on the basis of its measured gain response.

The magnitude is prescribed over a finite frequency interval. The function `find_phase2.m` takes frequency and magnitude vectors as inputs and returns the phase.

Behavior of the magnitude outside the specified frequency interval is assumed to be an asymptotic extension of the response within the specified interval. That is, a system that has −40 dB/decade roll off within the specified frequency range is assumed rolling off at the same rate at higher frequencies. The integrals of the tails are precomputed and stored in a lookup table created by `table_maker.m`. While these could be computed when needed, the lookup table yields faster code. The errors introduced by such handling of the tails might become dominant when the high-frequency and low-frequency asymptotes are not well developed within the prescribed frequency interval.

The accuracy of the finite difference approximation of the integral used in the routines is sufficient. The functions have been tested on a variety of transfer functions. When the input is 100 logarithmically spaced points spread over 4 decades, typical errors are under 0.1 degrees. As the number of data points increases, accuracy improves. By placing data points unevenly, with more of them where the known function is rapidly changing, accuracy can be improved with minimal added computational cost.

The programs' listing follows:

Function *find_phase2*

```
function [phase] = find_phase2(magdb, natfreq)
% function [phase] = find_phase2(magdb ,freq)
% This routine uses the Bode Integral to generate phase data.
% from a magnitude response of a m.p. function.
% magdb:   row gain vector given in dB
% freq:    row frequency vector given in rad/sec
% The Magnitude and Frequency vectors must be the same length.
% Before running this function prepare table with table_maker.

[row,col] = size(magdb); if col == 1, magdb = magdb'; end;
[row,col] = size(natfreq); if col == 1, natfreq = natfreq'; end;

%%% table_maker % calls the function creating the table
load table  % load data needed for toe and tail calculations:
            % table con hilimit lolimit numentries numintstep
points = length(natfreq);
numsteps = points - 1;

% The following variables are for the lookuptable (u domain)
ilnfreq = log(natfreq(1));
flnfreq = log(natfreq(points));
toeslope = (magdb(2) - magdb(1))/(log(natfreq(2))-ilnfreq);
tailslope = (magdb(points) - magdb(numsteps))/(flnfreq -
log(natfreq(numsteps)));
```

```
dmagdb = magdb(2:points) - magdb(1:numsteps);
dnatfreq = natfreq(2:points) - natfreq(1:numsteps);

deriv = dmagdb./dnatfreq;

nnfreq = natfreq(1:numsteps);
w1 = nnfreq';w2 = natfreq(2:points)';

for I = 1:points,

% The next lines perform the integration...
wc = natfreq(i)*ones(w1);
weights = (w2+ wc).*log(w2 + wc) - (w1 + wc).*log(w1 + wc) +...
   (wc-w2).*log(abs(wc-w2+eps)) - (wc-w1).*log(abs(wc-w1+eps));

looplnfreq = log(natfreq(i));
u = [flnfreq ilnfreq] - [looplnfreq]*[1 1] + [eps eps];
ind = (log10(abs(u))+[-log10(lolimit)-log10(lolimit)])/con+[1 1];
ind = max([1 1;ind]);ind=min([numentries numentries;ind]);
tailtoe = abs(pi^2/4*[1;1]-table(ind',2));

phase(i) = (deriv*weights + [tailslope toeslope]*tailtoe)/pi;
end
```

Function *table_maker.m*

```
function table_maker
% table_maker.m, called by function find_phase2. This routine
% creates the lookup table for calculation the toe and tail
% 1e-15 is used for zero, 650 is used for infinity
% reasonable limits:  1e-9 to 100
% This contains nearly all of the area (to 6 places).

lolimit = 1e-9;  hilimit = 100;
numentries = 4001;  % number of table entries, should be odd
numintstep = 100;   % number of steps for each integration

vector = logspace(log10(lolimit),log10(hilimit),numentries);
table(:,1) = vector';

[table(1,2)] = integral_u(1e-15,lolimit, numintstep);

for k = 2:length(vector),
[table(k,2)]=table(k-1,2)+integral_u(table(k-1,1),table(k,1),
numintstep);
end

clear vector k

con = log10(table(7,1))-log10(table(6,1));

save table con hilimit lolimit numentries numintstep table
clear con hilimit lolimit numentries numintstep table
```

Function *integral_u*

```
function [trap]=integral_u(u1,u2,numsteps)
% function [trap]=integral_u(u1,u2,numsteps)
```

```
% This routeens integrates in u domain.
% It may not handle some special cases, however.
% vector=linspace calculation(u1,u2,numsteps+1);

if (u1==0),
   u1=1e-15;
end
if (u2==0),
   u2=1e-15;
end
vector = logspace(log10(u1),log10(u2),numsteps+1);
% This next calculation is the logu_function

valvector = log(abs((exp(vector)+1)./(exp(vector)-1)));
delta = vector(2:numsteps+1)-vector(1:numsteps);

%up =    valvector(1:numsteps)*delta';
%down = valvector(2:numsteps+1)*delta';

trap=(.5*(valvector(1:numsteps)+valvector(2:numsteps+1)))*delta';
if trap==NaN,
   trap=0;
   disp('Warning:  NaN found in integration, result set to 0')
end
```

Different MATLAB programs for Bode integral calculations are listed in [6], QFT.

For the function without high-Q resonance peaks, the phase response corresponding to a gain response can be found using function `boniqas.m` from Appendix 14.

Appendix 6 Generic single-loop feedback system

When a feedback system cannot be easily broken into a connection of links, a more general description of the feedback system can be used as a connection of a unilateral two-port w to a passive four-port B as displayed in Fig. A6.1.

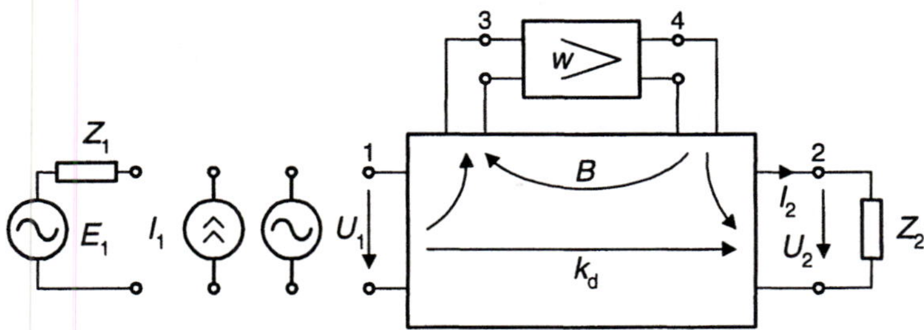

Fig. A6.1 Single-loop feedback system

The two-port w is assumed to have zero reverse transmission and either infinite or infinitesimal input and output impedances. When the circuit diagram in Fig. A6.1 is applied to the analysis of circuits with physical amplifiers, the amplifier input and output impedances may be imitated by connecting passive two-poles in parallel or in series to the amplifier's input and output. Further, these two-poles have to be integrated into the B-network.

In principle, the dimensionality of the signals at the input and output of the amplifier and at the input and output terminals of the whole system does not influence the following analysis. However, to simplify the exposition, we select one of the possible versions and characterize the signal at the input to the active element by the voltage E_1, and the signal at its output by the current I_4. The active element with transadmittance $w = I_4/E_3$ is therefore assumed to have high input and output impedances.

When an external emf E_1 is applied to the input of the amplifier disconnected from the B-circuit, as shown in Fig. A6.2, the return voltage U_3 appears at the port 3′. The return ratio is

$$T = -\frac{U_3}{E_3}. \tag{A6.1}$$

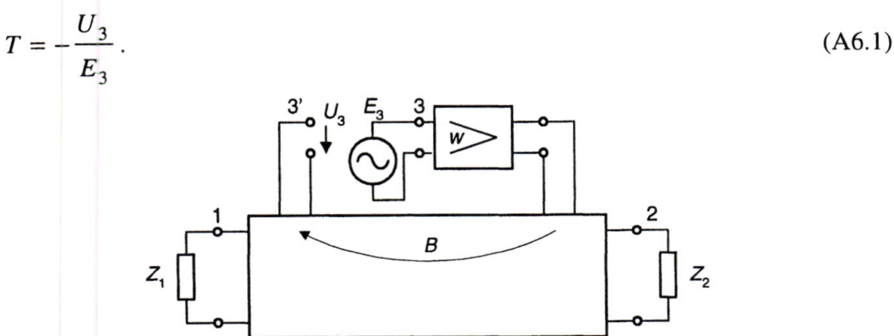

Fig. A6.2 Disconnected feedback loop

When the return ratio T is being measured, the emf E_1 of the source connected to the system's input 1 must be replaced by a short circuit, in accordance with the superposition principle. Thus, the source impedance Z_1 is connected to the port 1. The return ratio therefore depends on Z_1, i.e., $T = T(Z_1)$. In particular, $T(0)$ denotes the value of T measured while the input is shorted, and $T(\infty)$, while the input is open.

Fig. A6.3 shows the ***cross-sectioned feedback circuit***. The external two-poles Z' and Z" connected to the input and output of the broken loop provide appropriate loading for the disconnected parts of the B-circuit in order to keep their transfer coefficients unchanged. The emf E_6 applied to the input of the broken loop produces return signal U_5. The ratio

$$-\frac{U_5}{E_6} = -\frac{U_3}{E_6}\frac{I_4}{U_s}\frac{U_5}{I_4}$$

equals T. When T is measured this way, the two-pole Z' need not be connected since it is shunted by the emf E_3.

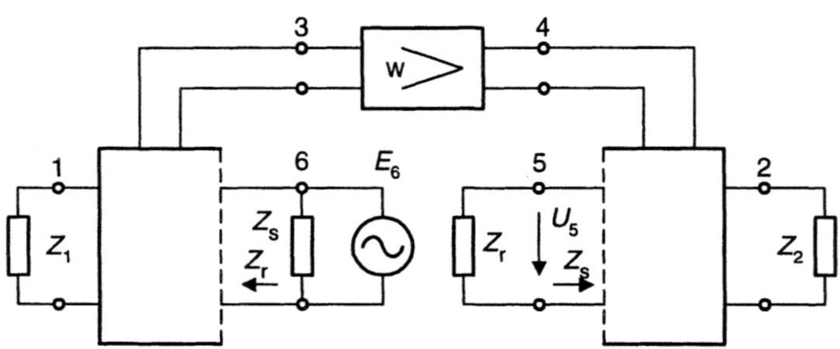

Fig. A6.3 Disconnecting feedback loop in the feedback path

Generally, the voltage output-input ratio of a linear fourpole can always be presented as the product of two ratios: the current output-input ratio, and the ratio of the load impedance to the fourpole input impedance. The latter ratio is found to be 1 when T is calculated. Therefore, T may be measured arbitrarily as the ratio of either voltages or currents.

In the closed-loop system, the signal U_3 is formed by superposition of the effects U_3° and $-U_3T$ produced respectively by the signal source and the output of the amplifier. Thus $U_3 = U_3^{\circ} - U_3T$, whence

$$U_3 = \frac{U_3^{\circ}}{F}; \tag{A6.2}$$

here, $F = T + 1$ is the return difference, i.e., the difference between the signals U_3 and E_3 relative to E_3.

The output of the feedback system in Fig. A6.1 is a linear combination of two signal sources: the output of the amplifier and the signal source. By virtue of (A6.2), the signal from the amplifier output is reduced F times. Hence, the closed-loop system transfer coefficients in voltage, current, and as the ratio of the output voltage to the signal emf are, respectively,

$$K_{OL} = \frac{U_2}{U_1} = \frac{K}{F(0)} + k_d\,, \tag{A6.3}$$

$$K_{OL\,I} = \frac{I_2}{I_1} = \frac{K_I}{F(\infty)} + k_{Id}\,, \tag{A6.4}$$

$$K_{OL\,E} = \frac{U_2}{E_1} = \frac{K_E}{F} + k_{Ed}\,. \tag{A6.5}$$

Here, K_{OL}, $K_{OL\,I}$, and $K_{OL\,E}$ are the open loop system transmission functions, measured while the feedback path is disconnected; $F(0)$, $F(\infty)$, and $F = F(Z_1)$ are the return differences measured under the conditions of connecting zero impedance, infinite impedance, or, respectively, impedance Z_1 to the system's input terminals; and k_d, k_{Id}, and k_{Ed} are the coefficients of direct signal propagation through the *B*-circuit and are determined under the same set of loading conditions at the input terminals.

Let Z designate the input impedance, and Z_o, the input impedance in the system without feedback (with a cross-sectioned feedback path, or with $w = 0$, i.e., with the active element killed out). Then

$$K = K_{OL}\frac{Z_2}{Z}\,, \tag{A6.6}$$

$$k_d = k_{Id}\frac{Z_2}{Z}\,, \tag{A6.7}$$

$$K_{OL\,I} = K_I\frac{Z_2}{Z_o}\,. \tag{A6.8}$$

By substituting (A6.7), (A6.8) into (A6.3), and (A6.4) into (A6.6), we get

$$\frac{K_I Z_2/Z_o}{F(0)} + k_{Id}\frac{Z_2}{Z} = \left(\frac{K_I}{F(\infty)} + k_{Id}\right)\frac{Z_2}{Z}$$

from which Blackman's formula follows:

$$Z = Z_o\frac{F(0)}{F(\infty)} = Z_o\frac{T(0)+1}{T(\infty)+1}\,. \tag{A6.9}$$

The formula expresses Z through the three easily calculated functions: Z_o and return ratios $T(0)$ and $T(\infty)$.

Since, in principle, any two nodes of the *B*-circuit can be regarded as input terminals of the feedback system, the formula (A6.9) can be used for calculation of driving point impedance at any port *n*, provided that $F(0)$ is understood as F measured with the port *n* shorted, and $F(\infty)$, with the port terminals open.

If the terminals are shorted, the voltage between them vanishes, but not the current. For this reason, the feedback is called ***current-mode*** (or ***series***) with respect to the terminals n if with respect to these terminals $T(0) = 0$ and $T(\infty) = 0$. Analogously, the feedback is ***voltage-mode*** (or ***parallel***) if $T(0) = 0$ and $T(\infty) = 0$. If neither of them equals 0, the feedback is called ***compound***.

If the feedback is infinite, the input impedance is

$$Z = \lim_{w \to \infty} Z = Z_o \frac{wB(0)}{wB(\infty)} = Z_o \frac{B(0)}{B(\infty)} . \tag{A6.10}$$

It depends exclusively on the B-circuit and not on w.

More detail along these lines is given in [2, 9].

Appendix 7 Effect of feedback on mobility

The following derivation of Blackman's formula follows Blackman's original proof, however in mechanical terms [21].

Consider a mechanical system with an actuator accessed via a single mechanical port. That is, consider an active structural member including an actuator. To the active member, a structure is connected. The structure's mobility is Z_L. The force F is measured between the active member and the structure.

For the purpose of analysis, consider disconnecting the feedback loop at the input to the actuator and applying signal E to the actuator input. The relative velocity V across the active member and the feedback return signal E_r at the end of the disconnected feedback loop can be each expressed as a linear function of the force F and the signal E:

$$V = aE + bF, \quad E_r = cE + dF \tag{A7.1}$$

Here a, b, c, and d are the constants to be determined from boundary conditions. Notice that when the feedback loop is closed, $E = E_r$, and $E = 0$ when it is open.

First, find the expressions for the mobility without and with feedback. The case without feedback, i.e., $E = 0$, gives $V = bF$. Thus, the active member mobility without feedback, Z_o, becomes

$$Z_o = V/F = b \tag{A7.2}$$

When the loop is closed, i.e., $E_r = E$, (A7.1) gives $V = [b + ad/(1 - c)]F$ so that the active member mobility with feedback, Z, becomes

$$Z = b + ad/(1 - c) \tag{A7.3}$$

Second, find the expressions for the return ratio for the two different loading conditions, when the active member is clamped and when it is free to expand. The return ratio T of the active member feedback loop is defined as the ratio of the return signal to the negative of the input signal, i.e., $T = -E_r/E$. This ratio is certainly a function of the mobility of the structure Z_L, i.e., $T(Z_L)$.

When $Z_L = 0$, the active member is rigidly constrained, i.e., $V = 0$. In this case, (A7.1) gives $F = -a/(bE)$ and $E_r = (c - ad/b)E$. Hence, the return ratio becomes

$$T(0) = -c + ad/b\,. \tag{A7.4}$$

On the other hand, when $Z_L = \infty$, i.e., the active member is free to expand, zero force is induced in the active member. In this case $E_r = cE$ and the return ratio is

$$T(\infty) = -c\,. \tag{A7.5}$$

Comparing the obtained equations (A7.2–A7.5) results in Blackman's formula as

$$Z = Z_o \frac{T(0) + 1}{T(\infty) + 1}.$$

The formula expresses active member mobility with feedback, Z, in terms of three other functions, Z_o, $T(0)$, and $T(\infty)$, that do not depend on the structural system to which the active member is connected.

Appendix 8 Dependence of a function on a parameter

The formulas derived in the two previous sections are bilinear, i.e., can be presented as a ratio of two linear functions. Generally, a Laplace transform transfer function of a physical linear system can be presented as Δ/Δ_o, where the main determinant Δ and the minor Δ_o are linear functions of the value of an element of the system (e.g., for an electrical system, resistance of a resistor, capacitance of a capacitor, inductance of an inductor, gain coefficient of an amplifier). The proof can be found in [2].

For example, return ratio is a bilinear function of the feedback system load impedance Z_L. This function can be expressed as

$$T(Z_L) = \frac{Z_o T(0) + Z_L T(\infty)}{Z_o + Z_L}$$

and can be used to analyze the effects of the plant uncertainty.

Appendix 9 Balanced bridge feedback

As is seen from Blackman's formula (A6.9), each one of the following equations

$$\left.\begin{aligned} T(0) &= T(\infty) \\ F(0) &= F(\infty) \\ Z &= Z_o \end{aligned}\right\} \qquad \text{(A9.1)}$$

validates the other two. This are the conditions of ***balanced bridge***. In such a system the feedback is not dependent on the external impedance, and therefore the value of the feedback is not limited by this impedance variations.

An example of a balanced bridge circuit is shown in Fig. A9.1. The Wheatstone bridge is balanced if $Z_a/Z_b = Z_c/Z_d$. In this case, the transmission from the output of the amplifier to each of the system output terminals is the same, the voltage across the bridge diagonal is zero, and connecting these terminals by any external impedance will not change the return ratio. Therefore, the first condition from (A9.1) is satisfied.

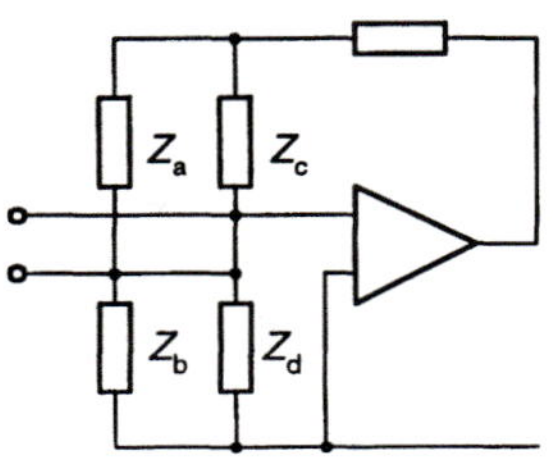

Fig. A9.1 Balanced bridge feedback

Zero transmission between the system's output terminals and the input of the amplifier is of similar use. An example can be drawn by inverting the direction of the amplifier in Fig A9.1. The desired output impedance of the amplifier (actuator) Z_d can be implemented by using a nested actuator feedback loop.

Appendix 10 Phase-gain relation for describing functions

For a describing function of a system composed of linear links and a nonlinear nondynamic link, the relation between the gain and phase frequency responses $B(A)$ is the same as in linear systems if the DF of the nonlinear link stays the same for all frequencies, i.e., the signal amplitude at the input to the nonlinear link does not depend on frequency. This is the case of the iso-*w* curves. When, however, the nonlinear link depends on frequency, the relation $B(A)$ becomes more complicated.

Assume that the conjunction of filter is applicable to the problem considered. Let us determine the relations between the real and imaginary components of the describing function of a nonlinear link:

$$H = \operatorname{Re} H + j\varphi = \ln U_2 - \ln U_1$$

Here, U_1 is the amplitude of the sinusoidal input signal, and U_2 is the complex amplitude of the output signal fundamental. The derivative is

$$\frac{dN}{d\ln U_1} = \frac{U_1\, dU_2}{U_2\, dU_1} - 1 = \exp(\theta - H) - 1$$

where the logarithmic transfer function for increments

$$\theta = A + jB = \ln dU_2 - \ln dU_1,$$

and dU_1 is real. Therefore,

$$\frac{d\operatorname{Re} H}{d\ln U_1} = \exp(A - \operatorname{Re} H)\cos(B - \varphi) - 1,$$

and, by applying the Bode relation to the differentially linearized stable m.p. link, we see that

$$\cos\left[\varphi - \frac{1}{\pi}\int_{-\infty}^{\infty}\frac{dA(U_1, u)}{du}\ln\coth\frac{|u|}{2}\,d|u|\right] = \left(1 + \frac{d\operatorname{Re} H}{d\ln U_1}\right)\exp(\operatorname{Re} H - A) \qquad \text{(A10.2)}$$

Using (A10.2) the phase shift $\varphi(U_1, \omega_o)$ can be calculated from the measurements of: the real component of the describing function $\operatorname{Re} H$, its derivative on the level of the input signal $d\operatorname{Re} H/d\ln U_1$, and the frequency response of the real part of the differentially linearized circuit A. The latter must be measured while simultaneously applying to the link's input the main signal with amplitude U and frequency ω_o. The magnitude A is easy to measure with a frequency-selective voltmeter if at this frequency no nonlinear product has a substantial amplitude. Since the power of the set of frequencies of these components is small in comparison with the power of all real frequencies, for the continuously differentiable function A (the differentially linearized circuit must be m.p. and not containing singularities on the $j\omega$-axis, which will be clearly seen on the results of the measurements), the integral in (A10.2) can be determined with sufficient accuracy from the measurement results of A on a limited number of frequencies.

The relation (A10.2) is rather complicated. For practical design applications, iso-*w* responses are preferable and should be used whenever possible.

Appendix 11 Discussions

The function of the "Discussions" sections is to anticipate questions and objections, and also to address certain persistent misconceptions. In most cases, these are condensed transcripts of conversations which took place with the authors' colleagues and students, to whom we extend our sincere gratitude.

A11.1 Compensator implementation

Q: In the Preface, it is stated that modern technology somehow relates to high-order compensators. What does the word "modern" have to do with the compensator's order? Isn't it used here just as an advertisement gimmick?

A: It is more than the gimmick. The old way of compensator implementation used the same technology as that for the actuation. There exist mechanical, hydraulic, and pneumatic gadgets whose outputs are proportional to the integral or to the derivative of the inputs. These devices were combined to make the desired transfer function of the compensator. Since the devices were relatively expensive and rather difficult to tune, the compensators remained low-order. The most complicated among them were the *PID* controllers using a proportional device, an integrator device, and a differentiator device, all connected in parallel.

The way of the mechanical controller implementation impacted the control theory. Convenient methods have been developed for low-order controller design – however, they are not suitable to handle high-order controllers.

On the other hand, feedback amplifier technology was not limited to low-order compensators since small capacitors, resistors and chokes used in the compensators were cheap. The Bell Laboratories scientists developed synthesis theory for high-order compensators to approximate the theoretical best. These theory and design methods only partially propagated into the so-called classical automatic control.

The sensor technology changed completely in 1980s and 1990s. Nowadays, in mechanical, pneumatic, and hydraulic systems the outputs of the sensors are electrical, the inputs of the actuators are electrical, and the compensators are electrical. Therefore, high-order compensators can be economically implemented in hardware or software, and the control can be made close to the best possible. Teaching the control at the elementary level, however, lagged, still practicing some tools that had become not only obsolete but plain counterproductive and confusing when they had been used to try to make a precision controller design.

Q: The specifications on the closed-loop performance are often formulated in the time domain. Is it worthwhile to convert them into frequency-domain specifications and then design the compensator with frequency-domain methods?

A: Definitely. The benefits of frequency-domain design far outweigh the trouble of converting the time-domain specifications. The conversion of most practical specifications is simple and transparent.

Q: For initial design, is it advisable to optimize the performance while ignoring some details like noise, nonlinearities, and plant parameter variations?

A: That sort of approach typically produces disastrous results in engineering practice. Even in the initial design, the system should be addressed in its entirety, or else the time and resources allocated for design will be wasted on dead ends.

A11.2 Feedback: positive and negative

Q: Does the minus sign at the summer mean that the feedback is negative?

A: No. At any specified frequency, the sign of the feedback also depends on the modulus and phase of the elements in the loop.

Q: From the transfer functions *CAP* and *B*, the closed-loop transfer function *CAP*/(1 + *CAPB*) can be calculated, and vice-versa. Is that all there is to it?

A: It's not quite that simple. When the feedback is large, only the first calculation will be accurate. Even rather large errors in the knowledge of *C*, *A*, and *P* do not cause much uncertainty in the closed-loop transfer function. The result is that the inverse problem is ill-conditioned, with small errors in the observed or calculated closed-loop response possibly mapping into large errors in the calculation of the loop transmission. Therefore, the calculations or measurements of the closed-loop response are commonly inadequate to determine the loop transfer function.

A11.3 Tracking systems

Q: Isn't a common definition of a tracking system that the feedback path transmission is 1?

A: $B = 1$ is required for frequencies much lower than the crossover, and B can be identically 1 in a system with a prefilter. However, with $B = 1$ and without a prefilter, the disturbance rejection and the compensator design need to be compromised to provide acceptable transient responses for the nominal plant. (This may be unavoidable for homing-type systems.)

A11.4 Elements (links) of the feedback system

Q: Should the feedback path and the compensator be analog or digital?

A: There are many factors to consider here. A digital compensator has the advantage that it can be reprogrammed in response to known changes in the plant; however, it should be pointed out that changing software is often neither trivial nor cheap. A primary disadvantage of digital compensators is that they reduce the available feedback when the computational delay in the loop is significant, as it often is. For this reason, analog compensators are the first choice for fast feedback loops. If necessary, the parameters of an analog compensator can be made "programmable" by employing multiplying D/A converters. The summing device and the feedback path can be analog as long as the required relative accuracy is not better than about 0.0001. To achieve much higher accuracy, they should probably be digital. Many other factors will affect this decision, including power requirements, cost, production quantity, and so forth.

Q: The compensator is optimized for disturbance rejection, and the feedback link (and/or prefilter) is designed to provide the desired closed-loop response with the nominal plant. But what happens to the closed-loop response when the plant deviates from the nominal?

A: A methodology has been developed, called quantitative feedback theory (QFT), that designs the system so as to satisfy the closed-loop performance specifications in the presence of the worst-case parameter variations. This procedure is discussed in Chapter 8. Although the methods involve a significant amount of calculation, they are useful when parameter variations are very large, i.e., more than 10 dB. For most applications, it is sufficient to first design the closed-loop response with the nominal plant, and then determine the changes caused by the plant variations, and, if necessary, modify the prefilter.

Q: What is wrong with design methods that use the closed-loop response as the objective for the compensator design, like the pole placement method?

A: With such methods, the appropriate value of feedback and the stability margins are not observed and preserved. This can cause many problems: the system stability may be only conditional, or wind-up might occur, the disturbance rejection may not be optimized, and the plant parameter variations may cause larger closed-loop response variations than what is achievable with feedback maximization methods based on the open-loop frequency response.

A11.5 Plant transfer function uncertainty

Q: How accurate must the gain response approximation be for the accuracy of the related phase response to be 5° ?

A: About 1/2 dB.

Q: The compensator transfer function must be high order to approximate well the required transcendental response. But, the poles and zeros of a high-order transfer function may be very sensitive to the polynomial coefficients. Then, rounding errors will make the system not robust. Is this correct?

A: Not exactly. Variations in the values of poles and zeros is not important as long as the frequency response doesn't change much. The sensitivity of the frequency response to polynomial coefficients in typical cases is not high, and the compensator order is in this aspect irrelevant. For example, in analog telecommunication systems, equalizers have been routinely employed with 20th order transfer functions, with the accuracy of implementation of the desired response of ±0.001 dB. The requirements for the control system compensators are by a long way easier.

Q: What if the plant is non-minimum-phase?

A: The non-minimum-phase component of the loop phase lag must be compensated for by reducing the minimum-phase component of the phase lag. This can be done by reducing the frequency f_b to increase the length of the Bode step. Simple formulas for these calculations are given in Chapter 4.

Q: The command, the prefilter, and the feedback path are known, and the sensor is measuring the plant's output. This data suffices to calculate *CAP*. Further, if *C* is known, *AP* can be determined exactly, and this information can be used to modify *C* such that the loop transmission function is as desired. Right?

A: Before deciding to do this, let's consider the errors in the calculation. The sensor is not ideal (noisy), the command is not well-suited to the goal of characterizing the plant, and the problem is ill-conditioned for all frequency components where the feedback is large. This approach might be used for the frequency band where the feedback is positive, i.e., in the neighborhood of the crossover frequency f_b. However, for this application, the high frequency components of the command need to be sufficiently large, larger than the noise, which is not usually the case.

Q: Can the effects of the plant resonances on the potentially available feedback be calculated in advance?

A: Sure. See Chapter 4, Problem 4 and the Answer to the Problems.

A11.6 The Nyquist stability criterion

Q: Is the distance of the poles of the closed-loop transfer function from the $j\omega$-axis a better robustness measure then the stability margins on the Nyquist diagram?

A: No. A practical counterexample is an active *RC* notch filter. Its closed-loop poles are very close to the $j\omega$-axis, but the stability margins in this feedback system are sufficient, the system is globally stable, reliable and widely used. Conversely, placing the poles of a closed-loop system far away from the $j\omega$-axis does not guarantee global stability, process stability, or robustness.

Q: Is it convenient to define the boundaries on the Nyquist diagram to reflect necessary and sufficient stability requirements?

A: Yes.

Q: How are the values for the stability margins determined?

A: Global and process stability must be assured in the presence of the plant and actuator parameter variations. (Often, process stability is not strictly achieved, but the satisfaction of a certain norm on the errors in the nonlinear state of operation is acceptable.) Also, the sensitivity of the closed loop response to plant parameter variations must be limited.

Q: Why should the area surrounding the critical point be defined by gain- and phase-margins, and not by mathematically simpler circular boundary?

A: The shape of the margin boundary is defined by global and process stability requirements and by the plant parameter variations. In the majority of plants, variations of the plant gain and plant phase in the neighborhood of the crossover are not well correlated. This necessitates defining gain- and phase-margins independently.

Q: Is much performance lost if we simplify the design methods and use circular stability margin boundary?

A: Changing the stability margins from those appropriate to circular would substantially reduce the integral of positive feedback in this frequency range and, therefore, the integral of negative feedback in the functional frequency band.

Q: Is the definition of *Nyquist-stable* the same as that of *conditionally stable*?

A: No. The Lyapunov definition for *conditional stability* is that the nonlinear system stability depends on the initial conditions. A Nyquist-stable system with the usual actuator saturation nonlinearity may be conditionally stable but can be rendered globally stable by the addition of nonlinear dynamic compensation.

A11.7 Actuator's output impedance

Q: Whatever the internal (output) impedance of the actuator, we can always apply to its input such a signal that the system output will be as required. Therefore, it doesn't matter what is the actuator output impedance. Right?

A: Wrong. For the nominal plant, the statement is right, but the effects of the plant parameter deviations from the nominal are to a large extent dependent on the actuator output impedance.

Q: Compound feedback produces some finite output impedance for the driver. But, this can be achieved more simply, by making voltage feedback and placing a resistor in series.

A: The result will be the same as long as the system remains linear. This method can be used in the laboratory for testing various control schemes. However, there will be power losses on this series resistor, and as a result, the actuator needs to be more powerful, with a larger saturation level.

Q: But, isn't the power dissipated anyway on the internal impedance of the actuator when the impedance is made finite by compound feedback?

A: No. Compound feedback causes no power losses and doesn't reduce the actuator efficiency.

A11.8 Integral of feedback

Q: At what frequencies is the feedback positive?

A: The feedback is positive within the circle of unit radius centered at (–1,0) on the T-plane. When the loop gain decreases monotonically with frequency, the major part of the area of positive feedback falls in the band $0.6f_b$ to $4f_b$. The integral of feedback at much higher frequencies is negligibly small (see bibl. to [9]).

Q: Well, the feedback makes the system better at some frequencies, but at others even worse than it was before. So, what is gained from the application of feedback?

A: Negative feedback is used to cover the frequency range where the plant parameter variations and the effects of disturbances are critical, and positive feedback is confined to the frequency range where the noise and the disturbances are so small that even after being increased by the positive feedback, they still will be acceptable.

A11.9 Bode integrals

Q: How many Bode relations are there, total?

A: There are about 30 different integral relations in Bode's book.

Q: Have new relations or expansions of these relations been made since then?

A: Yes. The most important are the R. Fano expansions of the Bode integral of the real part, which use not one but several terms in the Laurent expression. They are usable for RF and microwave circuits design. Also, Bode relations were expanded on unstable systems, discrete systems, multivariable systems, parallel paths of the signal propagation, and the transfer function for disturbance isolation. An example of an expansion on certain nonlinear systems is given in Appendix 10. However, these expansions are less important for practice, compared with the four fundamental Bode relations described in this book.

A11.10 The Bode phase-gain relation

Q: Why is the slope of the Bode diagram expressed here in dB/oct, not in dB/dec?

A: The same reason that supermarkets sell milk by gallons, not barrels. Gain samples one octave apart typically suffice to calculate the phase response with less than 5° error as needed for sound feedback system design. And it is certainly convenient that the slope of a well-designed Bode diagram is nearly –10 dB/oct.

Q: Does it make sense to consider a loop gain response having a constant slope of –10 dB/oct? How can such a response be implemented in a physical system?

A: It can be closely approximated by a rational function of s.

Q: Are the Bode integrals applicable to transcendental plant transfer functions?

A: Yes.

Q: At a university where I have taught, the teaching relies heavily on the root locus method. Afterward I started to work in aircraft industry and was surprised that here the engineers prefer Bode diagrams. Why is it so?

A: Bode noticed that for an engineer striving to make the best of the design, it is difficult to cope with three scalar variables: gain, phase, and frequency, and he *reduced* the number of the variables to only *two scalars*: gain and frequency. This approach allows handling high-order systems (including MIMO and multiloop systems). In contrast, applying the root locus method to an nth order system *increases* the number of the variables to *n complex variables* (roots) crawling all over the s-plane in a strange, threatening, and unmanageable manner.

A11.11 What limits the feedback?

Q: What are the physical factors which limit the feedback?

A: The major factors are: the uncertainties in the plant transfer function, the clipping of higher frequency components of the sensor noise in the actuator, and the level of noise in the system's output.

Q: Does it pay to exploit as much as possible the available knowledge about the plant when designing the control law?

A: Certainly. This increases the available feedback. The available feedback is infinite for hypothetical full-state feedback, and no feedback is available for a completely unpredictable plant.

Q: A typical concern of the project manager: the system design trade-offs require knowledge of the available performance of its subsystems, and these subsystems often employ feedback. How can the system trade-offs be made without designing the feedback subsystems?

A: The manager should request data about available subsystem performance, i.e., the performance that can be achieved with the optimal compensator. Using the Bode approach this can be calculated without actually designing the compensator.

Q: What kind of data do we need for this estimation? What kind of mathematics is involved?

A: The following data is generally required: the sensor noise spectral density over three octaves in the vicinity of the planned feedback bandwidth, the available (nondistorted) output signal amplitude from the actuator, ranges of frequencies where plant structural modes can fall, up to four octaves over the estimated feedback bandwidth, and sometimes more. This sort of data is typically available, although it may not always be very precise. The available disturbance rejection can then be estimated using the Bode integrals.

A11.12 Feedback maximization

Q: I want to develop and market a product. How can I be sure that my competitor will not enter the market, soon after me, with a superior product that uses the same actuator, plant, and sensor, but has bigger and faster feedback?

A: The Bode integral approach allows the determination of the best theoretically available system performance.

Q: Is it necessary to approximate a physical plant transfer function (which is sometimes transcendental) by a rational function and/or to perform plant "model reduction" when designing a feedback system with the Bode method?

A: No. The calculated or measured plant and actuator transfer functions are subtracted from the desired loop response to obtain the desired compensator response, and the rational function approximation is introduced only in the final stage of the compensator design.

Q: What is the required accuracy of the approximation of the desired loop gain response?

A: The accuracy must be such that the related phase response will approximate the desired phase response with an accuracy of about 5°.

Q: Why 5° ?

A: In this case, the *average* loop phase lag must stay 5° away from the stability margin boundary. Then, the average slope of the Bode diagram will be less than the maximum acceptable by 1/3 dB/oct (from the proportion: 90° for 6 dB/oct). Therefore, over each of the 3 octaves of the cut-off, 1 dB of feedback is lost. Typically, these losses are marginally acceptable. The small remaining possible feedback increase may not justify a further increase in the complexity of the compensator.

Q: How accurate must the gain response approximation be for the accuracy of the related phase response to be 5° ?

A: About 1/2 dB.

Q: The compensator transfer function must be high-order for accurate approximation of the required transcendental response. But, poles and zeros of a high-order transfer function may be very sensitive to the polynomial coefficients. Then, the rounding errors will make the system not robust. Is this correct?

A: Not exactly. Variations in the values of poles and zeros are not important as long as the frequency response doesn't change much. The typical compensator response has no sharp peaks and notches; the sensitivity of the such function modulus to the polynomial coefficients is typically less than 1, and the compensator order is in this aspect irrelevant. Therefore, in order for the compensator to be accurate within 0.5 dB, i.e., 6% in the magnitude of the transfer function, the polynomial coefficient accuracy need not be better than 1 to 5%, and the effect of rounding is insignificant.

Q: What is the best available accuracy of analog compensators?

A: Equalizers for analog telecommunication systems have been routinely designed with up to 20th order transfer function, with accuracy of implementation of the desired response of ±0.001 dB.

Q: How were these equalizers designed?

A: By interpolation, by cut-and-try procedures, by using asymptotic Bode diagrams, by adjustments in the element domain, by using Chebyshev polynomial series, by using the Second Remez Algorithm, and by using the Simplex Method.

Q: What if the plant is non-minimum-phase?

A: The non-minimum-phase component of the loop phase lag must be compensated for by reducing the minimum-phase component of the phase lag. This can be done by reducing the frequency f_b to increase the length of the Bode step. Simple formulas for these calculations are given in Chapter 4.

Q: Only analog systems have been discussed. What is the difference with regard to digital controllers?

A: Sampled data systems, which include an analog-to-digital converter, a digital filter, and a digital-to-analog converter are time-variable and have an inherent delay that reduces the available feedback.

Q: What about a proportional-integral-derivative (*PID*) compensator having the transfer function $C(s) = k_p + k_i/s + k_d s$?

A: With appropriate gains, the *PID* compensator roughly approximates the response of the optimal controller; however, the approximation error can be quite large. *PID* control falls short of the performance of a system with an appropriate high-order compensator.

Q: But how much better off than a *PID*'s is the optimal controller?

A: It depends on the frequency response of the plant transfer function and the disturbance spectrum density. For smooth plant responses, the expected improvement in disturbance rejection is 5 to 10 dB when a *PID* controller is replaced by a higher-order controller. For a plant with structural resonances, especially when it is used in conjunction with a nonlinear dynamic compensator, a 5 to 30 dB improvement can be expected from the high-order controller.

Q: Why aren't the Ziegler-Nichols conditions for tuning *PID* controllers presented in this book?

A: This method does not use a prefilter (or a command feedforward), and is suitable only for specific plants. More general and better results can be achieved with the frequency domain approach.

A11.13 Feedback maximization in multi-loop systems

Q: What about loop response shaping for multi-loop systems?

A: Basically, the technique is the same. Multi-loop systems can be designed one loop at a time, as long as loop coupling is taken care of by adjusting the responses and providing extra stability margins within certain frequency bands.

Q: Does this technique work well even when the number of the loops is large?

A: Yes, typically.

A11.14 Nonminimum phase functions

Q: Is there a case where it is advantageous to apply compensators with non-minimum-phase transfer functions?

A: Probably not. Sometimes, when the plant has high-frequency structural modes, a compensator must introduce a phase delay in the loop to make the system stable. However, this phase delay can always be achieved with a minimum-phase low-pass filter-type function – with the extra advantage of increasing the amplitude stability margin and reducing the noise effect at the actuator's input.

A11.15 Feedback control design procedure

Q: Isn't it awkward to go back and forth from the Nyquist diagram on the *L*-plane to the Bode diagram? Isn't one type of diagram enough?

A: It's a case of the right tool for the right job: global and process stability characterization, stability margin definitions, and nonlinear dynamic compensation design all require the use of the *L*-plane, but the trade-off resolution and the compensator design are simplified by using the Bode diagram due to a reduced number of variables.

Q: Do the design phases discussed in the following Appendix 12 apply for the design of single-loop or multi-loop systems?

A: Single-loop. But multi-loop system design is a natural extension of this procedure.

Q: Still, if you have to design, say, a three-input, three-output system, will it take 3, 9, or 27 times longer than a single-loop system of comparable complexity for each channel?

A: 3 times longer, typically.

A11.16 Global stability and absolute stability

Q: Is it convenient to use Lyapunov functions in the design of high performance control systems?

A: Devising appropriate Lyapunov functions is yet unmanageable for systems with a high-order linear part and several nonlinear elements.

Q: At the 1st Congress of the International Federation of Automatic Control (IFAC) in 1960 in Moscow, researchers from Russian academia presented important theoretical results using time-domain and state variables. This research impressed American professors attending the conference and shifted to a large extent the direction of research in American academia. Were these methods the methods Russian engineers employed to design control laws for their rockets and satellites?

A: No. Russian rockets' control systems have been designed with frequency domain methods in very much the same way as American rockets (see A13.9).

A11.17 Describing function and nonlinear dynamic compensation

Q: DF analysis fails to predict instability in certain systems. Would it be right therefore to discard the DF approach altogether?

A: No. It would be foolish to discard a Phillips screwdriver just because it cannot drive all screws. Similarly, although DF analysis is not a universal, foolproof tool, it is fairly accurate when employed for analysis and synthesis of *well designed* control loops with steep monotonic low-pass responses.

Q: What is the purpose of using DF now, when nonlinear systems can be simulated well with computers?

A: The DF advantage is seen when it is used not for the purpose of analysis only, but for the purpose of design, and especially, conceptual design of control systems with several nonlinear links. In many cases, the accuracy of the analysis need not be high. The phase of the DF has an error of up to 20° compared with the phase calculated with exact analysis – but, this phase is not of critical importance since NDCs can easily provide the required phase advance.

Q: DF analysis does not account for additional phase shifts created by higher harmonics interacting in the nonlinear link. So does the DF design guarantee robustness?

A: Interaction of high-order harmonics can typically yield up to 15–20° of extra phase lag for the fundamental. The NDC DF phase advance must be increased by this amount, which is rather easy to do since the typical NDC phase advance exceeds 100–200°.

Q: How much improvement in disturbance rejection can be expected from using an NDC?

A: It depends on the desired frequency shaping of the disturbance rejection – in some cases, up to 30 dB.

Q: Is it advantageous to use NDCs in MIMO systems?

A: Yes. In addition to providing the same advantages as for SISO systems, NDCs reduce the effects of nonlinear coupling between the loops on the system stability.

A11.18 Multi-loop systems

Q: Why is it the number of saturation elements that matters when defining what is the multi-loop system?

A: Because the DF of a saturation link changes from 1 down to the inverse of the loop gain coefficient, i.e., 100 or 1000 times. This effect needs to be accounted for during the stability analysis and the loop transfer functions synthesis. Compared with this effect, even the effects of the plant parameter variations are small.

A11.19 MIMO systems

Q: Where is the better place to implement the decoupling matrix – in the forward path or in the feedback path?

A: In the forward path, since in this case the decoupling matrix doesn't need to be precise. In this case, commands are formulated in the sensors' readings. When commands should be formulated in actuators' actions (and sensors are not aligned with the actuators), the decoupling matrix should be placed in the feedback path.

Q: Is the design of MIMO and multi-loop systems difficult?

A: It is not simple, but several oft-ignored techniques greatly simplify the design:

- The Bode technique of making trade-offs between the loops without designing the compensators,
- verifying the m.p. character of parallel signal propagation paths,
- considering the effects of loop coupling only in the narrow frequency ranges where the coupling is critical,
- using appropriate output impedances for the actuators, and
- using NDCs to provide global stability and good performance in the nonlinear mode of operation, without sacrificing the disturbance rejection.

Q: How does coupling between the loops affect the system stability and robustness?

A: The effects of loop coupling can be analyzed by considering the change in one loop transfer function caused by the plant and the actuator transfer function variations in another loop. The robustness can be provided for by correspondingly increasing the stability margins.

Q: At what frequencies is the coupling most dangerous?

A: In the nonlinear mode, at lower frequencies, where the loop gain is large and changes in it caused by saturation could affect other loops. However, this can be taken care of by using an NDC. In the linear mode, the coupling is more dangerous near the crossover frequencies where the feedback in the loops is positive.

A11.20 The Bode's book

Q: In the bibliography, the book is characterized as a book on synthesis, and not on analysis. However, Bode himself named it as the book on analysis. Why?

A: The book describes the tools for analysis also, but mostly for synthesis. The choice of tools for these two purposes are, generally speaking, quite different.

Q: What about the terminology in the Bode's book?

A: Bode developed several powerful approaches valid for a wide range of applications, and he optimized the terminology to serve this wide range of applications.

For example, in his definition the transfer function is the ratio of the input to the output, and the sensitivity is the ratio of the relative change of the cause to the that of the result. These definitions allowed integrating feedback theory with driving point functions analysis and synthesis.

Following these definitions, he defined certain functions as minimum phase as the functions having the smallest phase shift among all transfer functions with the same gain response. Today, however, we use the inverse definition of the transfer function which is, maybe, intuitively better for the narrower area of applications, but are not as elegant and general. With the definition employed today of the transfer function as the ratio of the response to the source, m.p. function has in fact *maximum* phase shift. To bring some sense to this, m.p. function must be called minimum *phase lag* functions – but still, in most books it is inconsistently called minimum phase function.

Bode didn't use the word "compensator". Instead, as a telecommunication engineer, he used the word "corrector", for feedback loop or for any other signal transmission channel. The meaning of this word is: we know the theoretically best response, we have measured the response (of the loop) as is, then, we have to correct the response for it to be as prescribed by the theory, to be the optimal.

Bode used separate definitions for the sensitivity and for the return difference, while in some contemporary works the return difference is formally defined as "sensitivity" – which may be acceptable if we only limit the use of this term to a very narrow class of problems, too narrow even for the control engineers.

Bode formulas has been later modified and generalized for the purposes of certain research goals and for the convenience of teaching. Nonetheless, the authors of the *Classical Feedback Control* found no versions in the existing literature better or equally usable for the engineering design purposes than those employed in the Bode's book. This is why *Classical Feedback Control*, with very few exceptions, uses original Bode's notations, definitions, and expressions.

Appendix 12 Design sequence

In the following, an estimate of the average time is given for completion of each of the steps by an experienced designer, and for a not-unusually-complex problem.

Phase I: Calculation of the available performance (up to 24 hours)

- determine the feedback bandwidth f_b (about 30 minutes);
- decide what kind of nonlinear dynamic compensation (if any) will be employed, and define the stability bounds on the L-plane (5 minutes);
- draw the desired Nyquist diagram on the L-plane which maximizes the feedback (5 minutes);
- find and draw the desired Bode diagram (5 to 30 minutes);
- make all trade-offs required by the frequency responses of the disturbance spectrums, by other feedback loops in the system, and by the plant transfer function variation, reshape the Bode diagram (10 minutes to 22 hours), determine the available feedback and report the system performance to your manager;

Phase II: Design of the linear compensators and prefilters (up to 5 days)

- *given the nominal plant response, or a plant model,* subtract the nominal plant gain response from the desired nominal loop gain response to obtain the required compensator gain response (10 to 30 minutes);
- approximate the required compensator response using one of the methods discussed in Chapter 6 and obtain the compensator transfer function and specifications for the compensator hardware/software (a day or two);
- make a linear system model and calculate the closed-loop frequency response (several hours);
- decide on the type of nominal closed-loop gain frequency response that is close to the optimum (2 minutes to several hours);
- subtract the calculated closed-loop gain response from the desired response to obtain the prefilter gain response (10 to 30 minutes); or, alternatively, find the desired response for the feedback path, B;
- find the prefilter (or feedback path link) to approximate the required response (10 minutes to 3 hours);
- determine the frequency and time responses for the linear system (1 to 24 hours);

Phase III: Nonlinear compensator design and time-response simulations

- create a nonlinear model for the system and find the responses for large-level signals (nonlinear mode) (up to 10 days);
- introduce nonlinear elements in the compensator and/or design a specialized nonlinear dynamic compensator, test and tune the system performance (up to 30 days);

Subtotal time for Phase III: up to 40 days.

Note that only Phase I is in the critical path for the global engineering system design (of which your feedback system is a subsystem). Phase I might be repeated many times for different configurations of the global system before a final decision is made, so it is important that this procedure be short, simple, and reliable.

Appendix 13 Examples

In addition to the examples in the book's chapters, several examples are given here of single-loop and multiloop systems designed with the frequency domain approach.

A13.1 Industrial furnace temperature control

Temperature control of a precision industrial furnace, Fig. A13.1, is implemented as a single-loop SISO system depicted in Fig. A13.2. The plant includes a sink representing heat dissipation into the environment, two thermal capacitances, the furnace and the payload, connected by the heat flux conductance as illustrated in Fig. A13.1(a). The plant equivalent electrical circuit is shown in Fig. A13.1(b), and the plant transfer function is the ratio of the payload temperature to the difference of the power of the heater and that dissipated to the environment. The plant gain frequency response is shown in Fig. A13.1(c).

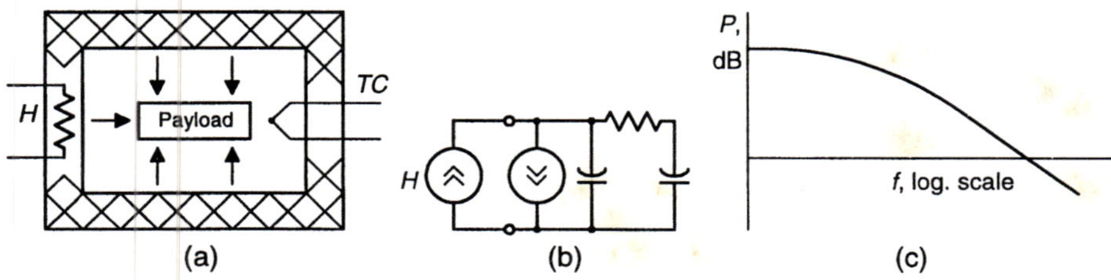

Fig. A13.1 (a) Industrial furnace with payload, (b) its equivalent thermal schematic diagram, and (c) the plant frequency response

The actuator is a pulse-width modulated electrical heater H, the carrier being 60 Hz. The signal amplitude of the thermocouple sensor TC is rather small, so that the error amplifier noise limits the system's accuracy. The response of the PID compensator can be adjusted by three manual regulators. The regulator adjustments in relation to the different payloads are easy to make.

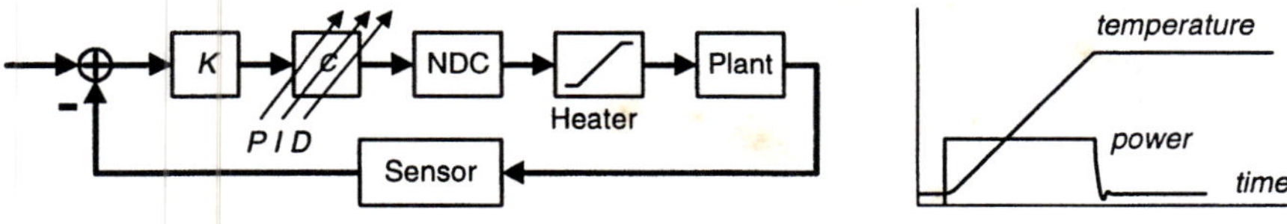

Fig. A13.2 Temperature control system

Fig. A13.3 Closed-loop transient response

The payload must be kept in the furnace for not less than a specified time (say, 3 hours) at a prescribed temperature (e.g., 1150° ± 2°). The furnace operation during the settling time is not paid for by the customer; therefore, the settling time needs to be minimized. The settling time depends on the heater's available output power, the feedback bandwidth, and the value of the feedback.

The feedback bandwidth is limited by the effects of the sensor noise and the amplifier noise on the actuator's input, and by the carrier frequency of the pulse-width modulation. The amplitude stability margin is 10 dB. Because of the low-order compensator employed, rather big variations in the loop phase shift are expected over the frequency range with positive loop gain. Then, for the minimum phase stability margin to be 40°, the average phase stability margin must be about 60°. The average slope of the Bode diagram must therefore be about –8 dB/oct. The average slope cannot be made steeper, since in this case the compensator adjustments for each type of payload need to be of higher precision than is practical.

With such stability margins, wind-up cannot be large, but it can be a problem if the system is not well tuned. To improve the system robustness, wind-up was eliminated with a simple NDC that reduces the loop phase lag for large-level signals.

The system transient response is shown in Fig. A13.3. The control looks nearly time-optimal because the settling time in the linear mode of operation is many times shorter than the full-power preheating time.

Digital implementation of the controller simplified the interfacing and made the controller cheaper in mass production.

A13.2 Scanning mirror of a mapping spectrometer

The mapping spectrometer of the Cassini spacecraft is an instrument producing an object image in the light of specified wavelengths. The spectrometer's scanning mirror is attached to a yoke suspended on a flexure as shown in Fig. A13.4.

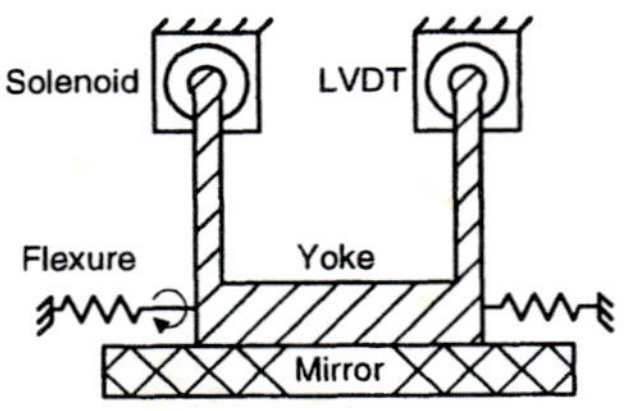

Fig. A13.4 Scanning mirror suspension with actuator (solenoid) and sensor (LVDT)

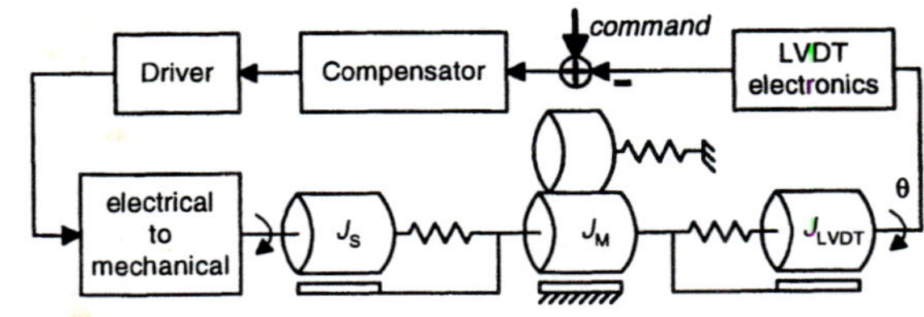

Fig. A13.5 Mechanical schematic diagram for scanning mirror positioning with feedback control

The yoke is rotated by a solenoid actuator. The mirror angle is measured by a linear voltage differential transformer (LVDT). The plant's mechanical schematic diagram in Fig. A13.5 shows the cores of the solenoid and the LVDT, and the mirror characterized respectively by the moments of inertia J_S, J_M, andJ_{LVDT}, the stiffness and the damping in the yoke's arms, and the suspension flexure stiffness and damping. The analytical model of the system dynamics was derived and the system was simulated as a block diagram with unidirectional blocks. Later, however, for the convenience of the analog controller design, the mechanical plant was modeled as an equivalent electrical circuit (the response for the electrical model was compared with that of the mechanical model and found to be exactly the same) and integrated with the controller's model in SPICE.

The control goal is to increase the exposure time, i.e., to scan the mirror in such a way that the light beam will remain longer in each of the focal plane photocells, with a staircase-type command as shown in Fig. A13.6. The mirror angle should track the command with a rise-time less than 2 msec. Thus, the feedback bandwidth must exceed 1 kHz.

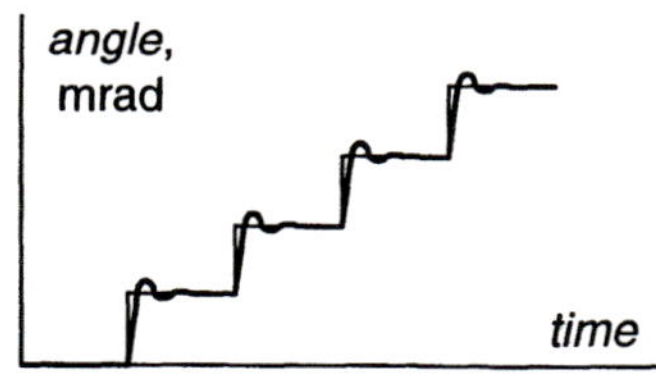

Fig. A13.6 Closed-loop transient response of the scanning mirror

The size of a pixel, angle-wise, is 0.9 mrad. The plant parameter variations are such that for the beam to stay reliably within the pixel, the feedback about the plant must exceed 60 dB at lower frequencies. The overshoot must not exceed 30%.

The feedback bandwidth is limited by the structural resonance of the solenoid core on its arm. The nominal resonance frequency is 6 kHz with possible deviations of ±20%. The damping of the resonance is known with the accuracy of ±20% (for damping, this is excellent accuracy). Therefore, the maximum plant gain in the 6 kHz neighborhood is known pretty well, with the accuracy of ±2 dB. The plant phase, however, is quite uncertain due to the effects of the LVDT resonance (at a higher frequency) and the stray inductances and capacitances in the coils. Therefore, the loop must be gain-stabilized at the frequency of the solenoid core resonance, i.e., the gain at the resonance must be below the amplitude stability margin of −10 dB.

The plant and controller simulation has been performed in SPICE. The chosen stability boundary on the *L*-plane is shown in Fig. A13.7(a) and has a 45° phase margin. The Bode diagram for the design is shown in Fig. A13.7(b). Here is seen the resonance of the mirror on its suspension at 8 Hz, and the solenoid core resonance on its arm at 6 kHz. The resonance frequency of the LVDT core on its arm is higher than 10 kHz.

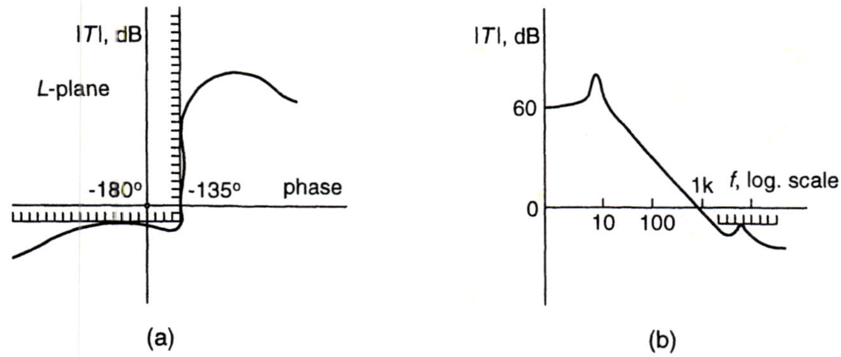

Fig. A13.7 (a) Nyquist and (b) Bode diagrams for scanning mirror control

It is seen that variations in the frequency of the plant resonances will not cause the system to oscillate. The stability margin of 45° is preserved over all frequencies where the loop gain is more than −10 dB.

The feedback path transfer coefficient (*B*) was chosen to be a constant. The response of a system with such a *B* to a small-amplitude step input is shown in Fig. A13.8, curve (a). The overshoot is about 30%, which is marginally acceptable. The overshoot can be efficiently reduced (to about 10%) by either putting a lead filter in the feedback path, or by an equivalent low-pass prefilter. However, this was not done since the actuator is not intended to operate in the linear mode.

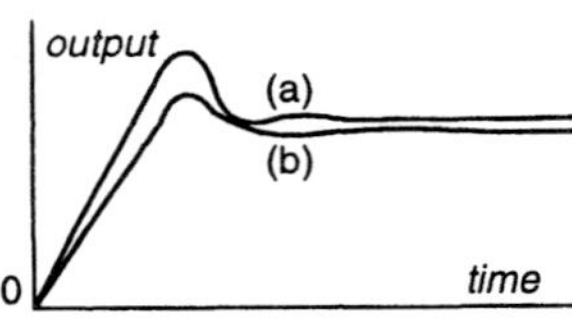

Fig. A13.8 Closed-loop transient response (a) without and (b) with saturation in the actuator

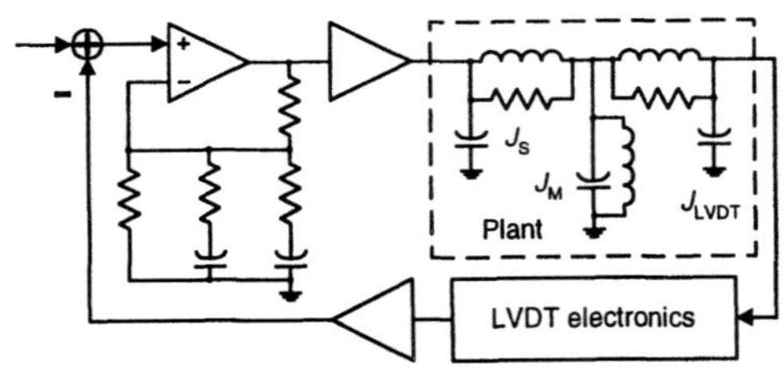

Fig. A13.9 Analog controller and plant model for scanning mirror (simplified)

It is economical to employ a small actuator whose maximum available torque is a small fraction of the maximum torque that would be seen in an idealized, linear system (without saturation).

Due to the 45° phase stability margin, large wind-up should not be expected. Therefore, application of an NDC is not required. This was confirmed by computer simulation of the output response to the step command with different actuator saturation thresholds. In Fig. A13.8, curve (a) relates to the linear system, and curve (b) to the threshold 5 times smaller than the maximum torque calculated for the linear system. Even with this small output power actuator, the output transient response is quite good, and the overshoot is reduced.

Because of the rather large feedback bandwidth, small production quantities, and required low power consumption, it was decided to make the entire loop analog. The system block diagram and the simplified schematic diagram of the compensator are shown in Fig. A13.9.

A13.3 Rocket booster nutation control

It was found that the nutation angle of a rocket booster shown in Fig. A13.10 increases exponentially during the main engine thrusting. To fight this, small auxiliary thrusters were employed to produce stabilizing moments. The thrust is pulse-width modulated (with the carrier frequency of 10 Hz) by a feedback loop using a gyro as a sensor. Since the nutation frequencies are in the range of 0.2 to 0.8 Hz, the feedback loop is a band-pass one.

It was shown that the larger the feedback is, the smaller is the real part of the nutation exponent. The loop gain and phase shift frequency responses of the controller's original version are shown by curves (o) in Fig. A13.11. The control loop compensator consists of a two-pole low-pass filter having a corner frequency of 4 Hz and a high-pass filter with a corner frequency of 0.1 Hz. The low-pass filter reduces the loop gain at the 10 Hz carrier frequency to the acceptable level of –27 dB. The high-pass filter attenuates dc and low-frequency components of the error to prevent them from overloading the thrusters. The filters are made analog as a cost-effective solution to low-quantity production.

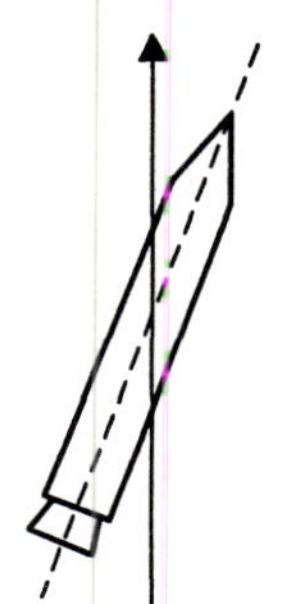

Fig. A13.10 Rocket booster nutation

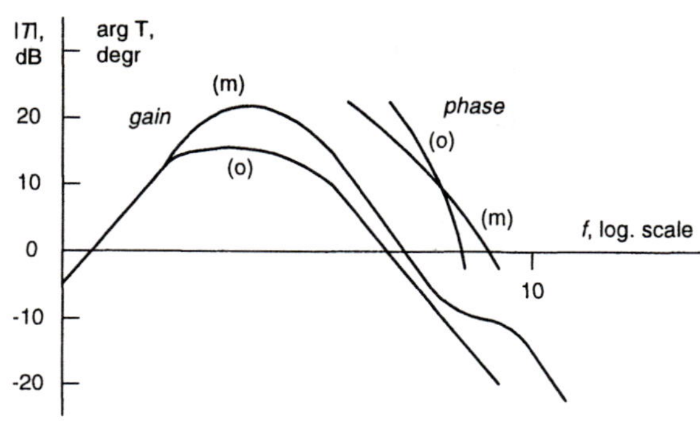

Fig. A13.11 Bode diagrams for nutation controller

The loop responses with the modified higher-order compensator, employing a notch filter to attenuate the carrier instead of the low-pass filter, are shown by the curves marked (m). Both the feedback over the range of the nutation frequencies and the amplitude stability margin are increased. The controller is globally stable; however, a substantial jump-resonance could be expected because of the phase stability margin of only 30° (the same as in the older system). Although the process instability might not create a problem, it is safer to eliminate even the possibility of jump-resonance and wind-up by introducing an NDC whose frequency response for small signal levels is shown in Fig. A13.12(a). The NDC schematic is shown in Fig. A13.12(b). For large signals, the diodes are open and the NDC gain is 0 dB.

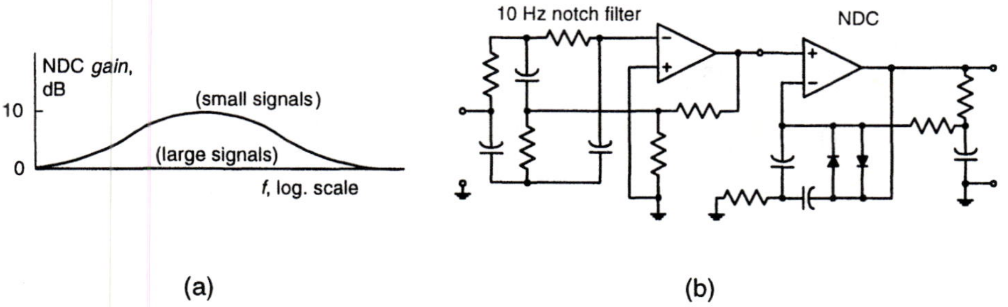

Fig. A13.12 Nonlinear dynamic compensator: (a) gain responses, (b) schematic diagram

Computer simulations showed that after 12 seconds of thrusting, the nutation amplitude in the modified system was 50 times smaller.

A13.4 Telecommunication repeater with an NDC

Linearity of repeaters for telecommunication systems with the frequency division multiplex should be very good, to reduce the channel intermodulation to an acceptable level. The linearity of this repeater designed in Russia in the 1960s is comparable to the linearity of a concurrent American counterpart, despite the fact that the bandwidth of the transistors was smaller and the linearity of the transistor in the ultimate stage was much worse than that of the special transistors developed in The Bell Laboratories. This was achieved by the increase in the feedback obtained by using a Nyquist-stable system with the Bode diagram shown in Fig. A13.13, upper curve.

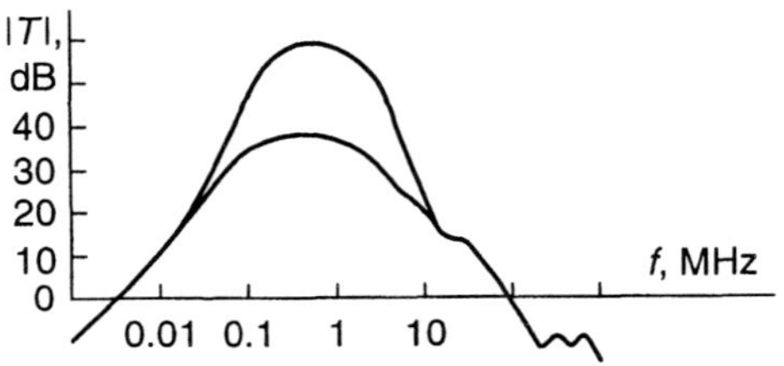

Fig. A13.13 Bode open-loop diagrams for a telecommunication repeater feedback amplifier

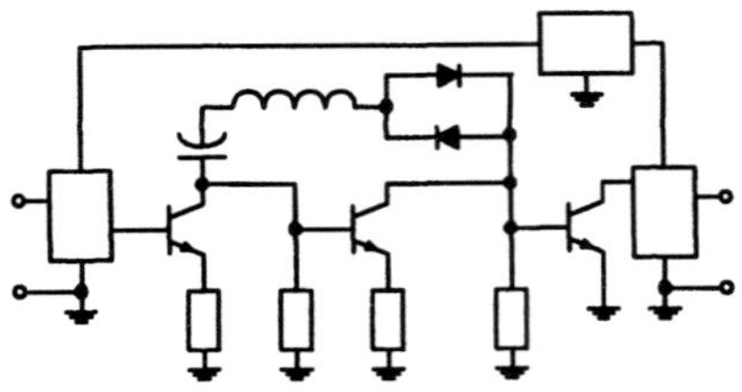

Fig. A13.14 Nonlinear dynamic compensation in a telecommunication feedback amplifier

The system global stability was provided by a nonlinear dynamic local feedback loop in the penultimate stage. The simplified schematic of the amplifier with the NDC is shown in Fig. A13.14. The diodes play the role of a dead-zone link. When the signal amplitude exceeds their threshold, the diodes start conducting and introduce negative feedback over the functional bandwidth, thus changing the gain in the common loop as shown by the lower curve in Fig. A13.13. This Bode diagram is less steep and the related phase lag is smaller, so that according to the DF method, further saturation in the ultimate stage does not cause periodical oscillation.

Many initial conditions were tried during experiments and none of them elicited oscillation. The repeater was put in mass production in large quantities and no stability problems were ever observed.

A13.5 Attitude control of a flexible plant

Consider the problem of pointing a spacecraft using a reaction wheel that resonates on a flexible shaft at 30 Hz in the system shown in Fig. A13.15. The reaction wheel is rotated by the motor M.

We consider here only the planar rotation about the center of rotation for the system. The moment of inertia of the spacecraft is J_{SC} = 100 kgm^2, and of the reaction wheel, J_R = 1 kgm^2.

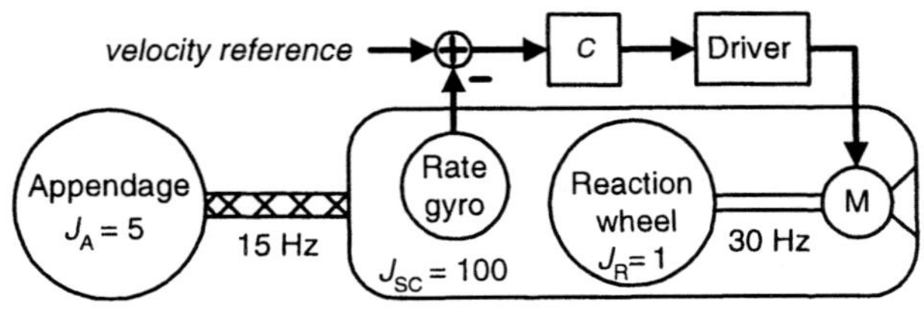

Fig. A13.15 Attitude control of a flexible plant

The amplitude stability margin was chosen to be 8 dB. The gyro noise limits the feedback bandwidth to less than 10 Hz. The flexible appendage with the moment of inertia $J_A = 5$ kg m^2, resonating somewhere in the 13–29 Hz region (nominally at 15 Hz) with very low damping, produces an area of uncertainty on the loop gain and phase responses, which further limits the feedback bandwidth as illustrated (not to scale) in Fig. A13.16(a). The loop at the resonance must be gain-stabilized. The gain peak depends on the value of the output mobility of the actuator.

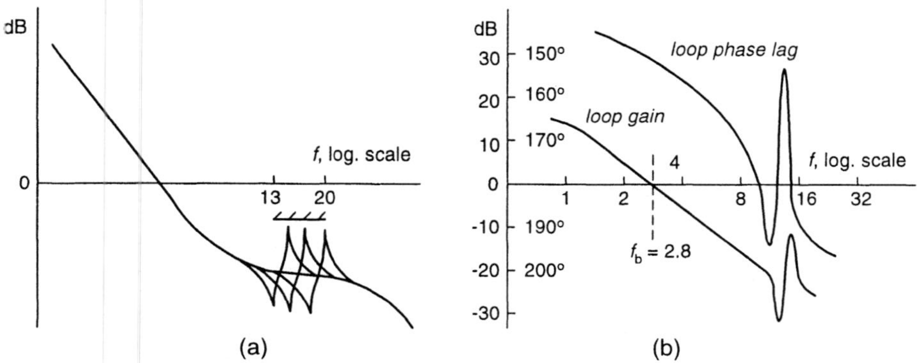

Fig. A13.16 Bode loop gain diagrams (a) for the flexible plant control with uncertain resonance frequency, not to scale, and (b) for the nominal plant

A local feedback loop about the driver (not shown in Fig. A13.15) makes the actuator output mobility equal to 0.03 (rad/sec)/Nm. This reduces the peak-to-notch swing from 54 dB to 28 dB compared with the case of the actuator being a pure torque source. Correspondingly, the available feedback bandwidth increases by almost an octave. The resulting Bode diagram is shown in Fig. A13.16(b).

A13.6 Voltage regulator with main, vernier, and local loops

The power supply voltage for a klystron transmitter is produced by a motor-generator whose output is rectified and passed through a low-pass filter. The –20 kV output voltage is the difference between the –22 kV at the filter's output and the 2 kV voltage drop on the bypass tube. The output voltage is regulated by (a) changing the current in the field winding of the generator (main actuator) and (b) changing the grid bias of the bypass tube (vernier actuator), as shown in Fig. A13.17. The goal (the reference) for the main (slow) loop is to keep the voltage on the by-pass tube to be on average 2 kV for the tube to be operational. The goal (the reference) for the fast vernier loop is to fine-tune the voltage drop on the by-pass tube for the regulator output voltage to be –20 kV (or any other specified value within the 0 to –40 kV range).

The range of the output voltage of the generator/rectifier/filter subsystem is from 0 to –42 kV. The feedback bandwidth in the main loop is limited to 10 Hz by the time-constant of the generator, the maximum voltage from the power amplifier that supplies the current to the field winding of the generator, and the noise of the voltage divider and the first stage of the amplifier in the compensator.

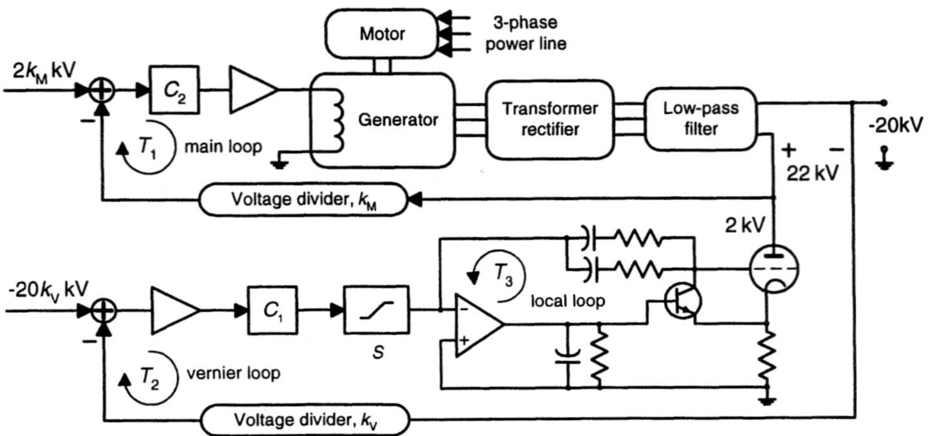

Fig. A13.17 High-voltage power supply controller

The vernier loop bandwidth was chosen to be 60 Hz to achieve the required disturbance rejection. This regulator can only regulate within the range of 4 kV, so that its average output voltage needs to be maintained at the 2 kV level (to be de-saturated) by the slower main loop.

The plant is high-order and nonlinear (because of the nonlinearity of the magnetization curve of the field winding). The loop compensator's order is rather high. The Bode diagrams and the Nyquist diagrams for the main loop return ratio T_1 and the vernier loop return ratio T_2 are presented in Fig. A13.18.

Experiments showed that the system was asymptotically stable following any initial conditions that take place in practice. However, initially, the saturation link S was not included in the system, and a substantial jump-resonance was observed. Analysis and experiments indicated that the reason for the jump-resonance was a nonlinear interaction between the vernier loop and the local loop T_3.

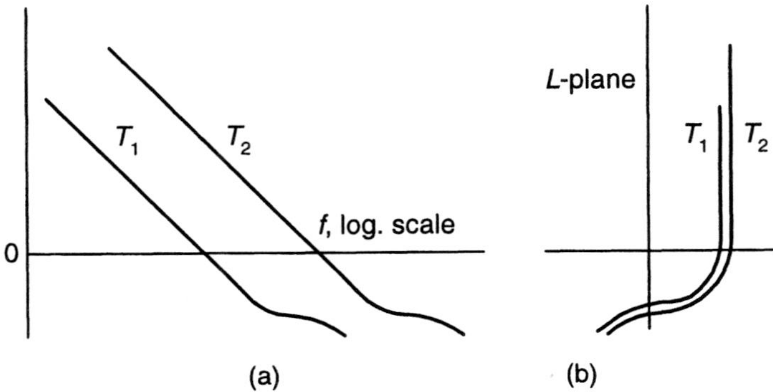

Fig. A13.18 (a) Bode and (b) Nyquist diagrams for a high-voltage power supply controller

This local loop is about the high-voltage transistor driving the bypass tube, to stabilize its transfer function, with 1 MHz feedback bandwidth. This loop contains an amplifier which is nearly an integrator; a power transistor which is nearly an integrator; and the feedback path which is nearly a differentiator, so that the phase stability margin in this loop is quite big and the stand-alone loop is globally stable. However, when the transistor is overloaded by a large-amplitude input signal, the feedback in this loop reduces. This increases the closed-loop phase lag of this subsystem, thus increasing the phase lag in the vernier loop.

The stability margins in the transistor loop have been increased and a signal amplitude limiter S was installed at the input to this loop. Since then, substantial jump resonance has not been observed.

A13.7 Telecommunication repeater

An electrical or a mechanical multiport can be characterized by transmission coefficients between the ports and the input impedances (mobilities) at the ports. MIMO feedback can be used to make all these parameters of an active multiport as desired, stable in time, and linear.

An example of such a system is a repeater designed in the Bell Laboratories in the 1930s for a 12-channel frequency division multiplex telecommunication system over open wires [2, pp. 499 – 502]. The repeater amplifier must have large feedback over the operational band to reduce the intermodulation between the channels. Over the operational frequency band, the repeater's gain must be continuously adjusted by automatic level control to match the attenuation of the open wire span between the repeaters for all weather conditions, which can cause the line attenuation to vary within 40 dB at higher frequencies and within 20 dB at lower frequencies. The nominal input and output impedances of the repeater equal to $Z = 150\,\Omega$. The provision of the desired impedances and the input-output gain can be illustrated with the flowchart shown in Fig. A13.19. This is a MIMO system having three inputs and three outputs (not counting the slower gain response control loops).

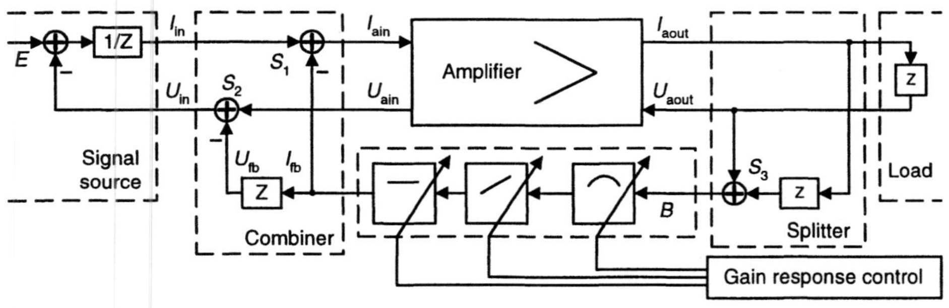

Fig. A13.19 Flowchart of a MIMO system

The amplifier is a two-port described by linear relations between the two input and two output signals: I_{ain}, U_{ain}, and I_{aout}, U_{aout}. The three-ports splitter and combiner connect the signal source and the load to the amplifier and the feedback path B. The ratios of the signals (variables) to be controlled are the amplifier transfer coefficient, the input impedance, and the output impedance. The input and output impedances are specified (commanded) by the transfer functions of the blocks Z. In other words, with the input current I_{in} and the outputs of the blocks Z being the command signals, the output current I_{out} is controlled using the summer S_1, the input voltage U_{in}, with the summer S_2, and the output voltage U_{out}, with the summer S_3.

The simplified schematic diagram including the gain control loops is shown in Fig. A13.20. The system is single-loop in the Bode sense since it includes only one actuator with a nonlinear link: the saturation in the ultimate stage of the amplifier. This diagram illustrates the benefits of structural design: over the functional frequency range, the input impedance depends exclusively on the combiner, and the output impedance, on the splitter. The repeater input-output gain depends on these two blocks and on the feedback path B. Thus, gain regulations in B do not affect the input and output impedances.

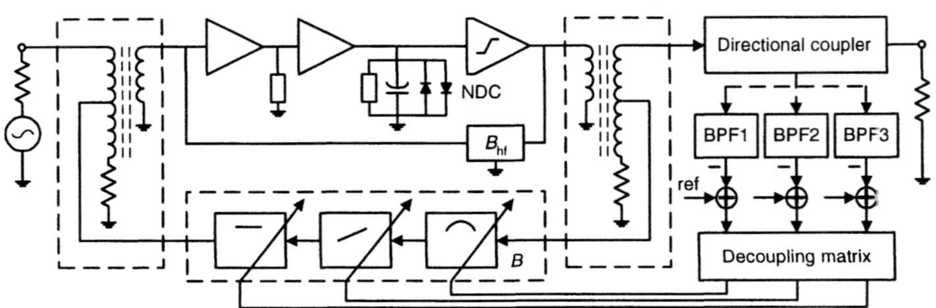

Fig. A13.20 Telecommunication repeater simplified schematic diagram

The combiner and the splitter are made with transformers having band-limiting stray reactive elements and multiple resonances. They are more complicated than the idealized ones shown in Fig. A13.19: unlike in the block diagrams, splitting a signal in electrical circuits means splitting the incoming power into two (or many) loads, and there is a trade-off between how much power goes to one output at the expense of the second output. The optimal solution to this trade-off is different at different frequencies. The stray elements bound the available signal-to-noise ratio, reflection attenuations, and the feedback. During the design, the trade-offs between these parameters were first resolved using the Bode integrals for the real part of an impedance that defines the available signal-to-noise ratio, for the reflection attenuation, and for the feedback. The frequency responses for the splitter and combiner in different directions of the signal propagation were chosen to be not overly complex. Then, these responses were approximated with high-order correctors and compensators that are implemented by incorporating extra electrical elements in the combiner and splitter.

The functional bandwidth over which the feedback sufficiently rejects the intermodulation was limited to 50 kHz by the quality of vacuum tubes available at the time. Today, the bandwidth of FET feedback amplifiers can reach several GHz – but the design principles remain essentially the same.

The loop gain response with a crisp Bode step and maximum available feedback over the operational bandwidth was implemented with the accuracy better than 0.5 dB, with the order of the compensators over 25. The feedback path was arranged in two parallel paths. The higher frequencies bypassed the feedback path B (whose transformers' attenuate higher frequencies) via the high-pass filter B_{hf}. The minimum-phase property of the total feedback path $B + B_{hf}$ was preserved by appropriate shaping of the frequency responses B and B_{hf}.

A simple NDC – a back-parallel pair of diodes shunting the capacitive interstage load – provided some phase advance for large-level signals and significantly reduced the jump amplitude of the jump resonance.

The slope of the Bode diagram and the available feedback over the operational band could be increased with Nyquist-stable loop design. However, in those days (and until the 1960's) the methods of providing global stability with NDCs were not yet well developed, and without such an NDC, the system would not be globally stable.

Three Bode variable equalizers in the feedback path have the following gain regulation frequency responses: flat, slanted, and convex, as depicted in Fig. A13.21. Three pilot signals with frequencies at the ends and in the center of the functional frequency band were being sent continuously. An analog computer in the form of a resistor matrix decoupled the automatic gain control loops. The automatic gain control kept the pilot levels at the output of the amplifier constant thus assuring that the total gain of the open wire line and the repeater remained 0 dB at these frequencies and, therefore, approximately 0 dB over the entire frequency band of operation.

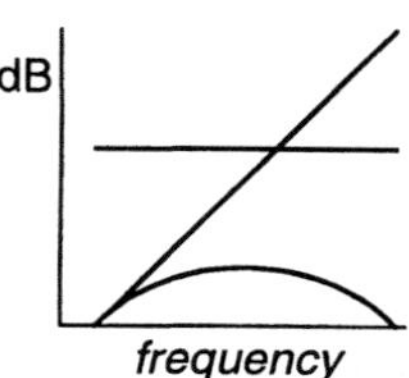

Fig. A13.21 Gain regulation frequency responses

A13.8 Distributed regulators

The signal level along the telecommunication trunk can be controlled by sending a small-amplitude single-frequency pilot signal and changing the attenuation of the feedback paths of the repeaters such that the levels of the pilot signal selected by appropriate bandpass filters will be equal to the references, as shown in Fig. A13.22.

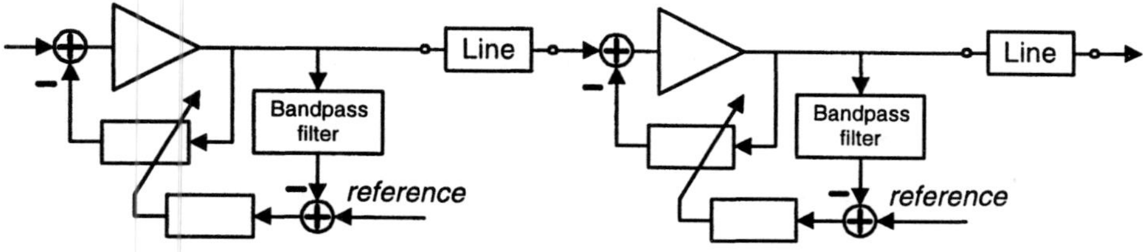

Fig. A13.22 Cascaded signal level regulation in telecommunication system

When the number of such regulators in a long telecommunication trunk is large, there exists substantial interaction between them. When, say, the signal level on the nth repeater output suddenly becomes larger than the reference, the levels at the outputs of all repeaters starting with the nth become larger than their references, and all these automatic gain control system start acting to correct the signal level.

The analysis of such systems shows that even a small overshoot in a single regulator transient response causes a large overshoot at the output of the trunk as illustrated in Fig. A13.23.

If all regulators are identical, then to make small the overshoot of the total response, a 90° stability margin must be preserved in each regulator over the frequency band where the regulator loop gain is bigger than -60 dB, which, although feasible, presents substantial implementation difficulties.

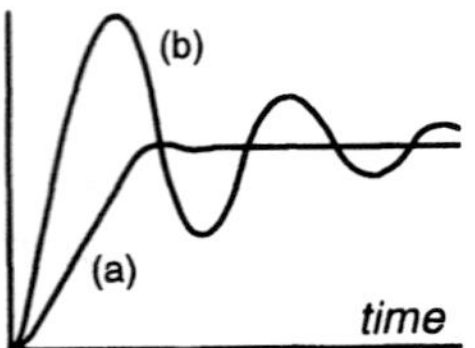

Fig. A13.23 Transient responses for (a) a single regulator and (b) a chain of regulators

Alternatively, NDCs can be introduced in the regulator loops, or the regulators can be made differently, with a faster regulator following several slower regulators, thus reducing the coupling between the regulators.

Chains or networks of regulation are employed for shape correction of flexible bodies (optical mirrors, for example) with multiple spatially distributed regulators.

A13.9 Saturn V S-IC flight control system

For the flight control system for the first stage of the Saturn V launch vehicle, the attitude is sensed using an inertially stabilized platform, and the attitude rate is available from a rate gyro package. The torque is provided by gimballing four of the five F-1 rocket engines.

The controller block diagram is shown in Fig. A13.24. It is part digital and part analog. The digital computer compares the programmed attitude command history to the attitude feedback sampled from the inertial platform. The attitude error is passed through a D/A converter to the analog flight control computer. The analog feedback from the rate gyros goes directly to the flight control computer. This computer consists of the high-order *RC* compensators C_1 and C_2. Note that the rate feedback and the compensators are analog because they must have a fast response, while the generation of the attitude errors in the digital computer can be rather slow.

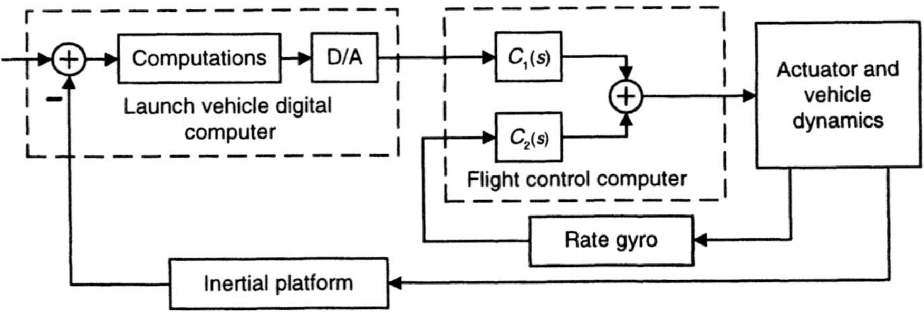

Fig. A13.24 Control block diagram for the Saturn V

Fig. A13.25 is a frequency map of the transfer function from the gimbal actuators to the sensors. It can be seen that the plant contains many resonances due to structural dynamics and propellant slosh, the frequencies of which are uncertain.

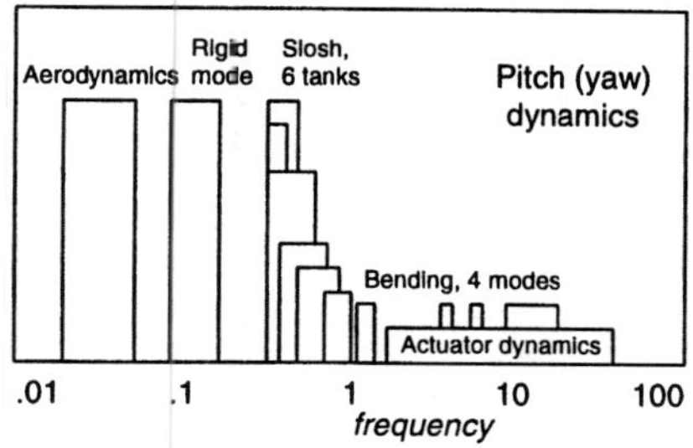

Fig. A13.25 Frequency map of Saturn V dynamics

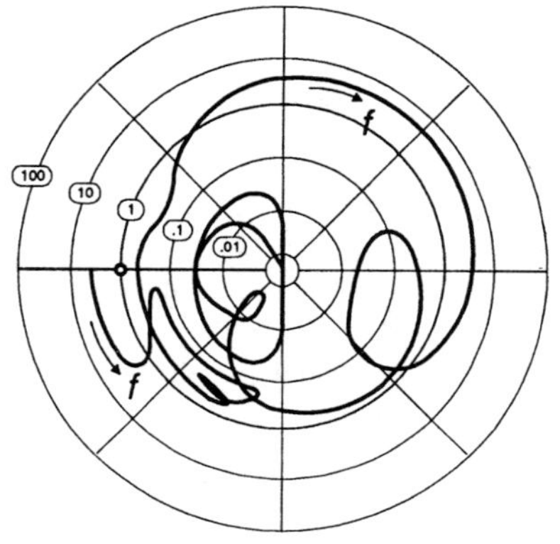

Fig. A13.26 Nyquist diagram for Saturn V pitch control

The Nyquist diagram for the pitch channel is shown in Fig. A13.26. The resonance modes of the structure and propellant appear as loops. The control system has sufficient stability margins for all possible frequencies of the flexible modes within the boundaries shown in Fig. A13.25.

A13.10 PLL computer clock with duty cycle adjustments

The 20 MHz computer clock of the Cassini spacecraft radar must have 50% duty cycle and must be synchronized with a 10 MHz ultra-stable quartz generator.

A nearly sinusoidal 20 MHz signal is generated by an *LC* VCO. The VCO is followed by a Schmitt trigger to make the clock shape rectangular. To stabilize the frequency and the duty cycle of the clock, two feedback loops are employed. The coupling between the two loops is small and can be neglected.

The first loop is a PLL. The clock frequency is halved by a D flip-flop and applied to one of the inputs of an exclusive OR gate which is used as a phase detector. To the other input of the gate is applied the reference 10 MHz signal. The averaged phase detector output (phase error) is applied via an op-amp with *RC* compensation to the VCO tuning diode and corrects the clock frequency.

The duty-cycle loop uses as a sensor a simple *RC* low-pass filter. The filter's output voltage is the average value of the clock signal and is proportional to the duty cycle. This voltage is subtracted from a half of the VCC (obtained by a voltage divider), and the difference (duty cycle error) adds to the dc bias of the transistor in the VCO, via an op-amp with *RC* compensation, thus changing the degree of asymmetry of the generated signal. After being clamped by the Schmitt trigger, the output is forced by the feedback to have the required 50% duty cycle.

A13.11 Attitude control of solar panels

The DS-1 (Deep Space 1) spacecraft attitude is bang-bang controlled by short bursts fired from the attitude control thrusters, with the cycle length of approximately 1000 sec. Between the firings, the spacecraft attitude changes linearly with time (if the solar wind and the out-gassing are neglected) and the maximum deviation from the nominal exceeds 1°. Making the error smaller would mean firing the thrusters more frequently and using more propellant, which is unacceptable. However, the required accuracy of pointing the solar panels outfitted with Fresnel lens light concentrators is 0.5°. Therefore, the panels' attitude toward the sun must be continuously corrected by motors. The chosen stepper motors rotate the panels via a gear without backlash and make up to 40 steps per second.

When the motors are commanded open-loop, the action is fast. However, the possibility is not excluded that the motors might slip over one or several steps during extreme conditions of the thruster firing and excitation of structural modes in the large panels (the panels supply the power to the ion engine). To prevent this from affecting the steady state accuracy, the system must be controlled closed-loop using the data from the encoder placed on the solar panel shaft.

The sampling frequency is 0.5 Hz. Therefore, $f_b \approx 0.05$ Hz, the closed-loop bandwidth ≈ 0.07 Hz, and the rise time is about 5 sec. With this rise time, the settling time of the system is about 10 seconds, which is too long. A better solution is to combine the advantages of the open-loop control and the closed-loop control by using a feedback-feedforward scheme. A simplified block diagram of such a controller is shown in Fig. A13.27. Here, `alpha_com` is the commanded angle, `alpha_enc` is the encoder readings, `pulse_num` is the number of steps the motor must make during the sampling period, `fforw_num` is the number of pulses commanded by the feedforward path, `fback_num` is the number of pulses commanded by the feedback compensator.

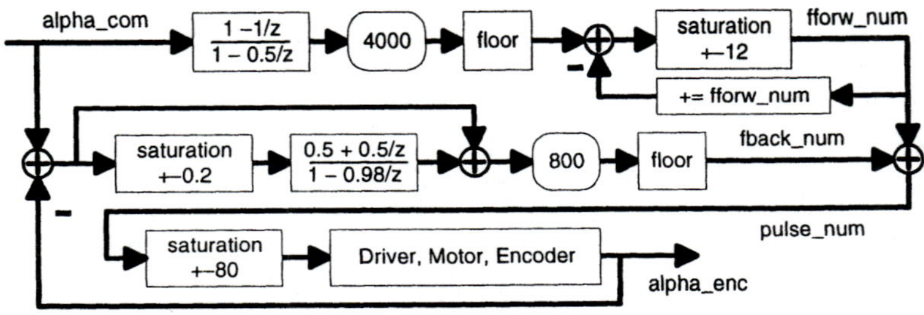

Fig. A13.27 Simplified block diagram for solar panels' attitude control

The feedback controller has a common configuration of parallel connection of proportional control path and a low-pass link with saturation in front of it.

The upper path in the block diagram is the feedforward. It begins with a high-pass filter that does not pass dc. Therefore, in seven or so seconds after a step command this path becomes disabled and the feedback control takes over, but for small times it supplies up to 12 pulses to the motor to speed it up and to reduce the rise time and the settling time.

A13.12 Conceptual design of an antenna attitude control

The Microwave Limb Sounder (MLS) of the Chemistry Spacecraft includes a TeraHerz radiometer. The radiometer antenna mirror scans the Earth horizon with 6 arcsec pointing accuracy. After each 20 sec, 2° scan, the antenna must be turned away by nearly 180° for calibration, and then returned to start the next scan. The required settling accuracy after the return is 10^{-5} (the ratio of 6 arcsec to 180°), and the settling time must be less than 0.8 sec, which is not easy to achieve.

The radiometer pointing direction can be shifted by 180° with an additional mirror. In this case, there will be two separate mechanisms: a scanning mirror and a switching mirror. The design of attitude controllers in the dual mirror system is simple since the switching mirror can be small and can be switched completely off the optical path of the main mirror during the scan, and the scanning mechanism can include a gear.

An economic alternative is to use a single mirror directly driven by a motor, and a single servo loop for both scanning and switching. In the attitude controller for this mechanism, the small settling time requirement necessitates a wide feedback bandwidth; the bandwidth might be limited, however, by structural resonances. The question regarding feasibility of the single-mirror servo has to be resolved at the initial stages of the project when accurate mathematical models of the mechanisms' dynamics have not yet been developed. Therefore, for the purposes of conceptual design and the control performance estimation, a simple specialized model of the system's dynamics has to be developed by the control system designer.

The single-mirror mechanism is shown in Fig. A13.28(a). The angular encoder disk is used as the angle feedback sensor. It is placed close to the motor rotor, thus making the control collocated. The antenna mirror is connected to the motor rotor via a shaft. The frequency of torsional resonance of the antenna on the shaft is 300 Hz. The entire mechanism is mounted on the spacecraft with a thrust structure. On the same instrument base, a heavy 50 kg laser is installed.

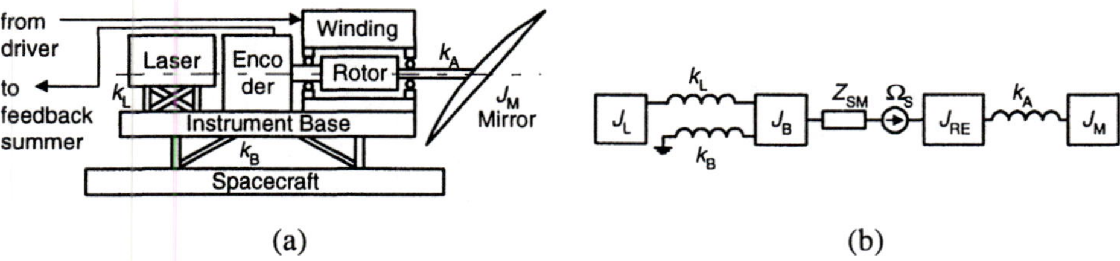

(a) (b)

Fig. A13.28 (a) The TeraHerz antenna mechanism and (b) its simplified mechanical schematic diagram

The major structural modes of the laser and the base are rocking modes. There is a requirement that no structural mode can be lower in frequency than 50 Hz. The lowest frequency of the modes of the base is about 50 Hz, and of the laser modes, about 60 Hz. The modes are coupled with the torsional motion of the rotor. The coupling coefficients are not yet known at this stage of the project.

A simplified mechanical schematic diagram for the rotor torsional motion shown in Fig. A13.28(b) is obtained by projecting the rocking structural modes onto rotation about the motor rotation axis. We will consider the worst case: the rocking mode motion aligned with the torsional modes. These projections use the moments of inertia of the rigid bodies about the rotation axis. The spring coefficients are chosen such that the torsional modes resulting from these projections have the same frequencies as the original rocking modes.

The employed models have no provisions for modal damping. The damping can be introduced in the model, but in this example there is no need for this, since the modes will be already substantially damped by the dissipative output mobility of the motor driven by a voltage source.

The moment of inertia of the laser about the rotor rotation axis $J_L \approx 0.5$ kgm^2. The instrument base, the stator, and the encoder constitute a rigid body with the moment of inertia about the motor rotation axis $J_B \approx 0.2$ kgm^2. The antenna mirror moment of inertia $J_M = 0.01$ kgm^2. The actuator output is shown as a source of velocity Ω_S with internal mobility Z_{SM}.

The initial rough estimation of the plant dynamics' effects on the control loop based on the diagrams in Fig. A13.28 is the following: The base moment of inertia is larger by more than by an order of magnitude than the mirror's moment of inertia. Therefore, we can assume in the course of the system analysis that the base moment of inertia is infinite. This simple model is valid at frequencies up to half of the frequency of the lowest structural resonance, i.e., up to 25 Hz. The resonances at 50 and 60 Hz of the base and the laser will not profoundly affect the loop because the base mobility is rather low relative to the antenna mobility – as long as the frequency range of interest is up to 25 Hz, i.e., the control bandwidth is below 15 Hz.

For extending the control bandwidth beyond 15 Hz, the worst-case stability analysis can be made using the plant model in Fig. A13.28(b).

The equivalent electrical schematic diagram is shown in Fig. A13.29(a), with the 1-to-1 ratio in the analogy of the mechanical to electrical parameters and variables, e.g., with J_M replaced by $C_M = 0.01\,F$. The torsional stiffness coefficient of the mirror shaft k_M is replaced by $L_M = 1/k_M$. The electrical contour impedance of the driver's output is the winding resistance r_w (the motor is driven by a voltage driver); it is converted to mobility and to the mobility 1-to-1 electrical impedance representation $Z_S = k^2 r_w$, where k is the electromechanical motor constant.

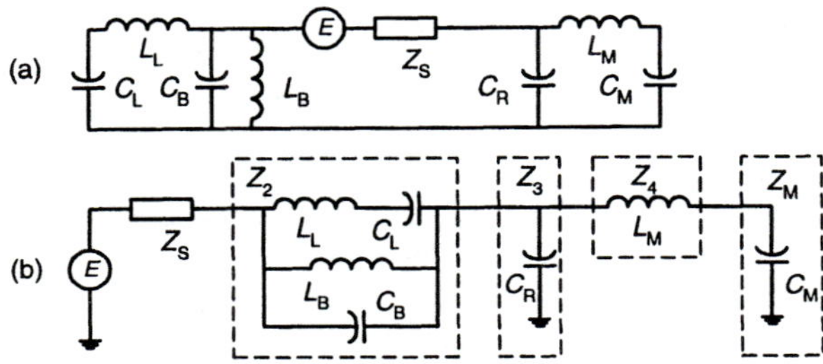

Fig. A13.29 (a) Electrical equivalent schematic diagram for the plant dynamics of the TeraHerz antenna attitude control, (b) modified diagram

The schematic diagram modified by changing the order of series connection of the two-poles about the contour is shown in Fig. A13.29(b). Here,

$k = 0.85$, $r_w = 4\,\Omega$, $C_B = 0.2$ F, $C_L = 0.5$ F, $C_R = 0.0002$ F, $C_M = 0.01$F,

the resonance frequencies $\omega_L = 400$ rad/sec, $\omega_B = 320$ rad/sec, $\omega_A = 2000$ rad/sec,

$1/Z_S = 1/(k^2 r_w) = 0.346$,

$1/Z_3 = sC_R = 0.002s$,

$Z_4 = sL_A = s/(\omega_A^{\,2} C_A) = 0.000025s$,

$Z_M = 1/(sC_M) = 100/s$,

$$Z_2 = \frac{s^3 L_L C_L L_B + sL_B}{s^4 L_L C_L L_B C_B + s^2 (L_B C_B + L_L C_L + C_L L_B) + 1}.$$

Since $L_i C_i = 1/\omega_i^2$, we can express Z_2 as

$$Z_2 = \frac{s^3 / (C_B \omega_L^2 \omega_B^2) + s / (C_B \omega_B^2)}{s^4 / (\omega_L^2 \omega_B^2) + s^2 [1/\omega_B^2 + 1/\omega_L^2 + C_L / (C_B \omega_B^2)] + 1},$$

or

$$Z_2 = \frac{305 \times 10^{-6} s^3 + 48.8s}{61.035 \times 10^{-6} s^4 + 31.64s^2 + 10^6}.$$

This function is an impedance of a lossless two-pole, and its zeros 0; $\pm j400$ and poles $\pm j184$; $\pm j700$ alternate along the $j\omega$-axis.

The schematic diagram in Fig. A13.29 could be analyzed with SPICE; in this case the expressions for the impedances do not even need to be derived. However, for the convenience of system design and integration, the plant model is worth building in SIMULINK. Fig. A13.30 shows the block diagram model of the plant dynamics following the approach described in Fig. 7.18(d) and (f). On the SIMULINK block diagram, mechanical variables are listed.

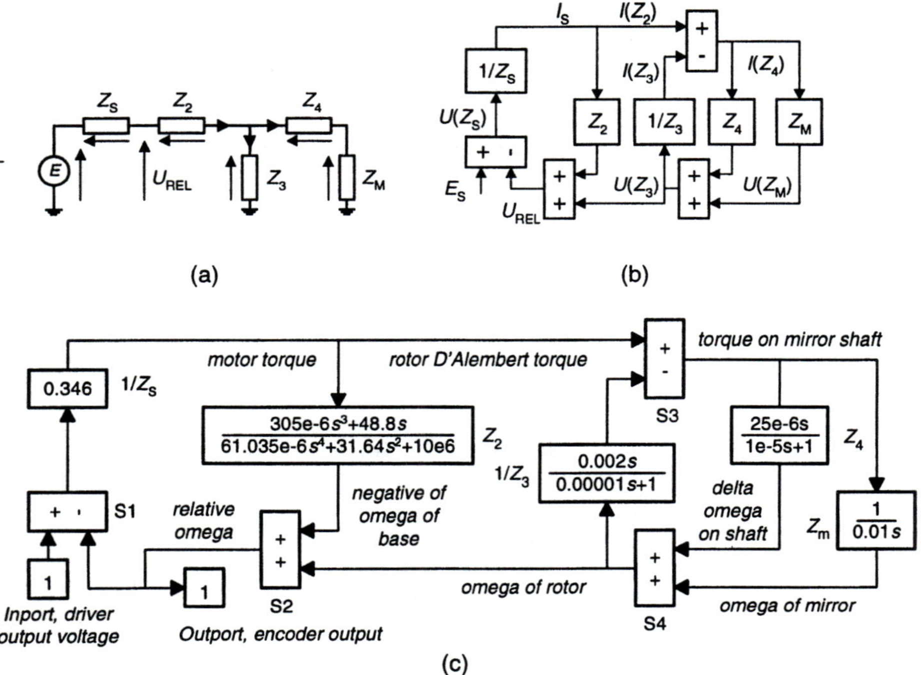

Fig. A13.30 (a) Ladder network plant model of the TeraHerz antenna attitude control, (b) Block diagram model of the plant, (c) SIMULINK model

Notice that the diagrams (b) and (c) in Fig. A13.28 are very convenient for troubleshooting and investigating the effects of the series and parallel branches in (a): disconnecting the input of a link pointing down replaces the related series two-pole by a short circuit, and disconnecting the input to a link pointing up, disconnects the corresponding shunting branch.

The Bode diagram in Fig. A13.31 for the plant transfer function $U(Z_M)/E_S$ is plotted with

```
w = logspace(1,3);
[A,B,C,D] = linmod('mlsplan');
bode(A,B,C,D,1,w)
hold on
% remove the connection to the input of block Z2
bode(A,B,C,D,1,w)
```

for two cases: for the nominal values of the plant parameters, and for $Z_2 = 0$, i.e., for an infinitely massive base.

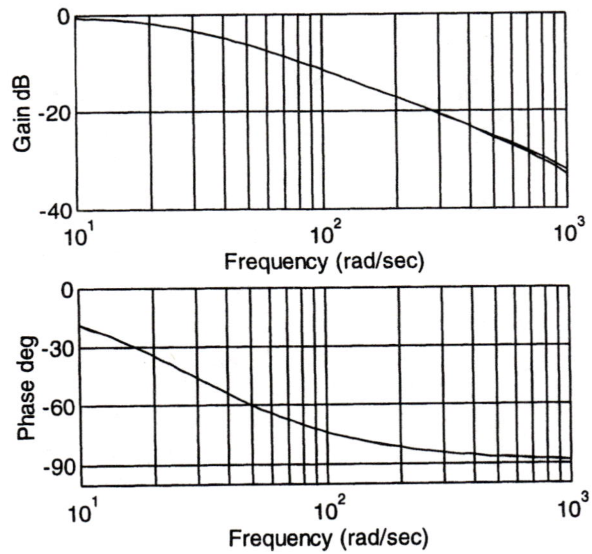

Fig. A13.31 Bode diagram for the plant of the TeraHerz antenna attitude control loop

For a much larger primary mirror of the MLS GHz radiometer, single-mirror attitude-and-switching control is not feasible. This device must be made with two separate mechanisms, a scanning mirror and a switching mirror.

The two gain and phase responses in the picture are very close to each other. Hence, the structural resonances of the laser and of the base do not constrain the feedback bandwidth which can be therefore 30 Hz. With an appropriate prefilter, a multiwindow controller, command profiling (see Section 5.11) and loop shaping, this feedback bandwidth ensures the required accuracy with substantial margins. Based on this analysis, a decision can be made to use the economic option of a single-mirror mechanism.

A13.13 Pathlength control of an optical delay line

System description

The optical delay line of the stellar interferometer (SIM) is placed in the path of the light gathered from one of the widely separated optical elements called siderostats. The delay is regulated to provide an interference pattern on the focal plane where this light is combined with the light from another siderostat. The delay line is positioned on a truss structure as shown in Fig. A13.32(a). A reaction wheel assembly is employed for the interferometer attitude control.

The delay line includes two mirrors. The larger, spherical mirror is actuated by a voice coil (VC) capable of a maximum displacement of 1 cm at low frequencies. The smaller, flat mirror is moved by a piezoactuator (PZT) with a maximum stroke of 30 μm. The light bouncing between the mirrors along the variable pathlength experiences a controllable delay. The pathlength must be controlled with better than 5 nm accuracy.

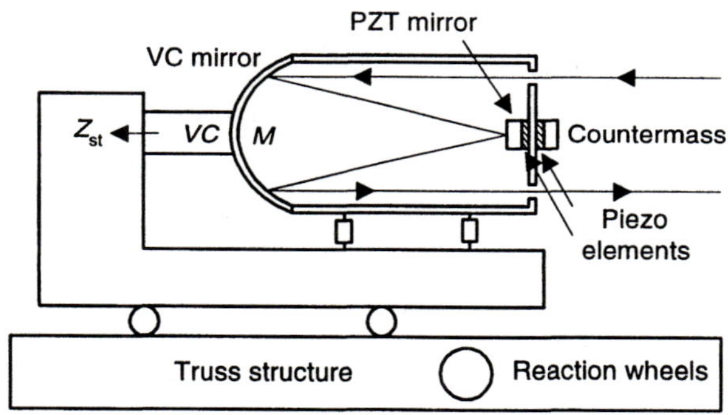

Fig. A13.32 Optical delay line

The PZT consists of two piezo elements and applies a force between the smaller mirror and the countermass. Since the forces applied to the supporting structure balance out, the PZT does not excite the structure, and the feedback bandwidth in this loop is not limited by the structural modes.

The VC de-saturates the PZT. The VC is, in turn, de-saturated by placing it on a cart; the cart motor control loop will not be discussed here.

The voice coil moves the mirror having mass $M = 0.5$ kg against the flexible structure with mobility Z_{st}. The equivalent electrical schematic diagrams for the VC actuation of the mirror are shown in Fig. A13.33. Here, C_M reflects the mass of the mirror, and L_{susp} and R_{susp} reflect the VC suspension, with the resonance at 5 Hz. The VC mobility $Z_{VC} = (r_{vc} + Z_D)/k^2$ where r_{vc} is the coil resistance, Z_D is the driver amplifier output impedance, and k is the force/current

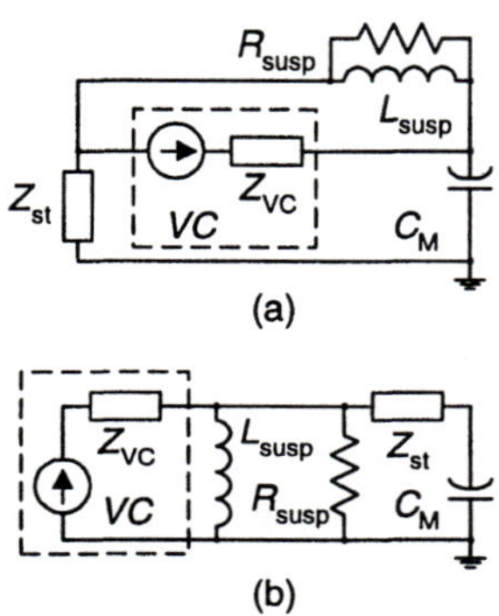

Fig. A13.33 Equivalent electrical circuit for the voice coil drive: (a) following the structure, (b) modified

electromechanical coefficient of the voice coil. Over the frequency range of interest, $|Z_{st}|$ is substantially smaller than the mobility of the contour it is in and does not much affect the velocity of the mirror (the voltage on C_M).

Numerical design constrains

The PZT maximum displacement is 30 μm over the entire frequency range of interest.

The VC maximum displacement amplitude is

$$D_{max} = (2\pi f)^{-2}\, kI_{max}/M$$

With the VC constant $k = 0.3$ N/A and the current saturation threshold of the VC driver amplifier $I_{max} = 3$ A, the frequency at which the VC maximum displacement amplitude equals that of the PZT, $D_{max} = 0.00003$ m, is

$$f_{cross} = \sqrt{kI_{max}/(MD_{max})}/(2\pi) \approx 40\text{Hz}.$$

The feedback in the VC and PZT loops must be sufficiently large to reject the vibrational disturbances over the bandwidth up to 500 Hz, caused by the reaction wheels. The feedback must be at least 60 dB at 16 Hz. The disturbance forces' spectral density responses are assumed to be flat over the 500 Hz bandwidth, so that the vibration amplitudes are, on the average, inversely proportional to the square of the frequency. Therefore, both the feedback and the maximum actuator output amplitude can decrease inversely proportional to the square of the frequency. Hence, the loop gain slope must be close to –12 dB/oct.

The available feedback in the VC loop is affected by the mobility of the structure Z_{st}, as seen in Fig. A13.33. As can be estimated using an asymptotic Bode-step response, the structural modes and their uncertainties prevent the VC loop bandwidth from exceeding 100 Hz – if the loop is designed as stand-alone-stable. In this case, the gain in the VC loop at 16 Hz is only 26 dB, less than the required 60 dB.

Therefore, the VC loop has to be designed as stand-alone-unstable; in this case, the gain and the slope of the loop Bode diagram can be substantially increased. The loop can be designed as a self-oscillating dithering system as described in Section 9.7.

The sampling frequency for the PZT loop is 5 kHz. For the VC loop, the sampling frequency can be chosen to be 1 kHz.

Higher-level design objectives

The control system performance index is the mean square error in the delay line pathlength. The principle design objective is to keep the mean square error below 5 nm.

In addition to the principle design objective, the following characteristics are required or highly desirable:

- system robustness,
- good output responses to commands of different shapes and amplitudes,
- transient responses to large-amplitude vanishing disturbances that are neither excessively large in amplitude nor duration, or
- a large disturbance/command triggering threshold of nonlinear phenomena, if the nonlinear phenomena responses are prolonged and violent and cannot be excluded by the design, in order that these nonlinear phenomena happen infrequently.

Design approach

The controller for this nonlinear, flexible, and uncertain plant must be reasonably close to the best achievable, but not overly complicated. We choose the controller to be multiloop, high-order, and nonlinear, but not time variable (i.e., not adaptive).

The conceptual design employing Bode integrals and the Bode asymptotic diagrams should produce a solution in the vicinity of the globally optimal time-invariable controller (the "optimality" here means best satisfaction of the higher-level design objectives, according to the opinion of the customers, i.e., system engineers).

The design begins with making some reasonable assumptions and translating the higher-level objectives into a set of lower-level guidelines. The latter consists of the design objectives and the design considerations. If possible, these objectives and considerations should be formulated in a mutually decoupled (orthogonal) form to simplify the system trade-offs and to speed up the design. This is more easily accomplished using frequency-domain specifications.

Lower Level Design Objectives

The lower level design objectives for the feedback system under consideration can be formulated as the following:

- To effectively reject the vibrational disturbances of rather large amplitudes, the VC loop gain must exceed 50 dB at 40 Hz and the feedback must increase toward the lower frequencies.
- To effectively reject vibrations at higher frequencies, however with smaller amplitudes, the feedback in the PZT loop must have wide bandwidth; it has been calculated that 600 Hz bandwidth will suffice.
- The transient response to the commands of different amplitudes must be good.
- The system must be globally stable.
- If the VC loop is made unstable when stand-alone, the frequency of oscillation in this loop should be as high as possible in order for the amplitude of the limit-cycle oscillation in this loop to be as small as possible and not to overload the PZT.

The design considerations are the following:

- The VC and PZT loops are nearly m.p., and the combined loop must be m.p. For this, the parallel channels must be shaped appropriately.
- At frequencies below $f_{cross} \approx 40$ Hz the VC provides a larger stroke and is therefore the main actuator; at higher frequencies, PZT is the main actuator. Therefore, rejection of the large amplitude disturbances depends at $f < f_{cross}$ on the loop gain in the VC loop and at $f > f_{cross}$, on the PZT loop. Outside of these ranges, the only requirements for the VC and PZT loop gain shaping is the provision of stability and robustness.

  (Generally, making the loop cross at f_{cross} is not necessary. Extra gain in the loop which is not the main one does not hurt the system performance – but it unnecessarily consumes a part of the area specified by a related Bode integral; therefore, this is not worth doing.)
- At the frequency of 5 Hz, the VC driver output impedance should be small to damp the VC suspension resonance. At frequencies over 50 Hz, the impedance can be

made higher, thus making $|Z_{VC}|$ higher in order for the loop transfer function at these frequencies to be less affected by the system structural modes.

Conventional design approach

Most commonly, industrial main-vernier systems do not include NDCs or only include simple NDCs for improving the responses to the commands. Such systems are made globally stable by making all stand-alone loops globally stable.

When the feedback and the disturbance rejection that are achievable in such systems are not sufficient, the disturbance rejection can be improved at the price of making the main-vernier system only conditionally stable [25,50]. This, however, causes the command generator to be complicated and the reaction to the commands sluggish. It also necessitates a procedure to recover the system from the limit cycle which can be quite violent and damaging to the hardware. For this reason, this option is normally not used in industrial systems.

The chosen design options

The chosen design options for the system with an NDC are the following:

- The VC loop is stand-alone-unstable. In this case the slope of the Bode diagram of the VC loop can be made steeper thus improving the disturbance rejection and the handling of large-amplitude disturbances.
- An NDC provides global stability.
- The PZT loop bandwidth is 600 Hz wide, with a Bode step response, with the Bode diagram slope –10 dB/oct at frequencies 150 to 1200 Hz, and –12 dB/oct over the range 40 to 150 Hz. At 40 Hz the gain in the PZT loop is therefore 12 dB/oct$\times$ $\log_2(150/40) + 20 = 43$ dB. The gain may gradually roll down at lower frequencies.
- We choose the asymptotic Bode diagrams shown in Fig. A13.34(a), since for this particular design task the advantages of using the better shaped diagrams exemplified in (b) do not justify higher complexity of the design.
- The VC loop Bode diagram crosses the PZT loop Bode diagram at 40 Hz.
- The PZT loop phase margin at 40 Hz is approximately 40°.
- The VC loop Bode diagram slope is –18 dB/oct down to 20 rad/sec, and – 6 dB/oct below this frequency. The gain at 40 Hz must be 43 dB, i.e., 141 times. From here, $T_{VC} = a/[s(s + 20)^2]$ where $a = 141(2\pi 40)^3 = 2.24\times10^9$.
- At 40 Hz, the phase difference between the loops is 270° – 140° = 130°. This ensures minimum phase character of the total loop with 50° safety margin.
- To ensure the system's global stability and good transient responses to the commands, an NDC will be included in the combined loop, with the dead zone equal to the threshold of the PZT actuator. As the linear block of the NDC, the PZT loop transfer function is employed.

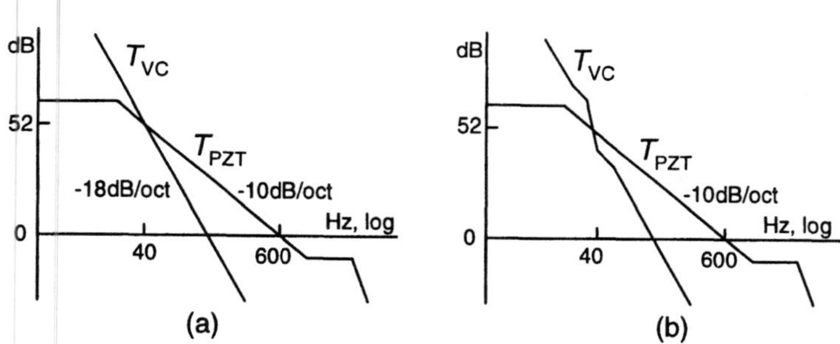

Fig. A13.34 Asymptotic Bode diagrams for stand-alone loop transfer functions; (a) simpler and (b) higher performance

Block diagram with rational transfer functions

The asymptotic (transcendental) responses shown in Fig. A13.34(a) need to be approximated with rational function responses.

The frequency-normalized Bode-step-type loop response (including certain n.p. lag of sampling) is obtained with MATLAB function `bostep` (from the Bode Step toolbox) and shifted in frequency with `lp2lp` for the crossover frequency to be $2\pi 600$ rad/sec. This forms the stand-alone PZT loop transfer function. The stand-alone VC loop transfer function has been already defined as $2.24 \times 10^9/[s(s + 20)^2]$.

The Bode diagrams for the stand-alone VC and PZT loops and the summed loop Bode diagram are obtained with the following script:

```
npzt = [2.33e-10 6.7e-6 -5.14e11 1.26e17 1.5e21 9.5e24 1.2e28];
dpzt = [1   2.83e5   7e9   9.18e13   6.5e17   1.684e21   0   0];
nvc = [2.5e9]; dvc = [1 40 400 0];

ntot1 = conv(npzt,dvc);
ntot2 = conv(dpzt,nvc);
adz = zeros(1,length(ntot1)-length(ntot2));
ntot = ntot1 + [adz ntot2];
dtot = conv(dvc,dpzt);

w = logspace(1,4.5);
bbode(npzt,dpzt,w);
hold on
bode(nvc,dvc,w);
bode(ntot,dtot,w);
hold off
zoom on
```

The logarithmic plane Nyquist diagrams are plotted with `nyqlog` from the Bode Step toolbox:

```
nyqlog(3700, nvc,dvc)
hold on
nyqlog(3700, npzt,dpzt)
hold on
nyqlog(3700, ntot,dtot)
hold off
```

The diagrams shown in Fig. A13.35 indicate that the system in the linear mode of operation is stable, robust, and provides 69 dB disturbance rejection at 16 Hz.

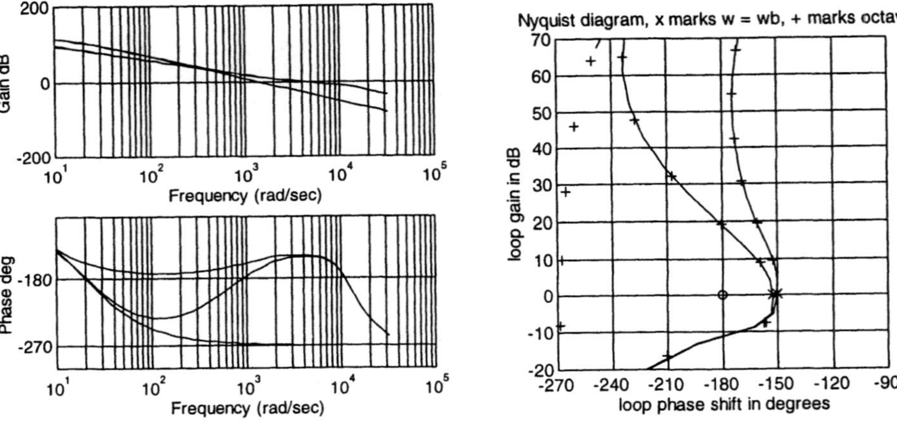

Fig. A13.35 Bode and Nyquist diagrams for the rational transfer functions approximating the asymptotic diagrams for the VC, the PZT, and the combined loop. (The combined response is the top one on the gain response, and the intermediate one on the phase response and the Nyquist diagram.)

The SIMULINK block diagram using these transfer functions is shown in Fig. A13.36. The pathlength is the sum of the displacements produced by the PZT and VC actuators.

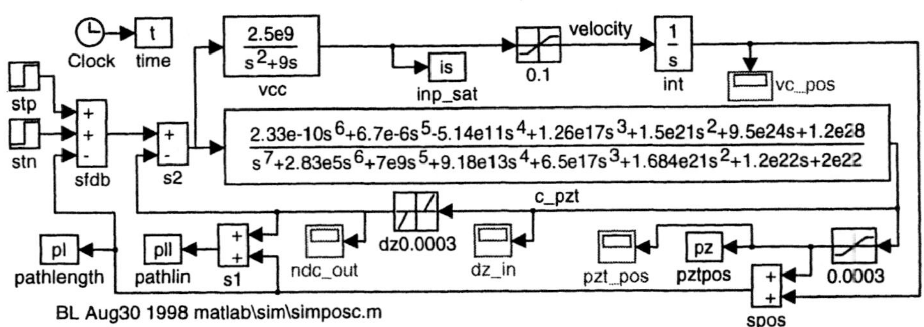

Fig. A13.36 Simplified SIMULINK block diagram of pathlength control

In this block diagram and in the simulations, the thresholds of the saturation, the dead zone, and the signal amplitudes have been increased 10 times relative to the real system. This increases the numerical stability of simulation. This scaling certainly does not affect the theory of operation and the responses' shapes.

To simplify the analysis, VC is represented by a velocity source with velocity saturation; using a detailed VC actuator model will not change the principle character of the results.

The feedback path via the dead zone dz0.0003 implements nonlinear dynamic compensation (NDC). The dead zone equals the PZT saturation threshold. It does not pass small amplitude signals and has no effect on the system performance in the linear state of operation. The NDC effect on the global stability and the transient responses to large amplitude commands will be discussed later.

Self-oscillation in the system without an NDC

With frequency responses shown in Fig. A13.353, the system is stable in the linear mode of operation when both the PZT loop and the VC loop are closed. The stand-alone VC loop is unstable.

Without an NDC, the system is not globally stable. Fig. A13.37 shows the pathlength oscillation triggered by a 30 msec, 3 mm pulse command. The oscillation approaches a limit cycle with the period longer than 5 sec.

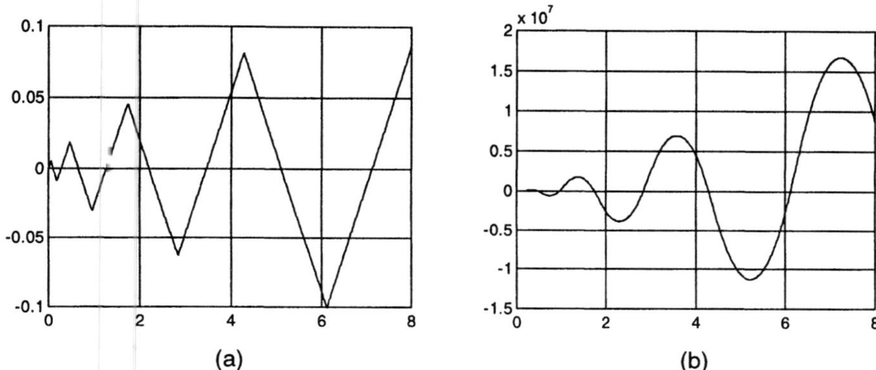

Fig. A13.37 Oscillation triggered by a short pulse command in the system without an NDC (a) of the VC output pathlength component and (b) at the input to the saturation link

Compared with the amplitude of this oscillation, the PZT output is negligibly small. Therefore, the system can be analyzed as having a single nonlinear link, the saturation in the VC actuator, and the analysis can be performed with the describing function method. The accuracy of such analysis is sufficient since the VC loop is a low-pass and the signal at the input to the saturation link becomes nearly sinusoidal when approaching the limit cycle as seen in Fig. A13.37(b).

In other words, when the vibrations' amplitudes are large enough to saturate the PZT, the VC loop is left alone and the system bursts into the limit cycle self-oscillation at a low frequency with large signal amplitude. When the high-amplitude high-frequency vibrations vanish and no longer saturate the PZT, the limit cycle oscillation does not change substantially since the amplitude of the output of the PZT is negligibly small compared with the displacement generated by the VC. Thus, the limit cycle of the stand-alone VC loop is in fact also the limit cycle of the system as a whole. Large-amplitude vanishing signals with substantial frequency components of the limit cycle oscillation belong to the basin of attraction for the limit cycle. The system is conditionally stable.

Limit cycle of the VC loop with the NDC

Since the stand-alone VC loop is unstable and the NDC dead zone does not pass the signals of small amplitudes, the stand-alone VC loop is unstable even with the NDC. It is important to emphasize, however, that the NDC radically changes the amplitude and the frequency of the limit cycle oscillation.

The amplitude of the limit cycle oscillation is determined by the dead zone in the NDC, since for signals much larger than the dead zone, the dead zone describing function approaches 1, the system becomes close to linear, and such a system, as has been shown before, is stable with sufficient stability margins. Therefore, oscillation can only take place with amplitudes not much higher than the dead zone.

Since the oscillation amplitude is small, during the describing function stability analysis the saturation link in the VC actuator can be replaced by 1, and the system can be viewed as that shown in Fig. A13.38(a). This system's linear part is, however, unstable; for the conventional describing function methods to be applicable, the diagram (a) is transformed equivalently to the diagram (b) by replacing the dead zone with parallel connection of a unity link and a saturation link with the threshold equal to the dead zone. The stand-alone linear part of the latter (the dashed box) is stable since this is a feedback system with parallel connection of VC and PZT loops. The transfer function of the box is

$$\frac{T_{\mathrm{PZT}}}{1+T_{\mathrm{VC}}+T_{\mathrm{PZT}}}.$$

Notice that the signal about the loop via the saturation is applied in phase.

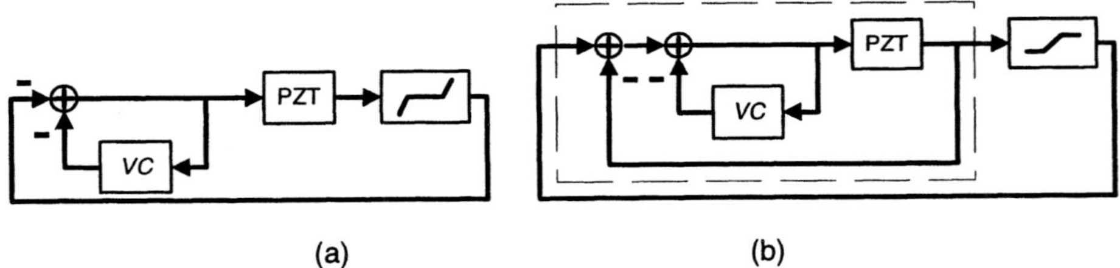

Fig. A13.38 Equivalent block diagrams for describing function analysis, (a) with the unstable linear part and (b) with the stable linear part

The Bode diagram for this transfer function shown in Fig. A13.39 is obtained with the script

```
npzt = [2.33e-10 6.7e-6 -5.14e11 1.26e17 1.5e21 9.5e24
1.2e28];
dpzt = [1  2.83e5  7e9  9.18e13  6.5e17  1.684e21  0  0];
nvc = [2.5e9]; dvc = [1 40 400 0];

n1 = conv(npzt,dvc);
d1 = conv(dvc,dpzt);
d11 = [0 n1] + d1;

d2 = conv(dpzt,nvc);
deq = d11 + [0 0 0 d2];

w = logspace(0,4);
% nyqlog(1260,-n1,deq)
bbode(n1,deq,w)
```

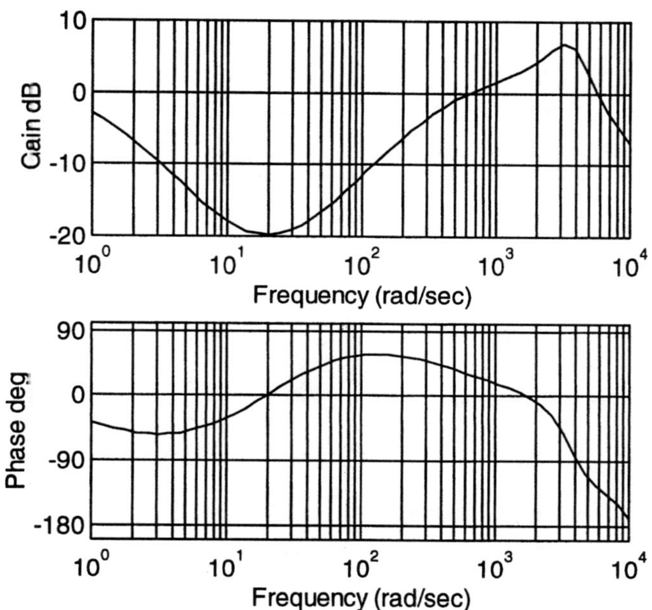

Fig. A13.39 Bode diagram for the oscillation describing function analysis

It is seen that the condition of oscillation (zero loop phase shift while the loop gain is positive) occurs at approximately 1600 rad/sec, or, approximately, at 250 Hz. The loop gain at this frequency is 3.5 dB, i.e., the gain coefficient is 1.5. The oscillation amplitude E can be found by equating the describing function to the inverse of the loop gain coefficient, i.e., using (11.8),

$$1.27(E/e_S)^{-1} - 0.27(E/e_S)^{-4} \approx 1/1.5,$$

wherefrom $E \approx 1.8e_S \approx 0.00054$.

With this E, the describing function of the PZT equals 0.67, i.e., is only 3.5 dB less than the gain in the linear state of operation. Since in the linear state of operation the stability and safety margins of order of 10 dB are provided, the system is asymptotically globally stable.

Therefore, when the PZT actuator recovers from being saturated by disturbances or commands, it is capable of delivering an output signal comparable with the self-oscillation in the VC loop. Application of this signal makes the system stable, and the oscillation in the VC asymptotically dies down.

The time-history of the pathlength in this case is shown in Fig. A13.40, (a) without and (b) with saturation in the VC that limits the slew rate.

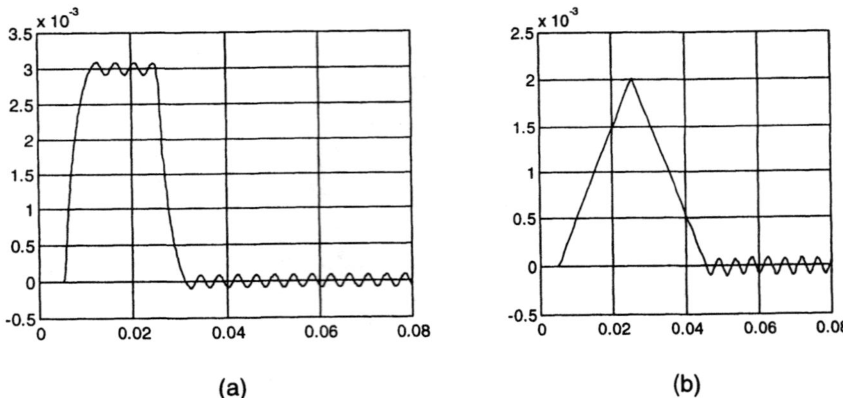

Fig. A13.40 Pathlength time-history after a pulse command in a system with the NDC and without a PZT actuator, (a) without limiting the VC velocity and (b) with a ±0.1 m/sec limiter of the VC velocity

Global stability of a system with the NDC

The dead-zone branch is in parallel with the PZT saturation branch. These two parallel branches equal a linear branch with the same transfer function. (Augmenting a nonlinear system with an extra nonlinear subsystem, in our case by the dead zone, so that the total system properties are those of a linear system, is sometimes called ***exact linearization***.)

The system therefore includes only one nonlinear link: the VC saturation. The transfer function of the equivalent linear link from the VC saturation output to the VC saturation input is $T_{VC}/(1 + T_{PZT})$. The Bode diagram of this transfer function found with the script

```
npzt = [2.33e-10 6.7e-6 -5.14e11 1.26e17 1.5e21 9.5e24 1.2e28];
dpzt = [1  2.83e5  7e9  9.18e13  6.5e17  1.684e21  0  0];
nvc = [2.5e9]; dvc = [1 40 400 0];

n1 = conv(nvc,dpzt);
d1 = conv(dvc,dpzt);

d2 = conv(npzt,dvc);
deq = d1 + [0 d2];

w = logspace(1,4);
bode(n1,deq,w)
```

is plotted in Fig. A13.41. Application of the Popov criterion to this system shows that the system is asymptotically globally stable.

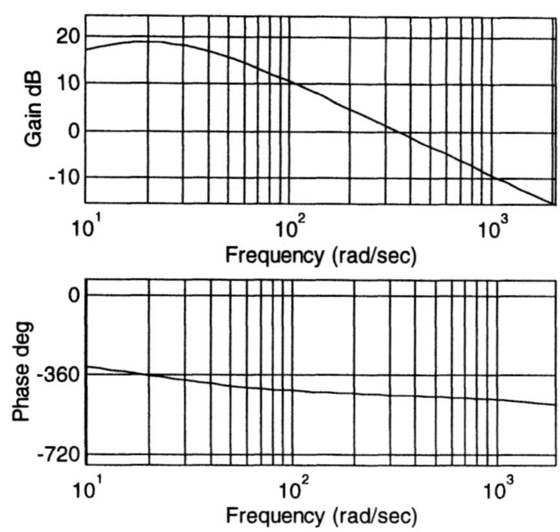

Fig. A13.41 Bode diagram for transfer function $T_{VC}/(1 + T_{PZT})$

Transient responses to commands

The output transient response to a 3 mm pulse command in a system, assuming no saturation in the VC actuator, is shown in Fig. A13.42.

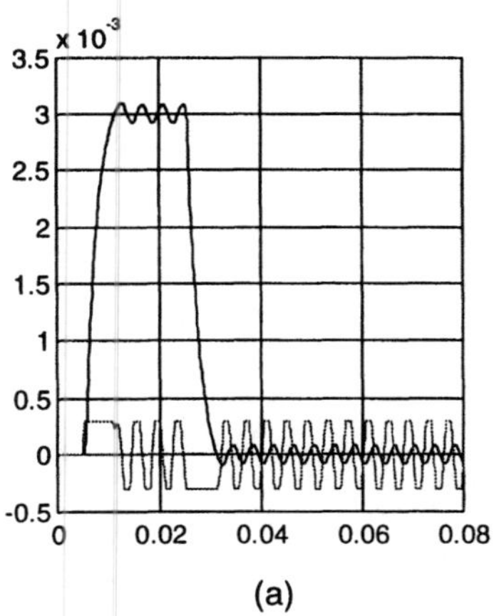

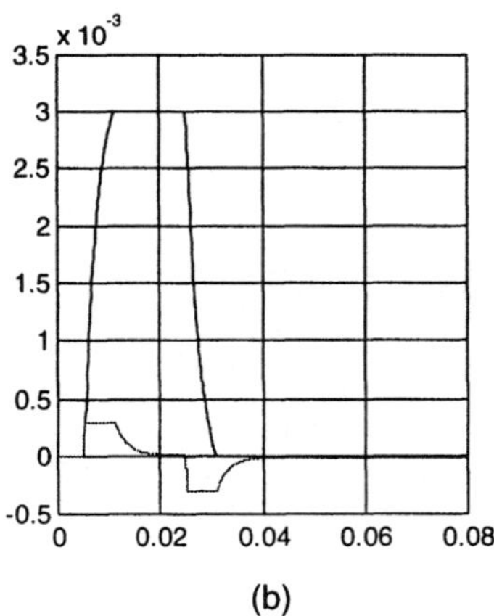

Fig. A13.42 Closed-loop transient responses to 20 ms, 3 mm pulse command, without saturation in the VC actuator; pathlength, upper curves, and PZT mirror displacement time-response, lower curves; (a) without PZT, ending in a small-amplitude 250 Hz limit cycle oscillation, and (b) with PZT, asymptotically globally stable, rapidly settling

The zoomed-on parts of the responses are shown in Fig. A13.43. After the PZT becomes desaturated, the NDC dead zone stops passing the signal, the pathlength response coincides with the response of the equivalent linear system, and the error dies down rapidly.

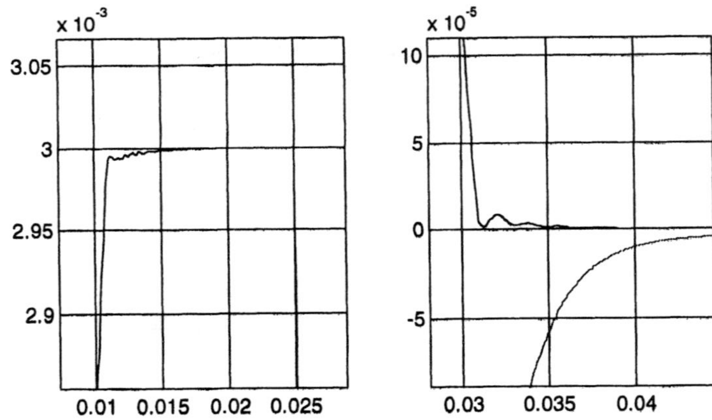

Fig. A13.43 Zoomed-on pieces of the closed-loop transient response to a 20 ms, 3 mm pulse command

The output transient response to a 50 ms pulse command in a system with saturation in the VC actuator with threshold 0.01 is shown in Fig. A13.44. The saturation limits the slew rate of the output, and in other aspects the responses are similar to those in Fig. A13.42.

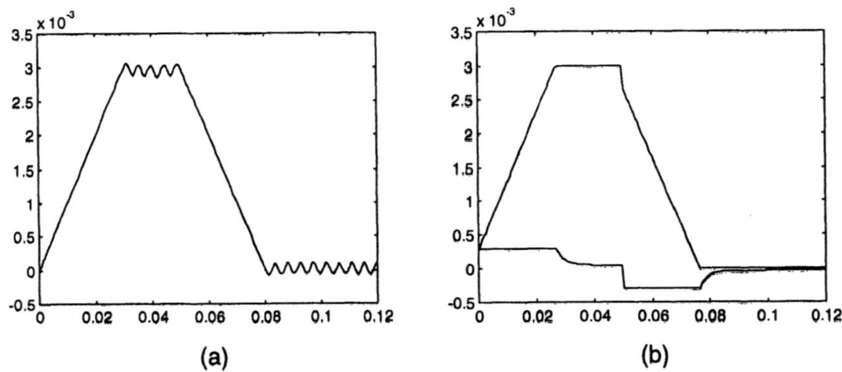

Fig. A13.44 Closed-loop transient responses to 50 ms, 3 mm pulse command, with saturation in the VC actuator; pathlength, solid lines, and PZT mirror displacement time-response, dashed lines; (a) without PZT, ending in a small-amplitude 250 Hz limit cycle oscillation, and (b) with PZT, asymptotically globally stable, rapidly settling

Settling time to high accuracy is that of the linear system. The tail of the response is that of a linear system. Fig. A13.45 shows the responses of the equivalent linear system (dotted line) and the response of the real system (solid line).

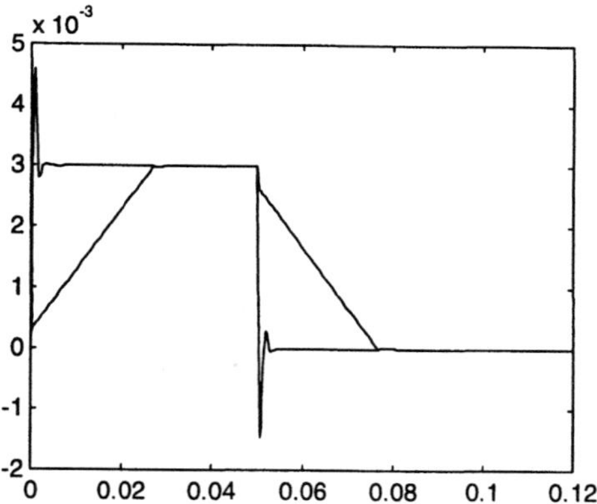

Fig. A13.45 Transient responses to step commands, of the linear system (Π-shaped with overshoots) and the main-vernier nonlinear system (with finit slew rate)

These figures demonstrate the function of an NDC: not only does it ensure global stability but also it provides good transient response to large signals.

In the large-signal (nonlinear) mode of operation, the loop response is, approximately, $T_{vc}/(1+T_{pzt})$. This loop response shown in Fig. A13.41 is close to a single integrator response over the frequency range near the crossover, from 1 Hz to 15 Hz. Such a loop response results in a single-pole closed loop transfer function which is known to produce a rather good closed-loop transient response (although not as good as a higher-order Bessel filter response, so there remains some room for improvement).

Fig. A13.46 shows the output response to the (a) small and (b) large step commands applied at the instant of 50 msec. The 75% overshoot in the small-signal (linear) mode of operation results from the stability margins that are justifiably chosen reasonably narrow (since the plant parameter uncertainty is small). If needed, the small-signal overshoot can be reduced by a prefilter or a command feedforward, but these are not needed because in absolute values the overshoot is small.

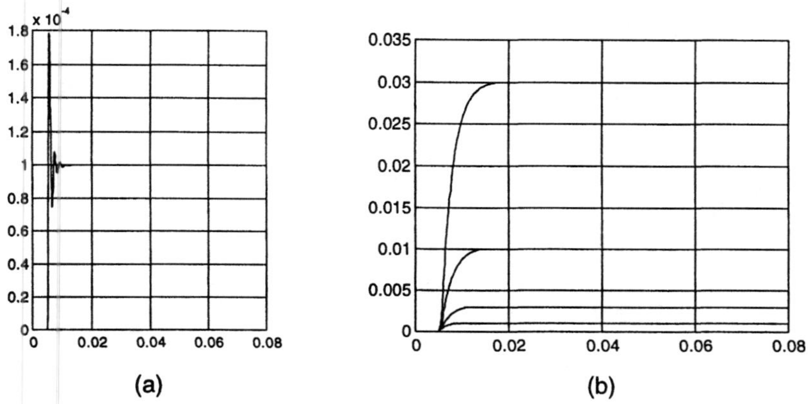

Fig. A13.46 Transient responses to step commands of (a) 0.001 m and (b) larger amplitudes' commands

Robustness

As long as the threshold in the PZT saturation is exactly equal to the dead zone in the NDC, the system is equivalent to a system with only one nonlinear element, the VC actuator. This system satisfies the Popov criterion with conventional stability margins, and is therefore robust against variations in the linear links' transfer functions and in the nonlinear characteristic of the VC actuator.

However, it would be desirable to make the dead zone wider than the threshold of the PZT saturation by 10 to 30% so as not to impair the PZT output stroke. Therefore, the system must be also robust against variations in the threshold of the saturation in the PZT. To such systems including more than one nonlinear link, the Popov criterion is not directly applicable. Instead, we rely on computer simulations. When the PZT saturation threshold is changed from 0.0003 to 0.00015, the system remains stable and the response to the pulse command shown in Fig. A13. 47 remains acceptable. Further reduction of the threshold to 0.0001 makes the system unstable and causes a small amplitude oscillation similar to that shown in Fig. A13.42.

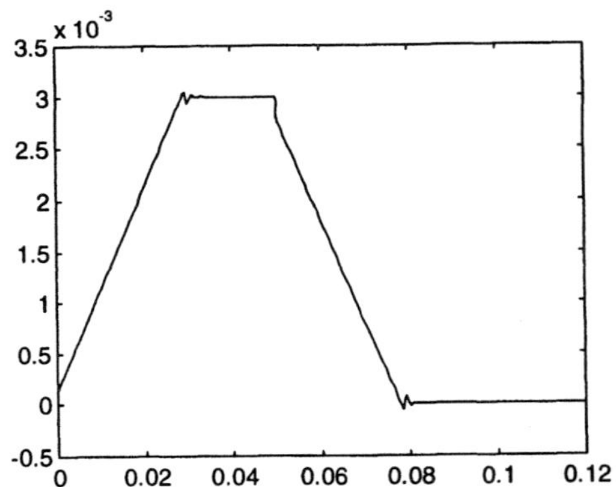

Fig. A13.47 Transient responses to the pulse command in the system with PZT saturation threshold 0.00015

A13.14 MIMO motor control having loop responses with Bode steps

An approximation to the theoretical loop response with Bode step described in Section 4.2.3 has been employed in precision SISO motor control systems for retroreflector carriage motion control and for gyroscope attitude control of an instrument on the *Chemistry* spacecraft.

It was also employed for several less precise attitude, position, and motion controls of robot mounted drills: for a robot designed to take samples of concrete in Chernobyl nuclear station, for a rover to take samples of Mars rocks, and for a device for taking material samples of a comet or an asteroid in nearly zero-gravity environment.

In these MIMO control systems the feedback loops' bandwidths are limited by the rover and drill structural resonances. The same nominal loop response was used in most of the critical feedback loops. The *i*-th loop response was therefore fully defined by the crossover frequency f_{bi}. The trade-offs between the different performance requirements for the loops of these MIMO systems was reduced to finding the set of scalar numbers f_{bi} that provides good system performance. The iterative adjustments of the crossover frequencies with computer simulations and from the experiments converged rapidly.

After the control system becomes operational (at least in simulations), the loop response at frequencies much lower than f_{bi} can be reshaped following Bode relation (3.12). In most cases there was no reason to do this and it was not done, but in some cases the required accuracy of the control loop was able to be well satisfied with much lesser feedback, and the feedback at lower frequencies was reduced. Within the limitations imposed by the Bode integrals, this loop gain reduction was traded for bigger phase stability margins and, therefore, better transient responses.

The last design step was the introduction of nonlinear dynamic compensation in some of the loops that improved the transient responses to large amplitude commands and disturbances.

A13.15 Mechanical snake control

A gigantic animated mechanical snake employed in the movie *Anaconda* was constructed of chain-connected identical mechanical links, each with the ability to bend in pitch and yaw directions. Position feedback allowed achieving good transient response of a single link. However, transient responses of several such links in a chain were very oscillatory. The oscillation was caused by the interference of mechanical waves propagating along the snake.

To correct the problem, force feedback was added and, using the methods detailed in Chapter 7, the output mobilities of the actuators were made dissipative. The damping introduced into the system reduced the effects of the waves' interference, the transient response of the chain of several links was drastically improved as exemplified in Fig. A13.48 for the 5 link chain, and the movements of the entire mechanical snake became agile and very impressive.

This example was contributed by JAS Company, La Crescenta, California, <schier@compuserve.com>.

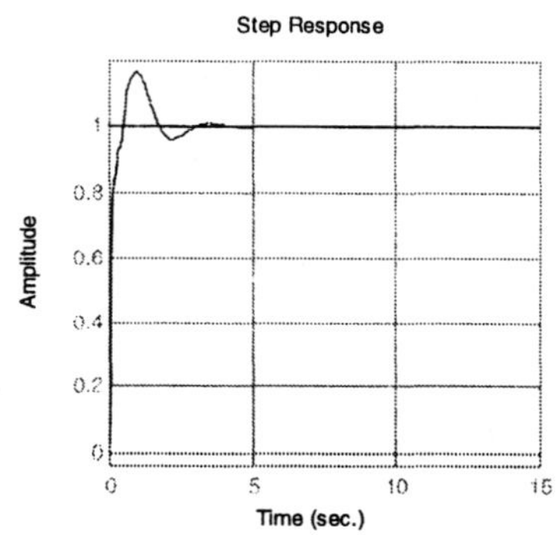

Fig. A13.48 Transient responses of 5 links in a chain with position and force feedback

Appendix 14 Bode Step toolbox

The toolbox includes the following MATLAB functions:

1. NYQLOG (Nyquist plot on logarithmic scale with frequency in octave marks)
2. BOLAGNYQ (Bode diagram for the gain and the lag margin, and the logarithmic Nyquist plot)
3. TFSHIFT (Frequency transform, frequency de-normalization, similar to lp2lp)
4. BONYQAS (Asymptotic Bode diagram, phase plot, and logarithmic Nyquist plot)
5. BOSTEP (Rational function approximation to optimal Bode step response)
6. BOCLOS (Prefilter design for closed-loop low overshoot and fast settling)
7. BOINTEGR (Breaking compensator into two parallel paths, one of which is dominant at low frequencies.)
8. BOCOMP (Calculation of the compensator function for a dc motor control)
9. NDCP (Nyquist plot for describing functions of NDC with parallel paths)
10. BNDCP (Bode plot for describing functions of NDC with parallel paths)
11. NDCB (Nyquist plot for describing functions of NDC with feedback path)
12. BNDCB (Bode plot for describing functions of NDC with feedback path)

These MATLAB functions simplify designing linear/nonlinear servo loops with Bode optimal loop responses and prefilters. The m-functions (and MATLAB scripts for the examples and the problems listed in the book) are available from the author's web page `www.luriecontrol.com`. The functions do the following:

A. Plotting routines

1 NYQLOG plots a Nyquist diagram on the logarithmic plane with octave marks. The marks allow reading the slope of the Bode diagram without plotting it. The loop shaping can be performed iteratively using this plot instead of the Bode diagrams.

2 BOLAGNYQ plots frequency responses of the gain and the phase lag stability margin, and plots the Nyquist diagram on the logarithmic plane with octave marks. This plots' arrangement is convenient for loop shaping. Using the plot of the lag margin allows one to pay attention to only a small area of the plot where the gain response crosses the 0-dB level, and to read there the value of the guard-point phase lag margin.

B. Calculations

3 TFSHIFT de-normalizes an initially normalized frequency response similarly to the MATLAB function `lp2lp` but does this by a polynomial transform in the *s*-domain instead of matrix manipulation in the *t*-domain (`lp2lp` gives complex numbers as the answers for the gain and, sometimes, produces inaccurate answers). It can be used, for example, in problems like that described in Example 3 in Section 4.2.3.

C. Plant modeling, asymptotic diagrams, and conceptual design

4 BONYQAS simplifies the system conceptual design. In particular, it helps to accomplish the following two tasks:

I. Calculating and plotting the phase shift of the plant when only the plant gain has been estimated or measured (since the measurements of the plant gain are often less time-consuming then the measurements of the phase shift).

II. Completing the conceptual design of the feedback loop using asymptotic piece-linear Bode diagrams. The function generates the asymptotic Bode diagram and calculates and plots the phase and the Nyquist diagrams. It allows for easy shaping of the responses by changing the corner frequencies, the gain at the corner frequencies, the linear nonminimum phase lag coefficient, and the asymptotic slopes at low frequencies (the Type of the loop) and at high frequencies. The obtained parameters of the Bode step can be further used as the input file parameters for the BOSTEP function.

D. Feedback loop design

5 BOSTEP generates a rational function approximation to the theoretical transcendental (or irrational, obtained with BONYQAS) loop response with a Bode step for specified margins, asymptotic slope, non-minimum phase, and the feedback Type.

6 BOCLOS generates a normalized closed-loop transfer function without and with a prefilter, and the frequency response of a 4th-order Bessel filter. Also, it generates the step response. It is usable for designing a prefilter rendering a good closed-loop transient response.

7 BOINTEGR converts the linear compensator transfer function into a sum of two transfer functions, the low-pass one (a generalization of an integrator of a PID controller) and the rest. Then, saturation can be introduced into the low-frequency path to improve the transient response and to provide global stability.

E. Compensator design

8 BOCOMP calculates the compensator transfer function for the loop response with a Bode step of a servo with a dc permanent magnet motor. The input file includes the loop transfer function generated by BOSTEP and the dc motor and the inertial load parameters.

F. Nonlinear dynamic compensator design

9 NDCP plots iso-*E* describing functions on the logarithmic Nyquist plane for an NDC with parallel paths, one of which includes a variable or a nonlinear (typically, a saturation) link. A series link is also included that can imitate the rest of the feedback loop so that the loop Nyquist diagrams can be plotted, with the values of the describing function of the nonlinear element from 0 to 1.

10 BNDCP plots iso-*E* describing functions Bode diagrams for an NDC with parallel paths, one of which includes a variable or a nonlinear link (typically, a saturation). A series link is also included that can imitate the rest of the feedback loop so that the loop Bode diagrams can be plotted, with the values of the describing function of the nonlinear element from 0 to 1.

11 NDCB plots iso-*E* describing functions on the logarithmic Nyquist plane for an NDC with a feedback path which includes a variable or a nonlinear link (typically, a dead zone). A series link is also included that can imitate the rest of the feedback loop so that the loop Nyquist diagrams can be plotted, with the values of the describing function of the nonlinear element from 0 to 1.

12 BNDCB plots iso-*E* describing functions Bode diagrams for an NDC with a feedback path which includes a variable or a nonlinear link (typically, a dead zone). A series link is also included that can imitate the rest of the feedback loop so that the loop Bode diagrams can be plotted, with the values of the describing function of the nonlinear element from 0 to 1.

The functions' HELP files include default/demo that generate plots when the name of the function is typed in without arguments. The plots are shown in Figs. A14.1–8 and A14.9–12.

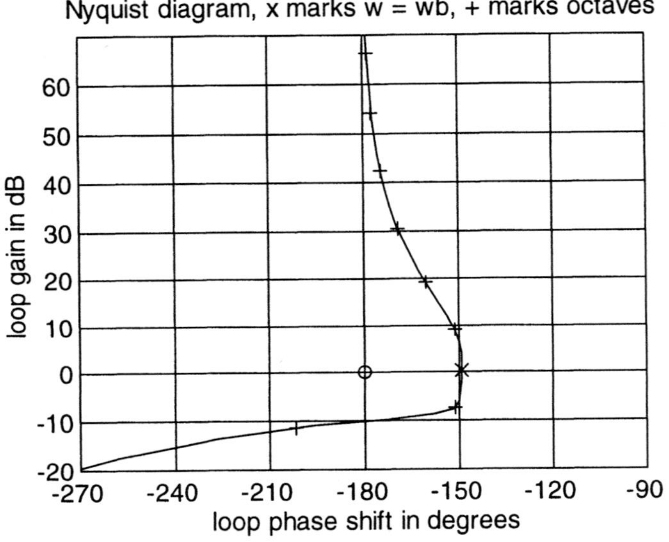

Fig. A14.1 Default/demo for `nyqlog` and `bostep`

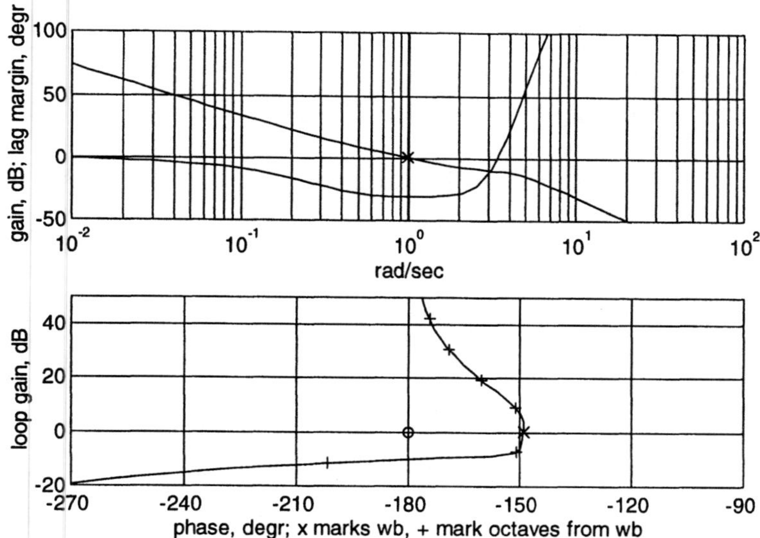

Fig. A14.2 Default/demo for `bolagnyq`

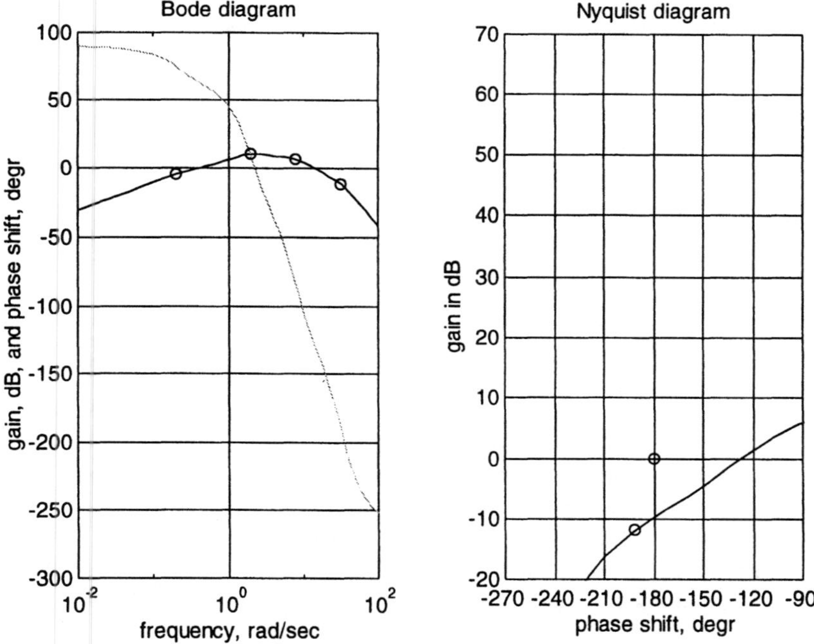

Fig. A14.3 Default/demo 1 for `bonyqas`, calculating and plotting a band-pass plant phase. Circles on the gain response mark the corner frequencies of the piece-linear approximation

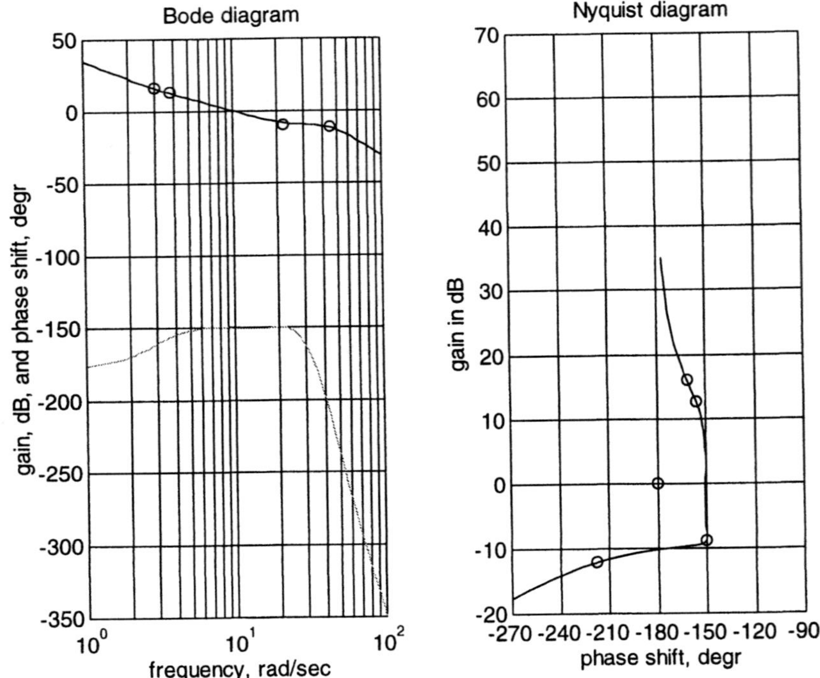

Fig. A14.4 Default/demo 2 for `bonyqas`, asymptotic loop response. Circles on the gain response mark the corner frequencies of the piece-linear asymptotic Bode diagram for loop response with Bode step

Dc motor servo design

A dc motor control system (like that described in Example 2 in Section 4.2.3) can be designed using the following sequence of the toolbox functions.

First, with `bonyqas`, an asymptotic Bode diagram with a Bode step is chosen such that the feedback satisfies the disturbance rejection requirements, and the attenuation in the feedback loop is sufficient at the frequencies of structural modes to guarantee system stability with the chosen stability margins. An example of the diagram is shown in Fig. A14.4. The stability margins, the frequencies of the beginning and the end of the Bode step, the asymptotic slope, the nonminimum phase lag, and the system Type are determined during this conceptual feedback loop design.

Second, with the function `bostep` a rational function approximation for the loop response is obtained, and a compensator function for the nominal single-integrator plant with a small nonminimum phase lag is calculated (as was done in Example 1 in Section 4.2.3). The responses at this point are normalized in frequency, i.e., the loop gain response has unity crossover frequency, like that shown in Fig. A14.1.

Third, with the function `boclos` the closed-loop frequency response is plotted, the prefilter is calculated, and the closed-loop with the prefilter response is plotted, like that shown in Fig. A14.5. The MATLAB function can be used to calculate the closed-loop response with a prefilter comprising two notches, one peak, and a low-pass – or with any subset of these.

The goal for the closed-loop response with the prefilter is a response close to a de-normalized Bessel filter response. The de-normalized filter response can be obtained by shifting the nominal filter response along the logarithmic frequency axis.

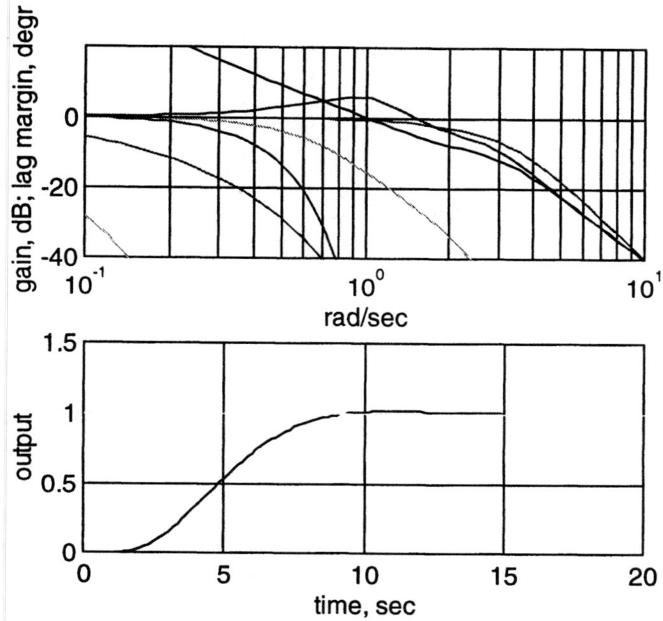

Fig. A14.5 Default/demo for `boclos` plots several responses: an open-loop gain response (the one with large gain at low frequencies); closed-loop responses, without a prefilter (the one with a 7 dB hump) and with a prefilter that consists of a notch, a peak, and a 3rd-order Bessel filter (the response with the widest bandwidth); 4th-order Bessel filter response with nominal bandwidth 1 rad/sec that has 15 dB attenuation at this frequency. The phase responses (the lower three) are of the Bessel filter, of the closed loop with the prefilter, and of closed-loop without a prefilter. The lower plot shows the closed-loop transient response with the prefilter.

Fourth, using the motor and load parameters, the compensator and the plant transfer functions are calculated and plotted with `bocomp`, as were those in Fig. A14.6.

Fifth, with the function `bointegr`, the compensator transfer function is split into a parallel connection of the low-frequency path and the second path (for the medium and higher frequencies), as exemplified by the plots in Fig. A14.7.

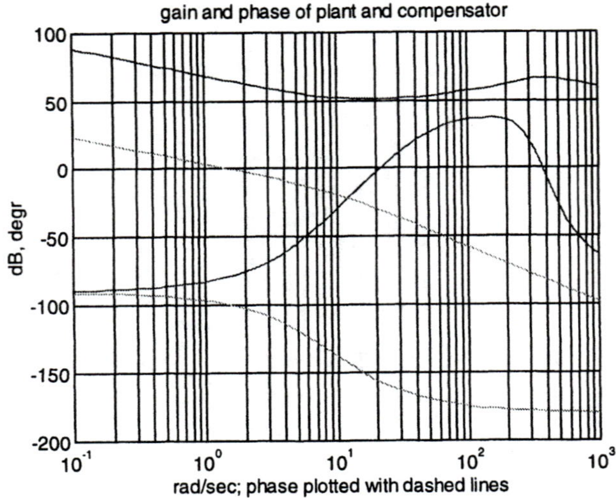

Fig. A14.6 Default/demo for `bocomp`, the gain and phase responses for the compensator and the plant of a dc motor control system

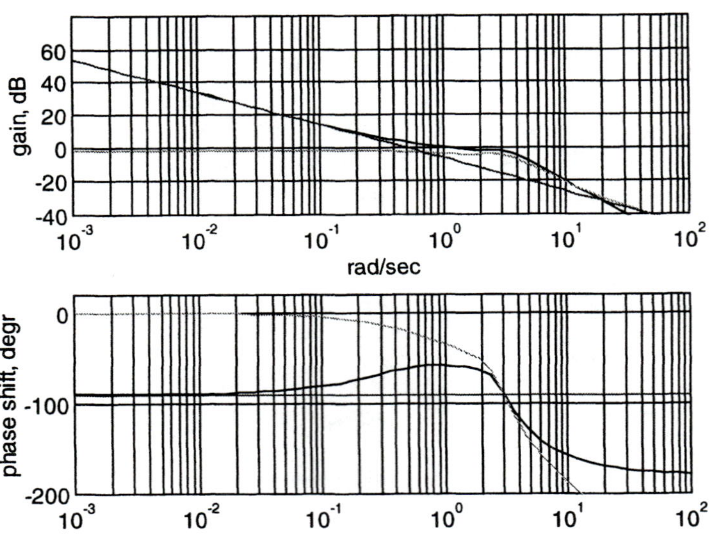

Fig. A14.7 Default/demo for `bointegr`, the gain and phase responses for the entire compensator and its two parallel paths

Next, a SIMULINK model of the system should be built like that shown in Fig. 7.26. The model must include flexible modes of the load and the nonlinear elements in the actuator (saturation in current and voltage), in the friction model, and in the compensator (saturation is introduced in front of the low-frequency path of the compensator). With the model, the controller is fine-tuned to perform well over the specified range of the plant parameter variations, and the system performance and robustness are evaluated.

NDC iso-*E* describing functions

The iso-*E* responses for signal paths shown in Fig. A14.8 are exemplified in Figs. A14.9–12 with the DF values 0; 0.2; 0.4; 0.6; 0.8; 1. These values correspond to the signal levels $E/e_s = \infty$; 6.3; 3; 2; 1.5; < 1 for the saturation and E/e_{dz} < 1; 1.5; 2; 3.5; 8; ∞ for the dead zone.

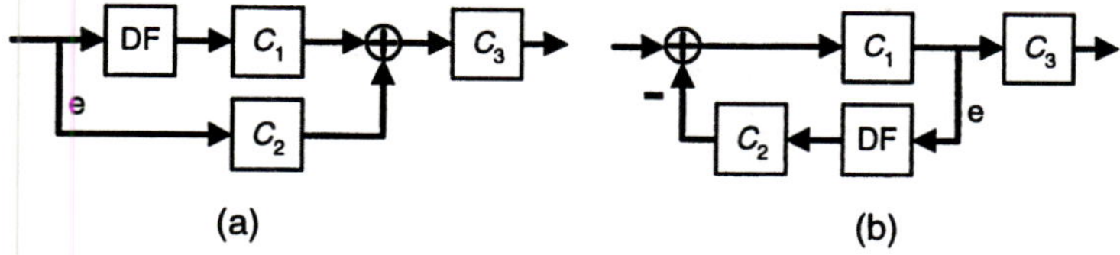

Fig. A14.8 NDCs with (a) parallel paths and (b) with a feedback path

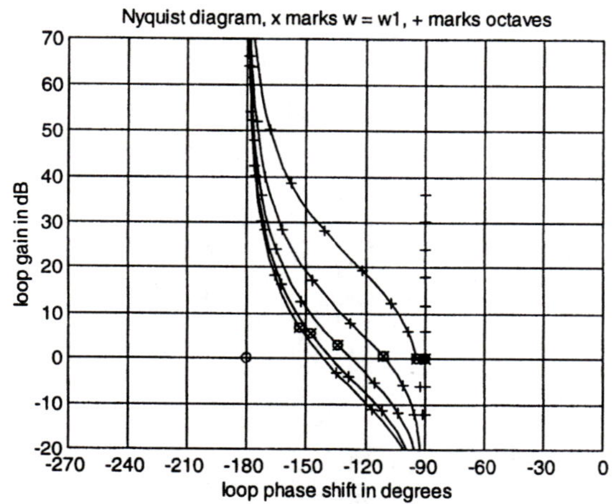

Fig. A14.9 Default/demo for `ndcp`

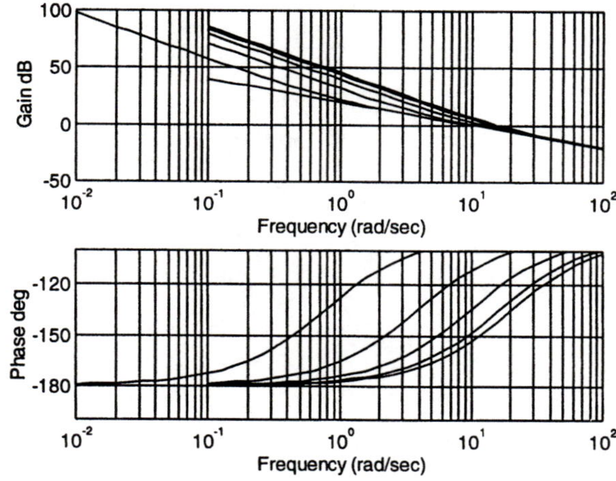

Fig. A14.10 Default/demo for `bndcp`

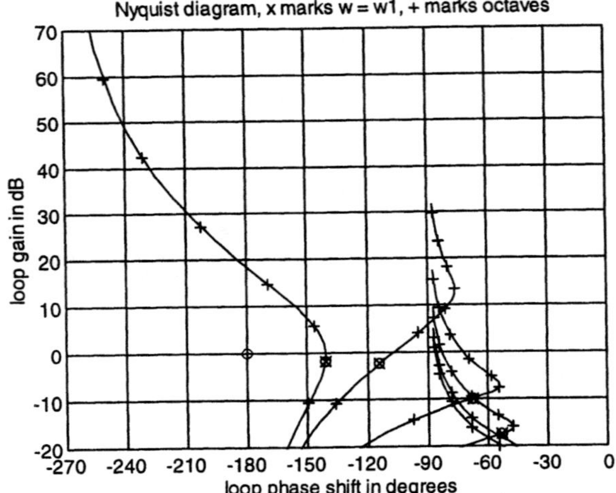

Fig. A14.11 Default/demo for `ndcb`

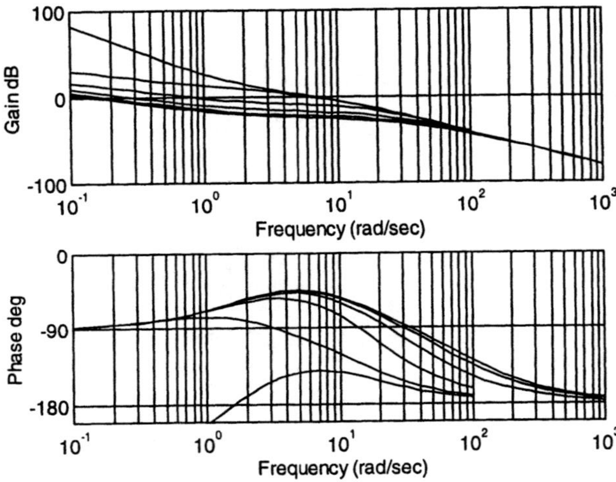

Fig. A14.12 Default/demo for `bndcb`

BIBLIOGRAPHY

Textbooks

1 G. Biernson, *Principles of Feedback Control*, v. 1 and 2. John Wiley and Sons, NY: 1988 (the book contains many practical examples of frequency-domain design).

2 H. W. Bode, *Network Analysis and Feedback Amplifier Design*. Van Nostrand, NY: 1945 (or any of numerous later editions).

The book establishes frequency-domain theory of feedback maximization in causal single-loop and multi-loop systems, which is of immense importance for both theorists and practical engineers.

While concentrating on linear systems, the book develops design methods applicable to practical systems with large parameter uncertainties and with nonlinear actuators, and abstains from discussing those design approaches that has been found non-practical or inferior. When the reader becomes confused why the book doesn't follow up on some apparently attractive idea, why some theory is not further generalized or, conversely, simplified, or even why the employed notations are not what the reader would consider appropriate, he should be advised to suppress the urge to correct the classic and humbly do his homework of finding the reason why Bode did so. Some of these issues are clarified in [6, 9] and in the present book.

The book cannot be recommended as a first book on feedback systems, although the transparent and informative introductory chapter is worth reading for everybody interested in feedback. But afterward, the difficulty rises steeply since the book was written on the basis of lectures Bode gave for his colleagues at The Bell Laboratories where he later headed the mathematical and physical development.

3 P. A. Crafton, *Shock and Vibration in Linear Systems*. Harper and Brothers, NY: 1961 (one of the best books on using electromechanical analogies).

4 R. C. Dorf and R. H. Bishop, *Modern Control Systems*. Addison-Wesley, Menlo Park, CA: 1997.

5 P. Garnell, *Guided Weapon Control Systems*, 2nd ed. Pergamon Press, NY: 1980.

6 I. M. Horowitz, *Synthesis of Feedback Systems*. Academic Press, NY: 1963. Also: *Quantitative Feedback Design Theory (QFT)*. QFT Publications, Boulder Co, 1993.

7 J. V. Jose, E. J. Saletan, *Classical Dynamics: A Contemporary Approach*. Cambridge University Press, NY: 1998.

8 B. C. Kuo, *Automatic Control Systems*, 5th ed. Prentice Hall, Upper Saddle River, NJ: 1995.

9 B. J. Lurie, *Feedback Maximization*. Artech House, Dedham, MA: 1986. (The book describes the theory and the methods for the analysis and design of certain linear and nonlinear feedback systems, see www.luriecontrol.com.)

10 K. Ogata, *Modern Control Engineering*, 3rd ed. Prentice Hall, Englewood Cliffs, NJ: 1997.

11 C. E. Rohrs, J. L. Melsa, D. G. Schultz, *Linear Control Systems*. McGraw-Hill, NY: 1993.

12 M. J. Sidi, *Spacecraft Dynamics and Control*. Cambridge University Press, NY: 1997.

13 M. Vidyasagar, *Nonlinear Systems Analysis*, 2nd ed. Prentice Hall, Upper Saddle River, NJ: 1993.

Additional bibliography

Classical control history is well reflected in the bibliography to [11]. The bibliographies in [6, 9] cover most topics specific to the present book. Many important papers are collected in [9.128] (reference no. 128 in [9]). Transfer function design approach is compared with state-variable methods in [9.71].

The following is a short list of selected recent publications and patents (most of them written by the authors' colleagues at the Jet Propulsion Laboratory or by the authors) reflected in or related to the book material. Some of these sources contain extensive additional bibliographies.

14 D. S. Bayard, Necessary and sufficient conditions for LTI Representation of Adaptive Systems with Sinusoidal Regressors. Proc. American Control Conference, Albuquerque, NM, June 1997.

15 D. S. Bayard, An Algorithm for State-Space Frequency Domain Identification Without Windowing Distortions. IEEE Trans. AC, vol. 39, no. 9, 1994.

16 D. S. Bernstein, Four and a Half Control Experiments and What I learned From Them: A Personal Journey. Proc. of the American Control Conference, Albuquerque, NM, June 1997.

17 J. O'Brien, B. Lurie, Integral Force Feedback for Structural Damping. First International Conf. on Adaptive Structures, San Diego, Nov. 1992.

18 J. F. O'Brien and G. W. Neat, Micro-Precision Interferometer: Pointing Control System. 4th Conference on Control Applications, Albany, NY, September 1995.

19 G.- S. Chen, B. J. Lurie, Bridge Feedback for Active Damping Augmentation. AIAA paper 90-1243, 1989.

20 G.- S. Chen, C. R. Lawrence, B. J. Lurie, Active Member Vibration Control Experiment in a KC-135 Reduced Gravity Environment. First US/Japan Conference on Adaptive Structures. Maui, Hawaii, Nov. 1990.

21 G.- S. Chen and B. J. Lurie, Active Member Bridge Feedback Control for Damping Augmentation. "J. of Guidance, Control and Dynamics", v. 15, no. 5, 1992.

22 B. P. Dolgin, F. T. Hartley, B. J. Lurie, and P. M. Zavracky, Electrostatic Actuation of a Microgravity Accelerometer, 43rd National Symposium of American Vacuum Society, Microelectronical Mechanical Systems Topical Conference, Philadelphia, PA, 14-18 Oct. 1996.

23 P. J. Enright, F. Y. Hadaegh, and B. J. Lurie, Nonlinear Multi-window Controllers. AIAA Guidance, Navigation and Control Conference, San Diego, July 1996.

24 J. L. Fanson, Cheng-Chih Chu, and B. J. Lurie, Damping and Structural Control of the JPL Phase 0 Testbed Structure. "J. of Intelligent Material Systems and Structures", v. 2, no. 3, 1991.

25 J. S. Freudenberg and D. P. Looze, *Frequency Domain Properties of Scalar and Multivariable Feedback Systems, Lecture notes in control and information science*, vol. 104. Berlin: Springer Verlag, 1987.

26 R. L. Grogan, G. H. Blackwood, and R. J. Calvet, Optical Delay Line Nanometer Level Pathlength Control Law Design For Space-Based Interferometry, 1998.

27 J. J. Hench, B. J. Lurie, R. Grogan, R. Johnson. Implementation of nonlinear control laws for the RICST optical delay line. IEEE Aerospace Conference, March 2000, Big Sky, Montana.

28 Y.-H. Lin, B. J. Lurie, Nonlinear Dynamic Compensation for Control System with Digital Sensor. US patent 5,119,003, June 2, 1992.

29 B. J. Lurie, Three loop balanced bridge feedback pointing control. "Proc. of the American Control Conference", Atlanta, 1988.

30 B. J. Lurie, Global Stability of Balanced Bridge Feedback. ICCON '89, IEEE International Conference on Control and Applications. Jerusalem, April 1989.

31 B. J. Lurie, Balanced Bridge Feedback Control System. USA Patent 4, 912,386, March 27, 1990.

32 B. J. Lurie, Multiloop Balanced Bridge Feedback in Application to Precision Pointing. "Intern. J. Control", 1990, vol. 51, no. 4.

33 B. J. Lurie, Tunable TID controller. US patent 5,371,670, December 6, 1994.

34 B. J. Lurie, a letter to Bob's Mailbox, Electronic Design, March/4 1996.

35 B. J. Lurie, Integral Relations for Disturbance Isolation. J. Guidance, Control and Dynamics, vol. 20, no. 3, May-June 1997.

36 B. J. Lurie, J. Daegas, An Improved High-Voltage DC Regulator for a Radar and Communication Transmitter. "Proc. 18th Power Modulator Symposium", Hilton-Head, 1988.

37 B. J. Lurie, J. A. Schier, M. M. Socha, Torque Sensor Having a Spoked Sensor Element Support Structure. USA Patent 4,932,270, June 12, 1990.

38 B. J. Lurie, J. L. Fanson and R. A. Laskin, Active Suspensions for Vibration Isolation. 32nd SDM Conf., Baltimore, April 1991.

39 B. Lurie, J. O'Brien, S. Sirlin, and J. Fanson, The Dial-a-Strut Controller for Structural Damping. ADPA/AIAA/ASME/SPIE Conference on Active Materials and Adaptive Structures, Alexandria, VA, Nov. 1991.

40 B. J. Lurie and F. Y. Hadaegh, Applications of Multiwindow Control in Spacecraft Systems, NASA URC Tech. Conf., Albuquerque, Febr. 1997.

41 B. J. Lurie, A. Ahmed and F. Y. Hadaegh, Asymptotically Globally Stable Multiwindow Controllers, AIAA Conf. on Guidance, Navigation and Control, New Orlean, Aug. 1997.

42 B. J. Lurie, A. Ahmed and F. Y. Hadaegh, Acquisition and Tracking with High-Order Plant and Nonlinear Regulation of Bode Diagram. Control Conference, Pullman, Wa, Sept. 1998.

43 B. J. Lurie, A. Ghavimi, F.Y. Hadaegh and E. Mettler, System Trades with Bode Step Control Design. AIAA Conf. on Guidance, Navigation and Control, Boston, 1998.

44 B. J. Lurie, J. J. Hench, A. Ahmed, and F. Y. Hadaegh, Nonlinear Control of the Optical Delay Line Pathlength. AeroSense '99, SPIE conference, Orlando, FL, 5-7 April 1999.

45 B. J. Lurie, B. Dolgin, F. Y. Hadaegh. Motor Control with Active Impedance of the Driver. Space 2000, Jan-Feb 2000, Albuquerque, NM.

46 J. W. Melody, G. W. Neat, Integrated Modeling Methodology Validation Using the Micro-Precision Interferometer Testbed: Assessment of Closed-Loop Performance Prediction Capability. American Control Conference, Albuquerque, NM, June 1997.

47 S. Mitter and A. Tannenbaum, The Legacy of George Zames. IEEE Transaction Automatic Control, v. 43, no. 5, 1998.

48 A. Murray, DSP Motor Control Boosts Efficiency In Home Appliances, Electronic Design, pp. 102-110, May 25, 1998.

49 G. Neat, J. O'Brien, B. Lurie, A. Garnica, W. Belvin, J. Sulla, J. Won, Joint Langley Research Center/Jet Propulsion Laboratory CSI Experiment. 15th Annual AAS Guidance and Control Conference, Keyston, Colorado, Feb. 1992.

50 G. W. Neat, A. Abramovichi, J. W. Melody, R. J. Calvet, N. M. Nerheim, and J. F. O'Brien, Control Technology Readiness for Spaceborn Optical Interferometer Missions. SMACS 2, Toulouse, France, May 1997.

51 G. W. Neat, J. W. Melody, B. J. Lurie, Vibration Attenuation Approach for Spaceborn Optical Interferometers. IEEE Transaction Control Technology, v. 6, no. 6, 1998.

52 Z. Rahman, J. Spanos, and D. Bayard, Multi-tone Adaptive Vibration Isolation of Engineering Structures. 36th AIAA/ASME/ASCE/AHS/ASC Structures, Structural Dynamics and Material Conference, New Orleans, LA, April 1995.

53 J. T. Spanos and M. C. O'Neal, Nanometer level optical control of the JPL phase B testbed. ADPA/AIAA/ASME/SPIE Conf. on Active Materials and Adaptive Structures, 1992.

54 J. Spanos, Z. Rahman, A. von Flotov, Active Vibration Isolation on Experimental Flexible Structure. Smart Structures and Intelligent Systems, SPIE Proc. 1917-60, Albuquerque, 1993.

55 J. Spanos and Z. Rahman, Narrow-band Control Experiments in Active Vibration Isolation. Vibration Monitoring and Control, SPIE Proc. 2264-01, San Diego, CA, July 1994.

56 A. Talwar, Noise and Distortion Reduction in Amplifiers using Adaptive Cancellation. Microwave Journal, August 1997.

References to chapters

References [2], [6], [9] and [9.113] are generic for those topics covered in this book which are not conventionaly included in contemporary textbooks. The references listed below exemplify and expand on selected topics, or indicate earlier publications.

Ch. 1 **1.7** [9.24]; **1. 9** [9.65]

Ch. 2 **2.4** [9.24, 9.176, 55]; **2.6** [29-32]; **2.7** [26, 36, 53]

Ch. 3 **3.** [41]; **3.9** [1, 9, 9.113, 9.44, 17, 25, 50] **3.9.2** [35]; **3.9.3** [17]; **3.12** [9.99, 9.103, 9]; **3.13** [9.99, 9.103, 9.113]

Ch. 4 **4.1**[9.81, 33]; **4.2.3** [34]; **4.5** [9.99]

Ch. 6 **6.6.2** [33]; **6.7.2** [9.21, 9.109]

Ch. 7 **7.1.1** [3, 38, 39], **7.4** [49, 51]; **7.4.5** [19, 21, 29, 38, 39, 51]; **7.5** [29, 30, 31, 32]; **7.9.1.1** [37]; **7.9.1.3** [22]; **7.9.2** [28, 9.122]

Ch. 8 **8.1**[9.128, 9.71]; **8.4** [9.69, 9.71]; **8.5** [47]

Ch. 9 **9.2** [54]; **9.7** [14,15, 42]

Ch. 10 **10. 3.1**[9, 9.105, 9.113] **10.5** [9, 9.51, 9.95, 9.110, 9.111, 9.118, 9.119, 9.120,]; **10.7** [9, 9.111, 41]

Ch. 11 **11.2** [9.57]; **11.3** [9, 9.113, 9.162, 9.158, 9.108, 9.102, 9.126, 28]; **11.9** [30]; **11.3.2** [9, 9.113] **11. 6, 11.7, 11.8** [9.113]

Ch. 12 **12.2** [9.86, 9.105, 9.107, 9.106]; **12.5** [9.4, 9.105, 9.106, 9.107]

Ch. 13 **13** [23, 40, 41]

NOTATION

A, $A(s)$ – actuator transfer function, 1
A, $A(s)$ – even part of $\theta(s)$, 74
$A(j\omega) = \text{Re}\,\theta(j\omega)$, 74
A – system matrix, 249
A_o – maximum available feedback, 106
A_∞ – value of A at $s = \infty$, 75
B, $B(s)$ – feedback path transfer function, 1
B, $B(s)$ – odd part of $\theta(s)$, 74
B – control-input matrix, 249
$B(0)$ – feedback path transfer function with two specified nodes connected, 382
$B(\infty)$ – feedback path transfer function with two specified nodes disconnected, 382
$B(j\omega) = \text{Im}\,\theta(j\omega)$, 74
B_1, B_3 – coefficients in Laurent expansion of $\theta(s)$, 75
B_n – nonminimal phase lag, 85-86
C – capacitance, 76
C, $C(s)$ – compensator transfer function, 1
C – output matrix, 249
E – emf of the signal source, 212
E – amplitude of the fundamental at the input to a nonlinear link, 291
E_1 – amplitude of the fundamental at the input to NDC, 301
E_2 – amplitude of the fundamental at the input to the actuator, 297
E_a' – amplitude of the fundamental after the jump down, 325
E_a'' – amplitude of the fundamental after the jump up, 325
E_b' – amplitude of the fundamental before the jump down, 325
E_b'' – amplitude of the fundamental before the jump up, 325
$\overline{E}$ – mean square thermal noise amplitude, 171
E_{1C}, E_{2C} – values of E_1, E_2 causing $|T| = 1$, 301
E_{sb} – amplitude of a subharmonic, 327
F – force, 207, 207
$F = T + 1$ – return difference, 2
$F(0)$ – value of return difference when two specified terminals shorted, 381
$F(\infty)$ – value of return difference when two specified terminals open, 381
$|F|$, or $20 \log|F|$ – feedback, 2
$G(s)$, G – return ratio of linear links in NDC, 276
G – plant noise distribution matrix, 254
H – describing function, 287
H – measurement matrix, 254
H_∞ – norm, 255
J – moment of inertia, 12, 207-208
J – quadratic functional, 253
K – closed loop transmission coefficient in voltage, 381
K – gain matrix, 249
K_E – closed loop ratio of the output voltage to the source emf, 381

K_E – estimator gain matrix, 254
K_I – closed loop transmission coefficient in current, 381
K_{OL} – open loop system transmission coefficient in voltage, 381
$K_{OL\,I}$ – open loop system transmission coefficient in current, 381
$K_{OL\,E}$ – open loop system ratio of the output voltage to the source emf, 381
L – inductance, 77, 207-208
M – mass of a rigid body, 116, 204-205
$M = T/F$, 4
M, N – coefficients in bilinear function, 302
N_A – noise at the actuator input, 112
$P, P(s)$ – plant transfer function, 1
$P_o, P_o(s)$ – transfer function of nominal plant, 31,
Q – regulation of a symmetrical regulator, 195
$Q = 1/(2\zeta)$ – quality factor of a resonance, 137
Q – weighting matrix, 253
$R, R(s)$ – prefilter transfer function, 33
R – weighting matrix, 253
R_L – load resistance, 212-213
R_T – thermal resistance, 208
S – sensitivity, 17
S_H – Horowitz sensitivity, 19
T – return ratio, 2
T – absolute temperature, 209
$T(0)$ – return ratio in a system with the two specified nodes connected, 216, 381-382
$T(\infty)$ – return ratio when two specified nodes are disconnected, 216, 381-382
T_P – return ratio about the plant, 276
T_E – return ratio in an equivalent system, 277
T_S – sampling period, 147
U – voltage, 212
U – amplitude of sinusoidal signal at the system's input, 324
U', U'' – threshold values of U causing jumps in E, 320-321
$U_{2(n)}$ – amplitude of nth harmonic at a system's output, 16
V – amplitude of fundamental at the output of nonlinear link, 324
V – velocity, 205, 207-208
V – voltage, 207-208
W – transfer function, or immitance, 194
$W(w)$ – transfer function (immitance) of a regulator, 194
$W(0), W(\infty)$ – regulator functions with zero or infinite values of the regulating element w, 194
Y – admittance, 77
Y' – admittance of a two-pole that is not zero at infinite frequency, 77
Z – impedance, 76
Z' – impedance of a two-pole that is not zero at infinite frequency, 77
Z_L – load impedance, 212, 216
Z_o – impedance (mobility) of a system without feedback, 216, 381, 383

Z_S – source impedance, 212
$a_1, a_2, ...$ – polynomial coefficients, 151-152
$b_1, b_2, ...$ – polynomial coefficients, 151-152
e, $e(t)$ – signal at the input to a nonlinear link, 323
e_d – dead zone, 293
e_s – saturation threshold, 293
f – frequency in Hz, 7
f_b – frequency at which loop gain is 0 dB, 8
f_c – lowest frequency at which loop gain is $-x$ dB, 98
f_c – central frequency of a segment of constant slope of a Bode diagram, 82
f_d – highest frequency at which loop gain is $-x$ dB, 98
f_g – lowest frequency at which loop gain is x_1 dB, 109
f_h – highest frequency at which loop gain is x_1 dB, 109
f_p – frequency of a pole, 132
f_S – sampling frequency, 148
f_T – unity gain bandwidth, 170-171
f_z – frequency of a zero, 132
k – gain coefficient, 294, 303
k – motor constant, 223
k – spring stiffness coefficient, 207-208
k_o – coefficient of forward propagation in voltage, 379, 381
k_{oE} – coefficient of forward signal propagation relative to the source emf, 381
k_{oI} – coefficient of forward signal propagation, 381
k_c – coupling coefficient, 120, 313
n – slope coefficient of an asymptotic Bode diagram, 97, 131
n – transformation coefficient, 226
$p_1, p_2, ...$ – polynomial coefficients, 147
q – Popov's criterion coefficient, 271-272
$q_1, q_2,$ – coefficients in modified Popov criterion, 275
$q_1, q_2, ...$ – polynomial coefficients, 147
r – reference, 249
$s = \sigma + j\omega$ – operational variable, 362
t – time, 54
t_d – delay time, 54
t_r – rise time, 54
t_s – settling time, 54
$u = \ln \omega/\omega_c$, 79
u, $u(t)$ – signal at system's input, command, 320
u – control vector, 249
u_{th} – dead beat threshold, 298
u_n – signal sample at the input to trapezoid integrator, 147
v, $v(t)$ – signal at the output of a nonlinear link, 323
v – vector of process noise, 254
v_n – signal sample at the output of trapezoid integrator, 147
w – impedance, admittance, or transfer coefficient of a variable element, 194
w – amplifier transfer immitance or coefficient in a feedback system, 379
w – width in octaves of a trapezium gain response segment, 82
w – vector of sensor noise, 254

w_o – nominal value of variable element in Bode symmetrical regulator, 195
x – displacement, 208, 312
x – lower amplitude stability margin in dB, 64
x – vector of state variables, 249
x_1 – upper amplitude stability margin in dB, 65
x_E – state estimate, 254
y, or $y\pi$ – phase stability margin, 63-64
y – output vector, 249
y – displacement along y-axis, 312
z – measurements, 254
z^{-1} – delay operator, 147-148
Δ – difference, 78
Δ – main determinant, 384
Δ_o – minor, 384
Ω – angular velocity, rad/sec, 207
Ω_f – free run angular velocity, rad/sec, 213
θ, $\theta(s)$, $\theta(j\omega)$ – transfer function, 74
θ – angle of rotation, 208
ϕ – nonlinear function, 271
ρ – wave resistance (mobility), 231
τ – torque, 12
τ_b – brake torque, 213
$\zeta = 1/(2Q)$ – damping coefficient, 137
ω – frequency, rad/sec, 7
ω_c – frequency at which phase shift is calculated, 79
– denotes parallel connection of impedances, e.g., $Z_1 \# Z_2 = Z_1Z_2/(Z_1 + Z_2)$

INDEX